SYSTEMS

ANALYSIS

DESIGN

Second Edition

ALAN DENNIS

Indiana University

BARBARA WIXOM

University of Virginia

John Wiley & Sons, Inc.

http://www.wiley.com/college

CREDITS

ACQUISITIONS EDITOR	Beth Golub
MARKETING MANAGER	Gitti Lindner
SENIOR PRODUCTION EDITOR	Patricia McFadden
SENIOR DESIGNER	Dawn Stanley
PRODUCTION MANAGEMENT SERVICES	Hermitage Publishing Services

This book was set in 10.5/12 Times New Roman by Hermitage Publishing Services and printed and bound by Von Hoffman Press, Inc. The cover was printed by Von Hoffman Press, Inc.

This book is printed on acid-free paper. ∞

ISBN 0-471-07322-9

Printed in the United States of America

10 9 8 7 6 5 4 3 2 1

To Eileen and Alec

To Chris, Haley, and Linda

BRIEF CONTENTS

CONTENTS

CHAPTER 10   USER INTERFACE DESIGN   **303**

PREFACE

PURPOSE OF THIS BOOK

Systems Analysis and Design (SAD) is an exciting, active field in which analysts continually learn new techniques and approaches to develop systems more effectively and efficiently. However there is a core set of skills that all analysts need to know—no matter what approach or methodology is used. All information systems projects move through the four phases of planning, analysis, design, and implementation; all projects require analysts to gather requirements, model the business needs, and create blueprints for how the system should be built; and all projects require an understanding of organizational behavior concepts like change management and team building.

This book captures the dynamic aspects of the field by keeping students focused on doing SAD while presenting the core set of skills that we feel every systems analyst needs to know today and in the future. This book builds on our professional experience as systems analysts and on our experience in teaching SAD in the classroom.

This book will be of particular interest to instructors who have students do a major project as part of their course. Each chapter describes one part of the process, provides clear explanations on how to do it, gives a detailed example, and then has exercises for the students to practice. In this way, students can leave the course with experience that will form a rich foundation for further work as a systems analyst.

OUTSTANDING FEATURES

A Focus on Doing SAD

The goal of this book is to enable students to do SAD—not just read about it, but understand the issues so they can actually analyze and design systems. The book introduces each major technique, explains what it is, explains how to do it, presents an example, and provides opportunities for students to practice before they do it for real in a project. After reading each chapter, the student will be able to perform that step in the system development life cycle (SDLC) process.

Rich Examples of Success and Failure

The book includes a running case about a fictitious company called CD Selections. Each chapter shows how the concepts are applied in situations at CD Selections. Unlike running cases in other books we have tried to focus these examples on planning, managing, and executing the activities described in the chapter, rather than on detailed dialogue between fictitious actors. In this way, the running case serves as a template that students can apply to their own work. Each chapter also includes numerous Concepts in Action boxes that describe how real companies succeeded—and failed—in performing the activities in the chapter. Many of these examples are drawn from our own experiences as systems analysts.

Incorporation of Object-Oriented Concepts and Techniques

The field is moving toward object-oriented concepts and techniques, both through UML, the new standard for object-oriented analysts and design, as well as by gradually incorporating object-oriented concepts into traditional techniques. We have taken two approaches to incorporating object-oriented analysis and design into the book. First, we have integrated several object-oriented concepts into our discussion of traditional techniques, although this may not be noticed by the students because few concepts are explicitly labeled as object-oriented concepts. For example, we include the development of use cases as the first step in process modeling (i.e., data flow diagramming) in Chapter 5, the use (and reuse) of standard interface templates and use scenarios for interface design in Chapter 10.

Second, and more obvious to students, we include a final chapter on the major elements of UML that can be used as an introduction to object-oriented analysts and design. This chapter can be used at the end of a course—while students are busy working on projects—or can be introduced after or instead of Chapters 6 and 7.

Real World Focus

The skills that students learn in a systems analysis and design course should mirror the work that they ultimately will do in real organizations. We have tried to make this book as "real" as possible by building extensively on our experience as professional systems analysts for organizations such as Arthur Andersen, IBM, the U.S. Department of Defense, and the Australian Army. We have also worked with diverse industry advisory boards of IS professionals and consultants in developing the book and have incorporated their stories, feedback, and advice throughout. Many students who use this book will eventually use the skills on the job in a business environment, and we believe they will have a competitive edge in understanding what successful practitioners feel is relevant in the real world.

Project Approach

We have presented the topics in this book in the SDLC order in which an analyst encounters them in a typical project. Although the presentation is necessarily linear (because students have to learn concepts in the way in which they build on each other), we emphasize the iterative, complex nature of SAD as the book unfolds. The presentation of the material should align well with courses that encourage students to work on projects because it presents topics as students need to apply them.

Graphic Organization

The underlying metaphor for the book is doing SAD through a project. We have tried to emphasize this graphically throughout the book so that students better understand how the major elements in the SDLC are related to each other. First, at the start of every major phase of the system development life cycle, we have a graphic illustration that shows the major deliverables that will be developed and added to the "project binder" during that phase. Second, at the start of each chapter, we present a checklist of key tasks or activities that will be performed to produce the deliverables associated with this chapter. These graphic elements—the binder of deliverables tied to each phase and the task checklist tied to chapter—can help students better understand how the tasks, deliverables, and phases are related and flow from one to another.

Finally, we have highlighted important practical aspects throughout the book by marking boxes and illustrations with a "pushpin." These topics are particularly important in the practical day-to-day life of systems analysts and are the kind of topics that junior analysts should pull out of the book and post on the bulletin board in their office to help them avoid costly mistakes!

WHAT'S NEW IN THIS EDITION

The second edition contains three fundamental improvements, including reorganized chapters, new topics, and expanded detail. First, chapters have been reorganized to combine the analysis process and requirements gathering techniques; to combine interface design structure and components; and to separate use cases from process modeling. Second, a half chapter about an organization's project selection process was added to Chapter 3, along with other new topics that are described later. Third, more detail was added to economic feasibility, process modeling, data modeling, and IT architecture topics to make them easier to learn and understand.

Several changes have been made across all chapters in the second edition. The Concepts in Action boxes have been updated when appropriate, and the running case has been changed to include issues involving integration with a company's existing IT infrastructure. We interviewed Chief Information Officers of seven different organizations regarding project selection and project management at the organizational level, and the results of these interviews have been incorporated into the general content, the Concepts in Action boxes, and the instructor's manual. New practitioner war stories also have been incorporated into the textbook and supplemental materials. A new chapter organization graphic strengthens the linkages between the components of the SDLC.

The first chapter, Introduction, now includes a discussion of Agile Development and Extreme Programming. It also provides more coverage of the skills of an analyst.

Part 1, Planning, has been reorganized to provide more depth within several important topics. Chapter 2, Project Initiation, includes more detailed coverage of economic feasibility analysis, and a section on project selection was added to emphasize organizational project considerations, such as trade-offs and portfolio management.

Chapter 3, Project Management, now contains more detailed descriptions of Gantt charts, PERT charts, and CASE tools.

Part 2, Analysis, has been reorganized to emphasize business requirements for a system and how requirements drive the systems analysis and design process. Chapter 4, Requirements Determination, combines earlier chapters about process improvement and requirements-gathering techniques, and it contains an additional section that defines and illustrates business requirements. The discussions of use cases and process modeling have been broken into two separate chapters, chapter 5 and Chapter 6. Much more detail has been added to Chapters 6 (Process Modeling) and 7 (Data Modeling), and both chapters include more examples. More coverage of structured English, decision tables, and decision trees have been added to Chapter 6. Chapter 7 now includes the topic of normalization and more coverage of the data dictionary.

Part 3, Design, has been reorganized to expand coverage on IT architecture (Chapter 9) and to integrate interface concepts into one chapter (Chapter 10) without losing substance. Chapter 8, System Design, now includes a discussion of CRUD matrix.

Part 4, Implementation, and the final chapter, The Movement to Objects, have been only lightly revised.

ORGANIZATION OF THIS BOOK

This book is organized by the phases of the Systems Development Life Cycle (SDLC). Each chapter has been written to teach students specific tasks that analysts need to accomplish over the course of a project, and the deliverables that will be produced from the tasks. As students complete the book, tasks will be "checked off" and deliverables will be completed and filed in a Project Binder. Along the way, students will be reminded of their progress using road maps that indicate where their current task fits into the larger context of SAD.

Chapter 1 introduces the SDLC and describes the roles and skills needed for a project team. Part 1 contains Chapters 2 and 3, which describe the first phase of the SDLC, the Planning Phase. Chapter 2 presents Project Initiation, with a focus on the System Request, Feasibility Analysis, and Project Selection. In Chapter 3 students learn about Project management, with emphasis on the Workplan, Staffing Plan, Project Charter, and Risk Assessment that are used to help manage and control the project.

Part Two presents techniques needed during the Analysis Phase. In Chapter 4, students are introduced to requirements determination and learn a variety of analysis techniques to help with Business Process Automation, Business Process Improvement, and Business Process Reengineering. Chapter 5 focuses on Use Cases, while Chapter 6 covers Process Models, and Chapter 7 explains Data Models and Normalization.

The Design Phase is covered in Part 3 of the textbook. In Chapter 8, students learn how to convert existing process and data models into physical representations of the To-Be system. They create an Alternative Matrix that compares custom, packaged, and outsourcing alternatives. Chapter 9 focuses on architecture design, which includes the Architecture Design, Hardware/Software Specification, and Security Plan. Chapter 10 focuses on the user interface and presents interface design, and students learn how to create the Interface Structure, Interface Standards, User Interface Template, and User Interface Design. Finally, the data storage and program designs are illustrated in Chapters 11 and 12, which contain information regarding the Data Storage Design, Program Structure Chart, and Program Specification.

The Implementation Phase is presented in Chapters 13 and 14. Chapter 13 focuses on system construction, and students learn how to build and test the system. It includes information about the Test Plan and User Documentation. Conversion is covered in Chapter 14, and students learn about the Conversion Plan, Change Management Plan, Support Plan, and the Project Assessment.

Chapter 15 provides a background of object orientation and explains several key object concepts supported by the standard set of object-modeling techniques used by systems analysts and developers. Then, we explain how to draw four of the most effective models in UML: the use-case diagram, the sequence diagram, the class diagram, and the statechart diagram.

SUPPLEMENTS
(www.wiley.com/college/dennis)

Online Instructors Manual

The Instructors manual, prepared by Roberta M. Roth of University of Northern Iowa, provides resources to support the instructor both inside and out of the classroom:

- Short experiential exercises that instructors can use to help students experience and understand key topics in each chapter
- Short stories have been provided by people working in both corporate and consulting environments for instructors to insert into lectures to make concepts more colorful and real
- Additional mini-cases for every chapter allow students to perform some of the key concepts that were learned in the chapter.
- Answers to end of chapter questions and exercises are provided

Online Instructor's Resources

- PowerPoint slides, prepared by Roberta M. Roth of University of Northern Iowa, that instructors can tailor to their classroom needs and that students can use to guide their reading and studying activities
- Test Bank, prepared by Tom Dillon of James Madison University, that includes a variety of questions ranging from multiple choice to essay style questions. A computerized version of the Test Bank will also be available.

Instructor's Resources CD

Includes electronic files for the Instructor's Manual, Powerpoint slides and Test Bank. It also includes a computerized version of the Test Bank.

WebCT and Blackboard Courses

These online course management systems are tools that facilitate the organization and delivery of course materials on the Web. Easy to use, they provide powerful communication, loaded content, flexible course administration and sophisticated online testing and diagnostic systems.

Student Website

- Web Resources, prepared by Lenore Horowitz of Schenectady Community College, provide instructors and students with weblinks to resources that reinforce the major concepts in each chapter. See http://www.wiley.com/college/dennis.
- Web Quizzes, prepared by Bruce White of Quinnipiac University, help students prepare for class tests.

CASE Software

Three CASE (Computer-Aided Software Engineering) tools can be purchased with the text:

1. Oracle's Enterprise Development Suite, comprising Oracle 8i Personal Edition 8.1.5, Oracle Developer 6.0 and Oracle Designer 6.0. This software is available under a "Development Sublicense" for personal development purposes only, and has no time restrictions or limitations.
2. Visible Systems Corporation's Visible Analyst Student Edition.
3. Microsoft's Visio 2002

Contact your local Wiley sales representative for details, including pricing and ordering information.

Project Management Software

A 120-Day Trial Edition of Microsoft Project 2002 can be purchased with the textbook. Contact your local Wiley sales representative for details.

ACKNOWLEDGMENTS

Our thanks to the many people who contributed to the preparation of this second edition. We are indebted to the staff at John Wiley & Sons for their support, including Beth Lang Golub, Executive Editor, Lorraina Raccuia, Assistant Editor, Ailsa Manny, Editorial Assistant, Trish McFadden, Production Editor, Dawn L. Stanley, Senior Designer, Norman Christensen, Freelance Illustrator.

We would like to thank the following reviewers and focus group participants for their helpful and insightful comments:

Murugan Anandarajan	*Drexel University*
John Baron	*University of Illinois, Ph.D. Student*
Meral Binbasioglu	*Hofstra University*
Thomas Case	*Georgia Southern University*
Manoj Choudhary	*DeVry Institute of Technology, Scarborough, CA*
Subhasish Dasgupta	*George Washington University*
Mark Dishaw	*University of Wisconsin, Oshkosh*
Raol Freeman	*California State University*
Mark N. Frolick	*University of Memphis*
Rick Gibson	*American University*
Peter C. Johnson	*California State University, Sacramento*
Bill Hardgrave	*University of Arkansas*

Fred G. Harold	*Florida Atlantic University*
Jeffrey S. Harper	*Indiana State University*
Albert Harris	*Appalachian State University*
Monica C. Holmes	*Central Michigan University*
Rebecca Horner	*University of Virginia*
Ron Kelly	*Nova Scotia Community College, Burridge Campus*
Deepak Khazanchi	*Northern Kentucky University*
Chung S. Kim	*Southwest Missouri State University*
Chang Koh	*University of North Texas*
George M. Marakas	*Indiana University*
Vicki McKinney	*University of Wisconsin, Milwaukee*
Eric Meier	*University of Virginia*
Michael Morris	*University of Virginia*
Fred Niederman	*University of Baltimore*
Maggie O'Hara	*East Carolina University*
Richard O'Lander	*St. John's University-St. Vincent's College*
Elizabeth Perry	*SUNY Binghamton*
Tom Pettay	*DeVry Institute of Technology, Columbus, OH*
Alan M. Przyworski	*DeVry Institute of Technology, Decatur, GA*
Thomas C. Richards	*University of North Texas*
Cynthia Ruppel	*University of Toledo*
Nancy L. Russo	*Northern Illinois University*
Linda Salchenberger	*Loyola University, Chicago*
Stephen L. Shih	*Southern Illinois University*
Ulrike Schultze	*Southern Methodist University*
Tony Scime	*State University of New York, College at Brockport*
John B. Schwartz	*University of Maryland, Baltimore County*
Anne Marie Smith	*LaSalle University*
Ted Strickland	*University of Louisville*
James Suleiman	*University of Colorado, Colorado Springs*
Ron Thompson	*University of Vermont*
Jonathan Trower	*Baylor University*
Duane P. Truex III	*Georgia State University*
William J. Vachula	*University of Pennsylvania*
David Vance	*Southern Illinois University*
Bruce White	*Quinnipiac University*
Rosann Webb Collins	*University of South Florida*

We would like to thank the practitioners for helping us add a real world component to this project:

Matthew Anderson	*Andersen Consulting*
Adam Beck	*Capital One*
James Chapman	*Bell Atlantic*
David Dierolf	*Crutchfield Corporation*
Jeff Elgin	*Capital One*
Graig Galef	*American Management Systems*
Jane Griffin	*Arthur Andersen & Co.*
Mary Haffey	*Arthur Andersen & Co.*
Don Hallacy	*Sprint Corporation*

Fred Hanson	*Ferguson*
Rixey Jones	*Home Depot*
Randeen Klarin	*EDS*
Diane Krehmeyer	*PriceWaterhouseCoopers*
Theresa Leahy	*Willis Company*
Pavan Marpaka	*Ascential Software*
Brice Marsh	*Computer Sciences Corporation*
Lyn McDermid	*Dominion Virginia Power*
Darrell Piatt	*First USA Bank*
David Price	*CoBank*
Don Stoller	*Owens & Minor*
Troy Venis	*CheckFree Corporation*
Carl Wilson	*Marriott Corporation*
Gillen Young	*Ernst & Young*

Thanks also to our families and friends for their patience and support along the way, especially Christopher Wixom, Haley Wixom, Linda Estey, Eileen Dennis, and Alec Dennis.

Alan Dennis
ardennis@indiana.edu

Barb Wixom
bwixom@mindspring.com

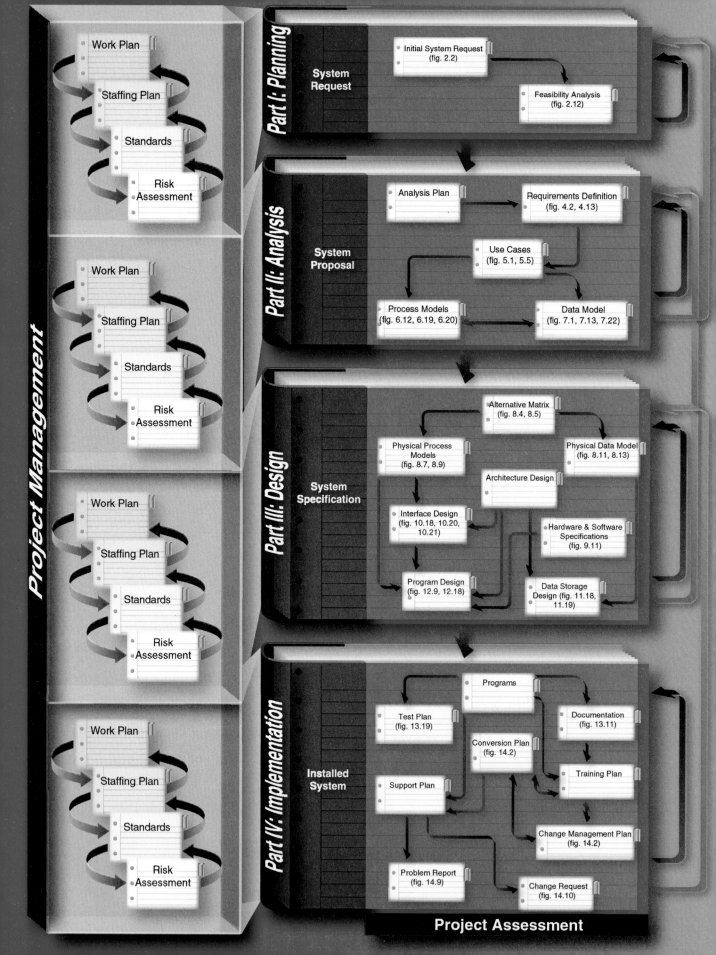

CHAPTER 1

INTRODUCTION

This chapter introduces the systems development life cycle, the fundamental four-phase model (planning, analysis, design, and implementation) that is common to all information system development projects. It then examines several commonly used methodologies that differ in their focus and approach to each of these phases. The chapter closes with a discussion of the skills and roles within the project team.

OBJECTIVES

- Understand the fundamental systems development life cycle and its four phases.
- Understand several different types of methodologies and how to choose among them.
- Be familiar with the different skills and roles on the project team.

CHAPTER OUTLINE

INTRODUCTION

The *systems development life cycle* (SDLC) is the process of understanding how an information system (IS) can support business needs, designing the system, building it, and delivering it to users. If you have taken a programming class or have programmed on your own, this probably sounds pretty simple. Unfortunately, it is not. A 1996 survey by the Standish Group found that 42% of all corporate IS projects were abandoned before completion. A similar study done in 1996 by the General Accounting Office found 53% of all U.S. government IS projects were abandoned. Unfortunately, many of the systems that aren't abandoned are delivered to the users significantly late, end up costing far more than expected, and have fewer features than originally planned.

Most of us would like to think that these problems only occur to other people or other organizations, but they happen in most companies. Even Microsoft has a history of failures and overdue projects (e.g., Windows 1.0, Windows 95).[1]

We would like to promote this book as a silver bullet that will keep you from IS failures, but a silver bullet that guarantees IS development success does not exist. Instead, this book will provide you with several fundamental concepts and many practical techniques that you can use to improve the probability of success.

The key person in the SDLC is the systems analyst, who analyzes the business situation, identifies opportunities for improvements, and designs an IS to implement them. Being a systems analyst is one of the most interesting, exciting, and challenging jobs around. As a systems analyst, you will work with a variety of people and learn how they conduct business. You will work with a team of systems analysts, programmers, and others on a common mission. You will feel the satisfaction of seeing systems that you designed and developed make a significant business impact, and you will know that you contributed your unique skills to make that happen.

It is important to remember that the primary objective of the systems analyst is not to create a wonderful system. The primary goal is to create value for the organization, which for most companies means increasing profits (government agencies and not-for-profit organizations measure value differently). Many failed systems were abandoned because the analysts tried to build a wonderful system without clearly understanding how the system would fit with the organization's goals, current business processes, and other information systems to provide value. An investment in an information system is like any other investment, such as a new machine tool. The goal is not to acquire the tool, because the tool is simply a means to an end; the goal is to enable the organization to perform work better so it can earn greater profits or serve its constituents more effectively.

This book will introduce you to the fundamental skills you need to be a systems analyst. This is a pragmatic book that discusses best practices in systems development; it does not present a general survey of systems development that exposes you to everything about the topic. By definition, systems analysts *do things* and challenge the current way that organizations work. To get the most out of this book, you will need to actively apply the ideas and concepts in the examples, those

---

[1] For more information on the problem, see Capers Jones, *Patterns of Software System Failure and Success,* London: International Thompson Computer Press, 1996; Capers Jones, *Assessment and Control of Software Risks,* Englewood Cliffs, NJ: Yourdon Press, 1994; Julia King, "IS Reins in Runaway Projects," *Computerworld.* February 24, 1997.

CONCEPTS

IN ACTION

1-A AN EXPENSIVE FALSE START

A real-estate group in the federal government cosponsored a data warehouse with the IT department. A formal proposal was written by IT in which costs were estimated at $800,000, the project duration was estimated to be eight months, and the responsibility for funding was defined as the business unit's. The IT department proceeded with the project before hearing whether the proposal was ever accepted.

The project actually lasted two years because requirements gathering took nine months instead of one and a half, the planned user base grew from 200 to 2,500, and the approval process to buy technology for

the project took a year. Three weeks prior to technical delivery, the IT Director canceled the project. This failed endeavor cost the organization $2.5 million.

Source: "Data Warehousing Failure: Case Studies and Findings" *The Journal of Data Warehousing,* by Hugh J. Watson et al., 4 (1), 1999, pp. 44–54.

QUESTION:

Why did this system fail? Why would a company spend money and time on a project and then cancel it? What could have been done to prevent this?

in the "Your Turn" exercises that are presented throughout, and, ideally, from those you develop in your own systems development project. This book will guide you through all the steps for delivering a successful information system. Also, we will illustrate how one organization (which we call CD Selections) applies the steps in one project (developing a Web-based CD sales system). By the time you finish the book, you won't be an expert analyst, but you will be ready to start building systems for real.

In this chapter, we first introduce the basic SDLC that IS projects follow. This life cycle is common to all projects, although the focus and approach to each phase of the life cycle may differ. In the next section, we discuss three fundamentally different types of methodologies (structured design, rapid application development, and agile development). Finally, we discuss one of the most challenging aspects of systems development—the depth and breadth of skills that are required. Today, most organizations use project teams that contain members with unique but complementary skills. This chapter closes with a discussion of the key roles played by members of the systems development team.

THE SYSTEMS DEVELOPMENT LIFE CYCLE

In many ways, building an information system is similar to building a house. First, the house (or the information system) starts with a basic idea. Second, this idea is transformed into a simple drawing that is shown to the customer and refined (often through several drawings, each improving on the other) until the customer agrees that the picture depicts what he or she wants. Third, a set of blueprints is designed that present much more detailed information about the house (e.g., the type of water faucets, where the telephone jacks will be placed). Finally, the house is built following the blueprints—and often with some changes and decisions made by the customer as the house is erected.

The SDLC has a similar set of four fundamental *phases*: planning, analysis design, and implementation (Figure 1-1). Different projects may emphasize differ-

Phase	Chapter	Step	Technique	Deliverable
Planning (Why build the system?) System Request	2 2	Identify Opportunity Analyze Feasibility	Project Identification Technical Feasibility Economic Feasibility Organizational Feasibility	System Request Feasibility Analysis
	3	Develop Workplan	Time Estimation Timeboxing Task Identification Work Breakdown Structure Pert Chart GANTT Chart Scope Management	Workplan
	3	Staff Project	Project Staffing Project Charter	Staffing Plan
	3	Control and Direct Project	CASE Repository Standards Documentation Risk Management	Standards List Risk Assessment
Analysis (Who, what, when, where will the system be?) System Proposal	4	Develop Analysis Strategy	Business Process Automation Business Process Improvement Business Process Reengineering	System Proposal
	4	Determine Business Requirements	Interview JAD session Questionnaire Document Analysis Observation	Requirements Definition
	5	Create Use Cases	Use-Case Analysis	Use Cases
	6	Model Processes	Data Flow Diagramming	Process Models
	7	Model Data	Entity Relationship Modeling Normalization	Data Model
Design (How will the system Work?) System Specification	8	Design Physical System	Design Selection Data Flow Diagramming Entity Relationship Modeling	Alternative Matrix Physical Process Models Physical Data Model System Specification
	9	Design Architecture	Architecture Design Hardware & Software Selection	Architecture Report Hardware & Software Specification
	10	Design Interface	Use Scenario Interface Structure Interface Standards Interface Prototype Interface Evaluation	Interface Design
	11	Design Databases and Files	Data Format Selection Denormalization Performance Tuning Size Estimation	Data Storage Design
	12	Design Programs	Transform Analysis Program Structure Chart Program Specification	Program Design
Implementation (System delivery) Installed System	13	Construct System	Programming Software Teasting Performance Testing	Test Plan Programs Documentation
	14	Install System	Conversion Style Selection Training	Conversion Plan Training Plan
	14	Maintain System	Support Selection System Maintenance Project Assessment	Support Plan Problem Report Change Request
	14	Postimplementation	Postimplementation Audit	Postimplementation Audit Report

FIGURE 1-1
Systems Development Life Cycle Phases

ent parts of the SDLC or approach the SDLC phases in different ways, but all projects have elements of these four phases. Each phase is itself composed of a series of *steps,* which rely on *techniques* that produce *deliverables* (specific documents files that provide understanding about the project).

For example, when you apply for admission to a university, there are several phases that all students go through: information gathering, applying, and accepting. Each of these phases has steps: information gathering includes such steps as searching for schools, requesting information, and reading brochures. Students then use techniques (e.g., Internet searching) that can be applied to steps (e.g., requesting information) to create deliverables (e.g., evaluations of different aspects of universities).

Figure 1-1 suggests that the SDLC phases and steps proceed in a logical path from start to finish. In some projects, this is true, but in many projects, the project teams move through the steps consecutively, iteratively, or in other patterns. In this section, we describe the phases and steps and some of the techniques that are used to accomplish the steps at a very high level. We should emphasize that not all organizations follow the SDLC in exactly the way described next. As we shall shortly see, there are many variations on the overall SDLC.

For now, there are two important points to understand about the SDLC. First, you should get a general sense of the phases and steps that IS projects move through and some of the techniques that produce certain deliverables. Second, it is important to understand that the SDLC is a process of *gradual refinement.* The deliverables produced in the analysis phase provide a general idea of the shape of the new system. These deliverables are used as input to the design phase, which then refines them to produce a set of deliverables that describe in much more detailed terms exactly how the system will be built. These deliverables in turn are used in the implementation phase to produce the actual system. Each phase refines and elaborates on the work done previously.

Planning

The *planning phase* is the fundamental process of understanding *why* an information system should be built and determining how the project team will go about building it. The first step is to identify opportunity, during which the system's business value to the organization is identified—how will it lower costs or increase profits? Most ideas for new systems come from outside the IS area (from the marketing department, accounting department, etc.) in the form of a *system request.* A system request presents a brief summary of a business need, and it explains how a system that supports the need will create business value. The IS department works together with the person or department that generated the request (called the *project sponsor*) to conduct a *feasibility analysis* that examines the idea's technical feasibility (i.e., can we build it?), the economic feasibility (i.e., will it provide business value?), and the organizational feasibility (i.e., if we build it, will it be used?).

The system request and feasibility analysis are presented to an IS *approval committee* (sometimes called a steering committee), which decides whether the project should be undertaken. If the committee approves the project, then the next step of planning occurs—*project management.* During project management, the *project manager* creates a *workplan,* staffs the project, and puts techniques in place to help him or her control and direct the project through the entire SDLC. The deliverable for project management is a *project workplan* that describes how the project team will go about developing the system. This workplan includes all of the project management deliverables.

Analysis

The *analysis phase* answers the questions of *who* will use the system, *what* the system will do, and *where* and *when* it will be used. During this phase, the project team investigates any current system(s), identifies improvement opportunities, and develops a concept for the new system. See Figure 1-1.

Analysis begins with the development of an *analysis strategy* that guides the project team's efforts. Such a strategy usually includes an analysis of the current system (called the *as-is system*) and its problems and then an analysis of ways to design a new system (called the *to-be system*). The next step is gathering business requirements (e.g., through interviews or questionnaires). The analysis of this information—in conjunction with input from the project sponsor and many other people—leads to the development of a list of business requirements for the new system. This Requirements Definition is then used as a basis to develop use cases and a *process model* that describe how the business will operate if the new system were developed. Figure 1-2 shows an example process model, with processes (the tasks that someone performs) shown as boxes with rounded corners, inputs and outputs of the processes as arrows, and data storage containers as boxes with an open side. Finally, a *data model* is developed to describe the information that is needed to support the process.

The analyses, Requrements Definition Use Cases, process model, and data model are combined into a document called the *system proposal,* which is presented to the project sponsor and other key decision makers (e.g., members of the approval committee) who decide whether the project should continue to move forward. See Figure 1-1.

The system proposal is the initial deliverable that describes what business requirements the new system should meet. Because it is really the first step in the design of the new system, some experts argue that it is inappropriate to use the term *analysis,* as the name for this phase, some argue a better name would be *analysis*

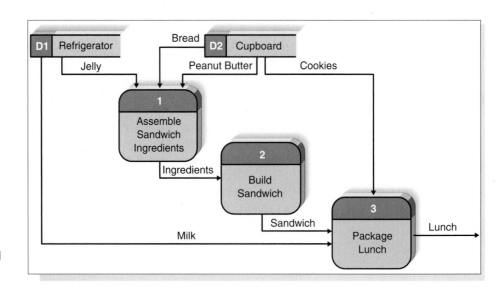

FIGURE 1-2
A Simple Process Model for Making Lunch

and initial design. Because most organizations continue to use the name *analysis* for this phase, we will use it in this book as well, but it is important to remember that the deliverable from the analysis phase is both an analysis and a high-level initial design for the new system.

Design

The *design phase* decides *how* the system will operate, in terms of the hardware, software, and network infrastructure; the user interface, forms, and reports that will be used; and the specific programs, databases, and files that will be needed. Although most of the strategic decisions about the system were made in the development of the system concept during the analysis phase, the steps in the design phase determine exactly how the system will operate.

The first step in the design phase is to perform the *design selection:* will the system will be developed by the company's own programmers, will it will be outsourced to another firm (usually a consulting firm), or will the company will buy an existing software package. This leads to the development of the basic *architecture design* for the system that describes the hardware, software, and network infrastructure that will be used. In most cases, the system will add or change the infrastructure that already exists in the organization. The *interface design* specifies how the users will move through the system (i.e., navigation methods such as menus and on-screen buttons) and the forms and reports that the system will use. Next, the *data storage design* that defines exactly what and how data will be stored is developed. Finally, the analysis team develops the *program design,* which defines the programs that need to be written and exactly what each program will do.

This collection of deliverables (architecture design, interface design, database and file specifications, and program design) is the *system specification* that is handed to the programming team for implementation. At the end of the design phase, the feasibility analysis and workplan are reexamined and revised, and another decision is made by the project sponsor and approval committee about whether to terminate the project or continue. See Figure 1-1.

Implementation

The final phase in the SDLC is the *implementation phase,* during which the system is actually built (or purchased, in the case of a packaged software design). This is the phase that usually gets the most attention, because for most systems it is the longest and most expensive single part of the development process.

The first step in implementation is system *construction,* during which the system is built and tested to ensure it performs as designed. Testing is one of the most critical steps in implementation, because the cost of bugs can be immense; most organizations spend more time and attention on testing than on writing the programs in the first place. Once the system has passed a series of tests, it is installed. *Installation* is the process by which the old system is turned off and the new one turned on, and it may include a direct cutover approach (in which the new system immediately replaces the old system), a parallel conversion approach (in which both the old and new systems are operated for a month or two until it is clear that there are no bugs in the new system), or a phased conversion strategy (in which the new system is installed in one part of the organization as an initial trial and then gradually installed in others). One of the most important aspects of conversion is the development of a *training plan* to teach users how to use the new system and help manage the changes caused by the new system.

Once the system has been installed, the analyst team establishes a *support plan* for the system. This plan usually includes a formal or informal postimplementation review as well as a systematic way for identifying major and minor changes needed for the system.

SYSTEMS DEVELOPMENT METHODOLOGIES

A *methodology* is a formalized approach to implementing the SDLC (i.e., it is a list of steps and deliverables). There are many different systems development methodologies, and each one is unique because of its emphasis on processes versus data and the order and focus it places on each SDLC phase. Some methodologies are formal standards used by government agencies, whereas others have been developed by consulting firms to sell to clients. Many organizations have their own internal methodologies that have been refined over the years, and they explain exactly how each phase of the SDLC is to be performed in that company.

There are many ways to categorize methodologies. One way is by looking at whether they focus on business processes or on the data that supports the business. Methodologies are *process centered* if they emphasize process models (Chapter 6) as the core of the system concept. In Figure 1-2, for example, process-centered methodologies would focus first on defining the processes (e.g., assemble sandwich ingredients). Methodologies are *data centered* if they emphasize data models (Chapter 7) as the core of the system concept. In Figure 1-2, for example, data-centered methodologies would focus first on defining the contents of the storage areas (e.g., refrigerator) and how the contents were organized. Other methodologies, such as *object-oriented methodologies* (Chapter 15), attempt to balance the focus between process and data by incorporating both into one model.[2] In Figure 1-2, these methodologies would focus first on defining the major elements of the system (e.g., sandwiches, lunches) and look at the processes and data (i.e., storage items) that were involved with each.

Another important factor in categorizing methodologies is the sequencing of the SDLC phases and the amount of time and effort devoted to each.[3] In the early days of computing, the need for formal and well-planned life cycle methodologies was not well understood. Programmers tended to move directly from a very simple planning phase right into the construction step of the implementation phase; in other words, they moved directly from a very fuzzy, not-well-thought-out system request into writing code, which is the same approach that you sometimes use when writing programs for a programming class. This approach can work for small programs that

---

[2] The classic modern process-centered methodology is that by Edward Yourdon, *Modern Structured Analysis,* Englewood Cliffs, NJ: Yourdon Press, 1989. An example of a data-centered methodology is information engineering by James Martin, *Information Engineering,* volumes 1-3, Englewood Cliffs, NJ: Prentice Hall, 1989. Many new object-oriented methodologies are based on the Unified Modeling Language defined in *UML Document Set,* Santa Clara, CA: Relational Software Corp., 1997. A widely accepted standardized methodology that balances processes and data is IDEF; see FIPS 183, *Integration Definition for Function Modeling,* Federal Information Processing Standards Publications, Washington, D.C.: U.S. Department of Commerce, 1993.

[3] A good reference for comparing systems development methodologies is Steve McConnell, *Rapid Development,* Redmond, WA: Microsoft Press, 1996.

require only one programmer, but if the requirements are complex or unclear, you may miss important aspects of the problem and have to start all over again, throwing away part of the program (and the time and effort spent writing it). This approach also makes teamwork difficult because members have little idea about what needs to be accomplished and how to work together to produce a final product.

Structured Design

The first category of systems development methodologies is called *structured design.* These methodologies became dominant in the 1980s, replacing the previous ad hoc and undisciplined approach. Structured design methodologies adopt a formal step-by-step approach to the SDLC that moves logically from one phase to the next. There are numerous process-centered methodologies and data-centered methodologies that follow the basic approach of the two structured design categories outlined next.

Waterfall Development   The original structured design methodology (that is still used today) is *waterfall development.* With waterfall development, the analysts and users proceed in sequence from one phase to the next; see Figure 1-3. The key deliverables for each phase are typically produced on paper (often hundreds of pages in length) and are presented to the project sponsor for approval as the project moves from phase to phase. Once the sponsor approves the work that was conducted for a phase, the phase ends and the next one begins. This approach is called waterfall development because it moves forward from phase to phase in the same manner as a waterfall. Although it is possible to go backward in the SDLC (e.g., from design back to analysis), it is extremely difficult. (Imagine yourself as a salmon trying to swim upstream in a waterfall as shown in Figure 1-3.)

The two key advantages of waterfall development are that it identifies system requirements long before programming begins and that it minimizes changes to the

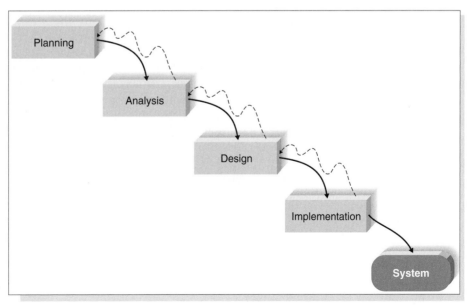

FIGURE 1-3
The Waterfall Development Methodology

requirements as the project proceeds. The two key disadvantages are that the design must be completely specified on paper before programming begins and that a long time elapses between the completion of the system proposal in the analysis phase and the delivery of the system (usually many months or years). A paper document is often a poor communication mechanism, so important requirements can be overlooked in the hundreds of pages of documentation. Users rarely are prepared for their introduction to the new system, which occurs long after the initial idea for the system was introduced. If the project team misses important requirements, expensive postimplementation programming may be needed. (Imagine yourself trying to design a car on paper. How likely would you be to remember to include interior lights that come on when the doors open or to specify the right number of valves on the engine?)

A system also may require significant rework because the business environment has changed from the time that the analysis phase occurred. When changes do occur, it means going back to the initial phases and following the change through each of the subsequent phases in turn.

Parallel Development The *parallel development* methodology attempts to address the problem of long delays between the analysis phase and the delivery of the system. Instead of doing the design and implementation in sequence, it performs a general design for the whole system and then divides the project into a series of distinct subprojects that can be designed and implemented in parallel. Once all subprojects are complete, there is a final integration of the separate pieces, and the system is delivered (Figure 1-4).

The primary advantage of this methodology is that it can reduce the schedule time required to deliver a system; thus there is less chance of changes in the business environment causing rework. However, the approach still suffers from problems caused by paper documents. It also adds a new problem: sometimes the subprojects are not completely independent; design decisions made in one subproject may affect another, and the end of the project may require significant integrative efforts.

Rapid Application Development

Rapid application development (RAD) is a newer approach to systems development that emerged in the 1990s. RAD attempts to address both weaknesses of the structured development methodologies: long development times and the difficulty in understanding a system from a paper-based description. RAD methodologies adjust the SDLC phases to get some part of the system developed quickly and into the hands of the users. In this way, the users can better understand the system and suggest revisions that bring the system closer to what is needed.[4]

Most RAD methodologies recommend that analysts use special techniques and computer tools to speed up the analysis, design, and implementation phases, such as CASE (computer-aided software engineering) tools (see Chapter 3), JAD (joint application design) sessions (see Chapter 5), fourth-generation/visual programming languages that simplify and speed up programming (e.g., Visual Basic), and code generators that automatically produce programs from design specifica-

---

[4] One of the best RAD books is that by Steve McConnell, *Rapid Development,* Redmond. WA: Microsoft Press, 1996.

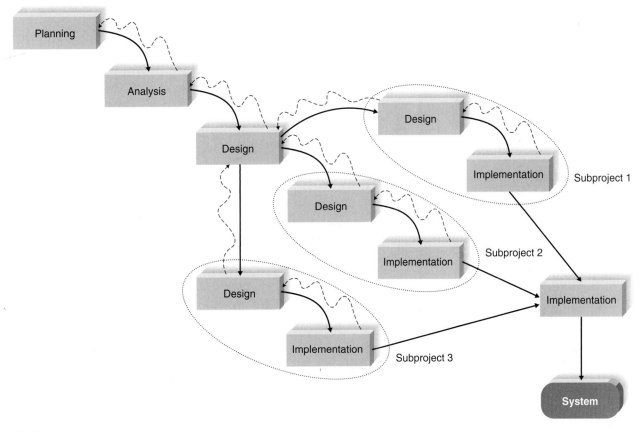

FIGURE 1-4
The Parallel Development Methodology

tions. It is the combination of the changed SDLC phases and the use of these tools and techniques that improves the speed and quality of systems development. There are process-centered, data-centered, and object-oriented methodologies that follow the basic approaches of the three RAD categories described next.

Phased development The *phased development* methodology breaks the overall system into a series of *versions* that are developed sequentially. The analysis phase identifies the overall system concept, and the project team, users, and system sponsor then categorize the requirements into a series of versions. The most important and fundamental requirements are bundled into the first version of the system. The analysis phase then leads into design and implementation, but only with the set of requirements identified for version 1 (Figure 1-5).

Once version 1 is implemented, work begins on version 2. Additional analysis is performed on the basis of the previously identified requirements and combined with new ideas and issues that arose from users' experience with version 1 . Version 2 then is designed and implemented, and work immediately begins on the next version. This process continues until the system is complete or is no longer in use.

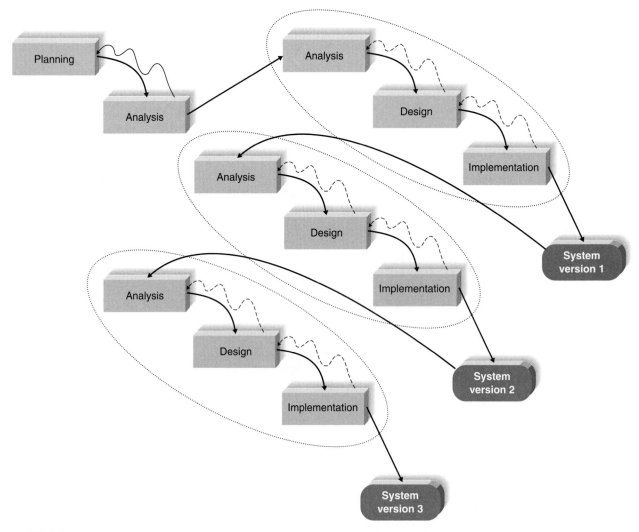

FIGURE 1-5
The Phased Development Methodology

Phased development has the advantage of quickly getting a useful system into the hands of the users. Although it does not perform all the functions the users need at first, it begins to provide business value sooner than if the system were delivered after completion, as is the case with the waterfall or parallel methodologies. Likewise, because users begin to work with the system sooner, they are more likely to identify important additional requirements sooner than with structured design situations.

The major drawback to phased development is that users begin to work with systems that are intentionally incomplete. It is critical to identify the most important and useful features and include them in the first version while managing users' expectations along the way.

Prototyping The *prototyping* methodology performs the analysis, design, and implementation phases concurrently, and all three phases are performed repeatedly

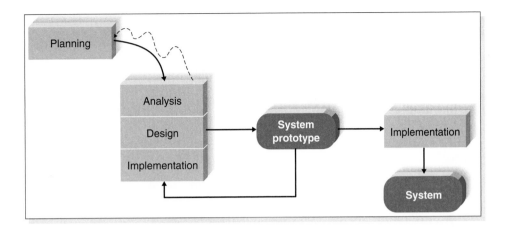

FIGURE 1-6
The Prototyping Methodology

in a cycle until the system is completed. With this approach, the basics of analysis and design are performed, and work immediately begins on a *system prototype,* a "quick-and-dirty" program that provides a minimal amount of features. The first prototype is usually the first part of the system that the user will use. This is shown to the users and the project sponsor, who provide comments, which are used to reanalyze, redesign, and reimplement a second prototype that provides a few more features. This process continues in a cycle until the analysts, users, and sponsor agree that the prototype provides enough functionality to be installed and used in the organization. After the prototype (now called the "system") is installed, refinement occurs until it is accepted as the new system (Figure 1-6).

The key advantage of prototyping is that it *very* quickly provides a system for the users to interact with, even if it is not ready for widespread organizational use at first. Prototyping reassures the users that the project team is working on the system (there are no long delays in which the users see little progress), and the approach helps to more quickly refine real requirements. Rather than attempting to understand a system specification on paper, the users can interact with the prototype to better understand what it can and cannot do.

The major problem with prototyping is that its fast-paced system releases challenge attempts to conduct careful, methodical analysis. Often the prototype undergoes such significant changes that many initial design decisions become poor ones. This can cause problems in the development of complex systems because fundamental issues and problems are not recognized until well into the development process. Imagine building a car and discovering late in the prototyping process that you have to take the whole engine out to change the oil (because no one thought about the need to change the oil until after it had been driven 10,000 miles).

Throwaway Prototyping The *throwaway prototyping* methodology is similar to the prototyping methodology in that it includes the development of prototypes; however, throwaway prototypes are done at a different point in the SDLC. These prototypes are used for a very different purpose than ones previously discussed, and they have a very different appearance[5] (Figure 1-7).

---

[5] Our description of the throwaway prototyping methodology is a modified version of the spiral development methodology developed by Barry Boehm, "A Spiral Model of Software Development and Enhancement," *Computer,* May 1988, 21(5):61–72.

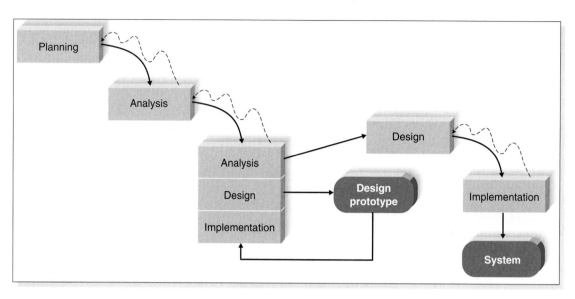

FIGURE 1-7
The Throwaway Prototyping Methodology

Throwaway prototyping has a relatively thorough analysis phase that is used to gather information and to develop ideas for the system concept. However, many of the features suggested by the users may not be well understood, and there may be challenging technical issues to be solved. Each of these issues is examined by analyzing, designing, and building a *design prototype*. A design prototype is not a working system; it only represents a part of the system that needs additional refinement and it contains only enough detail to enable users to understand the issues under consideration. For example, suppose users are not completely clear on how an order entry system should work. The analyst team might build a series of hypertext mark-up language (HTML) pages viewed using a Web browser to help the users visualize such a system. In this case, a series of mock-up screens *appear* to be a system, but they really do nothing. Alternately, suppose that the project team needs to develop a sophisticated graphics program in Java. The team could write a portion of the program with pretend data to ensure that they could do a full-blown program successfully.

A system that is developed using this approach typically uses several design prototypes during the analysis and design phases. Each of the prototypes is used to minimize the risk associated with the system by confirming that important issues are understood before the real system is built. Once the issues are resolved, the project moves into design and implementation. At this point, the design prototypes are thrown away, which is an important difference between this approach and prototyping, in which the prototypes evolve into the final system.

Throwaway prototyping balances the benefits of well-thought-out analysis and design phases with the advantages of using prototypes to refine key issues before a system is built. It may take longer to deliver the final system as compared with prototyping (because the prototypes do not become the final system), but the approach usually produces more stable and reliable systems.

Agile Development

The third category of systems development methodologies is still emerging today, and it is called *Agile Development*. These programming-centric methodologies have few rules and practices, all of which are fairly easy to follow. They focus on streamlining the SDLC by eliminating much of the modeling and documentation overhead, and the time spent on those tasks. Instead, projects emphasize simple, iterative application development. Examples of Agile Development methodologies include extreme programming, scrum, and the dynamic systems development method (DSDM). The Agile Development approach as described next typically is used in conjunction with object-oriented methodologies.

Extreme Programming   *Extreme programming (XP)*[6] uses continuous testing, simple coding performed by pairs of developers, and close interactions with end users to build systems very quickly. After a superficial planning process, projects perform analysis, design, and implementation phases iteratively (Figure 1-8). The system functionality grows over time.

Testing and efficient coding practices are core to XP. In fact, each day code is tested and placed into an integrative testing environment. If bugs exist, the code is backed out until it is completely free of errors. XP relies heavily on refactoring, which is a disciplined way to restructure code to keep it simple.

An XP project begins with user stories that describe what the system needs to do. Then, programmers code in small, simple modules and test to meet those needs. Users are required to be available to clear up questions and issues as they arise. Standards are very important to minimize confusion, so XP teams use a common set of names, descriptions, and coding practices.

XP projects deliver results sooner than even the RAD approaches, and they rarely get bogged down in gathering requirements for the system. XP works very well when requirements are undefined or rapidly changing, and it can be a good fit with object-oriented technologies. The methodology works well with projects that have very short schedules or time-critical requirements.

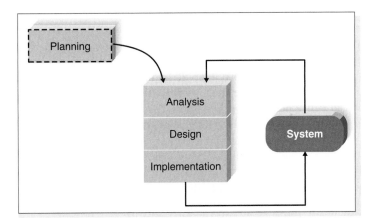

FIGURE 1-8
The Extreme Programming Methodology

[6] For more information, see *Extreme Programming Explained: Embrace Change*, City: Addison-Wesley, 1999.

Ability to Develop Systems	Structured Methodologies			RAD Methodologies		Agile Methodologies
	Waterfall	Parallel	Phased	Prototyping	Throwaway Prototyping	XP
with Unclear User Requirements	Poor	Poor	Good	Excellent	Excellent	Excellent
with Unfamiliar Technology	Poor	Poor	Good	Poor	Excellent	Poor
that are Complex	Good	Good	Good	Poor	Excellent	Poor
that are Reliable	Good	Good	Good	Poor	Excellent	Good
with a Short Time Schedule	Poor	Good	Excellent	Excellent	Good	Excellent
with Schedule Visibility	Poor	Poor	Excellent	Excellent	Good	Good

FIGURE 1-9
Criteria for Selecting a Methodology

XP requires a great deal of discipline, however. Otherwise, projects may become unfocused and chaotic. It is recommended for small groups of developers—no more than twelve developers working in pairs, and it is not advised for large applications. Also, the methodology needs a lot of on-site user input, something to which many business units cannot commit.

Selecting the Appropriate Development Methodology

Since there are many methodologies, the first challenge faced by analysts is to select which methodology to use. Choosing a methodology is not simple, because no one methodology is always best. (If it were, we'd simply use it everywhere!) Many organizations have standards and policies to guide the choice of methodology. You will find that organizations range from having one approved methodology to having several methodology options to having no formal policies at all.

Figure 1-9 summarizes some important methodology selection criteria. One important item not discussed in this figure is the degree of experience of the analyst team. Many of the RAD methodologies require the use of new tools and techniques that have a significant learning curve. Often these tools and techniques increase the complexity of the project and extra require time for learning. However, once they are adopted and the team becomes experienced, they can significantly increase the speed of the SDLC.

Clarity of User Requirements   When the user requirements for what the system should do are unclear or subject to change, it is difficult to understand them by talking about them and explaining them with written reports. Users normally need to interact with technology to really understand what the new system can do and how to best apply it to their needs. Prototyping, throwaway prototyping, and XP usually more appropriate when user requirements are unclear or unstable because they provide prototypes for users to interact with early in the SDLC.

Familiarity with Technology   When the system will use new technology with which the analysts and programmers are not familiar (e.g., the first Web development project with Java), early application of the new technology in the SDLC will improve the chance of success. If the system is designed without some familiarity with the base technology, risks increase because the tools may not be capable of doing what is needed. Throwaway prototyping is particularly appropriate for a lack of familiarity with technology because it explicitly encourages the developers to develop design prototypes for areas with high risks. Phased development is good as well because it creates opportunities to investigate the technology in some depth before the design is complete. Although one might think prototyping would also be appropriate, it is much less so, because the early prototypes that are built usually only scratch the surface of the new technology. Usually, it is only after several prototypes and several months that the developers discover weaknesses or problems in the new technology.

System Complexity   Complex systems require careful and detailed analysis and design. Throwaway prototyping is particularly well suited to such detailed analysis and design, but prototyping is not. The traditional structured methodologies can handle complex systems, but without the ability to get the system or prototypes into users' hands early on, some key issues may be overlooked. Even though the phased methodology enables users to interact with the system early in the process, we have observed that project teams who follow this methodology tend to devote less attention to the analysis of the complete problem domain than they might if they were using other methodologies.

System Reliability   System reliability is usually an important factor in system development. After all, who wants an unreliable system? However, reliability is just one factor among several. For some applications reliability is truly critical (e.g., medical equipment, missile control systems), whereas for other applications is it is merely important (e.g., games, Internet video). Throwaway prototyping is the most appropriate when system reliability is a high priority, because it combines detailed analysis and design phases with the ability for the project team to test many different approaches through design prototypes before completing the design. Prototyping is generally not a good choice when reliability is critical, because it lacks the careful analysis and design phases that are essential for dependable systems.

Short Time Schedules   Projects that have short time schedules are well suited for RAD and Agile methodologies because those methodologies are designed to increase the speed of development. Prototyping, phased development, and XP are excellent choices when timelines are short because they best enable the project team to adjust the functionality in the system on the basis of a specific delivery date, and if the project schedule starts to slip, it can be readjusted by removing functionality from the version or prototype under development. The waterfall methodology is the worst choice when time is at a premium because it does not allow for easy schedule changes.

Schedule Visibility   One of the greatest challenges in systems development is knowing whether a project is on schedule. This is particularly true of the structured methodologies because design and implementation occur at the end of the project. The RAD methodologies move many of the critical design decisions to an earlier point in the project to help project managers recognize and address risk factors and keep expectations in check.

PROJECT TEAM SKILLS AND ROLES

As should be clear from the various phases and steps performed during the SDLC, the project team needs a variety of skills. Project members are *change agents* who identify ways to improve an organization, build an information system to support them, and train and motivate others to use the system. Leading a successful organizational change effort is one of the most difficult jobs that someone can do. Understanding what to change, how to change it, and convincing others of the need for change requires a wide range of skills. These skills can be broken down into six major categories: technical, business, analytical, interpresonal, management, and ethical.

Analysts must have the technical skills to understand the organization's existing technical environment, the technology that will comprise the new system, and the way in which both can be fit into an integrated technical solution. Business skills are required to understand how IT can be applied to business situations and to ensure that the IT delivers real business value. Analysts are continuous problem solvers at both the project and the organizational level, and they put their analytical skills to the test regularly.

Often, analysts need to communicate effectively one-on-one with users and business managers (who often have little experience with technology) and with programmers (who often have more technical expertise than the analyst). They must be able to give presentations to large and small groups and write reports. Not only do they need to have strong interpersonal abilities, but also they need to manage people with whom they work and to manage the pressure and risks associated with unclear situations.

Finally, analysts must deal fairly, honestly, and ethically with other project team members, managers, and system users. Analysts often deal with confidential information or information that if shared with others could cause harm (e.g., dissent among employees); it is important to maintain confidence and trust with all people.

In addition to these six general skill sets, analysts require many specific skills that are associated with roles that are performed on a project. In the early days of systems development, most organizations expected one person, the analyst, to have all of the specific skills needed to conduct a systems development project. Some small organizations still expect one person to perform many roles, but because organizations and technology have become more complex, most large organizations now build project teams that contain several individuals with clearly defined responsibilities. Different organizations divide the roles differently, but Figure 1-10

YOUR TURN

1-1 SELECTING A METHODOLOGY

Suppose you are an analyst for the ABC Company, a large consulting firm with offices around the world. The company wants to build a new knowledge management system that can identify and track the expertise of individual consultants anywhere in the world on the basis of their education and the various consulting projects on which they have worked. Assume that this is a new idea that has never before been attempted in ABC or elsewhere. ABC has an international network, but the offices in each country may use somewhat different hardware and software. ABC management wants the system up and running within a year.

QUESTION:

What methodology would you recommend ABC Company use? Why?

Role	Responsibilities
Business analyst	Analyzing the key business aspects of the system
	Identifying how the system will provide business value
	Designing the new business processes and policies
Systems analyst	Identifying how technology can improve business processes
	Designing the new business processes
	Designing the information system
	Ensuring that the system conforms to information systems standards
Infrastructure analyst	Ensuring the system conforms to infrastructure standards
	Identifying infrastructure changes needed to support the system
Change management analyst	Developing and executing a change management plan
	Developing and executing a user training plan
Project manager	Managing the team of analysts, programmers, technical writers, and other specialists
	Developing and monitoring the project plan
	Assigning resources
	Serving as the primary point of contact for the project

FIGURE 1-10
Project Team Roles

presents one commonly used set of project team roles. Most IS teams include many other individuals, such as the *programmers* who actually write the programs that make up the system and *technical writers*, who prepare the help screens and other documentation (e.g., users manuals, systems manuals).

Business Analyst

The *business analyst* focuses on the business issues surrounding the system. These include identifying the business value that the system will create, developing ideas and suggestions for how the business processes can be improved, and designing the new processes and policies in conjunction with the systems analyst. This individual will likely have business experience and some type of professional training (e.g., the business analyst for accounting systems will likely be a CPA [certified public accountant in the United States] or a CA [chartered accountant in Great Britain and Canada]). He or she represents the interests of the project sponsor and the ultimate users of the system. The business analyst assists in the planning and design phases but is most active in the analysis phase.

Systems Analyst

The *systems analyst* focuses on the IS issues surrounding the system. This person develops ideas and suggestions for how IT can improve business processes, designs the new business processes with help from the business analyst, designs the new information system, and ensures that all IS standards are maintained. The systems

Suppose you decide to become an analyst after you graduate. What type of analyst would you most prefer to be? What type of courses should you take before you graduate? What type of summer job or internship should you seek?

QUESTION:
Develop a short plan that describes how you will prepare for your career as an analyst.

analyst likely will have significant training and experience in analysis and design, programming, and even areas of the business. He or she represents the interests of the IS department and works intensively throughout the project but perhaps less so during the implementation phase.

Infrastructure Analyst

The *infrastructure analyst* focuses on the technical issues surrounding how the system will interact with the organization's technical infrastructure (e.g., hardware, software, networks, and databases). The infrastructure analyst's tasks include ensuring that the new information system conforms to organizational standards and identifying infrastructure changes needed to support the system. This individual will likely have significant training and experience in networking, database administration, and various hardware and software products. He or she represents the interests of the organization and IS group that ultimately will have to operate and support the new system once it has been installed. The infrastructure analyst works throughout the project but perhaps less so during planning and analysis phases.

Change Management Analyst

The *change management analyst* focuses on the people and management issues surrounding the system installation. The roles of this person include ensuring that adequate documentation and support are available to users, providing user training on the new system, and developing strategies to overcome resistance to change. This individual likely will have significant training and experience in organizational behavior in general and change management in particular. He or she represents the interests of the project sponsor and users for whom the system is being designed. The change management analyst works most actively during the implementation phase but begins laying the groundwork for change during the analysis and design phases.

Project Manager

The *project manager* is responsible for ensuring that the project is completed on time and within budget and that the system delivers all benefits that were intended by the project sponsor. The role of the project manager includes managing the team members, developing the project plan, assigning resources, and being the primary point of contact when people outside the team have questions about the project. This individual likely will have significant experience in project management and likely has worked for many years as a systems analyst beforehand. He or she represents the interests of the IS department and the project sponsor. The project manager works intensely during all phases of the project.

SUMMARY

The System Development Life Cycle

All system development projects follow essentially the same fundamental process called the system development life cycle (SDLC). SDLC starts with a planning phase in which the project team identifies the business value of the system, conducts a feasibility analysis, and plans the project. The second phase is the analysis phase, in which the team develops an analysis strategy, gathers information, writes use cases, builds a process model, and builds a data model. In the next phase, the design phase, the team develops the physical design, architecture design, interface design, data storage design, and program design. In the final phase, implementation, the system is built, installed, and maintained.

Systems Development Methodologies

Structured design methodologies, such as waterfall and parallel development, use a formal step-by-step approach to the SDLC that moves logically from one phase to the next (and makes it hard to reverse direction). They produce a solid, well-thought-out system but can overlook requirements because users must specify them early in the design process before seeing the actual system. Rapid application development (RAD) methodologies attempt to speed up development and make it easier for users to specify requirements by having parts of the system developed sooner either by producing different versions (phased development) or by using prototypes (prototyping, throwaway prototyping). Agile development methodologies focus on streamlining the SDLC by eliminating many of the tasks and time associated with requirements definition and documentation. The choice of a methodology is influenced by the clarity of the user requirements, familiarity with the base technology, system complexity, the need for system reliability, time pressures, and the need to see progress on the time schedule.

Project Team Roles and Skills

The project team needs a variety of skills. All analysts need to have general skills, such as technical, business, analytical, interpersonal, management, and ethical. However, different kinds of analysts require specific skills in addition to these. Business analysts usually have business skills that help them to understand the business issues surrounding the system, whereas systems analysts also have significant experience in analysis and design and programming. The infrastructure analyst focuses on technical issues surrounding how the system will interact with the organization's technical infrastructure, and the change management analyst focuses on people and management issues surrounding the system installation. In addition to analysts, project teams will include a project manager, programmers, technical writers, and other specialists.

KEY TERMS

Agile development	Gradual refinement	Step
Analysis phase	Implementation phase	Structured design
Analysis strategy	Infrastructure analyst	Support plan
Approval committee	Interface design	System analyst
Architecture design	Methodology	System development life cycle
As-is system	Object-oriented methodology	(SDLC)
Business analyst	Parallel development	System proposal
Change agent	Phase	System prototype
Change managament analyst	Phased development	System request
Construction	Planning phase	System specification
Conversion	Process model	Technical writer
Data model	Process-centered methodology	Technique
Data storage design	Program design	Throwaway prototyping
Data-centered methodology	Programmer	To-be system
Deliverable	Project manager	Training plan
Design phase	Project binder	Version
Design prototype	Project sponsor	Waterfall development
Design selection	Prototyping	Workplan
Extreme programming (XP)	Rapid application development	
Feasibility analysis	(RAD)	

QUESTIONS

1. Compare and contrast phases, steps, techniques, and deliverables.
2. Describe the major phases in the systems development life cycle (SDLC).
3. Describe the principal steps in the planning phase. What are some major deliverables?
4. Describe the principal steps in the analysis phase. What are some major deliverables?
5. Describe the principal steps in the design phase. What are some major deliverables?
6. Describe the principal steps in the implementation phase. What are some major deliverables?
7. What does *gradual refinement* mean in the context of SDLC?
8. Compare and contrast process-centered methodologies, data-centered methodologies, and object-oriented methodologies.
9. Compare and contrast structured design methodologies in general to rapid application development (RAD) methodologies in general.
10. Compare and contrast extreme programming and throwaway prototyping.

11. Describe the major elements and issues with waterfall development.
12. Describe the major elements and issues with parallel development.
13. Describe the major elements and issues with phased development.
14. Describe the major elements and issues with prototyping.
15. What are the key factors in selecting a methodology?
16. What are the six general skills all project team members should have?
17. What are the major roles on a project team?
18. Compare and contrast the role of a systems analyst, business analyst, and infrastructure analyst.
19. Why do you think the most popular methodology in the 1970s and 1980s was waterfall development? Why do you think RAD approaches are more common today? Why are Agile approaches becoming popular?
20. Which phase in the SDLC is most important?

EXERCISES

A. Suppose you are a project manager using the water-fall development methodology on a large and complex project. Your manager has just read the latest article in *Computerworld* that advocates replacing the waterfall methodology with prototyping and comes to your office requesting you to switch. What do you say?

B. The six basic methodologies discussed in this chapter can be combined and integrated to form new hybrid methodologies. Suppose you were to combine throwaway prototyping with the use of parallel development. What would the methodology look like? Draw a picture (similar to Figure 1-7). How would this new methodology compare to the others? Develop a new column for Figure 1-8.

C. Suppose you were an analyst working for a small company to develop an accounting system. What methodology would you use? Why?

D. Suppose you were an analyst developing a new executive information system (EIS) intended to provide key strategic information from existing corporate databases to senior executives to help in their decision making. What methodology would you use? Why?

E. Suppose you were an analyst developing a new information system to automate the sales transactions and manage inventory for each retail store in a large chain. The system would be installed at each store and exchange data with a mainframe computer at the company's head office. What methodology would you use? Why?

F. Look in the classified section of your local newspaper. What kinds of job opportunities are available for people who want analyst positions? Compare and contrast the skills that the ads ask for to the skills that we presented in this chapter.

G. Think about your ideal analyst position. Write a newspaper ad to hire someone for that position. What requirements would the job have? What skills and experience would be required? How would applicants demonstrate that they have the appropriate skills and experience?

MINICASES

1. Barbara Singleton, manager of western regional sales at the WAMAP Company, requested that the IS department develop a sales force management and tracking system that would enable her to better monitor the performance of her sales staff. Unfortunately, due to the massive backlog of work facing the IS department, her request was given a low priority. After 6 months of inaction by the IS department, Barbara decided to take matters into her own hands. Based on the advice of friends, Barbara purchased a PC and simple database software and constructed a sales force management and tracking system on her own.

   Although Barbara's system has been "completed" for about 6 weeks, it still has many features that do not work correctly, and some functions are full of errors. Barbara's assistant is so mistrustful of the system that she has secretly gone back to using her old paper-based system, since it is much more reliable.

   Over dinner one evening, Barbara complained to a systems analyst friend, "I don't know what went wrong with this project. It seemed pretty simple to me. Those IS guys wanted me to follow this elaborate set of steps and tasks, but I didn't think all that really applied to a PC-based system. I just thought I could build this system and tweak it around until I got what I wanted without all the fuss and bother of the methodology the IS guys were pushing. I mean, doesn't that just apply to their big, expensive systems?"

   Assuming you are Barbara's systems analyst friend, how would you respond to her complaint?

2. Marcus Weber, IS project manager at ICAN Mutual Insurance Co., is reviewing the staffing arrangements for his next major project, the development of an expert system-based underwriters assistant. This new system will involve a whole new way for the underwriters to perform their tasks. The underwriters assistant system will function as sort of an underwriting supervisor, reviewing key elements of each application, checking for consistency in the underwriter's decisions, and ensuring that no critical factors have been overlooked. The goal of the new system is to improve the quality of the underwriters' decisions and to improve underwriter productivity. It is expected that the new system will substantially change the way the underwriting staff do their jobs.

Marcus is dismayed to learn that due to budget constraints, he must choose between one of two available staff members. Barry Filmore has had considerable experience and training in individual and organizational behavior. Barry has worked on several other projects in which the end users had to make significant adjustments to the new system, and Barry seems to have a knack for anticipating problems and smoothing the transition to a new work environment. Marcus had hoped to have Barry's involvement in this project.

Marcus's other potential staff member is Kim Danville. Prior to joining ICAN Mutual, Kim had considerable work experience with the expert system technologies that ICAN has chosen for this expert system project. Marcus was counting on Kim to help integrate the new expert system technology into ICAN's systems environment, and also to provide on-the-job training and insights to the other developers on this team.

Given that Marcus's budget will only permit him to add Barry or Kim to this project team, but not both, what choice do you recommend for him? Justify your answer.

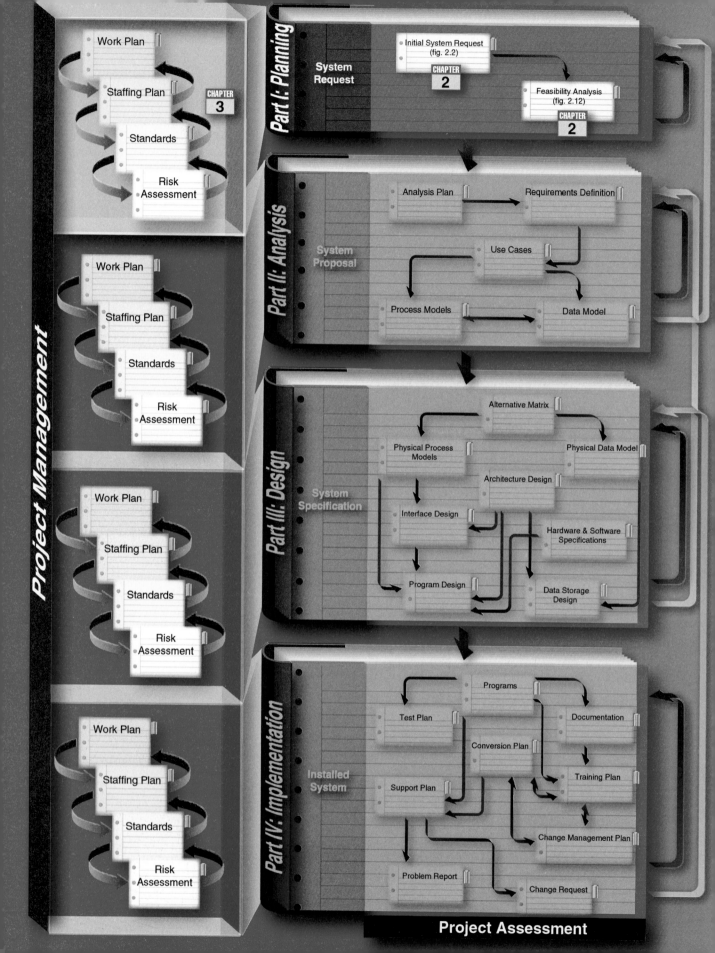

PART ONE

PLANNING PHASE

PROJECT BINDER

Project
Initiation

CHAPTER 2

The Planning Phase
is the fundamental process
of understanding why
an information system should be built,
and determining how
the project team will build it.

The deliverables from both steps
are combined into the system request
which is presented to the project sponsor
and approval committee at the end of the
Planning Phase. They decide whether it is
advisable to proceed with the system.

Project
Management

CHAPTER 3

PLANNING

- [ ] **Identify Project**
- [ ] **Analyze Technical Feasibility**
- [ ] **Analyze Economic Feasibility**
- [ ] **Analyze Organizational Feasibility**
- [ ] Estimate Time
- [ ] Identify Tasks
- [ ] Create Work Breakdown Structure
- [ ] Create PERT Charts
- [ ] Create Gantt Charts
- [ ] Manage Scope
- [ ] Staff Project
- [ ] Create Project Charter
- [ ] Set up CASE Respository
- [ ] Develop Standards
- [ ] Begin Documentation
- [ ] Manage Risk

TASK CHECKLIST

PLANNING   ANALYSIS   DESIGN

CHAPTER 2

PROJECT

INITIATION

T his chapter describes Project Initiation, the point at which an organization creates and assesses the original goals and expectations for a new system. The first step in the process is to identify a project that will deliver value to the business and to create a system request that provides basic information about the proposed system. Next the analysts perform a feasibility analysis to determine the technical, economic, and organizational feasibility of the system, and if appropriate, the system is selected, and the development project begins.

OBJECTIVES

- Understand the importanace of linking the information system to business needs.
- Be able to create a system request.
- Understand how to assess technical, economic, and organizational feasibility.
- Be able to perform a feasibility analysis.
- Understand how projects are selected in some organizations.

CHAPTER OUTLINE

IMPLEMENTATION

INTRODUCTION

The first step in any new development project is for someone—a manager, staff member, sales representative, or systems analyst—to see an opportunity to improve the business. New systems start first and foremost from some business need or opportunity. Many ideas for new systems or improvements to existing ones arise from the application of a new technology, but an understanding of technology is usually secondary to a solid understanding of the business and its objectives.

This may sound like common sense, but unfortunately, many projects are started without a clear understanding of how the system will improve the business. The IS field is filled with thousands of buzzwords, fads, and trends (e.g., customer relationship management [CRM], mobile computing, data mining). The promise of these innovations can appear so attractive that organizations begin projects even if they are not sure what value they offer, because they believe that the technologies are somehow important in their own right. A 1996 survey by the Standish Group found that 42% of all corporate IS projects were abandoned before completion; a similar 1996 study by the General Accounting Office found 53% of all U.S. government IS projects were abandoned. Most times, problems can be traced back to the very beginning of the SDLC where too little attention was given to the identifying business value and understanding the risks associated with the project.

Does this mean that technical people should not recommend new systems projects? Absolutely not. In fact, the ideal situation is for both IT people (i.e., the experts in systems) and the business people (i.e., the experts in business) to work closely to find ways for technology to support business needs. In this way, organizations can leverage the exciting technologies that are available while ensuring that projects are based upon real business objectives, such as increasing sales, improving customer service, and decreasing operating expenses. Ultimately, information systems need to affect the organization's bottom line (in a positive way!).

In general, a *project* is a set of activities with a starting point and an ending point meant to create a system that brings value to the business. *Project initiation* begins when someone (or some group) in the organization (called the *project sponsor*) identifies some business value that can be gained from using information technology. The proposed project is described briefly using a technique called the system request, which is submitted to an approval committee for consideration. The approval committee reviews the system request and makes an initial determination, based on the information provided, of whether to investigate the proposal or not. If so, the next step is the feasibility analysis.

The feasibility analysis plays an important role in deciding whether to proceed with an IS development project. It examines the technical, economic, and organizational pros and cons of developing the system, and it gives the organization a slightly more detailed picture of the advantages of investing in the system as well as any obstacles that could arise. In most cases, the project sponsor works together with an analyst (or analyst team) to develop the feasibility analysis for the approval committee.

Once the feasibility analysis has been completed, it is submitted back to the approval committee along with a revised system request. The committee then decides whether to approve the project, decline the project, or table it until additional information is available. Projects are selected by weighing risks and return, and by making trade-offs at the organizational level.

2-A   INTERVIEW WITH LYN MCDERMID, CIO, DOMINION VIRGINIA POWER

A CIO needs to have a global view when identifying and selecting projects for her organization. I would get lost in the trees if I were to manage on a project-by-project basis. Given this, I categorize my projects according to my three roles as a CIO, and the mix of my project portfolio changes depending on the current business environment.

My primary role is to **keep the business running.** That means every day when each person comes to work, they can perform his or her job efficiently. I measure this using various service level, cost, and productivity measures. Projects that keep the business running could have a high priority if the business were in the middle of a merger, or a low priority if things were running smoothly, and it were "business as usual."

My second role is to push **innovation that creates value for the business.** I manage this by looking at our lines of business and asking which lines of business create the most value for the company. These are the areas for which I should be providing the most value. For example, if we had a highly innovative marketing strategy, I would push for innovation there. If operations were running smoothly, I would push less for innovation in that area.

My third role is strategic, to look beyond today and find **new opportunities** for both IT and the business of providing energy. This may include investigating process systems, such as automated meter reading or looking into the possibilities of wireless technologies.

Lyn McDermid

PROJECT IDENTIFICATION

A project is identified when someone in the organization identifies a *business need* to build a system. This could occur within a business unit or IT, by a steering committee charged with identifying business opportunities, or evolve from a recommendation made by external consultants. Examples of business needs include supporting a new marketing campaign, reaching out to a new type of customer, or improving interactions with suppliers. Sometimes, needs arise from some kind of "pain" within the organization, such as a drop in market share, poor customer service levels, or increased competition. Other times, new business initiatives and strategies are created, and a system is required to enable them.

Business needs also can surface when the organization identifies unique and competitive ways of using IT. Many organizations keep an eye on *emerging technology*, which is technology that is still being developed and not yet viable for widespread business use. For example, if companies stay abreast of technology like the Internet, smart cards, and scent technology in their earliest stages, they can develop business strategies that leverage the capabilities of these technologies and introduce them into the marketplace as a *first mover*. Ideally, they can take advantage of this first-mover advantage by making money and continuing to innovate while competitors trail behind.

The *project sponsor* is someone who recognizes the strong business need for a system and has an interest in seeing the system succeed. He or she will work throughout the SDLC to make sure that the project is moving in the right direction from the perspective of the business. The project sponsor serves as the primary point of contact for the system. Usually the sponsor of the project is from a business function, such as Marketing, Accounting, or Finance; however, members of the IT area also can sponsor or cosponsor a project.

The size or scope of the project determines the kind of sponsor that is needed. A small, departmental system may only require sponsorship from a single manager; however, a large, organizational initiative may need support from the entire senior management team, and even the CEO. If a project is purely technical in nature (e.g., improvements to the existing IT infrastructure; research into the viability of an emerging technology), then sponsorship from IT is appropriate. When projects have great importance to the business, yet are technically complex, joint sponsorship by both the business and IT may be necessary.

The business need drives the high-level *business requirements* for the system. Requirements are what the information system will do, or what *functionality* it will contain. They need to be explained at a high level so that the approval committee and, ultimately, the project team understand what the business expects from the final product. Business requirements are what features and capabilities the information system will have to include, such as the ability to collect customer orders online or the ability for suppliers to receive inventory information as orders are placed and sales are made.

The project sponsor also should have an idea of the *business value* to be gained from the system, both in tangible and intangible ways. *Tangible value* can be quantified and measured easily (e.g., 2% reduction in operating costs). An *intangible* value results from an intuitive belief that the system provides important, but hard-to-measure benefits to the organization (e.g., improved customer service, a better competitive position).

Once the project sponsor identifies a project that meets an important business need, and he or she can identify the system's business requirements and value, it is time to formally initiate the project. In most organizations, project initiation begins with a technique called a system request.

YOUR TURN

2-1 IDENTIFY TANGIBLE AND INTANGIBLE VALUE

Dominion Virginia Power is one of the nation's 10 largest investor-owned electric utilities. The company delivers power to more than 2 million homes and businesses in Virginia and North Carolina. In 1997, the company overhauled some of its core processes and technology. The goal was to improve customer service and cut operations costs by developing a new workflow and geographic information system. When the project was finished, service engineers who used to sift through thousands of paper maps could pinpoint the locations of electricity poles with computerized searches. The project helped the utility improve management of all its facilities, records, maps, scheduling, and human resources. That, in turn, helped increase employee productivity, improve customer response times, and reduce the costs of operating crews.

Source: Computerworld November 11, 1997

QUESTIONS:
1. What kinds of things does Dominion Virginia Power do that requires it to know power pole locations? How often does it do these things? Who benefits if the company can locate power poles faster?
2. Based on your answers to question 1, describe three tangible benefits that the company can receive from its new computer system. How can these be quantified?
3. Based on your answers to question 1, describe three intangible benefits that the company can receive from its new computer system. How can these be quantified?

System Request

A *system request* is a document that describes the business reasons for building a system and the value that the system is expected to provide. The project sponsor usually completes this form as part of a formal system project selection process within the organization. Most system requests include five elements: project sponsor, business need, business requirements, business value, and special issues (see Figure 2-1). The sponsor describes the person who will serve as the primary contact for the project, and the business need presents the reasons prompting the project. The business requirements of the project refer to the business capabilities that the system will need to have, and the business value describes the benefits that the organization should expect from the system. *Special issues* are included on the document as a catchall for other information that should be considered in assessing the project. For example, the project may need to be completed by a specific deadline.

Element	Description	Examples
Project Sponsor	The person who initiates the project and who serves as the primary point of contact for the project on the business side.	Several members of the Finance department Vice President of Marketing IT Manager Steering committee CIO CEO
Business Need	The business-related reason for initiating the system.	Increase sales Improve market share Improve access to information Improve customer service Decrease product defects Streamline supply acquisition Processes
Business Requirements	The business capabilities that the system will provide.	Provide online access to information Capture customer demographic information Include product search capabilities Produce management reports Include online user support
Business Value	The benefits that the system will create for the organization.	3% increase in sales 1% increase in market share Reduction in headcount by 5*FTEs $200,000 cost savings from decreased supply costs $150,000 savings from removal of existing system
Special Issues or Constraints	Issues that are relevant to the implementation of the system and committee make decisions about the project.	Government-mandated deadline for May 30 System needed in time for the Christmas holiday season Top-level security clearance needed by project team to work with data

* = Full-time equivalent

FIGURE 2-1
Elements of the System Request Form

CONCEPTS

IN ACTION

2-B INTERVIEW WITH DON HALLACY, PRESIDENT, TECHNOLOGY SERVICES, SPRINT CORPORATION

At Sprint, network projects originate from two vantage points—IT and the business units. IT projects usually address infrastructure and support needs. The business-unit projects typically begin after a business need is identified locally, and a business group informally collaborates with IT regarding how a solution can be delivered to meet customer expectations.

Once an idea is developed, a more formal request process begins, and an analysis team is assigned to investigate and validate the opportunity. This team includes members from the user community and IT, and they scope out at a high level what the project will do; create estimates for technology, training, and development costs; and create a business case. This business case contains the economic value-add and the net present value of the project.

Of course, not all projects undergo this rigorous process. The larger the project, the more time is allocated to the analysis team. It is important to remain flexible and not let the process consume the organization. At the beginning of each budgetary year, specific capital expenditures are allocated for operational improvements and maintenance. Moreover, this money is set aside to fund quick projects that deliver immediate value without going through the traditional approval process. *Don Hallacy*

Project teams need to be aware of any special circumstances that could affect the outcome of the system.

The completed system request is submitted to the *approval committee* for consideration. This approval committee could be a company steering committee that meets regularly to make information systems decisions, a senior executive who has control of organizational resources, or any other decision-making body that governs the use of business investments. The committee reviews the system request and makes an initial determination, based on the information provided, of whether to investigate the proposal or not. If so, the next step is to conduct a feasibility analysis.

Applying the Concepts at CD Selections

Throughout the book, we will apply the concepts in each chapter to a fictitious company called CD Selections. For example, in this section, we will illustrate the creation of a system request. CD Selections is a chain of 50 music stores located in California, with headquarters in Los Angeles. Annual sales last year were $50 million, and they have been growing at about 3% to 5% per year for the past few years.

Background Margaret Mooney, Vice President of Marketing, has recently become both excited by and concerned with the rise of Internet sites selling CDs. The Internet has great potential, but Margaret wants to use it in the right way. Rushing into e-commerce without considering things like its effect on existing brick-and-mortar stores and the implications on existing systems at CD Selections could cause more harm than good.

CD Selections currently has a Web site that provides basic information about the company and about each of its stores (e.g., map, operating hours, phone number). The page was developed by an Internet consulting firm and is hosted by a prominent local Internet Service Provider (ISP) in Los Angeles. The IT department at CD Selections has become experienced with Internet technology as it has worked with the ISP to maintain the site; however, it still has a lot to learn when it comes to conducting business over the Web.

System Request   At CD Selections, new IT projects are reviewed and approved by a project steering committee that meets quarterly. The committee has representatives from IT as well as from the major areas of the business. For Margaret, the first step was to prepare a system request for the committee.

Figure 2-2 shows the system request she prepared. The sponsor is Margaret, and the business needs are to increase sales and to better service retail customers. Notice that the need does not focus on the technology, such as the need "to upgrade our Web page." The focus is on the business aspects: sales and customer service.

For now, the business requirements are described at a very high level of detail. In this case, Margaret's vision for the requirements includes the ability to help brick-and-mortar stores reach out to new customers. Specifically, customers should

---

System Request—Internet order project

Project sponsor: Margaret Mooney, Vice President of Marketing

Business Need: **This project has been initiated to reach new Internet customers and to better serve existing customers using Internet sales support.**

Business Requirements:

Using the Web, customers should be able to search for products and identify the brick-and-mortar stores that have them in stock. They should be able to put items on hold at a store location or place an order for items that are not carried or not in stock. The functionality that the system should have is listed below:

- Search through the CD Selections' inventory of products
- Identify the retail stores that have the product in stock
- Put a product on hold at a retail store and schedule a time to pick up the product
- Place an order for products not currently in stock or not carried by CD Selections
- Receive confirmation that an order can be placed and when it will be in stock

Business Value:

We expect that CD Selections will increase sales by reducing lost sales due to out-of-stock or nonstocked items and by reaching out to new customers through its Internet presence. We expect the improved services will reduce customer complaints, primarily because 50% of all customer complaints stem from out of stocks or nonstocked items. Also, CD Selections should benefit from improved customer satisfaction and increased brand recognition due to its Internet presence.

Conservative estimates of tangible value to the company includes:

- $750,000 in sales from new customers
- $1,875,000 in sales from existing customers
- $50,000 yearly reduction in customer service calls

Special Issues or Constraints:

- The Marketing Department views this as a strategic system. This Internet system will add value to our current business model, and it also will serve as a proof of concept for future Internet endeavors. For example, in the future, CD Selections may want to sell products directly over the Internet.
- The system should be in place for the holiday shopping season next year.

FIGURE 2-2
System Request for CD Selections

be able to search for products over the Internet, locate a retail store that contains the product, put a product on "hold" for later store pickup, and order products that are not currently being stocked.

The business value describes how the requirements will affect the business. Margaret found identifying intangible business value to be fairly straightforward in this case. The Internet is a "hot" area, so she expects the Internet to improve customer recognition and satisfaction. Estimating tangible value is more difficult. She expects that Internet ordering will increase sales in the retail stores, but by how much?

Margaret decides to have her marketing group do some market research to learn how many retail customers do not complete purchases because the store does not carry the item they are looking for. They learn that stores lose approximately 5% of total sales from "out-of-stocks and nonstocks." This number gives Margaret some idea of how much sales could increase from the existing customer base (i.e., about $50,000 per store), but it does not indicate how many new customers the system will generate.

Estimating how much revenue CD Selections should anticipate from new Internet customers was not simple. One approach was to use some of CD Selections' standard models for predicting sales of new stores. Retail stores average about $1 million in sales per year (after they have been open a year or two), depending upon location factors such as city population, average incomes, proximity to universities, and so on. Margaret estimated that adding the new Internet site would have similar effects of adding a new store. This would suggest ongoing revenues of $1 million, give or take several hundred thousand dollars, after the Web site had been operating for a few years.

Together, the sales from existing customer ($2.5 million) and new customers ($1 million) totaled approximately $3.5 million. Margaret created conservative and optimistic estimates by reducing and increasing this figure by 25%. This created a possible range of values from $2,625,000 to $4,375,000. Margaret is conservative, so she decided to include the lower number as her sales projection. See Figure 2-2 for the completed system request.

YOUR TURN **2-2 CREATE A SYSTEM REQUEST**

Think about your own university or college and choose an idea that could improve student satisfaction with the course enrollment process. Currently can students enroll for classes from anywhere? How long does it take? Are directions simple to follow? Is online help available?

Next, think about how technology can help support your idea. Would you need completely new technology? Can the current system be changed?

QUESTION:
Create a system request that you could give to the administration that explains the sponsor, business need, business requirements, and potential value of the project. Include any constraints or issues that should be considered.

FEASIBILITY ANALYSIS

Once the need for the system and its business requirements have been defined, it is time to create a more detailed business case to better understand the opportunities and limitations associated with the proposed project. *Feasibility analysis* guides the organization in determining whether to proceed with a project. Feasibility analysis also identifies the important *risks* associated with the project that must be addressed if the project is approved. As with the system request, each organization has its own process and format for the feasibility analysis, but most include three techniques: technical feasibility, economic feasibility, and organizational feasibility. The results of these techniques are combined into a *Feasibility Study* deliverable that is given to the approval committee at the end of Project Initiation. See Figure 2-3.

Although we will discuss feasibility analysis now within the context of Project Initiation, most project teams will revise their feasibility study throughout the SDLC and revisit its contents at various checkpoints during the project. If at any point the project's risks and limitations outweigh its benefits, the project team may decide to cancel the project or make necessary improvements.

Technical Feasibility

The first technique in the feasibility analysis is to assess the *technical feasibility* of the project, the extent to which the system can be successfully designed, developed,

---

Technical Feasibility: Can We Build It?

- Familiarity with Application: Less familiarity generates more risk
- Familiarity with Technology: Less familiarity generates more risk
- Project Size: Large projects have more risk
- Compatibility: The harder it is to integrate the system with the company's existing technology, the higher the risk

Economic Feasibility: Should We Build It?

- Development costs
- Annual operating costs
- Annual benefits (cost savings and revenues)
- Intangible costs and benefits

Organizational Feasibility: If We Build It, Will They Come?

- Project champion(s)
- Senior management
- Users
- Other stakeholders
- Is the project strategically aligned with the business?

FIGURE 2-3
Feasibility Analysis Assessment Factors

and installed by the IT group. Technical feasibility analysis is in essence a *technical risk analysis* that strives to answer the question: "*Can* we build it?"[1]

There are many risks that can endanger the successful completion of the project. First and foremost is the users' *and* analysts' *familiarity with the application*. When analysts are unfamiliar with the business application area, they have a greater chance of misunderstanding the users or missing opportunities for improvement. The risks increase dramatically when the users themselves are less familiar with an application, such as with the development of a system to support a new business innovation (e.g., Microsoft starting up a new Internet dating service). In general, the development of new systems is riskier than extensions to an existing system because existing systems tend to be better understood.

Familiarity with the technology is another important source of technical risk. When a system will use technology that has not been used before *within the organization,* there is a greater chance that problems will occur and delays will be incurred because of the need to learn how to use the technology. Risk increases dramatically when the technology itself is new (e.g., a new Java development toolkit).

Project size is an important consideration, whether measured as the number of people on the development team, the length of time it will take to complete the project, or the number of distinct features in the system. Larger projects present more risk, both because they are more complicated to manage and because there is a greater chance that some important system requirements will be overlooked or misunderstood. The extent to which the project is highly integrated with other systems (which is typical of large systems) can cause problems because complexity is increased when many systems must work together.

Finally, project teams need to consider the *compatibility* of the new system with the technology that already exists in the organization. Systems rarely are built in a vacuum—are built in organizations that have numerous systems already in place. New technology and applications need to be able to integrate with the existing environment for many reasons. They may rely on data from existing systems, they may produce data that feed other applications, and they may have to use the company's existing communications infrastructure. A new CRM system, for example, has little, value if it does not use customer data found across the organization, in existing sales systems, marketing applications, and customer service systems.

The assessment of a project's technical feasibility is not cut-and-dried because in many cases, some interpretation of the underlying conditions is needed (e.g., how large does a project need to grow before it becomes less feasible?). One approach is to compare the project under consideration with prior projects undertaken by the organization. Another option is to consult with experienced IT professionals in the organization or external IT consultants; often they will be able to judge whether a project is feasible from a technical perspective.

Economic Feasibility

The second element of a feasibility analysis is to perform an *economic feasibility* analysis (also called a *cost–benefit analysis*) that identifies the financial risk associated with the project. This attempts to answer the question: "*Should* we build the system?" Economic feasibility is determined by identifying costs and benefits asso-

---

[1] We use the words "build it" in the broadest sense. Organizations can also choose to buy a commercial software package and install it, in which case, the question might be: "Can we select the right package and successfully install it?"

1. Identify Costs and Benefits	List the tangible costs and benefits for the project. Include both one-time and recurring costs.
2. Assign Values to Costs and Benefits	Work with business users and IT professionals to create numbers for each of the costs and benefits. Even intangibles should be valued if at all possible.
3. Determine Cash Flow	Project what the costs and benefits will be over a period of time, usually 3 to 5 years. Apply a growth rate to the numbers, if necessary.
4. Determine Net Present Value	Calculate what the value of future costs and benefits are if measured by today's standards. You will need to select a rate of growth to apply the NPV formula.
5. Determine Return on Investment	Calculate how much money the organization will receive in return for the investment it will make using the ROI formula.
6. Calcultate Break-Even Point	Find the first year in which the system has greater benefits than costs. Apply the break-even formula using figures from that year. This will help you understand how long it will take before the system creates real value for the organization.
7. Graph Break-Even Point	Plot the yearly costs and benefits on a line graph. The point at which the lines cross is the break-even point.

FIGURE 2-4
Steps to Conduct Economic Feasibility

ciated with the system, assigning values to them, and then calculating the cash flow and return on investment for the project. The more expensive the project, the more rigorous and detailed the analysis. Figure 2-4 lists the steps to perform a cost–benefit analysis; each step will be described in the upcoming sections.

Identify Costs and Benefits The first task when developing an economic feasibility analysis is to identify the kinds of costs and benefits the system will have and list them along the left-hand column of a spreadsheet. Figure 2-5 lists examples of costs and benefits that may be included.

The costs and benefits can be broken down into four categories (1) development costs, (2) operational costs, (3) tangible benefits, and (4) intangibles. *Development costs* are those tangible expenses that are incurred during the construction of the system, such as salaries for the project team, hardware and software expenses, consultant fees, training, and office space and equipment. Development costs are usually thought of as one-time costs. *Operational costs* are those tangible costs that are required to operate the system, such the salaries for operations staff, software licensing fees, equipment upgrades, and communications charges. Operational costs are usually thought of as ongoing costs.

Revenues and cost savings are those *tangible benefits* that the system enables the organization to collect or tangible expenses that the system enables the organization to avoid. Tangible benefits may include increased sales, reductions in staff, and reductions in inventory.

Of course, a project also can affect the organization's bottom line by reaping *intangible benefits* or incurring *intangible costs*. Intangible costs and benefits are more difficult to incorporate into the economic feasibility because they are based

Development Costs	Operational Costs
Development Team Salaries	Software Upgrades
Consultant Fees	Software Licensing Fees
Development Training	Hardware Repairs
Hardware and Software	Hardware Upgrades
Vendor Installation	Operational Team Salaries
Office Space and Equipment	Communications Charges
Data Conversion Costs	User Training

Tangible Benefits	Intangible Benefits
Increased Sales	Increased Market Share
Reductions in Staff	Increased Brand Recognition
Reductions in Inventory	Higher Quality Products
Reductions in IT Costs	Improved Customer Service
Better Supplier Prices	Better Supplier Relations

FIGURE 2-5
Example Costs and Benefits for Economic Feasibility

on intuition and belief rather than "hard numbers." Nonetheless, they should be listed in the spreadsheet along with the tangible items.

Assign Values to Costs and Benefits   Once the types of costs and benefits have been identified, you will need to assign specific dollar values to them. This may seem impossible—how can someone quantify costs and benefits that haven't happened yet? And how can those predictions be realistic? Although this task is very difficult, you have to do the best you can to come up with reasonable numbers for all of the costs and benefits. Only then can the approval committee make an educated decision about whether or not to move ahead with the project.

The best strategy for estimating costs and benefits is to rely on the people who have the best understanding of them. For example, costs and benefits that are related

CONCEPTS   **2-C   INTANGIBLE VALUE AT CARLSON HOSPITALITY**

IN ACTION

I conducted a case study at Carlson Hospitality, a global leader in hospitality services, encompassing more than 1,300 hotel, resort, restaurant, and cruise ship operations in 79 countries. One of its brands, Radisson Hotels & Resorts, researched guest stay information and guest satisfaction surveys. The company was able to quantify how much of a guest's lifetime value can be attributed to his or her perception of the stay experience. As a result, Radisson knows how much of the collective future value of the enterprise is at stake given the per-

ceived quality of stay experience. Using this model, Radisson can confidently show that a 10% increase in customer satisfaction among the 10% of highest quality customers, will capture one-point market share for the brand. Each point in market share for the Radisson brand is worth $20 million in additional revenue. *Barbara Wixom*

QUESTION:
How can a project team use this information to help determine the economic feasibility of a system?

to the technology or the project itself can be provided by the company's IT group or external consultants, and business users can develop the numbers associated with the business (e.g., sales projections, order levels). You also can consider past projects, industry reports, and vendor information, although these approaches probably will be a bit less accurate. Likely, all of the estimates will be revised as the project proceeds.

What about the *intangible* costs and benefits? Sometimes, it is acceptable to list intangible benefits, such as improved customer service, without assigning a dollar value; whereas, other times estimates have to made regarding how much an intangible benefit is "worth." We suggest that you quantify intangible costs or benefits if at all possible. If you do not, how will you know if they have been realized? Suppose that a system is supposed to improve customer service. This is intangible, but let's assume that the greater customer service will decrease the number of customer complaints by 10% each year over 3 years and that $200,000 is spent on phone charges and phone operators who handle complaint calls. Suddenly we have some very tangible numbers with which to set goals and measure the original intangible benefit.

Figure 2-6 shows costs and benefits along with assigned dollar values. Notice that the customer service intangible benefit has been quantified based on

Benefits[a]	
Increased sales	500,000
Improved customer service[b]	70,000
Reduced inventory costs	68,000
Total benefits	**638,000**
Development costs	
2 servers @ $125,000	250,000
Printer	100,000
Software licenses	34,825
Server software	10,945
Development labor	1,236,525
Total development costs	**1,632,295**
Operational costs	
Hardware	54,000
Software	20,000
Operational labor	111,788
Total operational costs	**185,788**
Total costs	**1,818,083**

[a] An important yet intangible benefit will be the ability to offer services that our competitors currently offer.

[b] Customer service numbers have been based on reduced costs for customer complaint phone calls.

FIGURE 2-6
Assign Values to Costs and Benefits

fewer customer complaint phone calls. The intangible benefit of being able to offer services that competitors currently offer was not quantified, but it was listed so that the approval committee will consider the benefit when assessing the system's economic feasibility.

Determine Cash Flow　A formal cost–benefit analysis usually contains costs and benefits over a selected number of years (usually 3 to 5 years) to show cash flow over time (see Figure 2-7). When using this *cash flow method*, the years are listed across the top of the spreadsheet to represent the time period for analysis, and numeric values are entered in the appropriate cells within the spreadsheet's body. Sometimes fixed amounts are entered into the columns. For example, Figure 2-7

	2003	2004	2005	2006	2007	Total
2 Servers @ $125,000	250,000	0	0	0	0	
Printer	100,000	0	0	0	0	
Software licenses	34,825	0	0	0	0	
Server software	10,945	0	0	0	0	
Development labor	1,236,525	0	0	0	0	
Total Operational Costs:	1,632,295	0	0	0	0	
Hardware	54,000	81,261	81,261	81,261	81,261	
Software	20,000	20,000	20,000	20,000	20,000	
Operational labor	111,788	116,260	120,910	125,746	130,776	
Total Operational Costs:	185,788	217,521	222,171	227,007	232,037	
Total Costs:	1,818,083	217,521	222,171	227,007	232,037	
PV of Costs:	**1,765,129**	205,034	203,318	201,693	200,157	2,575,331
PV of All Costs:	**1,765,129**	1,970,163	2,173,481	2,375,174	2,575,331	
Increased sales	500,000	530,000	561,800	595,508	631,238	
Reduction in customer complaint calls	70,000	70,000	70,000	70,000	70,000	
Reduced inventory costs	68,000	68,000	68,000	68,000	68,000	
Total Benefits:	638,000	668,000	699,800	733,508	769,238	
PV of Benefits:	**619,417**	**629,654**	**640,416**	**651,712**	**663,552**	**3,204,752**
PV of All Benefits:	**619,417**	**1,249,072**	**1,889,488**	**2,541,200**	**3,204,752**	
Total Project Costs Less Benefits:	**(1,180,083)**	**450,479**	**477,629**	**506,501**	**537,201**	
Yearly NPV:	**(1,145,712)**	**424,620**	**437,098**	**450,019**	**463,395**	**629421**
Cumulative NPV:	**(1,145,712)**	**(721,091)**	**(283,993)**	**166,026**	**629,421**	
Return on Investment:	**24.44%**	(629,421/2,575,331)				
Break-even Point:	**4.63 years**	(break-even occurs in year 4; 450,019 + 166,026 / 450,019 = .63)				
Intangible Benefits:	This serivce is currently provided by competitors Improved customer satisfaction					

FIGURE 2-7
Cost–Benefit Analysis

lists the same amount for customer complaint calls and inventory costs for all 5 years. Usually amounts are augmented by some rate of growth to adjust for inflation or business improvements, as shown by the 6% increase that is added to the sales numbers in the sample spreadsheet. Finally, totals are added to determine what the overall benefits will be, and the higher the overall total, the more feasible the solution is in terms of its economic feasibility.

Determine Net Present Value and Return on Investment   There are several problems with the cash flow method because it does not consider the time value of money (i.e., a dollar today is *not* worth a dollar tomorrow), and it does not show the overall "bang for the buck" that the organization is receiving from its investment. Therefore, some project teams add additional calculations to the spreadsheet to provide the approval committee with a more accurate picture of the project's worth.

Net present value (NPV) is used to compare the present value of future cash flows with the investment outlay required to implement the project. Consider the table in Figure 2-8 that shows the future worth of a dollar investment today, given different numbers of years and different rates of change. If you have a friend who owes you a dollar today, but instead she gives you a dollar 3 years from now— you've been had! Given a 10% increase in value, you'll be receiving the equivalent of 75 cents in today's terms.

NPV can be calculated in many different ways, some of which are extremely complex. See Figure 2-9 for a basic calculation that can be used in your cash flow analysis to get more relevant values. In Figure 2-7, the present value of the costs and benefits are calculated first (i.e., they are shown at a discounted rate). Then NPV is calculated, and it shows the discounted rate of the combined costs and benefits.

The *return on investment (ROI)* is a calculation that is listed somewhere on the spreadsheet that measures the amount of money an organization receives in return for the money it spends—a high ROI results when benefits far outweigh costs. ROI is determined by taking the total benefits less the costs of the system and dividing that number by the total costs of the system (see Figure 2-9). ROI can be determined per year, or for the entire project over a period of time. One drawback of ROI is that it only considers the end points of the investment, not the cash flow in between, so it should not be used as the sole indicator of a project's worth. Find the ROI figure on the spreadsheet in Figure 2-7.

Determine Break-Even Point   If the project team needs to perform a rigorous cost–benefit analysis, they may need to include information about the length of

Number of years	6%	10%	15%
1	.943	.909	.870
2	.890	.826	.756
3	.840	.751	.572
4	.792	.683	.497

This table shows how much a dollar today is worth 1–4 years from now in today's terms across different interest rates.

FIGURE 2-8
The Value of a Future Dollar Today

Calculation	Definition	Formula
Present Value (PV)	The amount of an investment today compared to that same amount in the future, taking into account inflation and time.	$$\dfrac{\text{Amount}}{(1 + \text{interest rate})^n}$$ n = number of years in future
Net Present Value (NPV)	The present value of benefit less the present value of costs.	PV Benefits − PV Costs
Return on Investment (ROI)	The amount of revenues or cost savings results from a given investment.	$$\dfrac{\text{Total benefits} - \text{Total costs}}{\text{Total costs}}$$
Break-Even Point	The point in time at which the costs of the project equal the value it has delivered.	$$\dfrac{\text{Yearly NPV}^* - \text{Cumulative NPV}}{\text{Yearly NPV}^*}$$

*Use the Yearly NPV amount from the first year in which the project has a positive cash flow.

Add the above amount to the year in which the project has a positive cash flow.

FIGURE 2-9
Financial Calculations Used For Cost–Benefit Analysis

time before the project will break-even, or when the returns will match the amount invested in the project. The greater the time it takes to break-even, the riskier the project. The *break-even point* is determined by looking at the cash flow over time and identifying the year in which the benefits are larger than the costs (see Figure 2-7). Then, the yearly and cumulative NPV for that year are divided by the yearly NPV to determine how far into the year the break-even will occur. See Figure 2-9 for the break-even calculation.

The break-even point also can be depicted graphically as shown in Figure 2-10. The cumulative present value of the costs and benefits for each year are plotted on a line graph, and the point at which the lines cross is the break-even point.

CONCEPTS
IN ACTION

2-D RETURN ON INVESTMENT

In 1996, IDC conducted a study of 62 companies that had implemented data warehousing. They found that the implementations generated an average three-year return on investment of 401%. The interesting part is that while 45 companies reported results between 3% and 1,838%, the range varied from—1,857% to as high as 16,000%!

QUESTIONS:
1. What kinds of reasons would you give for the varying degrees of return on investment?
2. How would you interpret these results if you were a manager thinking about investing in data warehousing for your company?

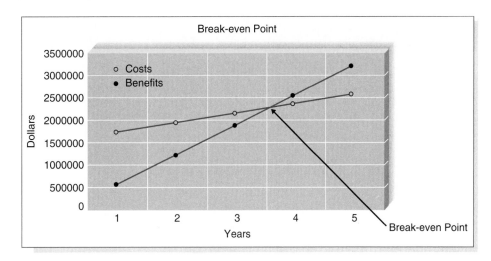

FIGURE 2-10

Organizational Feasibility

The final technique used for feasibility analysis is to assess the *organizational feasibility* of the system, how well the system ultimately will be accepted by its users and incorporated into the ongoing operations of the organization. There are many organizational factors that can have an impact on the project, and seasoned developers know that organizational feasibility can be the most difficult feasibility dimension to assess. In essence, an organizational feasibility analysis attempts to answer the question: "If we build it, will they come?"

One way to assess the organizational feasibility of the project is to understand how well the goals of the project align with business objectives. *Strategic alignment* is the fit between the project and business strategy — the greater the alignment, the less risky the project will be from an organizational feasibility perspective. For example, if the Marketing Department has decided to become more customer focused, then a CRM project that produces integrated customer information would have strong strategic alignment with Marketing's goal. Many IT projects fail when the IT department initiates them because there is little or no alignment with business unit or organizational strategies.

A second way to assess organizational feasibility is to conduct a *stakeholder analysis*.[2] A *stakeholder* is a person, group, or organization that can affect (or will be affected by) a new system. In general, the most important stakeholders in the introduction of a new system are the project champion, system users, and organizational management (see Figure 2-11), but systems sometimes affect other stakeholders as well. For example, the IS department can be a stakeholder of a system because IS jobs or roles may be changed significantly after its implementation. One key stakeholder outside of the champion, users, and management in Microsoft's project that embedded Internet Explorer as a standard part of Windows was the U.S. Department of Justice.

---

[2] A good book on stakeholder analysis that presents a series of stakeholder analysis techniques is R. O. Mason and I.I. Mittroff, *Challenging Strategic Planning Assumptions: Theory, Cases, and Techniques,* New York: John Wiley & Sons, 1981.

	Role	Techniques for improvement:
Champion	A champion: • Initiates the project • Promotes the project • Allocates his or her time to project • Provides resources	• Make a presentation about the objectives of the project and the proposed benefits to those executives who will benefit directly from the system • Create a prototype of the system to demonstrate its potential value
Organizational Management	Organizational Managers: • Know about the project • Budget enough money for the project • Encourage users to accept and use the system	• Make a presentation to management about the objectives of the project and the proposed benefits • Market the benefits of the system using memos and organizational newsletters • Encourage the champion to talk about the project with his or her peers
System Users	Users: • Make decisions that influence the project • Perform hands-on activities for the project • Ultimately determine whether the project is successful by using or not using the system	• Assign users official roles on the project team • Assign users specific tasks to perform with clear deadlines • Ask for feedback from users regularly (e.g., at weekly meetings)

FIGURE 2-11
Some Important Stakeholders for Organizational Feasibility

The champion is a high-level, IS executive who is usually but not always the project sponsor who created the system request. The champion supports the project by providing time, resources (e.g., money), and political support within the organization by communicating the importance of the system to other organizational decision makers. More than one champion is preferable because if the champion leaves the organization, the support could leave as well.

While champions provide day-to-day support for the system, organizational management also needs to support the project. Such management support conveys to the rest of the organization the belief that the system will make a valuable contribution and that necessary resources will be made available. Ideally, management should encourage people in the organization to use the system and to accept the many changes that the system will likely create.

A third important set of stakeholders is the system users who ultimately will use the system once it has been installed in the organization. Too often, the project team meets with users at the beginning of a project and then disappears until after the system is created. In this situation, rarely does the final product meet the expectations and needs of those who are supposed to use it because needs change and users become savvier as the project progresses. User participation should be promoted throughout the development process to make sure that the final system will be accepted and used by getting users actively involved in the development of the system (e.g., performing tasks, providing feedback, and making decisions).

The final feasibility study helps organizations make wiser investments regarding IS because it forces project teams to consider technical, economic, and organizational factors that can affect their projects. It protects IT professionals from criticism by keeping the business units educated about decisions and positioned as the leaders in the decision-making process. Remember—the feasibility study should be revised several times during the project at points where the project team makes critical decisions about the system (e.g., before the design begins). It can be used to support and explain the critical choices that are made throughout the SDLC.

Applying the Concepts at CD Selections

The steering committee met and placed the Internet Order project high on its list of projects. A senior systems analyst, Alec Adams, was assigned to help Margaret conduct a feasibility analysis because of his familiarity with CD Selections' sales and distribution systems. He also was an avid user of the Web and had been offering suggestions for the improvement of CD Selections' Web site.

Alec and Margaret worked closely together over the next few weeks on the feasibility analysis. Figure 2-12 presents the executive summary page of the feasibility analysis; the report itself was about 10 pages long, and it provided additional detail and supporting documentation.

As shown in Figure 2-12, the project is somewhat risky from a technical perspective. CD Selections has minimal experience with the proposed application and the technology because the ISP had been managing most of the Web site technology to date. One solution may be to hire a consultant with e-commerce experience to work with the IT department and to offer guidance. Further, the new system would have to exchange order information with the company's brick-and-mortar order system. Currently individual retail stores submit orders electronically, so receiving orders and exchanging information with the Internet systems should be possible.

The economic feasibility analysis includes refined assumptions that Margaret made in the system request. Figure 2-13 shows the summary spreadsheet that led to the conclusions on the feasibility analysis. Development costs are expected to be about $250,000. This is a very rough estimate, as Alec has had to make some assumptions about the amount of time it will take to design and program the system. These estimates will be revised after a detailed workplan has been developed and as the project proceeds.[3] Traditionally, operating costs include the costs of the computer operations. In this case, CD Selections has had to include the costs of business staff, because they are creating a new business unit, resulting in a total of about $450,000 each year. Margaret and Alec have decided to use a conservative estimate for revenues although they note the potential for higher returns. This shows that the project can still add significant business value, even if the underlying assumptions prove to be overly optimistic. The spreadsheet was projected over 3 years, and the ROI and break-even point were included.

The organizational feasibility is presented in Figure 2-12. There is a strong champion, well placed in the organization to support the project. The project originated in the business or functional side of the company, not the IS department, and Margaret has carefully built up support for the project among the senior management team.

This is an unusual system in that the ultimate end users are the consumers external to CD Selections. Margaret and Alec have not done any specific market

---

[3] Some of the salary information may seem high to you. Most companies use a "full cost" model for estimating salary cost in which all benefits (e.g., health insurance, retirement, payroll taxes) are included in salaries when estimating costs.

Internet Order Feasibility Analysis Executive Summary

Margaret Mooney and Alec Adams created the following feasibility analysis for the CD Selections Internet Order System Project. The System Proposal is attached, along with the detailed feasibility study. The highlights of the feasibility analysis are:

Technical Feasibility

The Internet Order System is feasible technically, although there is some risk.

CD Selections' risk regarding familiarity with Internet order applications is high
- The Marketing Department has little experience with Internet-based marketing and sales.
- The IT Department has strong knowledge of the company's existing order systems; however, it has not worked with Web-enabled order systems.
- Hundreds of retailers that have Internet Order applications exist in the marketplace.

CD Selections' risk regarding familiarity with the technology is medium
- The IT Department has relied on external consultants and an Information Service Provider to develop its existing Web environment.
- The IT Department has gradually learned about Web systems by maintaining the current Web site.
- Development tools and products for commercial Web application development are available in the marketplace, although the IT department has little experience with them.
- Consultants are readily available to provide help in this area.

The project size is considered medium risk
- The project team likely will include less than 10 people.
- Business user involvement will be required.
- The project timeframe cannot exceed a year because of the Christmas holiday season implementation deadline, and it should be much shorter.

The compatibility with CD Selections' existing technical infrastructure should be good
- The current Order System is a client-server system built using open standards. An interface with the Web should be possible.
- Retail stores already place and maintain orders electronically.
- An Internet infrastructure already is in place at retail stores and at the corporate headquarters.
- The ISP should be able to scale their services to include a new Order System.

Economic Feasibility

A cost–benefit analysis was performed; see attached spreadsheet for details. A conservative approach shows that the Internet Order System has a good chance of adding to the bottom line of the company significantly.

ROI over 3 years is: 229%
Total benefit after three years is: $3.5 million (adjusted for present value)
Break-even occurs: after 1.7 years

Intangible Costs and Benefits

- Improved customer satisfaction
- Greater brand recognition

Organizational Feasibility

From an organizational perspective, this project has low risk. The objective of the system, which is to increase sales, is aligned well with the senior management's goal of increasing sales for the company. The move to the Internet also aligns with Marketing's goal to become more savvy in Internet marketing and sales.

The project has a project champion, Margaret Mooney, Vice President of Marketing. Margaret is well positioned to sponsor this project and to educate the rest of the senior management team when necessary. To date, much of senior management is aware of and supports the initiative.

The users of the system, Internet consumers, are expected to appreciate the benefits of CD Selections' Web presence. And, management in the retail stores should be willing to accept the system, given the possibility of increased sales at the store level.

Additional Comments:

- The Marketing Department views this as a strategic system. This Internet system will add value to our current business model, and it also will serve as a proof of concept for future Internet endeavors. For example, in the future, CD Selections may want to sell products directly over the Internet.
- We should consider hiring a consultant with expertise in similar applications to assist with the project.
- We will need to hire new staff to operate the new system, from both the technical and business operations aspects.

FIGURE 2-12
Feasibility Analysis for CD Selections

	2003	2004	2005	Total
Labor: Analysis and Design	42,000	0	0	
Labor: Implementation	120,000	0	0	
Consultant Fees	50,000	0	0	
Training	5,000	0	0	
Office Space and Equipment	2,000	0	0	
Software	10,000	0	0	
Hardare	25,000	0	0	
Total Development Costs:	254,000	0	0	
Labor: Webmaster	85,000	87,550	90,177	
Labor: Network Technician	60,000	61,800	63,654	
Labor: Computer Operations	50,000	51,500	53,045	
Labor: Business Manager	60,000	61,800	63,654	
Labor: Assistant Manager	45,000	46,350	47,741	
Labor: 3 Staff	90,000	92,700	95,481	
Software Upgrades	1,000	1,000	1,000	
Software Licenses	3,000	1,000	1,000	
Hardware Upgrades	5,000	3,000	3,000	
User Training	2,000	1,000	1,000	
Communications Charges	20,000	20,000	20,000	
Marketing Expenses	25,000	25,000	25,000	
Total Operational Costs:	446,000	452,700	464,751	
Total Costs:	700,000	452,700	464,751	
PV of Costs:	**679,612**	**426,713**	**425,313**	**1,531,638**
PV of all Costs:	**679,612**	**1,106,325**	**1,531,638**	
Increased sales from new customers	0	750,000	772,500	
Increased sales from existing customers	0	1,875,000	1,931,250	
Reduction in customer complaint calls	0	50,000	50,000	
Total Benefits:	<u>0</u>	<u>2,675,000</u>	<u>2,753,750</u>	
PV of Benefits:	**0**	**2,521,444**	**2,520,071**	**5,041,515**
PV of All Benefits:	**0**	**2,521,444**	**5,041,515**	
Total Project Costs Less Benefits:	**(700,000)**	**2,222,300**	**2,288,999**	
Yearly NPV:	**(679,612)**	**2,094,731**	**2,094,758**	**3,509,878**
Cumulative NPV:	**(679,612)**	**1,415,119**	**3,509,878**	
Return on Investment:	**229.16%**	(3,509,878/1,531,638)		
Break-even Point:	<u>**1.68 years**</u>	(breakeven occurs in year 2; 2,094,731 + 1,415,119/2,094,731 = 1.68)		
Intangible Benefits:	Greater brand recognition Improved customer satisfaction			

FIGURE 2-13
Economic Feasibility Analysis for CD Selections

Think about the idea that you developed in "Your Turn 2.2" improve your university or college course enrollment.

QUESTIONS:

1. List three things that influence the technical feasibility of the system.

2. List three things that influence the economic feasibility of the system.

3. List three things that influence the organizational feasibility of the system.

4. How can you learn more about the issues that affect the three kinds of feasibility?

research to see how well potential customers will react to the CD Selections system, so this is a risk.

An additional stakeholder in the project is the management team responsible for the operations of the traditional stores, and the store managers. They should be quite supportive given the added service that they now can offer. Margaret and Alec need to make sure that they are included in the development of the system so that they can appropriately incorporate it into their business processes.

PROJECT SELECTION

Once the feasibility analysis has been completed, it is submitted back to the approval committee along with a revised system request. The committee then decides whether to approve the project, decline the project, or table it until additional information is available. At the project level, the committee considers the value of the project by examining the business need (found in the system request) and the risks of building the system (presented in the feasibility analysis).

Before approving the project, however, the committee also considers the project from an organizational perspective; it has to keep in mind the company's entire portfolio of projects. This way of managing projects is called *portfolio management*. Portfolio management takes into consideration the different kinds of projects that exist in an organization—large and small, high risk and low risk, strategic and tactical (see Figure 2-14 for the different ways of classifying projects). A good project portfolio will have the most appropriate mix of projects for the organization's needs. The committee acts as portfolio manager with the goal of maximizing the cost/benefit performance and other important factors of the projects in their portfolio. For example, a organization may want to keep high-risk projects to less than 20% of its total project portfolio.

The approval committee must be selective about where to allocate resources because the organization has limited funds. This involves *trade-offs* in which the organization must give up something in return for something else to keep its portfolio well balanced. If there are three potentially high-payoff projects, yet all have very high risk, then maybe only one of the projects will be selected. Also, there are times when a system at the project level makes good business sense, but it does not

Size	What is the size? How many people are needed to work on the project?
Cost	How much will the project cost the organization?
Purpose	What is the purpose of the project? Is it meant to improve the technical infrastructure? Support a current business strategy? Improve operations? Demonstrate a new innovation?
Length	How long will the project take before completion? How much time will go by before value is delivered to the business?
Risk	How likely is it that the project will succeed or fail?
Scope	How much of the organization is affected by the system? A department? A division? The entire corporation?
Return on investment	How much money does the organization expect to receive in return for the amount the project costs?

FIGURE 2-14
Ways to Classify Projects

at the organization level. Thus, a project may show a very strong ROI and support important business needs for a part of the company; however, it is not selected. This could happen for many reasons—because there is no money in the budget for another system, the organization is about to go through some kind of change (e.g., a merger, an implementation of a company-wide system like an ERP), projects that meet the same business requirements already are underway, or the system does not align well with current or future corporate strategy.

Applying the Concepts at CD Selections

The approval committee met and reviewed the Internet Order System project along with two other projects—one that called for the implementation of corporate Intranet and another that proposed in-store kiosks that would provide customers with information about the CDs that the store carried. Unfortunately, the budget would only allow for one project to be approved, so the committee carefully examined the costs, expected benefits, risks, and strategic alignment of all three projects. Currently, a primary focus of upper management is increasing sales in the retail stores, and the Internet system and kiosk project best aligned with that goal. Given that both projects had equal risk, but that the Internet Order project expected a much greater return, the committee decided to fund the Internet Order System.

YOUR TURN

2-4 To Select or Not To Select

It seems hard to believe that an approval committee would not select a project that meets real business needs, has a high potential ROI, and has a positive feasibility analysis. Think of a company you have worked for or know about. Describe a scenario in which a project may be very attractive at the project level, but not at the organization level.

CONCEPTS

IN ACTION

2-E INTERVIEW WITH CARL WILSON, CIO, MARRIOTT CORPORATION

At Marriott, we don't have IT projects—we have business initiatives and strategies that are enabled by IT. As a result, the only time a traditional "IT project" occurs is when we have an infrastructure upgrade that will lower costs or leverage better functioning technology. In this case, IT has to make a business case for the upgrade and prove its value to the company.

The way IT is involved in business projects in the organization is twofold. First, senior IT positions are filled by people with good business understanding. Second, these people are placed on key business committees and forums where the real business happens, such as finding ways to satisfy guests. Because IT has a seat at the table, we are able to spot opportunities to support business strategy. We look for ways in which IT can enable or better support business initiatives as they arise.

Therefore, business projects are proposed, and IT is one component of them. These projects are then evaluated the same as any other business proposal, such as a new resort—by examining the return on investment and other financial measures.

At the organizational level, I think of projects as must-do's, should-do's, and nice-to-do's. The "must do's" are required to achieve core business strategy, such as guest preference. The "should do's" help grow the business and enhance the functionality of the enterprise. These can be somewhat untested, but good drivers of growth. The "nice-to-do's" are more experimental and look farther out into the future.

The organization's project portfolio should have a mix of all three kinds of projects, with a much greater proportion devoted to the "must-do's." *Carl Wilson*

CONCEPTS

IN ACTION

2-F A PROJECT THAT DOES NOT GET SELECTED

Hygeia Travel Health is a Toronto-based health insurance company whose clients are the insurers of foreign tourists to the United States and Canada. Its project selection process is relatively straightforward. The project evaluation committee, consisting of six senior executives, splits into two groups. One group includes the CIO, along with the heads of operations and research and development, and it analyzes the costs of every project. The other group consists of the two chief marketing officers and the head of business development, and they analyze the expected benefits. The groups are permanent, and to stay objective, they don't discuss a project until both sides have evaluated it. The results are then shared, both on a spreadsheet and in conversation. Projects are then approved, passed over, or tabled for future consideration.

Last year, the marketing department proposed purchasing a claims database filled with detailed information on the costs of treating different conditions at different facilities. Hygeia was to use this information to

estimate how much money insurance providers were likely to owe on a given claim if a patient was treated at a certain hospital as opposed to any other. For example, a 45-year-old man suffering a heart attack may accrue $5,000 in treatment costs at hospital A, but only $4,000 at hospital B. This information would allow Hygeia to recommend the cheaper hospital to its customer. That would save the customer money and help differentiate Hygeia from its competitors.

The benefits team used the same three-meeting process to discuss all the possible benefits of implementing the claims database. Members of the team talked to customers and made a projection using Hygeia's past experience and expectations about future business trends. The verdict: The benefits team projected a revenue increase of $210,000. Client retention would rise by 2% And overall, profits would increase by 0.25%.

The costs team, meanwhile, came up with large estimates: $250,000 annually to purchase the database and an additional $71,000 worth of internal time to make the information usable. Put it all together and it was a financial loss of $111,000 in the first year.

The project still could have been good for marketing—maybe even good enough to make the loss acceptable. But some of Hygeia's clients were also in the claims information business and therefore potential competitors.

This, combined with the financial loss, was enough to make the company reject the project.

Source: "Two Teams Are Better Than One" *CIO Magazine,* July 15, 2001 by Ben Worthen.

YOUR TURN

2-5 PROJECT SELECTION

In April 1999, one of Capital Blue Cross's health-care insurance plans had been in the field for three years but hadn't performed as well as expected. The ratio of premiums to claims payments wasn't meeting historic norms. In order to revamp the product features or pricing to boost performance, the company needed to understand why it was underperforming. The stakeholders came to the discussion already knowing they needed better extraction and analysis of usage data in order to understand product shortcomings and recommend improvements.

After listening to input from the user teams, the stakeholders proposed three options. One was to persevere with the current manual method of pulling data from flat files via ad hoc reports and retyping it into spreadsheets.

The second option was to write a program to dynamically mine the needed data from Capital's customer information control system (CICS). While the system was processing claims, for instance, the program would pull out up-to-the-minute data at a given point in time for users to analyze.

The third alternative was to develop a decision-support system to allow users to make relational queries from a data mart containing a replication of the relevant claims and customer data.

Each of these alternatives was evaluated on cost, benefits, risks, and intangibles.

QUESTION:
1. What are three costs, benefits, risks, and intangibles associated with each project?
2. Based on your answer to question 1, which project would you choose?

Source: "Capital Blue Cross," *CIO Magazine,* February 15, 2000, by Richard Pastore.

SUMMARY

Project Initiation
Project initiation is the point at which an organization creates and assesses the original goals and expectations for a new system. The first step in the process is to identify the business value for the system by developing a system request that provides basic information about the proposed system. Next the analysts perform a feasibility analysis to determine the technical, economic, and organizational feasibility of the system, and if appropriate, the system is approved, and the development project begins.

System Request
The business value for an information system is identified and then described using a system request. This form contains the project's sponsor, business need, business

requirements, and business value of the information system, along with any other issues or constraints that are important to the project. The document is submitted to an approval committee who determines whether the project would be a wise investment of the organization's time and resources.

Feasibility Analysis

A feasibility analysis is then used to provide more detail about the risks associated with the proposed system, and it includes technical, economic, and organizational feasibilities. The technical feasibility focuses on whether the system *can* be built by examining the risks associated with the users' and analysts' familiarity with the application, familiarity with the technology, and project size. The economic feasibility addresses whether the system *should* be built. It includes a cost–benefit analysis of development costs, operational costs, tangible benefits, and intangible costs and benefits. Finally, the organizational feasibility assesses how well the system will be accepted by its users and incorporated into the ongoing operations of the organization. The strategic alignment of the project and a stakeholder analysis can be used to assess this feasibility dimension.

Proect Selection

Once the feasibility analysis has been completed, it is submitted back to the approval committee along with a revised system request. The committee then decides whether to approve the project, decline the project, or table it until additional information is available. The project selection process takes into account all of the projects in the organization using portfolio management. The approval committee weighs many factors and makes trade-offs before a project is selected.

KEY TERMS

Approval committee	Feasibility study	Return on/investment (ROI)
Break-even point	First mover	Risk
Business need	Functionality	Special issues
Business requirement	Intangible benefits	Stakeholder
Business value	Intangible costs	Stakeholder analysis
Cash flow method	Intangible value	Strategic alignment
Champion	Net present value (NPV)	System request
Compatibility	Operational cost	System users
Cost–benefit analysis	Organizational feasibility	Tangible benefits
Development cost	Organizational management	Tangible value
Economic feasibility	Portfolio management	Technical feasibility
Emerging technology	Project	Technical risk analysis
Familiarity with the application	Project initiation	Trade-offs
Familiarity with the technology	Project size	
Feasibility analysis	Project sponsor	

QUESTIONS

1. Give three examples of business needs for a system.
2. What is the purpose of an approval committee? Who is usually on this committee?
3. Why should the system request be created by a businessperson as opposed to an IS professional?
4. What is the difference between intangible value and tangible value? Give three examples of each.
5. What are the purposes of the system request and the feasibility analysis? How are they used in the project selection process?
6. Describe two special issues that may be important to list on a system request.
7. Describe the three techniques for feasibility analysis.
8. What factors are use to determine project size?
9. Describe a "risky" project in terms of technical feasibility. Describe a project that would *not* be considered "risky."
10. What are the steps for assessing economic feasibility? Describe each step.
11. List two intangible benefits. Describe how these benefits can be quantified.
12. List two tangible benefits and two operational costs for a system. How would you determine the values that should be assigned to each item?
13. Explain the net present value and return on investment for a cost–benefit analysis. Why would these calculations be used?
14. What is the break-even point for the project? How is it calculated?
15. What is stakeholder analysis? Discuss three stakeholders that would be relevant for most projects.

EXERCISES

A. Locate a news article in an IT trade magazine (e.g., *Computerworld*) about an organization that is implementing a new computer system. Describe the tangible and intangible value that the organization likely will realize from the new system.
B. Car dealers have realized how profitable it can be to sell automobiles using the Web. Pretend you work for a local car dealership that is part of a large chain such as CarMax. Create a system request you might use to develop a Web-based sales system. Remember to list special issues that are relevant to the project.
C. Suppose that you are interested in buying yourself a new computer. Create a cost–benefit analysis that illustrates the return on investment that you would receive from making this purchase. Computer-related Web sites (e.g., Dell Computers, Compaq Computers) should have real tangible costs that you can include in your analysis. Project your numbers out to include a 3-year period of time and provide the net present value of the final total.
D. Consider the Amazon.com Web site. The management of the company decided to extend their Web-based system to include products other than books (e.g., wine, specialty gifts). How would you have assessed the feasibility of this venture when the idea first came up? How "risky" would you have considered the project that implemented this idea? Why?
E. Interview someone who works in a large organization and ask him or her to describe the approval process that exists for approving new development projects. What do they think about the process? What are the problems? What are the benefits?
F. Reread the "Your Turn 2.1" box (Identify Tangible and Intangible Value). Create a list of the stakeholders that should be considered in a stakeholder analysis of this project.

MINICASES

1. The Amberssen Specialty Company is a chain of 12 retail stores that sell a variety of imported gift items, gourmet chocolates, cheeses, and wines in the Toronto area. Amberssen has an IS staff of three people who have created a simple but effective information system of networked point-of-sale registers at the stores, and a centralized accounting system at the company headquarters. Harry Hilman, the head of Amberssen's IS

group, has just received the following memo from Bill Amberssen, Sales Director (and son of Amberssen's founder).

> Harry—It's time Amberssen Specialty launched itself on the Internet. Many of our competitors are already there, selling to customers without the expense of a retail store-front, and we should be there too. I project that we could double or triple our annual revenues by selling our prod-ucts on the Internet. I'd like to have this ready by Thanks-giving, in time for the prime holiday gift-shopping season. Bill

After pondering this memo for several days, Harry scheduled a meeting with Bill so that he could clarify Bill's vision of this venture. Using the standard content of a system request as your guide, prepare a list of questions that Harry needs to have answered about this project.

2. The Decker Company maintains a fleet of 10 service trucks and crews that provide a variety of plumbing, heating, and cooling repair services to residential cus-tomers. Currently, it takes on average about 6 hours before a service team responds to a service request. Each truck and crew averages 12 service calls per week, and the average revenue earned per service call is $150. Each truck is in service 50 weeks per year. Due to the difficulty in scheduling and routing, there is considerable slack time for each truck and crew during a typical week.

In an effort to more efficiently schedule the trucks and crews and improve their productivity, Decker man-agement is evaluating the purchase of a prewritten routing and scheduling software package. The benefits of the system will include reduced response time to service requests and more productive service teams, but management is having trouble quantifying these benefits.

One approach is to make an estimate of how much service response time will decrease with the new sys-tem, which then can be used to project the increase in the number of service calls made each week. For exam-ple, if the system permits the average service response time to fall to 4 hours, management believes that each truck will be able to make 16 service calls per week on average—an increase of 4 calls per week. With each truck making 4 additional calls per week and the aver-age revenue per call at $150, the revenue increase per truck per week is $600 (4 × $150). With 10 trucks in service 50 weeks per year, the average annual revenue increase will be $300,000 ($600 × 10 × 50).

Decker Company management is unsure whether the new system will enable response time to fall to 4 hours on average, or will be some other number. Therefore, management has developed the following range of out-comes that may be possible outcomes of the new sys-tem, along with probability estimates of each outcome occurring.

New Response Time	# Calls/Truck/Week	Likelihood
2 hours	20	20%
3 hours	18	30%
4 hours	16	50%

Given these figures, prepare a spreadsheet model that computes the expected value of the annual revenues to be produced by this new system.

- ☑ **Identify Project**
- ☑ **Analyze Technical Feasibility**
- ☑ **Analyze Economic Feasibility**
- ☑ **Analyze Organizational Feasibility**
- ☐ **Estimate Time**
- ☐ **Identify Tasks**
- ☐ **Create Work Breakdown Structure**
- ☐ **Create PERT Charts**
- ☐ **Create Gantt Charts**
- ☐ **Manage Scope**
- ☐ **Staff Project**
- ☐ **Create Project Charter**
- ☐ **Set up CASE Repository**
- ☐ **Develop Standards**
- ☐ **Begin Documentation**
- ☐ **Manage Risk**

TASK CHECKLIST

PLANNING   ANALYSIS   DESIGN

This chapter describes the important steps of project management, which begins in the Planning Phase and continues throughout the systems development life cycle (SDLC). First, the project manager estimates the size of the project and identifies the tasks that need to be performed. Next, he or she staffs the project and puts several activities in place to help coordinate project activities. These steps produce important project management deliverables, including the workplan, staffing plan, and standards list.

OBJECTIVES

- Become familiar with estimation.
- Be able to create a project workplan.
- Understand why project teams use timeboxing.
- Become familiar with how to staff a project.
- Understand how computer-aided software engineering, standards, and documentation improve the efficiency of a project.
- Understand how to reduce risk on a project.

CHAPTER OUTLINE

IMPLEMENTATION

INTRODUCTION

Think about major projects that occur in people's lives, such as throwing a big party like a wedding or graduation celebration. Months are spent in advance identifying and performing all of the tasks that need to get done, such as sending out invitations and selecting a menu, and time and money are carefully allocated among them. Along the way, decisions are recorded, problems are addressed, and changes are made. The increasing popularity of the party planner, a person whose sole job is to coordinate a party, suggests how tough this job can be. In the end, the success of any party has a lot to do with the effort that went into planning along the way. System development projects can be much more complicated than the projects we encounter in our personal lives—usually, more people are involved (e.g., the organization), the costs are higher, and more tasks need to be completed. Therefore, you should not be surprised to learn that "party planners" exist for information system projects—they are called project managers.

Project management is the process of planning and controlling the development of a system within a specified time frame at a minimum cost with the right functionality.[1] A *project manager* has the primary responsibility for managing the hundreds of tasks and roles that need to be carefully coordinated. Nowadays, project management is an actual profession, and analysts spend years working on projects prior to tackling the management of them. In a 1999 *Computerworld* survey, more than half of 103 companies polled said they now offer formal project management training for IT project teams. There also is a variety of *project management software* available like Microsoft Project, Plan View, and PMOffice that support project management activities.

Although training and software are available to help project managers, unreasonable demands set by project sponsors and business managers can make project management very difficult. Too often, the approach of the holiday season, the chance at winning a proposal with a low bid, or a funding opportunity pressures project managers to promise systems long before they are able to deliver them. These overly optimistic timetables are thought to be one of the biggest problems that projects face; instead of pushing a project forward faster, they result in delays.

Thus, a critical success factor for project management is to start with a realistic assessment of the work that needs to be accomplished and then manage the project according to that assessment. This can be achieved by carefully following the four steps that are presented in this chapter: identifying the project size, creating and managing the workplan, staffing the project, and coordinating project activities. The project manager ultimately creates a workplan, staffing plan, and standards list, which are used and refined throughout the entire SDLC.

IDENTIFYING PROJECT SIZE

The science (or art) of project management is in making *trade-offs* among three important concepts: the size of the system (in terms of what it does), the time to complete the project (when the project will be finished), and the cost of the project.

---

[1] A good book on project management is by Jack R. Meredith and Samuel J. Mantel, *Project Management: A Managerial Approach,* New York: John Wiley & Sons, 1995.

Think of these three things as interdependent levers that the project manager controls throughout the SDLC. Whenever one lever is pulled, the other two levers are affected in some way. For example, if a project manager needs to readjust a deadline to an earlier date, then the only solution is to decrease the size of the system (by eliminating some of its functions) or to increase costs by adding more people or having them work overtime. Often, a project manager will have to work with the project sponsor to change the goals of the project, such as developing a system with less functionality or extending the deadline for the final system, so that the project has reasonable goals that can be met.

Therefore, in the beginning of the project, the manager needs to estimate each of these levers and then continuously assess how to roll out the project in a way that meets the organization's needs. *Estimation*[2] is the process of assigning projected values for time and effort, and it can be performed manually or with the help of an estimation software package like Costar or Construx—there are over fifty available on the market. The estimates developed at the start of a project are usually based on a range of possible values (e.g., the design phase will take 3 to 4 months) and gradually become more specific as the project moves forward (e.g., the design phase will be completed on March 22).

The numbers used to calculate these estimates can come from several sources. They can be provided with the methodology that is used, taken from projects with similar tasks and technologies, or provided by experienced developers. Generally speaking, the numbers should be conservative. A good practice is to keep track of the actual time and effort values during the SDLC so that numbers can be refined along the way, and the next project can benefit from real data. One of the greatest

---

CONCEPTS **3-A TRADE-OFFS**

IN ACTION

I was once on a project to develop a system that should have taken a year to build. Instead, the business need demanded that the system be ready within 5 months—impossible!

On the first day of the project, the project manager drew a triangle on a white board to illustrate some trade-offs that he expected to occur over the course of the project. The corners of the triangle were labeled Functionality, Time, and Money. The manager explained, "We have too little time. We have an unlimited budget. We will not be measured by the bells and whistles that this system contains. So over the next several weeks, I want you as developers to keep this triangle in mind and do everything it takes to meet this 5-month deadline."

At the end of the 5 months, the project was delivered on time; however, the project was incredibly over budget, and the final product was "thrown away" after it was used because it was unfit for regular usage. Remarkably, the business users felt that the project was very successful because it met the very specific business needs for which it was built. They believed that the trade-offs that were made were worthwhile. *Barbara Wixom*

QUESTIONS:
1. What are the risks in stressing only one corner of the triangle?
2. How would you have managed this project? Can you think of another approach that might have been more effective?

---

[2] A good book for further reading on software estimation is that by Capers Jones, *Estimating Software Costs*, New York: McGraw-Hill, 1989.

strengths of systems consulting firms is the past experience that they offer to a project; they have estimates and methodologies that have been developed and honed over time and applied to hundreds of projects.

There are two basic ways to estimate the time required to build a system. The simplest method uses the amount of time spent in the planning phase to predict the time required for the entire project. The idea is that a simple project will require little planning and a complex project will require more planning, so using the amount of time spent in the planning phase is a reasonable way to estimate overall project time requirements.

With this approach, you take the time spent in (or estimated for) the planning phase and use industry standard percentages (or percentages from the organization's own experiences) to calculate estimates for the other SDLC phases. Industry standards suggest that a "typical" business application system spends 15% of its effort in the planning phase, 20% in the analysis phase, 35% in the design phase, and 30% in the implementation phase. This would suggest that if a project takes 4 months in the planning phase, then the rest of the project likely will take a total of 22.66 person-months ($4 \div .15 = 22.66$). These same industry percentages are then used to estimate the amount of time in each phase (Figure 3-1). The obvious limitation of this approach is that it can be difficult to take into account the specifics of your individual project, which may be simpler or more difficult than the "typical" project.

Function Point Approach

The second approach to estimation, sometimes called the *function point approach*, uses a more complex—and, it is hoped, more reliable—three-step process (Figure 3-2). First, the project manager estimates the size of the project in terms of the number of lines of code the new system will require. This size estimate is then converted into the amount of effort required to develop the system in terms of the number of person-months. The estimated effort is then converted into an estimated schedule time in terms of the number of months from start to finish.

Step 1: Estimate System Size The first step is to estimate the size of a project using function points, a concept developed in 1979 by Allen Albrecht of IBM.[3] A

	Planning	**Analysis**	**Design**	**Implementation**
Typical industry standards for business applications	15%	20%	35%	30%
Estimates based on actual figures for first stages of SDLC	Actual: 4 person-months	Estimated: 5.33 person-months	Estimated: 9.33 person-months	Estimated: 8 person-months

SDLC = systems development life cycle.

FIGURE 3-1

Estimating Project Time Using the Planning Phase Approach

[3] Albrecht's original article is out of print, but much additional research has been done and can be found at www.ifpug.org.

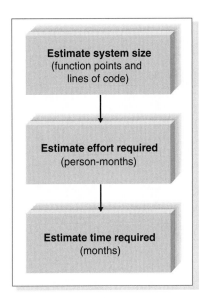

FIGURE 3-2
Estimating Project Time Using the Function Point Approach

function point is a measure of program size that is based on the system's number and complexity of inputs, outputs, queries, files, and program interfaces.

To calculate the function points for a project, components are listed on a worksheet to represent the major elements of the system. For example, data-entry screens are kinds of inputs, reports are outputs, and database queries are kinds of queries (see Figure 3-3). The project manager records the total number of each component that the system will include, and then he or she breaks down the number to show the number of components that have low, medium, and high complexity. In Figure 3.3, there are 19 outputs that need to be developed for the system, 4 of which have low complexity, 10 that have medium complexity, and 5 that are very complex. After each line is filled in, a total number of points are calculated per line by multiplying each number by a complexity index. The line totals are added up to determine the *total unadjusted function points (TUFP)* for the project.

The *complexity* of the overall system is greater than the sum of its parts. Things like the familiarity of the project team with the business area and the technology that will be used to implement the project also may influence how complex a project will be. A project that is very complex for a team with little experience might have little complexity for a team with lots of experience. To create a more realistic size for the project, a number of additional system factors, such as end-user efficiency, reusability, and data communications are assessed in terms of their effect on the project's complexity (see Figure 3-3). These assessments are totaled and placed into a formula to calculate an *adjusted project complexity (APC)* score. The TUFP value is multiplied by the APC value to determine the ultimate size of the project in terms of *total adjusted function points(TAFP)*. This number should give the project manager a reasonable idea as to how big the project will be.

Sometimes a shortcut is used to determine the complexity of the project. Instead of calculating the complexity for the 14 factors listed in Figure 3-3, project managers choose to assign an APC value that ranges from 0.65 for very simple systems to 1.00 for "normal" systems to as much as 1.35 for complex systems and multiply the value to the TUFP score. For example, a very simple system that has 200 unadjusted function points would have a size of 130 adjusted function points

System Components:

Description	Total Number	Complexity			Total
		Low	Medium	High	
Inputs	6	3 × 3	2 × 4	1 × 6	23
Outputs	19	4 × 4	10 × 5	5 × 7	101
Queries	10	7 × 3	0 × 4	3 × 6	39
Files	15	0 × 7	15 × 10	0 × 15	150
Program Interfaces	3	1 × 5	0 × 7	2 × 10	25
Total Unadjusted Function Points (TUFP):					338

Overall System:

Data communications	3
Heavy use configuration	0
Transaction rate	0
End-user efficiency	0
Complex processing	0
Installation ease	0
Multiple sites	0
Performance	0
Distributed functions	2
Online data entry	2
Online update	0
Reusability	0
Operational ease	0
Extensibility	0
Total Processing Complexity (PC):	7

(0 = no effect on processing complexity; 3 = great effect on processing complexity)

Adjusted Processing Complexity (APC):

.65 + (0.01 x 7) = .72

Total Adjusted Function Points (TAFP):

.72 (APC) x 338 (TUFP) = 243 **(TAFP)**

FIGURE 3-3
Function Point-Estimation Worksheet

CONCEPTS

IN ACTION

3-B FUNCTION POINTS AT NIELSEN

Nielsen Media used function point analysis (FPA) for an upgrade to the Global Sample Management System (GSMS) for Nielsen Media/NetRatings, which keeps track of the Internet rating sample, a group of 40,000 homes nationwide that volunteer to participate in ongoing ratings.

In late fall of 1998, Nielsen Media did a FP count based on the current GSMS. (FPA is always easier and more accurate when there is an existing system.) Nielsen Media had its counters—three quality assurance staff—do their FPA, and then input their count into Knowledge-Plan, a productivity modeling tool. In early 1999, seven programmers began writing code for the system, which they were expected to complete in 10 months. As November approached, the project was adding staff to try to meet the deadline. When it became evident that the deadline would not be met, a new FP count was con-

ducted. The GSMS had grown to 900 FPs. Besides the original 500 plus 20%, there were 300 FPs attributable to features and functions that had crept into the project.

How did that happen? The way it always does: The developers and users had added a button here, a new feature there, and soon the project was much larger than it was originally. But Nielsen Media had put a stake in the ground at the beginning from which they could measure growth along the way.

The best practice is to run the FPA and productivity model at the project's launch and again when there is a full list of functional requirements. Then do another analysis anytime there is a major modification in the functional definition of the project.

Source:"Ratings Game," *CIO Magazine*, October 2000, by Bill Roberts.

$(200 \times .65 = 130)$. However, if the system with 200 unadjusted function points were very complex, its function point size would be 270 $(200 \times .35 = 270)$.

In the Planning Phase, the exact nature of the system has not yet been determined, so it is impossible to know *exactly* how many inputs, outputs, and so forth will be in the system. It is up to the project manager to make an intelligent guess. Some people feel that using function points early on in a project is not practical for this reason. We believe function points can be a useful tool for understanding a project's size at any point in the SDLC. Later in the project, once more is known about the system, the project manager will revise the estimates using this better knowledge to produce more accurate results.

Once you have estimated the number of function points, you need to convert the number of function points into the lines of code that will be required to build the system. The number of lines of code depends on the programming language you choose to use. Figure 3-4 presents a very rough conversion guide for some popular languages.

For example, the system in Figure 3-3 has 243 function points. If you were to develop the system in COBOL, it would typically require approximately 26,730 lines of code to write it. Conversely, if you were to use Visual Basic, it typically would take 7290 lines of code. If you could develop the system using a package such as Excel or Access, it would take between 2,430 and 9,720 lines of code. There is a great range for packages, because different packages enable you to do different things, and not all systems can be built using certain packages. Sometimes you end up writing lots of extra code to do some simple function because the package does not have the capabilities you need.

There is also a very important message from the data in this figure. Since there is a direct relationship between lines of code and the amount of effort and time

Language	Approximate Number of Lines of Code per Function Point
C	130
COBOL	110
Java	55
C++	50
Turbo Pascal	50
Visual Basic	30
PowerBuilder	15
HTML	15
Packages (e.g., Access, Excel)	10–40

Source: Capers Jones, Software Productivity Research, http://www.spr.com

HTML = hypertext mark-up language.

FIGURE 3-4
Converting from Function Points to Lines of Code

required to develop a system, the choice of development language has a significant impact on the time and cost of projects.

Step 2: Estimate Effort Required   Once an understanding is reached about the size of the system, the next step is to estimate the effort that is required to build it. *Effort* is a function of the system size combined with production rates (how much work someone can complete in a given time). Much research has been done on software production rates. One of the most popular algorithms, the COCOMO

YOUR TURN

3-1 CALCULATE SYSTEM SIZE

Imagine that job hunting has been going so well that you need to develop a system to support your efforts. The system should allow you to input information about the companies with which you interview, the interviews and office visits that you have scheduled, and the offers that you receive. It should be able to produce reports, such as a company contact list, an interview schedule, and an office visit schedule, as well as produce thank-you letters to be brought into a word processor to customize. You also need the system to answer queries, such as the number of interviews by city and your average offer amount.

QUESTIONS:
1. Determine the number of inputs, outputs, interfaces, files, and queries that this system requires. For each element, determine if the complexity is low, medium, or high. Record this information on a worksheet similar to the one in Figure 3-3.
2. Calculate the total function points for each line on your worksheet by multiplying the number of each element with the appropriate complexity score.
3. Sum up the total unadjusted function points.
4. Suppose the system will be built by you using Visual Basic (VB). Given your VB skills, multiply the TUFP score by the APC score that best estimates how complex the system will be for you to develop (.65 = simple, 1 = average, 1.35 = complex), and calculate a TAFP value.
5. Using the table in Figure 3-4, determine the number of lines of code that correspond to VB. Multiply this number with the TAFP to find the total lines of code that your system will require.

YOUR	3-2 CALCULATE EFFORT AND SCHEDULE TIME
TURN	

Refer to the project size and lines of code that you calculated in "Your Turn 3-1."

QUESTIONS:
1. Determine the effort of your project in person-months by effort multiplying your lines of code (in thousands) by 1.4.

2. Calculate the schedule time in months for your project using the formula, $3.0 \times \text{person-months}^{1/3}$.
3. Based on your numbers, how much time will it take to complete the project if you are the developer?

model,[4] was designed by Barry W. Boehm to convert a lines-of-code estimate into a person-month estimate. There are different versions of the *COCOMO model* that vary based on the complexity of the software, the size of the system, the experience of the developers, and the type of software you are developing (e.g., business application software such as the registration system at your university; commercial software such as Word; or system software such as Windows). For small to moderate-size business software projects (i.e., 100,000 lines of code and 10 or fewer programmers), the model is quite simple:

$$\text{effort (in person-months)} = 1.4 \times \text{thousands of lines of code}$$

For example, let's suppose that we were going to develop a business software system requiring 10,000 lines of code. This project would typically take 14 person-months to complete. If the system in Figure 3.3 were developed in COBOL (which equates to 26,730 lines of code), it would require about 37.42 person-months of effort.

Step 3: Estimate Time Required   Once the effort is understood, the optimal schedule for the project can be estimated. Historical data or estimation software can be used as aids, or one rule of thumb is to determine schedule using the following equation:

$$\text{schedule time (months)} = 3.0 \times \text{person-months}^{1/3}$$

This equation is widely used, although the specific numbers vary (e.g., some estimators may use 3.5 or 2.5 instead of 3.0). The equation suggests that a project that has an effort of 14 person-months should be scheduled to take a little more than 7 months to complete. Continuing the Figure 3.3 example, the 37.42 person-months would require a little over 10 months. It is important to note that this estimate is for the analysis, design, and implementation phases; it does not include the planning phase.

---

[4] The original COCOMO model is presented by Barry W. Boehm in *Software Engineering Economics*, Englewood Cliffs, NJ: Prentice Hall, 1981. Since then, much additional research has been done. For the latest updates, see http://sunset.usc.edu/COCOMOII/cocomo.html

CREATING AND MANAGING THE WORKPLAN

Once a project manager has a general idea of the size and approximate schedule for the project, he or she creates a *workplan*, which is a dynamic schedule that records and keeps track of all of the tasks that need to be accomplished over the course of the project. The workplan lists each task, along with important information about it, such as when it needs to be completed, the person assigned to do the work, and any deliverables that will result (Figure 3-5). The level of detail and the amount of information captured by the workplan depend on the needs of the project (and the detail usually increases as the project progresses). Usually, the workplan is the main component of the project management software that we mentioned earlier.

To create a workplan, the project manager first identifies the tasks that need to be accomplished and determines how long they will take. Then the tasks are organized within a workplan and presented graphically using Gantt and PERT charts. All of these techniques help a project manager understand and manage the project's progress over time, and they are described in detail in the next sections.

Identify Tasks

The overall objectives for the system should be listed on the system request, and it is the project manager's job to identify all of the tasks that need to be accomplished to meet those objectives. This sounds like a daunting task—how can someone know everything that needs to be done to build a system that has never been built before?

One approach for identifying tasks is to get a list of tasks that has already been developed and to modify it. There are standard lists of tasks, or methodologies, that are available for use as a starting point. As we stated in Chapter 1, a *methodology* is a formalized approach to implementing the SDLC (i.e., it is a list of steps and deliverables). A project manager can take an existing methodology, select the steps and deliverables that apply to the current project, and add them to the workplan. If an existing methodology is not available within the organization, methodologies can be purchased from consultants or vendors, or books like this textbook can serve as guidance. Using an existing methodology is the most popular way to create a workplan because most organizations have a methodology that they use for projects.

Workplan Information	Example
Name of the task	Perform economic feasibility
Start date	Jan 05, 2003
Completion date	Jan 19, 2003
Person assigned to the task	Project sponsor; Mary Smith
Deliverable(s)	Cost–benefit analysis
Completion status	Open
Priority	High
Resources that are needed	Spreadsheet software
Estimated time	16 hours
Actual time	14.5 hours

FIGURE 3-5
Workplan Information

If a project manager prefers to begin from scratch, he or she can use a structured, top-down approach whereby high-level tasks are first defined and then these are broken down into subtasks. For example, Figure 3-6 shows a list of high-level tasks that are needed to implement a new IT training class. Some of the main steps in the process include identifying vendors, creating and administering a survey, and building new classrooms. Each step is then broken down in turn and numbered in a hierarchical fashion. There are eight subtasks (i.e., 7.1–7.8) for creating and administering a survey, and there are three subtasks (7.2.1–7.2.3) that comprise the review initial survey task. A list of tasks hierarchically numbered in this way is called a *work breakdown structure*, and it is the backbone of the project workplan.

The number of tasks and level of detail depend on the complexity and size of the project. The larger the project, the more important it becomes to define tasks at a low level of detail so that essential steps are not overlooked.

The Project Workplan

The project workplan is the mechanism that is used to manage the tasks that are listed in the work breakdown structure. It is the project manager's primary tool for managing the project. Using it, the project manager can tell if the project is ahead or behind schedule, how well the project was estimated, and what changes need to be made to meet the project deadline.

Task Number	Task Name	Duration (in weeks)	Dependency	Status
1	Identify vendors	2		Complete
2	Review training materials	6	1	Complete
3	Compare vendors	2	2	In Progress
4	Negotiate with vendors	3	3	Open
5	Develop communications information	4	1	In Progress
6	Disseminate information	2	5	Open
7	Create and administer survey	4	6	Open
7.1	Create initial survey	1		Open
7.2	Review initial survey	1	7.1	Open
7.2.1	Review by Director of IT Training	1		Open
7.2.2	Review by Project Sponsor	1		Open
7.2.3	Review by Representative Trainee	1		Open
7.3	Pilot test initial survey	1	7.1	Open
7.4	Incorporate survey changes	1	7.2, 7.3	Open
7.5	Create distribution list	.5		Open
7.6	Send survey to distribution list	.5	7.4, 7.5	Open
7.7	Send follow-up message	.5	7.6	Open
7.8	Collect completed surveys	1	7.6	Open
8	Analyze results and choose vendor	2	4, 7	Open
9	Build new classrooms	11	1	In Progress
10	Develop course options	3	8, 9	Open

FIGURE 3-6

Identifying Tasks

Basically, the workplan is a table that lists all of the tasks in the work breakdown structure along with important task information, such as the people who are assigned to perform the tasks, the actual hours that the tasks took, and the variances between estimated and actual completion times (see Figure 3-6). At a minimum, the information should include the duration of the task, the current statuses of the tasks (i.e., open, complete), and the *task dependencies*, which occur when one task cannot be performed until another task is completed. For example, Figure 3-6 shows that incorporating changes to the survey (task 7.4) takes a week to perform, but it cannot occur until after the survey is reviewed (task 7.2) and pilot tested (task 7.3). Key *milestones*, or important dates, are also identified on the workplan. Presentations to the approval committee, the start of end-user training, a company retreat, and the due date of the system prototype are the types of milestones that may be important to track.

Gantt Chart

A Gantt chart is a horizontal bar chart that shows the same task information as the project workplan, but in a graphical way. Sometimes a picture really is worth a thousand words, and the Gantt chart can communicate the high-level status of a project much faster and easier than the workplan. Creating a Gantt chart is simple and can be done using a spreadsheet package, graphics software (e.g., Microsoft VISIO), or a project management package.

First, tasks are listed as rows in the chart, and time is listed across the top in increments based on the needs of the projects (see Figure 3-7). A short project may be divided into hours or days; whereas, a medium-sized project may be represented using weeks or months. Horizontal bars are drawn to represent the duration of each task; the bar's beginning and end mark exactly when the task will begin and end. As people work on tasks, the appropriate bars are filled in proportionately to how much of the task is finished. Too many tasks on a Gantt chart can become confusing, so it's best to limit the number of tasks to around 20 to 30. If there are more tasks, break them down into subtasks and create Gantt charts for each level of detail.

There are many things a project manager can see by looking quickly at a Gantt chart. In addition to seeing how long tasks are and how far along they are, the project manager also can tell which tasks are sequential, which tasks occur at the same time, and which tasks overlap in some way. He or she can get a quick view of tasks that are ahead of schedule and behind schedule by drawing a vertical line on today's date. If a bar is not filled in and is to the left of the line, that task is behind schedule.

There are a few special notations that can be placed on a Gantt chart. Project milestones are shown using upside-down triangles or diamonds. Arrows are drawn between the task bars to show task dependencies. Sometimes, the names of people assigned to each task are listed next to the task bars to show what human resources have been allocated to each task.

PERT Chart

A second graphical way to look at the project workplan information is the *PERT chart*, which lays the project tasks in a flowchart (see Figure 3-8). A box (also called a *node*) represents each task, and a line connecting two boxes represents the dependency between two tasks. Usually the partially completed tasks are displayed with a diagonal line through the node, and completed tasks contain crossed lines. Mile-

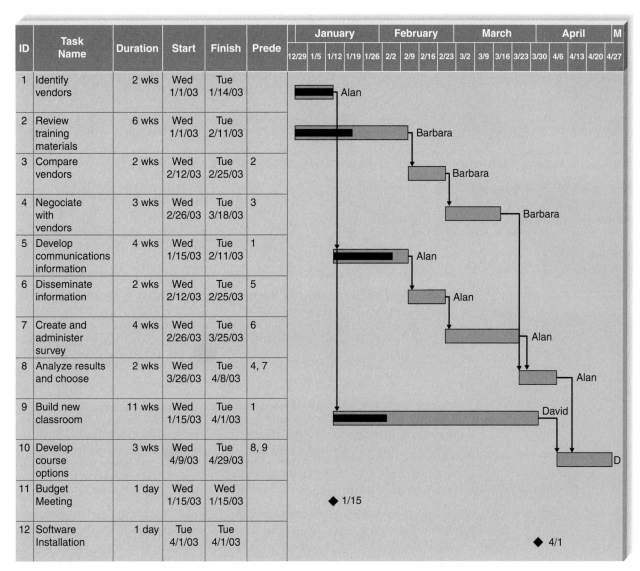

ID	Task Name	Duration	Start	Finish	Prede	January					February				March				April			M
						12/29	1/5	1/12	1/19	1/26	2/2	2/9	2/16	2/23	3/2	3/9	3/16	3/23	3/30	4/6	4/13	4/20	4/27
1	Identify vendors	2 wks	Wed 1/1/03	Tue 1/14/03																		
2	Review training materials	6 wks	Wed 1/1/03	Tue 2/11/03																		
3	Compare vendors	2 wks	Wed 2/12/03	Tue 2/25/03	2																	
4	Negociate with vendors	3 wks	Wed 2/26/03	Tue 3/18/03	3																	
5	Develop communications information	4 wks	Wed 1/15/03	Tue 2/11/03	1																	
6	Disseminate information	2 wks	Wed 2/12/03	Tue 2/25/03	5																	
7	Create and administer survey	4 wks	Wed 2/26/03	Tue 3/25/03	6																	
8	Analyze results and choose	2 wks	Wed 3/26/03	Tue 4/8/03	4, 7																	
9	Build new classroom	11 wks	Wed 1/15/03	Tue 4/1/03	1																	
10	Develop course options	3 wks	Wed 4/9/03	Tue 4/29/03	8, 9																	
11	Budget Meeting	1 day	Wed 1/15/03	Wed 1/15/03																		
12	Software Installation	1 day	Tue 4/1/03	Tue 4/1/03																		

FIGURE 3-7
Gantt Chart

stone tasks are emphasized in some way; in Figure 3-8, for example, the milestone tasks have solid blue borders.

PERT charts are the best way to communicate task dependencies because they lay out the tasks in the order in which they need to be completed. The longest path from the project inception to completion is referred to as the *critical path*. The critical path shows all of the tasks that must be completed on schedule for a project as a whole to finish on schedule. Each task on the critical path is a *critical task*, and usually they are depicted in a unique way; in Figure 3-8 they are shown with double borders.

The benefit of using project management software packages like Microsoft Project is that the workplan can be input once, and then the software can display the

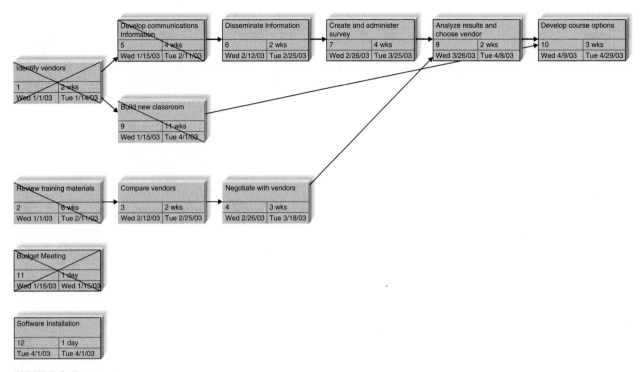

FIGURE 3-8
PERT Chart

information in many different formats. You can toggle between the workplan, a Gantt chart, and a PERT chart depending on your project management needs.

Refining Estimates

The estimates that are produced during the planning phase will need to be refined as the project progresses. This means not that estimates were poorly done at the start of the project but that it is virtually impossible to develop an exact assessment of the project's schedule before the analysis and design phases are conducted. A project manager should expect to be satisfied with broad ranges of estimates that become more and more specific as the project's product becomes better defined.

In many respects, estimating what an IS development project will cost, how long it will take, and what the final system will actually do follows a *hurricane model*. When storms and hurricanes first appear in the Atlantic or Pacific, forecasters watch their behavior and, on the basis of minimal information about them (but armed with lots of data on previous storms), attempt to predict when and where the storms will hit and what damage they will do when they arrive. As storms move closer to North America, forecasters refine their tracks and develop better predictions about where and when they are most likely to hit and their force when they do. The predictions become more and more accurate as the storms approach a coast, until they finally arrive.

In the Planning Phase when a system is first requested, the project sponsor and project manager attempt to predict how long the SDLC will take, how much it will cost, and what it will ultimately do when it is delivered (i.e., its functionality). However, the estimates are based on very little knowledge of the system. As the system moves into the Analysis Phase, more information is gathered, the system concept is developed, and the estimates become even more accurate and precise. As the system moves closer to completion, the accuracy and precision increase until the final system is delivered (Figure 3-9).

According to one of the leading experts in software development,[5] a well-done project plan (prepared at the end of the planning phase) has a 100% margin of error for project cost and a 25% margin of error for schedule time. In other words, if a carefully done project plan estimates that a project will cost $100,000 and take 20 weeks, the project will actually cost between $0 and $200,000 and take between 15 and 25 weeks. Figure 3-10 presents typical margins of error for other stages in the project. It is important to note that these margins of error apply only to well-done plans; a plan developed without much care has a much greater margin of error.

What happens if you overshoot an estimate (e.g., the Analysis Phase ends up lasting 2 weeks longer than expected)? There are number of ways to adjust future estimates. If the project team finishes a step ahead of schedule, most project managers shift the deadlines sooner by the same amount but do not adjust the promised completion date. The challenge, however, occurs when the project team is late in

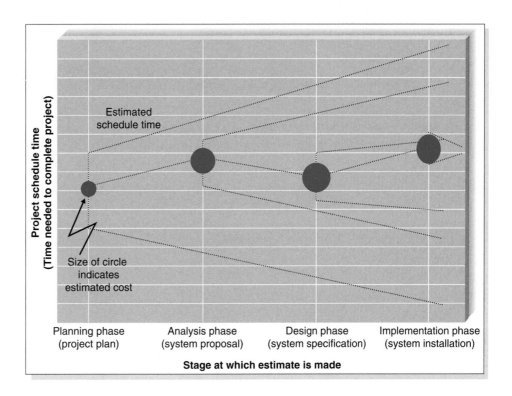

FIGURE 3-9

Hurricane Model

[5] Barry W. Boehm and colleagues, "Cost Models for Future Software Life Cycle Processes: COCOMO 2.0," in J. D. Arthur and S. M. Henry (editors), *Annals of Software Engineering: Special Volume on Software Process and Product Measurement*, Amsterdam: J. C. Baltzer AG Science Publishers, 1995.

		Typical Margins of Error for Well-Done Estimates	
Phase	**Deliverable**	**Cost (%)**	**Schedule Time (%)**
Planning phase	System request	400	60
	Project plan	100	25
Analysis phase	System proposal	50	15
Design phase	System specifications	25	10

Source: Barry W. Boehm and colleagues, "Cost Models for Future Software Life Cycle Processes: COCOMO 2.0," in J. D. Arthur and S. M. Henry (editors) *Annals of Software Engineering Special Volume on Software Process and Product Measurement,* Amsterdam: J. C. Baltzer AG Science Publishers, 1995.

FIGURE 3-10
Margins of Error in Cost and Time Estimates

meeting a scheduled date. Three possible responses to missed schedule dates are presented in Figure 3-11. We recommend that if an estimate proves too optimistic early in the project, do not expect to make up for lost time—very few projects end up doing this. Instead, change your future estimates to include an increase similar to the one that was experienced. For example, if the first phase was completed 10% over schedule, increase the rest of your estimates by 10%.

Assumptions	**Actions**	**Level of Risk**
If you assume the rest of the project is simpler than the part that was late and is also simpler than believed when the original schedule estimates were made, you can make up lost time	Do not change schedule.	High risk
If you assume the rest of the project is simpler than the part that was late and is no more complex than the original estimate assumed, you can't make up the lost time, but you will not lose time on the rest of the project.	Increase the entire schedule by the total amount of time that you are behind (e.g., if you missed the scheduled date by 2 weeks, move the rest of the schedule dates to 2 weeks later). If you included padded time at the end of the project in the original schedule, you may not have to change the promised system delivery date; you'll just use up the padded time.	Moderate risk
If you assume that the rest of the project is as complex as the part that was late (your original estimates too optimistic), then all the scheduled dates in the future underestimate the real time required by the same percentage as the part that was late.	Increase the entire schedule by the percentage of weeks that you are behind (e.g., if you are 2 weeks late on part of the project that was supposed to take 8 weeks, you need to increase all remaining time estimates by 25%). If this moves the new delivery date beyond what is acceptable to the project sponsor, the scope of the project must be reduced.	Low risk

FIGURE 3-11
Possible Actions When a Schedule Date Is Missed

Scope Management

You may assume that your project will be safe from scheduling problems because you carefully estimated and planned your project up front. However, the most common reason for schedule and cost overruns occurs after the project is underway—*scope creep*.

Scope creep happens when new requirements are added to the project after the original project scope was defined and "frozen." It can happen for many reasons: users may suddenly understand the potential of the new system and realize new functionality that would be useful; developers may discover interesting capabilities to which they become very attached; a senior manager may decide to let this system support a new strategy that was developed at a recent board meeting.

Unfortunately, after the project begins, it becomes increasingly difficult to address changing requirements. The ramifications of change become more extensive, the focus is removed from original goals, and there is at least some impact on cost and schedule. Therefore, the project manager plays a critical role in managing this change to keep scope creep to a minimum.

The keys are to identify the requirements as well as possible in the beginning of the project and to apply analysis techniques effectively. For example, if needs are fuzzy at the project's onset, a combination of intensive meetings with the users and prototyping could be used so that users "experience" the requirements and better visualize how the system could support their needs. In fact, the use of meetings and prototyping has been found to reduce scope creep to less than 5% on a typical project.

Of course, some requirements may be missed no matter what precautions you take, but several practices can be helpful to control additions to the task list. First, the project manager should allow only absolutely necessary requirements to be added after the project begins. Even at that point, members of the project team should carefully assess the ramifications of the addition and present the assessment back to the users. For example, it may require two more person-months of work to create a newly defined report, which would throw off the entire project deadline by several weeks. Any change that is implemented should be carefully tracked so that an audit trail exists to measure the change's impact.

Sometimes changes cannot be incorporated into the present system even though they truly would be beneficial. In this case, these additions to scope should be recorded as future enhancements to the system. The project manager can offer to provide functionality in future releases of the system, thus getting around telling someone no.

Timeboxing

Another approach to scope management is a technique called timeboxing. Up until now, we have described projects that are task oriented. In other words, we have described projects that have a schedule that is driven by the tasks that need to be accomplished, so the greater number of tasks and requirements, the longer the project will take. Some companies have little patience for development projects that take a long time, and these companies take a time-oriented approach that places meeting a deadline above delivering functionality.

Think about your use of word processing software. For 80% of the time, you probably use only 20% of the features, such as the spelling checker, boldfacing, and cutting and pasting. Other features, such as document merging and creation of mailing labels, may be nice to have, but they are not a part of your day-to-day needs.

The same goes for other software applications; most users rely on only a small subset of their capabilities. Ironically, most developers agree that typically 75% of a system can be provided relatively quickly, with the remaining 25% of the functionality demanding most of the time.

To resolve this incongruency, a technique called timeboxing has become quite popular, especially when using rapid application development (RAD) methodologies. This technique sets a fixed deadline for a project and delivers the system by that deadline no matter what, even if functionality needs to be reduced. Timeboxing ensures that project teams don't get hung up on the final "finishing touches" that can drag out indefinitely, and it satisfies the business by providing a product within a relatively fast time frame.

There are several steps to implement timeboxing on a project (Figure 3-12). First, set the date of delivery for the proposed goals. The deadline should not be impossible to meet, so it is best to let the project team determine a realistic due date. Next, build the core of the system to be delivered; you will find that timeboxing helps create a sense of urgency and helps keep the focus on the most important features. Because the schedule is absolutely fixed, functionality that cannot be completed needs to be postponed. It helps if the team prioritizes a list of features beforehand to keep track of what functionality the users absolutely need. Quality cannot be compromised, regardless of other constraints, so it is important that the time allocated to activities is not shortened unless the requirements are changed (e.g., don't reduce the time allocated to testing without reducing features). At the end of the time period, a high-quality system is delivered; likely, future iterations will be needed to make changes and enhancements, and the timeboxing approach can be used once again.

STAFFING THE PROJECT

Staffing the project includes determining how many people should be assigned to the project, matching people's skills with the needs of the project, motivating them to meet the project's objectives, and minimizing the conflict that will occur over time. The deliverable for this part of project management is a staffing plan, which describes the number and kinds of people who will work on the project, the overall reporting structure, and the project charter, which describes the project's objectives and rules.

Staffing Plan

The first step to staffing is determining the average number of staff needed for the project. To calculate this figure, divide the total person-months of effort by the opti-

1. Set the date for system delivery.
2. Prioritize the functionality that needs to be included in the system.
3. Build the core of the system (the functionality ranked as most important).
4. Postpone functionality that cannot be provided within the time frame.
5. Deliver the system with core functionality.
6. Repeat steps 3 through 5, to add refinements and enhancements.

FIGURE 3-12
Steps for Timeboxing

CONCEPTS

IN ACTION

3-C TIMEBOXING

DuPont was one of the first companies to use timeboxing. The practice originated in the company's fibers division, which was moving to a highly automated manufacturing environment. It was necessary to create complex application software quickly, and DuPont recognized that it is better to get a basic version of the system working, learn from the experience of operating with it, and then design an enhanced version than it is to wait for a comprehensive system at a later date.

DuPont's experience implies that:

- The first version must be built quickly.
- The application must be built so that it can be changed and added to quickly.

DuPont stresses that the timebox methodology works well for the company and is highly practical. It has resulted in automation being introduced more rapidly and effectively. DuPont quotes large cost savings from the methodology, and variations of timebox techniques have since been used in many other corporations.

QESTIONS:

1. Why do you think DuPont saves money by using this technique?
2. Are there situations in which timeboxing would not be appropriate?

Source: "Within the Timebox, Development Deadlines Really Work," *PC Week*, March 12, 1990, by James Martin.

mal schedule. So to complete a 40 person-month project in 10 months, a team should have an average of four full-time staff members, although this may change over time as different specialists enter and leave the team (e.g., business analysts, programmers, technical writers).

Many times, the temptation is to assign more staff to a project to shorten the project's length, but this is not a wise move. Adding staff resources does not translate into increased productivity; staff size and productivity share a disproportionate relationship, mainly because a large number of staff members is more difficult to coordinate. The more a team grows, the more difficult it becomes to manage. Imagine how easy it is to work on a two-person project team: the team members share a single line of communication. But adding two people increases the number of communication lines to six, and greater increases lead to more dramatic gains in communication complexity. Figure 3-13 illustrates the impact of adding team members to a project team.

One way to reduce efficiency losses on teams is to understand the complexity that is created in numbers and to build in a *reporting structure* that tempers its effects. The rule of thumb is to keep team sizes under 8 to 10 people; therefore, if more people are needed, create subteams. In this way, the project manager can keep the communication effective within small teams, which in turn communicate to a contact at a higher level in the project.

After the project manager understands how many people are needed for the project, he or she creates a *staffing plan* that lists the roles that are required for the project and the proposed reporting structure for the project. Typically, a project will have one project manager who oversees the overall progress of the development effort, with the core of the team comprising the various types of analysts described in Chapter 1. A *functional lead* usually is assigned to manage a group of analysts,

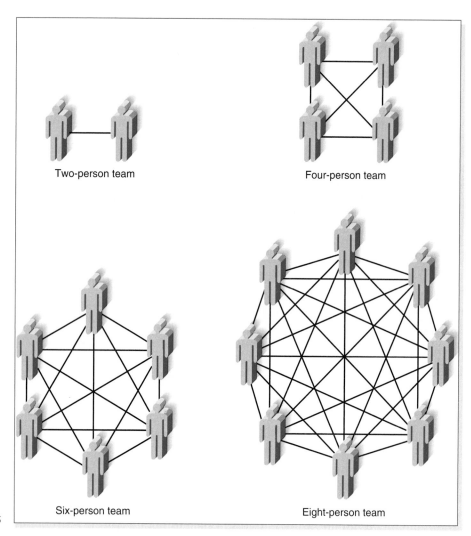

FIGURE 3-13
Increasing Complexity with Larger Teams

and *technical lead* oversees the progress of a group of programmers and more technical staff members.

There are many structures for project teams; Figure 3-14 illustrates one possible configuration of a project team. After the roles are defined and the structure is in place, the project manager needs to think about which people can fill each role. Often, one person fills more than one role on a project team.

When you make assignments, remember that people have *technical skills* and *interpersonal skills*, and both are important on a project. Technical skills are useful when working with technical tasks (e.g., programming in Java) and in trying to understand the various roles that technology plays in the particular project (e.g., how a Web server should be configured on the basis of a projected number of hits from customers).

Interpersonal skills, on the other hand, include interpersonal and communication abilities that are used when dealing with business users, senior management executives, and other members of the project team. They are particularly critical when performing the requirements-gathering activities and when addressing orga-

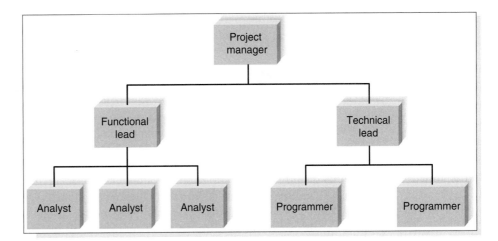

FIGURE 3-14
Possible Reporting Structure

nizational feasibility issues. Each project will require unique technical and interpersonal skills. For example, a Web-based project may require Internet experience or Java programming knowledge, or a highly controversial project may need analysts who are particularly adept at managing political or volatile situations.

Ideally, project roles are filled with people who have the right skills for the job; however, the people who fit the roles best may not be available; they may be working on other projects, or they may not exist in the company. Therefore, assigning project team members really is a combination of finding people with the appropriate skill sets and finding people who are available. When the skills of the available project team members do not match what is actually required by the project, the project manager has several options to improve the situation. First, people can be pulled off other projects, and resources can be shuffled around. This is the most disruptive approach from the organization's perspective. Another approach is to use outside help—such as a consultant or contractor—to train team members and start them off on the right foot. Training classes are usually available for both technical and interpersonal instruction, if time is available. Mentoring may also be an option;

a project team member can be sent to work on another similar project so that he or she can return with skills to apply to the current job.

Motivation

Assigning people to tasks isn't enough; project managers need to motivate the people to make the project a success. *Motivation* has been found to be the number-one influence on people's performance,[6] but determining how to motivate the team can be quite difficult. You may think that good project managers motivate their staff by rewarding them with money and bonuses, but most project managers agree that this is the last thing that should be done. The more often you reward team members with money, the more they expect it—and most times monetary motivation won't work.

Assuming that team members are paid a fair salary, technical employees on project teams are much more motivated by recognition, achievement, the work itself, responsibility, advancement, and the chance to learn new skills.[7] If you feel like you need to give some kind of reward for motivational purposes, try a pizza or free dinner, or even a kind letter or award. They often have much more effective results. Figure 3-15 lists some other motivational don'ts that you should avoid to ensure that motivation on the project is as high as possible.

Don'ts	Reasons
Assign unrealistic deadlines	Few people will work hard if they realize that a deadline is impossible to meet.
Ignore good efforts	People will work harder if they feel like their work is appreciated. Often, all it takes is public praise for a job well done.
Create a low-quality product	Few people can be proud of working on a project that is of low quality.
Give everyone on the project a raise	If everyone is given the same reward, then high-quality people will believe that mediocrity is rewarded—and they will resent it.
Make an important decision without the team's input	Buy-in is very important. If the project manager needs to make a decision that greatly affects the members of her team, she should involve them in the decision-making process.
Maintain poor working conditions	A project team needs a good working environment or motivation will go down the tubes. This includes lighting, desk space, technology, privacy from interruptions, and reference resources.

Source: Adapted *Rapid Development*, Redmond, WA: Microsoft Press, 1996, by from Steve McConnell.

FIGURE 3-15
Motivational Don'ts

---

[6] Barry W. Boehm, *Software Engineering Economics*, Englewood Cliffs, NJ: Prentice Hall, 1981. One of the best books on managing project teams is by Tom DeMarco and Timothy Lister, *Peopleware: Productive Projects and Teams*, New York: Dorset House, 1987.

[7] F. H. Hertzberg, "One More Time: How Do You Motivate Employees?" *Harvard Business Review,* 1968, January–February.

3-D HASTE SLOWS MICROSOFT

In 1984, Microsoft planned for the development of Microsoft Word to take one year. At the time, this was two months less than the most optimistic estimated deadline for a project of its size. In reality, it took Microsoft five years to complete Word. Ultimately, the overly aggressive schedule for Word slowed its development for a number of reasons. The project experienced high turnover due to developer burn-out from unreasonable pressure and work hours. Code was "finalized" prematurely, and the software spent much longer in "stabilization" (i.e., fixing bugs) than was originally expected (i.e., 12 months versus 3 months). And, the aggressive scheduling resulted in poor planning (the delivery date consistently was off by more than 60% for the first four years of the project).

QUESTION:

Suppose you take over as project manager in 1986 after the previous project manager has been fired. Word is now 1 year overdue. Describe three things you would do on your first day on the job to improve the project.

Source: Microsoft Corporation: Office Business Unit, *Harvard Business School Case Study 9-691-033,* revised May 31, 1994, Boston: Harvard Business School, 1994, by Marco Iansiti

Handling Conflict

The third component of staffing is organizing the project to minimize conflict among group members. *Group cohesiveness* (the attraction that members feel to the group and to other members) contributes more to productivity than do project members' individual capabilities or experiences.[8] Clearly defining the roles on the project and holding team members accountable for their tasks is a good way to begin mitigating potential conflict on a project. Some project managers develop a *project charter* that lists the project's norms and ground rules. For example, the charter may describe when the project team should be at work, when staff meetings will be held, how the group will communicate with each other, and the procedures for updating the workplan as tasks are completed. Figure 3-16 lists additional techniques that can be used at the start of a project to keep conflict to a minimum.

COORDINATING PROJECT ACTIVITIES

Like all project management responsibilities, the act of coordinating project activities continues throughout the entire project until a system is delivered to the pro-

---

- Clearly define plans for the project.
- Make sure the team understands how the project is important to the organization.
- Develop detailed operating procedures and communicate these to the team members.
- Develop a project charter.
- Develop schedule commitments ahead of time.
- Forecast other priorities and their possible impact on project.

Source: H. J. Thamhain and D. L. Wilemon, "Conflict Management in Project Life Cycles," *Sloan Management Review,* Spring 1975.

FIGURE 3-16
Conflict Avoidance Strategies

---

[8] B. Lakhanpal, "Understanding the Factors Influencing the Performance of Software Development Groups: An Exploratory Group-Level Analysis," *Information and Software Technology,* 1993, 35(8):468–473.

3-4 PROJECT CHARTER

Get together with several of your classmates and pretend that you are all staffed on the project described in "Your Turn 3-1" Discuss what would most motivate each of you to perform well on the project. List three potential sources of conflict that could surface as you work together.

QUESTION:
Develop a project charter that lists five rules that all team members will need to follow. How might these rules help avoid potential team conflict?

ject sponsor and end users. This step includes putting efficient development practices in place and mitigating risk. These activities occur over the course of the entire SDLC, but it is at this point in the project when the project manager needs to put them in place. Ultimately, these activities ensure that the project stays on track and that the chance of failure is kept at a minimum. The rest of this section will describe each of these activities in more detail.

CASE Tools

Computer-aided software engineering (CASE) is a category of software that automates all or part of the development process. Some CASE software packages are primarily used during the analysis phase to create integrated diagrams of the system and to store information regarding the system components (often called *upper CASE*), whereas others are design-phase tools that create the diagrams and then generate code for database tables and system functionality (often called *lower CASE*). *Integrated CASE*, or I-CASE, contains functionality found in both upper-CASE and lower-CASE tools in that it supports tasks that happen throughout the SDLC. CASE comes in a wide assortment of flavors in terms of complexity and functionality, and there are many good programs available in the marketplace, such as the Visible Analyst Workbench, Oracle Designer/2000, Rational Rose, and the Logic Works suite.

The benefits to using CASE are numerous. With CASE tools, tasks are much faster to complete and alter, development information is centralized, and information is illustrated through diagrams, which typically are easier to understand. Potentially, CASE can reduce maintenance costs, improve software quality, and enforce discipline, and some project teams even use CASE to assess the magnitude of changes to the project.

Of course, like anything else, CASE should not be considered a silver bullet for project development. The advanced CASE tools are complex applications that require significant training and experience to achieve real benefits. Often, CASE serves only as a glorified diagramming tool that supports the practices described in Chapter 6 (process modeling) and Chapter 7 (data modeling). Our experience has shown that CASE is a helpful way to support the communication and sharing of project diagrams and technical specifications—as long as it is used by trained developers who have applied CASE on past projects.

The central component of any CASE tool is the *CASE repository*, otherwise known as the information repository or data dictionary. The CASE repository stores

YOUR	3-5 COMPUTER-AIDED SOFTWARE ENGINEERING TOOL ANALYSIS
TURN	

Select a computer-aided software engineering (CASE) tool—either one that you will use for class, a program that you own, or a tool that you can examine over the Web. Create a list of the capabilities that are offered by the CASE tool.

QUESTION:
Would you classify the CASE as upper CASE, lower CASE, or integrated CASE (I-CASE)? Why?

the diagrams and other project information, such as screen and report designs, and it keeps track of how the diagrams fit together. For example, most CASE tools will warn you if you place a field on a screen design that doesn't exist in your data model. As the project evolves, project team members perform their tasks using CASE. As you read through the textbook, we will indicate when and how the CASE tool can be used so that you can see how CASE supports the project tasks.

Standards

Members of a project team need to work together, and most project management software and CASE tools provide access privileges to everyone working on the system. When people work together, however, things can get pretty confusing. To make matters worse, people sometimes get reassigned in the middle of a project. It is important that their project knowledge does not leave with them and that their replacements can get up to speed quickly.

One way to make certain that everyone is on the same page by performing tasks in the same way and following the same procedures is to create *standards* that the project team must follow. Standards can range from formal rules for naming files to forms that must be completed when goals are reached to programming guidelines. See Figure 3-17 for some examples of the types of standards that a project may create. When a team forms standards and then follows them, the project can be completed faster because task coordination becomes less complex.

Standards work best when they are created at the beginning of each major phase of the project and well communicated to the entire project team. As the team moves forward, new standards are added when necessary. Some standards (e.g., file-naming conventions, status reporting) are applied to the entire SDLC, whereas others (e.g., programming guidelines) are only appropriate for certain tasks.

Documentation

A final technique that project teams put in place during the planning phase is good *documentation*, which includes detailed information about the tasks of the SDLC. Often, the documentation is stored in *project binder(s)* that contain all the deliverables and all the internal communication that takes place—the history of the project.

A poor project management practice is waiting until the last minute to create documentation, and this typically leads to an undocumented system that no one understands. In fact, many problems that companies had updating their systems to

Types of Standards	Examples
Documentation standards	The date and project name should appear as a header on all documentation.
	All margins should be set to 1 inch.
	All deliverables should be added to the project binder and recorded in its table of contents.
Coding standards	All modules of code should include a header that lists the programmer, last date of update, and a short description of the purpose of the code.
	Indentation should be used to indicate loops, if-then-else statements, and case statements.
	On average, every program should include one line of comments for every five lines of code.
Procedural standards	Record actual task progress in the work plan every Monday morning by 10 A.M.
	Report to project update meeting on Fridays at 3:30 P.M.
	All changes to a requirements document must be approved by the project manager.
Specification requirement standards	Name of program to be created
	Description of the program's purpose
	Special calculations that need to be computed
	Business rules that must be incorporated into the program
	Pseudocode
	Due date
User interface design standards	Labels will appear in boldface text, left-justified, and followed by a colon.
	The tab order of the screen will move from top left to bottom right.
	Accelerator keys will be provided for all updatable fields.

FIGURE 3-17
A Sampling of Project Standards

handle the year 2000 crisis were the result of the lack of documentation. Good project teams learn to document the system's history as it evolves while the details are still fresh in their memory.

The first step to setting up your documentation is to get some binders and include dividers with which to separate content according to the major phases of the project. An additional divider should contain internal communication, such as the minutes from status meetings, written standards, letters to and from the business users, and a dictionary of relevant business terms. Then, as the project moves forward, place the deliverables from each task into the project binder with descriptions so that someone outside of the project will be able to understand it, and keep a table of contents up to date with the content that is added. Documentation takes time up front, but it is a good investment that will pay off in the long run.

Managing Risk

One final facet of project management is *risk management*, the process of assessing and addressing the risks that are associated with developing a project. Many

CONCEPTS 3-E POOR NAMING STANDARDS

IN ACTION

I once started on a small project (four people) in which the original members of the project team had not set up any standards for naming electronic files. Two weeks into the project, I was asked to write a piece of code that would be referenced by other files that had already been written. When I finished my piece, I had to go back to the other files and make changes to reflect my new work. The only problem was that the lead programmer decided to name the files using his initials (e.g., GGl.prg, GG2.prg, GG3.prg)—and there were over 200 files! I spent 2 days opening every one of those files because there was no way to tell what their contents were.

Needless to say, from then on, the team created a code for file names that provided basic information regarding the file's contents and they kept a log that recorded the file name, its purpose, the date of last update, and programmer for every file on the project. *Barbara Wixom*

QUESTION:
Think about a program that you have written in the past. Would another programmer be able to make changes to it easily? Why or why not?

things can cause risks: weak personnel, scope creep, poor design, and overly optimistic estimates. The project team must be aware of potential risks so that problems can be avoided or controlled well ahead of time.

Typically, project teams create a *risk assessment*, or a document that tracks potential risks along with an evaluation of the likelihood of the risk and its potential impact on the project (Figure 3-18). A paragraph or two is also included that explains potential ways that the risk can be addressed. There are many options: risks could be publicized, avoided, or even eliminated by dealing with its root cause. For example, imagine that a project team plans to use new technology but its members have identified a risk in the fact that its members do not have the right technical

CONCEPTS 3-F THE REAL NAMES OF THE SYSTEMS DEVELOPMENT LIFE CYCLE (SDLC) PHASES

IN ACTION

Dawn Adams, Senior Manager with Asymetrix Consulting, has renamed the SDLC phases:

1. Pudding (Planning)
2. Silly Putty (Analysis)
3. Concrete (Design).
4. Touch-this-and-you're-dead-sucker (Implementation)

Adams also uses icons, such as a skull and crossbones for the implementation phase. The funny labels lend a new depth of interest to a set of abstract concepts. But her names have had another benefit. "I had one participant who adopted the names wholeheartedly," she says, "including my icons. He posted an icon on his office door for the duration of each of the phases, and he found it much easier to deal with requests for changes from the client, who could see the increasing difficulty of the changes right there on the door."

QUESTION:
What would you do if your project sponsor demanded that an important change be made during the "touch-this-and-you're-dead-sucker" phase?

Source: Learning Technology Shorttakes (Wednesday, August 26, 1998, 1(2).

RISK ASSESSMENT

RISK #1:   The development of this system likely will be slowed considerably because project team members have not programmed in Java prior to this project.

Likelihood of risk: High probability of risk

Potential impact on the project: This risk likely will increase the time to complete programming tasks by 50%.

Ways to address this risk:
It is very important that time and resources are allocated to up-front training in Java for the programmers who are used for this project. Adequate training will reduce the initial learning curve for Java when programming begins. Additionally, outside Java expertise should be brought in for at least some part of the early programming tasks. This person should be used to provide experiential knowledge to the project team so that JAVA-related issues (of which novice Java programmers would be unaware) are overcome.

RISK #2:   ...

FIGURE 3-18
Sample Risk Assessment

skills. They believe that tasks may take much longer to perform because of a high learning curve. One plan of attack could be to eliminate the root cause of the risk—the lack of technical experience by team members—by finding time and resources that are needed to provide proper training to the team.

Most project managers keep abreast of potential risks, even prioritizing them according to their magnitude and importance. Over time, the list of risks will change as some items are removed and others surface. The best project managers, however, work hard to keep risks from having an impact on the schedule and costs associated with the project.

APPLYING THE CONCEPTS AT CD SELECTIONS

Alec Adams was very excited about managing the Internet Order System project at CD Selections, but he realized that his project team would have very little time to deliver at least some parts of the system because the company wanted the application developed in time for the holiday season. Therefore, he decided that the project should follow a RAD phased-development methodology, combined with the timeboxing technique. In this way, he could be sure that some version of the product would be in the hands of the users within several months, even if the completed system would be delivered at a later date.

As project manager, Alec had to estimate the project's size, effort, and schedule—some of his least favorite jobs because of how tough it is to do at the very beginning of the project. But he knew that the users would expect at least general

PRACTICAL	3-1 AVOIDING CLASSIC PLANNING MISTAKES
TIP	

As Seattle University's David Umphress has pointed out, watching most organizations develop systems is like watching reruns of *Gilligan's Island*. At the beginning of each episode, someone comes up with a cockamamie scheme to get off the island that seems to work for a while, but something goes wrong and the castaways find themselves right back where they started—stuck on the island. Similarly, most companies start new projects with grand ideas that seem to work, only to make a classic mistake and deliver the project behind schedule, over budget, or both. Here we summarize four classic mistakes in the planning and project management aspects of the project and discuss how to avoid them:

1. **Overly optimistic schedule:** Wishful thinking can lead to an overly optimistic schedule that causes analysis and design to be cut short (missing key requirements) and puts intense pressure on the programmers, who produce poor code (full of bugs).

   Solution: Don't inflate time estimates; instead, explicitly schedule slack time at the end of each phase to account for the variability in estimates, using the margins of error from Figure 3-10.

2. **Failing to monitor the schedule:** If the team does not regularly report progress, no one knows if the project is on schedule.

   Solution: Require team members to honestly report progress (or the lack or progress) every week. There is no penalty for reporting a lack of progress, but there are immediate sanctions for a misleading report.

3. **Failing to update the schedule:** When a part of the schedule falls behind (e.g., information gathering uses all of the slack in item 1 above plus 2 weeks), a project team often thinks it can make up the time later by working faster. It can't. This is an early warning that the entire schedule is too optimistic.

   Solution: Immediately revise the schedule and inform the project sponsor of the new end date or use timeboxing to reduce functionality or move it into future versions.

4. **Adding people to a late project:** When a project misses a schedule, the temptation is to add more people to speed it up. This makes the project take longer because it increases coordination problems and requires staff to take time to explain what has already been done.

   Solution: Revise the schedule, use timeboxing, throw away bug-filled code, and add people only to work on an isolated part of the project

Source: Adapted from *Rapid Development*, Redmond, WA: Microsoft Press, 1996, pp. 29–50, by Steve McConnell.

ranges for a product delivery date. He began by attempting to estimate the number of inputs, outputs, queries, files, and program interfaces in the new system. For the Web part of the system to be used by customers, he could think of four main queries (searching by artist, by CD title, by song title, and by ad hoc criteria), three input screens (selecting a CD, entering information to put a CD on hold, and a special order screen), four output screens (the home page with general information, information about CDs, information about the customer's special order, and the hold status), three files (CD information, inventory information, and customer orders), and two program interfaces (one to the company's special order system and one that communicates hold information to the retail store systems). For the part of the system to be used by CD Selections staff (to maintain the marketing materials), he identified three additional inputs, three outputs, four queries, one file, and one program interface. He believed most of these to be of medium complexity. He entered these numbers in a worksheet (Figure 3-19).

Rather than attempt to assess the complexity of the system in detail, Alec chose to use a value of 1.20 for APC. He reasoned that the system was of medium complexity and that the project team has little to moderate experience with Internet-base systems. This produced a TAFP of about 190.

System Components:

Description	Total Number	Complexity			Total
		Low	Medium	High	
Inputs	6	0 × 3	4 × 4	2 × 6	28
Outputs	7	2 × 4	4 × 5	1 × 7	35
Queries	8	3 × 3	4 × 4	1 × 6	31
Files	4	0 × 7	4 × 10	0 × 15	40
Program Interfaces	3	0 × 5	2 × 7	1 × 10	24
Total Unadjusted Function Points (TUFP):					158

Adjusted Project Adjusted Processing Complexicity (APC): 1.2

Total Adjusted Function Points (TAFP):

1.2 (APC) × 158 (TUFP) = 190 **(TAFP)**

FIGURE 3-19

Function Points for the Internet Order System

Converting function points into lines of code was challenging. The project would use a combination of C (for most programs) and HTML for the Web screens. Alec decided to assume that about 75% of the function points would be C and 25% would be HTML, which produced a total of about 19,200 lines of code [(.75 × 190 × 130) + (.25 × 190 × 15)].

Using the COCOMO formula, he found that this translated into about 27 person-months of effort (1.4 × 19.2). This in turn suggested a schedule time of about 9 months ($3.0 \times 27^{1/3}$). After much consideration, Alec decided to pad the estimate by 10% (by adding 1 extra month).

Once the estimation was underway, Alec began to identify the tasks that would be needed to complete the system. He used a RAD methodology that CD Selections had in-house, and he borrowed its high-level phases (e.g., analysis) and the major tasks associated with them (e.g., create requirements definition, gather requirements, analyze current system). These were recorded in a workplan using Microsoft Project. Alec expected to define the steps in much more detail at the beginning of each phase (Figure 3-20).

Staffing the Project

Alec next turned to the task of how to staff his project. On the basis of his earlier estimates, it appeared that about three people would be needed to deliver the system by the holidays (27 person-months over 10 months of calendar time means three people, rounded up).

First, he created a list of the various roles that he needed to fill. He thought he would need several analysts to work with the analysis and design of the system as well as an infrastructure analyst to manage the integration of the Internet order system with CD Selections' existing technical environment. Alec also needed people who had good programmer skills and who could be responsible for ultimately implementing the system. Ian, Anne, and K.C. are three analysts with strong tech-

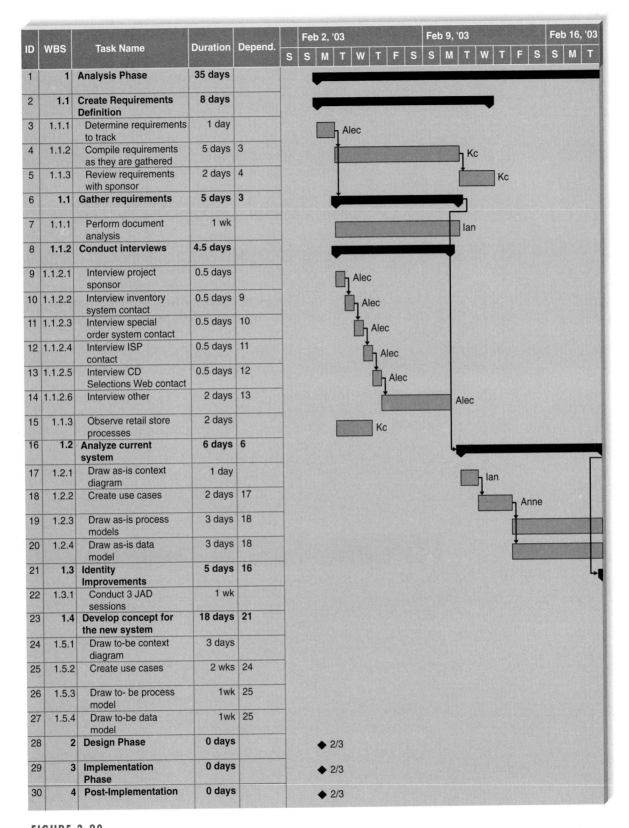

ID	WBS	Task Name	Duration	Depend.
1	1	**Analysis Phase**	35 days	
2	1.1	**Create Requirements Definition**	8 days	
3	1.1.1	Determine requirements to track	1 day	
4	1.1.2	Compile requirements as they are gathered	5 days	3
5	1.1.3	Review requirements with sponsor	2 days	4
6	1.1	**Gather requirements**	5 days	3
7	1.1.1	Perform document analysis	1 wk	
8	1.1.2	**Conduct interviews**	4.5 days	
9	1.1.2.1	Interview project sponsor	0.5 days	
10	1.1.2.2	Interview inventory system contact	0.5 days	9
11	1.1.2.3	Interview special order system contact	0.5 days	10
12	1.1.2.4	Interview ISP contact	0.5 days	11
13	1.1.2.5	Interview CD Selections Web contact	0.5 days	12
14	1.1.2.6	Interview other	2 days	13
15	1.1.3	Observe retail store processes	2 days	
16	1.2	**Analyze current system**	6 days	6
17	1.2.1	Draw as-is context diagram	1 day	
18	1.2.2	Create use cases	2 days	17
19	1.2.3	Draw as-is process models	3 days	18
20	1.2.4	Draw as-is data model	3 days	18
21	1.3	**Identity Improvements**	5 days	16
22	1.3.1	Conduct 3 JAD sessions	1 wk	
23	1.4	**Develop concept for the new system**	18 days	21
24	1.5.1	Draw to-be context diagram	3 days	
25	1.5.2	Create use cases	2 wks	24
26	1.5.3	Draw to- be process model	1wk	25
27	1.5.4	Draw to-be data model	1wk	25
28	2	**Design Phase**	0 days	
29	3	**Implementation Phase**	0 days	
30	4	**Post-Implementation**	0 days	

FIGURE 3-20

Gantt Chart

nical and interpersonal skills (although Ian is less balanced, having greater technical than interpersonal abilities), and Alec believed that they were available to bring onto this project. He wasn't certain if they had experience with the actual Web technology that would be used on the project, but he decided to rely on vendor training or an external consultant to build those skills later when they were needed. Because the project was so small, Alec envisioned all of the team members reporting to him because he would be serving as the project's manager.

Alec created a staffing plan that captured this information, and he included a special incentive structure in the plan (Figure 3-21). Meeting the holiday deadline was very important to the project's success, so he decided to offer a day off to the team members who contributed to meeting that date. He hoped that this incentive would motivate the team to work very hard. Alec also planned to budget money for pizza and sodas for times when the team worked long hours.

Before he left for the day, Alec drafted a project charter, to be fine-tuned after the team got together for its *kick-off meeting* (i.e., the first time the project team gets together). The charter listed several norms that Alec wanted to put in place from the start to eliminate any misunderstanding or problems that could come up otherwise (Figure 3-22).

Coordinating Project Activities

Alec wanted the Internet Order System project to be well coordinated, so he immediately put several practices in place to support his responsibilities. First, he acquired the CASE tool used at CD Selections and set up the product so that it could be used for the analysis-phase tasks (e.g., drawing the data flow diagrams). The team members would likely start creating diagrams and defining components of the system fairly early on. He pulled out some standards that he uses on all development projects

Role	Description	Assigned To
Project manager	Oversees the project to ensure that it meets its objectives in time and within budget	Alec
Infrastructure analyst	Ensures the system conforms to infrastructure standards at CD Selections; ensures that the CD Selections infrastructure can support the new system	Ian
Systems analyst	Designs the information system—with a focus on interfaces with the distribution system	Ian
Systems analyst	Designs the information system—with a focus on the process models and interface design	K.C.
Systems analyst	Designs the information system—with a focus on the data models and system performance	Anne
Programmer	Codes system	K.C.
Programmer	Codes system	Anne

Reporting structure: All project team members will report to Alec.

Special incentives: If the deadline for the project is met, all team members who contributed to this goal will receive a free day off, to be taken over the holiday season.

FIGURE 3-21
Staffing plan for the Internet Order System

Project objective: The Internet Order System project team will create a working Web-based system to sell CDs to CD Selections' customers in time for the holiday season.

The Internet Order System team members will

1. Attend a staff meeting each Friday at 2 P.M. to report on the status of assigned tasks.
2. Update the workplan with actual data each Friday by 5 P.M.
3. Discuss all problems with Alec as soon as they are detected.
4. Agree to support each other when help is needed, especially for tasks that could hold back the progress of the project.
5. Post important changes to the project on the team bulletin board as they are made.

FIGURE 3-22
Project Charter for the Internet Order System

and made a note to review them with his project team at the kick-off meeting for the system. He also had his assistant set up binders for the project deliverables that would start rolling in. Already he was able to include the system request, feasibility analysis, initial workplan, staffing plan, project charter, standards list, and risk assessment.

SUMMARY

Project Management
Project management is the second major component of the planning phase of the systems development life cycle (SDLC), and it includes four steps: identifying the project size, creating and managing the workplan, staffing the project, and coordinating project activities. Project management is important in ensuring that a system is delivered on time, within budget, and with the desired functionality.

Identifying Project Size
The project manager estimates the amount of time and effort that will be needed to complete the project. First, the size is estimated by relying on past experiences or industry standards or by calculating the function points, a measure of program size based on the number and complexity of inputs, outputs, queries, files, and program interfaces. Next, the project manager calculates the effort for the project, which is a function of size and production rates. Algorithms like the COCOMO model can be used to determine the effort value. Third, the optimal schedule for the project is estimated.

Creating and Managing the Workplan
Once a project manager has a general idea of the size and approximate schedule for the project, creates or she creates a workplan, which is a dynamic schedule that records and keeps track of all of the tasks that need to be accomplished over the course of the project. To create a workplan, the project manager first identifies the work breakdown structure, or the tasks that need to be accomplished, and then he or she determines how long the tasks will take. Important information about each task is entered into a workplan.

The workplan information can be presented graphically using Gantt and PERT charts. In the Gantt chart, horizontal bars are drawn to represent the duration of each task, and as people work on tasks, the appropriate bars are filled in proportionately to how much of the task is finished. PERT charts are the best way to communicate task dependencies because they lay out the tasks as a flowchart in the order in which they need to be completed. The longest path from the project inception to completion is referred to as the critical path.

Estimating what an IS development project will cost, how long it will take, and what the final system will actually do follows a hurricane model. The estimates become more accurate as the project progresses. One threat to the reliability of estimates is scope creep, which occurs when new requirements are added to the project after the original project scope was defined and "frozen." If the final schedule will not deliver the system in a timely fashion, timeboxing can be used. Timeboxing sets a fixed deadline for a project and delivers the system by that deadline no matter what, even if functionality must be reduced.

Staffing the Project

Staffing involves determining how many people should be assigned to the project, assigning project roles to team members, developing a reporting structure for the team, and matching people's skills with the needs of the project. Staffing also includes motivating the team to meet the project's objectives and minimizing conflict among team members. Both motivation and cohesiveness have been found to greatly influence performance of team members in project situations. Team members are motivated most by such nonmonetary things as recognition, achievement, and the work itself. Conflict can be minimized by clearly defining the roles on a project and holding team members accountable for their tasks. Some managers create a project charter that lists the project's norms and ground rules.

Coordinating Project Activities

Coordinating project activities includes putting efficient development practices in place and mitigating risk, and these activities occur over the course of the entire SDLC. Three techniques are available to help coordinate activities on a project: computer-aided software engineering (CASE), standards, and documentation. CASE is a category of software that automates all or part of the development process; standards are formal rules or guidelines that project teams must follow during the project; and documentation includes detailed information about the tasks of the SDLC. Often, documentation is stored in project binder(s) that contain all the deliverables and all the internal communication that takes place—the history of the project. A risk assessment is used to mitigate risk because it identifies potential risks and evaluates the likelihood of risk and its potential impact on the project.

KEY TERMS

Adjusted project complexity (APC)	Function point	Motivation
Complexity	Function point approach	Node
Computer-aided software engineering (CASE)	Functional lead	PERT chart
	Gantt chart	Project binder
CASE repository	Group cohesiveness	Project charter
COCOMO model	Hurricane model	Project management
Critical path	Integrated CASE	Project management software
Critical task	Interpersonal skills	Project manager
Documentation	Kick-off meeting	Reporting structure
Effort	Lower CASE	Risk assessment
Estimation	Methodology	Risk management
	Milestone	Scope creep

Staffing plan
Standards
Task dependency
Technical lead
Technical skills

Timeboxing
Total adjusted function points
  (TAFP)
Total unadjusted function points
  (TUFP)

Trade-offs
Upper CASE
Work breakdown structure
Workplan

QUESTIONS

1. Why do many projects end up having unreasonable deadlines? How should a project manager react to unreasonable demands?
2. What are the trade-offs that project managers must manage?
3. What are two basic ways to estimate the size of a project?
4. What is a function point, and how is it used?
5. Describe the three steps of the function point approach.
6. What is the formula for calculating the effort for a project?
7. Name two ways to identify the tasks that need to be accomplished over the course of a project.
8. What is the difference between a methodology and a workplan? How are the two terms related?
9. Compare and contrast the Gantt chart and the PERT chart.
10. Describe the hurricane model.
11. What is scope creep, and how can it be managed?
12. What is timeboxing, and why is it used?
13. Describe the differences between a technical lead and a functional lead. How are they similar?
14. Describe three technical skills and three interpersonal skills that would be very important to have on any project.
15. What are the best ways to motivate a team? What are the worst ways?
16. List three techniques to reduce conflict.
17. What is the difference between upper CASE (computer-aided software engineering) and lower CASE?
18. Describe three types of standards, and provide examples of each.
19. What belongs in the project binder? How is the project binder organized?
20. Create a list of potential risks that could affect the outcome of a project.
21. Some companies hire consulting firms to develop the initial project plans and manage the project, but use their own analysts and programmers to develop the system. Why do you think some companies do this?

EXERCISES

A. Visit a project management Web site, such as the Project Management Institute (www.pmi.org). Most have links to project management software products, white papers, and research. Examine some of the links for project management to better understand a variety of Internet sites that contain information related to this chapter.
B. Select a specific project management topic like computer-aided software engineering (CASE), project management software, or timeboxing and search for information on that topic using the Web. The URL listed in question A or any search engine (e.g., Yahoo!, Alta Vista, Excite, InfoSeek) can provide a starting point for your efforts.
C. Pretend that the Career Services office at your university wants to develop a system that collects student résumés and makes them available to students and recruiters over the Web. Students should be able to input their résumé information into a standard résumé template. The information then is presented in a résumé format, and it also is placed in a database that can be queried using an online search form. You have been placed in charge of the project. Develop a plan for estimating the project. How long do you think it would take for you and three other students to complete the project? Provide support for the schedule that you propose.
D. Refer to the situation in question C. You have been told that recruiting season begins a month from today and that the new system must be used. How would you approach this situation? Describe what you can do as the project manager to make sure that

your team does not burn out from unreasonable deadlines and commitments.

E. Consider the system described in question C. Create a workplan that lists the tasks that will need to be completed to meet the project's objectives. Create a Gantt chart and a PERT chart in a project management tool (e.g., Microsoft Project) or using a spreadsheet package to graphically show the high level tasks of the project.

F. Suppose that you are in charge of the project that is described in question C, and the project will be staffed by members of your class. Do your classmates have all of the right skills to implement such a project? If not, how will you go about making sure that the proper skills are available to get the job done?

G. Consider the application that is used at your school to register for classes. Complete a function point worksheet to determine the size of such an application. You will need to make some assumptions about the application's interfaces and the various factors that affect its complexity.

H. Read "Your Turn 3-1" near the beginning of this chapter. Create a risk assessment that lists the potential risks associated with performing the project, along with ways to address the risks.

I. Pretend that your instructor has asked you and two friends to create a Web page to describe the course to potential students and provide current class information (e.g., syllabus, assignments, readings) to current students. You have been assigned the role of leader, so you will need to coordinate your activities and those of your classmates until the project is completed. Describe how you would apply the project management techniques that you have learned in this chapter in this situation. Include descriptions of how you would create a workplan, staff the project, and coordinate all activities—yours and those of your classmates.

J. Select two project management software packages and research them using the Web or trade magazines. Describe the features of the two packages. If you were a project manager, which one would you use to help support your job? Why?

K. Select two estimation software packages and research them using the Web or trade magazines. Describe the features of the two packages. If you were a project manager, which one would you use to help support your job? Why?

L. In 1997, Oxford Health Plans had a computer problem that caused the company to overestimate revenue and underestimate medical costs. Problems were caused by the migration of its claims processing system from the Pick operating system to a UNIX-based system that uses Oracle database software and hardware from Pyramid Technology. As a result, Oxford's stock price plummeted, and fixing the system became the number-one priority for the company. Pretend that you have been placed in charge of managing the repair of the claims processing system. Obviously, the project team will not be in good spirits. How will you motivate team members to meet the project's objectives?

MINICASES

1. Emily Pemberton is an IS project manager facing a difficult situation. Emily works for the First Trust Bank, which has recently acquired the City National Bank. Prior to the acquisition, First Trust and City National were bitter rivals, fiercely competing for market share in the region. Following the acrimonious takeover, numerous staff were laid off in many banking areas, including IS. Key individuals were retained from both banks' IS areas, however, and were assigned to a new consolidated IS department. Emily has been made project manager for the first significant IS project since the takeover, and she faces the task of integrating staffers from both banks on her team. The project they are undertaking will be highly visible within the organization, and the time frame for the project is somewhat demanding. Emily believes that the team can meet the project goals successfully, but success will require that the team become cohesive quickly and that potential conflicts are avoided. What strategies do you suggest that Emily implement in order to help ensure a successfully functioning project team?

2. Tom, Jan, and Julie are IS majors at Great State University. These students have been assigned a class project by one of their professors, requiring them to develop a new Web-based system to collect and update information on the IS-program's alumni. This system will be used by the IS graduates to enter job and address information as they graduate, and then make changes to that information as they change jobs and/or addresses. Their professor also has a number of queries that she is interested in being able to implement. Based on their

preliminary discussions with their professor, the students have developed this list of system elements:

Inputs: 1 low complexity, 2 medium complexity, 1 high complexity

Outputs: 4 medium complexity

Queries: 1 low complexity, 4 medium complexity, 2 high complexity

Files: 3 medium complexity

Program Interfaces: 2 medium complexity

Assume that an adjusted program complexity of 1 2 is appropriate for this project. Calculate the total adjusted function points for this project.

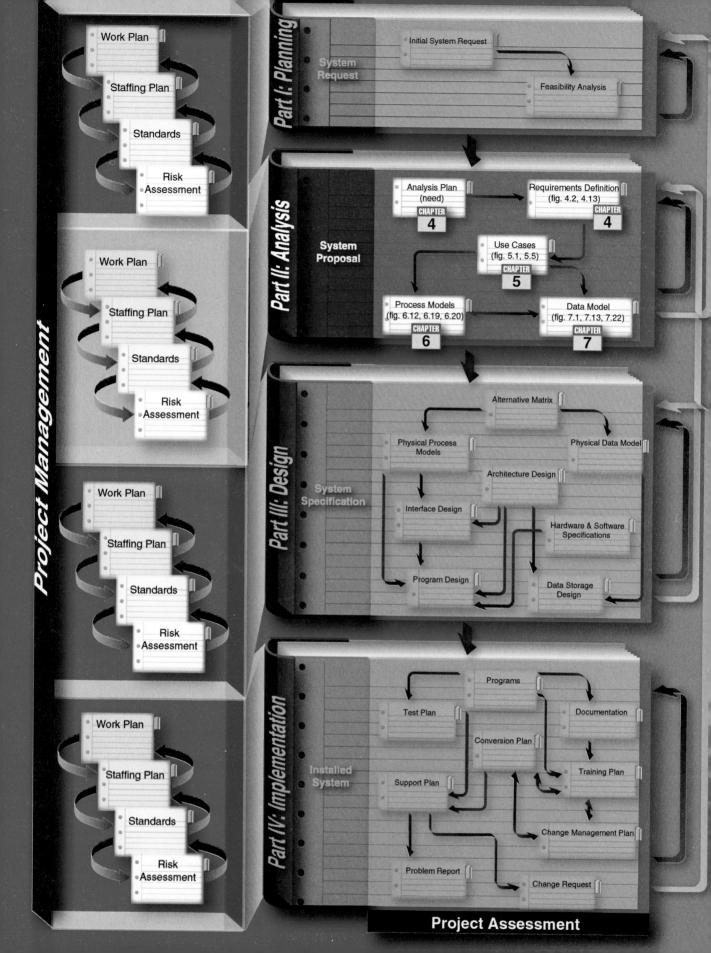

PART TWO
ANALYSIS PHASE

PROJECT BINDER

The Analysis Phase
answers the questions
of who will use the system,
what the system will do,
and
where and when it will be used.

All of the deliverables are combined
into a System Proposal,
which is presented to management
who decide whether the project should
continue to move forward.

Requirements
Determination

CHAPTER
4

Use Case
Analysis

CHAPTER
5

Process
Modeling

CHAPTER
6

Data
Modeling

CHAPTER
7

Analysis Plan

Requirements Definition

Use Cases

Process Models

Data Model

ANALYSIS

☐ **Apply Requirements Analysis Techniques (Business Process Automation, Business Process Improvement, or Business Process Reengineering)**

☐ **Use Requirements Gathering Techniques (Interview, JAD Session, Questionnaire, Document Analysis, or Observation)**

☐ Develop Use Cases

☐ Develop Data Flow Diagrams

☐ Develop Entity Relationship Model

☐ Normalize Entity Relationship Model

TASK CHECKLIST

PLANNING   ANALYSIS   DESIGN

CHAPTER 4

Requirements

Determination

During the analysis phase, the analyst determines the business requirements for the new system. This chapter begins by presenting the requirements definition, a document that lists the new system's capabilities. It then describes how to analyze requirements using business process automation, business process improvement, and business process reengineering techniques, and how to gather requirements using interviews, JAD sessions, questionnaires, document analysis, and observation.

OBJECTIVES

- Become familiar with the analysis phase of the SDLC.
- Understand how to create a requirements definition.
- Become familiar with requirements analysis techniques.
- Understand when to use each requirements analysis technique.
- Understand how to gather requirements using interviews, JAD sessions, questionnaires, document analysis, and observation.
- Understand when to use each requirements-gathering technique.

CHAPTER OUTLINE

IMPLEMENTATION

INTRODUCTION

The Systems Development Life Cycle (SDLC) is the process by which the organization moves from the current system (often called the *as-is system*) to the new system (often called the *to-be system*). The output of the Planning Phase that we discussed in Chapters 2 and 3 is the System Request, which provides general ideas for the to-be system, defines the project's scope, and provides the initial workplan. The analysis phase takes the general ideas in the system request and refines them into a detailed requirements definition (this chapter), use cases (Chapter 5), process models (Chapter 6) and a data model (Chapter 7) that together form the *System Proposal*. The system proposal also includes revised project management deliverables, such as the feasibility analysis (Chapter 2) and the workplan (Chapter 3).

The system proposal is presented to the approval committee who decides if the project is to continue. This usually happens at a system *walk-through*, a meeting at which the concept for the new system is presented to the users, managers, and key decision makers. The goal of the walk-through is to explain the system in moderate detail so that the users, managers, and key decision makers clearly understand it, can identify needed improvements, and are able to make a decision about whether the project should continue. If approved, the system proposal moves into the design phase, and its elements (requirements definition, use case, process models, and data model) are used as inputs to the steps in the design phase, which further refine them and define in much more detail how the system will be built.

The line between the analysis and design phases is very blurry, because the deliverables created in the analysis phase are really the first step in the design of the new system. Many of the major design decisions for the new system are found in the analysis deliverables: In fact, a better name for the analysis phase would really be "Analysis and Initial Design," but because this is rather long name, and because most organizations simply call this phase "Analysis," we will too. Nonetheless, it is important to remember that the deliverables from the analysis phase are really the first step in the design of the new system.

In many ways, the requirements determination step is the single most critical step of the entire SDLC, because it is here that the major elements of the system first begin to emerge. During requirements determination, the system is easy to change because little work has been done yet. As the system moves through the other phases in the SDLC, it becomes harder and harder to return to requirements determination and make major changes because of all of the rework that is involved. Several studies have shown that more than half of all system failures are due to problems with the requirements.[1] This is why the iterative approaches of many RAD and Agile methodologies are so effective—small batches of requirements can be identified and implemented in incremental stages, allowing the overall system to change and evolve over time.

In this chapter, we will focus on the requirements determination step of the analysis phase. We begin by explaining what a requirement is and the overall *analysis plan* of requirements gathering and requirements analysis. We then present a set of techniques that can be used to analyze and gather requirements.

---

[1] For example, see: *The Scope of Software Development Project Failures,* Dennis, MA: The Standish Group, 1995.

REQUIREMENTS DETERMINATION

The purpose of the *requirements determination* step is to turn the very high-level explanation of the business requirements stated in the System Request into a more precise list of requirements than can be used as input to the rest of the analysis phase (creating use cases, building process models, and building a data model), which further require and expand on the requirements and ultimately lead to design.

What is a Requirement?

A *requirement* is simply a statement of what the system must do or what characteristic it needs to have. During the analysis phase, requirements are written from the perspective of the businessperson, and they focus on "what" the system does. They focus on business user needs, so they usually are called *business requirements* (and sometimes user requirements). Later in the design phase, business requirements evolve to become more technical, and they describe "how" the system will be implemented. Requirements in the design phase are written from the developer's perspective, and they usually are called *system requirements*.

Before we continue, we want to stress that there is no black-and-white line dividing a business requirement and a system requirement—and some companies use the terms interchangeably. The important thing to remember is that a requirement is a statement of what the system must do, and requirements will change over time as the project moves from analysis to design to implementation. Requirements evolve from detailed statements of the business capabilities that a system should have to detailed statements of the technical way in which the capabilities will be implemented in the new system.

Requirements can be either functional or nonfunctional in nature. A *functional requirement* relates directly to a process the system has to perform or information it needs to contain. For example, requirements that state that the system must have the ability to search for available inventory or the ability to report actual and budgeted expenses are functional requirements. Functional requirements flow directly into the next steps of the analysis process (use cases, process models, data model) because they define the functions that the system needs to have.

Nonfunctional requirements refer to behavioral properties that the system must have, such as performance and usability. The ability to access the system using a Web browser would be considered a nonfunctional requirement. Nonfunctional requirements may influence the rest of the analysis process (use cases, process models, and data model) but often do so only indirectly; nonfunctional requirements are primarily used in the design phase when decisions are made about the user interface, the hardware and software, and the system's underlying architecture.

Figure 4-1 lists different kinds of nonfunctional requirements and examples of each kind. Notice that the nonfunctional requirements describe a variety of characteristics regarding the system: operational, performance, security, and cultural and political. These characteristics do not describe business processes or information, but they are very important in understanding what the final system should be like. For example, the project team needs to know if a system needs to be highly secure, requires subsecond response time, or has to reach a multilingual customer base. These requirements will affect design decisions that will be made in the design phase, particularly architecture design, so we will revisit them in detail in Chapter 9. The goal at this point is to identify any major issues.

Nonfunctional Requirement	Description	Examples
Operational	The physical and technical environments in which the system will operate	■ The system should be able to fit in a pocket or purse ■ The system should be able to integrate with the existing inventory system ■ The system should be able to work on any Web browser
Performance	The speed, capacity, and reliability of the system	■ Any interaction between the user and the system should not exceed 2 seconds ■ The system should receive updated inventory informataion every 15 minutes ■ The system should be available for use 24 hours per day, 365 days per year
Security	Who has authorized access to the system under what circumstances	■ Only direct managers can see personnel records of staff ■ Customers can only see their order history during business hours
Cultural and Political	Cultural, political factors and legal requirements that affect the system	■ The system should be able to distinguish between United States and European currency ■ Company policy says that we only buy computers from Dell ■ Country managers are permitted to authorize customed user interfaces within their units ■ The system shall comply with insurance industry standards

Source: The Atlantic Systems Guild, http;//www.systemsguild.com

FIGURE 4-1
Nonfunctional Requirements

YOUR TURN

4-1 IDENTIFYING REQUIREMENTS

One of the most common mistakes by new analysts is to confuse functional and nonfunctional requirements. Pretend that you received the following list of requirements for a sales system.

Requirements for Proposed System:
The system should...

1. be accessible to Web users
2. include the company standard logo and color scheme
3. restrict access to profitability information
4. include actual and budgeted cost information
5. provide management reports
6. include sales information that is updated at least daily
7. have 2-second maximum response time for predefined queries, and 10-minute maximum response time for ad hoc queries
8. include information from all company subsidiaries
9. print subsidiary reports in the primary language of the subsidiary
10. provide monthly rankings of salesperson performance

QUESTIONS:
1. Which requirements are functional business requirements? Provide two additional examples.
2. Which requirements are nonfunctional business requirements? What kind of nonfunctional requirements are they? Provide two additional examples.

CONCEPTS   4-A WHAT CAN HAPPEN IF YOU IGNORE NONFUNCTIONAL REQUIREMENTS

IN ACTION

I once worked on a consulting project in which my manager created a requirements definition without listing nonfunctional requirements. The project was then estimated based on the requirements definition and sold to the client for $5,000. In my manager's mind, the system that we would build for the client would be a very simple stand-alone system running on current technology. It shouldn't take more than a week analyze, design, and build.

Unfortunately, the client had other ideas. They wanted the system to be used by many people in three different departments, and they wanted the ability for any number of people to work on the system concurrently. The technology they had in place was antiquated, but nonetheless they wanted the system to run effectively on the existing equipment. Because we didn't set the project scope properly by including our assumptions about nonfunctional requirements in the requirements definition, we basically had to do whatever they wanted.

The capabilities they wanted took weeks to design and program. The project ended up taking four months, and the final project cost was $250,000. Our company had to pick up the tab for everything except the agreed upon $5,000. This was by far the most frustrating project situation I ever experienced. *Barbara Wixom*

Requirements Definition

The requirements definition report—usually just called the *requirements definition*—is a straightforward text report that simply lists the functional and nonfunctional requirements in an outline format. Figure 4-2 shows a sample requirements definition for a word processing program designed to compete against software, such as Microsoft Word.

The requirements are numbered in a legal or outline format so that each requirement is clearly identified. The requirements are first grouped into functional and nonfunctional requirements and then within each of those headings they are further grouped by the type of nonfunctional requirement or by function.

Sometimes, business requirements are prioritized on the requirements definition. They can be ranked as having "high," "medium," or "low" importance in the new system, or they can be labeled with the version of the system that will address the requirement (e.g., release 1, release 2, release 3). This practice is particularly important when using RAD methodologies that deliver requirements in batches by developing incremental versions of the system.

The most *obvious* purpose of the requirements definition is to provide the information needed by the other deliverables in the analysis phase, which include use cases, process models, and data models, and to support activities in the design phase. The most *important* purpose of the requirements definition, however, is to define the scope of the system. The document describes to the analysts exactly what the system needs to end up doing, and when discrepancies arise, the document serves as the place to go for clarification.

Determining Requirements

Determining requirements for the requirements definition is both a business task and an IT task. In the early days of computing, there was a presumption that the systems analysts, as experts with computer systems, were in the best position to define how

---

D. Nonfunctional Requirements

1. Operational Requirements
   1.1. The system will operate in Windows and Macintosh environments
   1.2. The system will be able to read and write Word documents, RTF, and HTML
   1.3. The system will be able to import Gif, Jpeg, and BMP graphics files

2. Performance Requirements
   2.1. Response times must be less than 7 seconds
   2.2. The Inventory database must be updated in real time

3. Security Requirements
   3.1. No special security requirements are anticipated

4. Cultural and Political Requirements
   4.1. No special cultural and political requirements are anticipated

---

C. Functional Requirements

1. Printing
   1.1. The user can select which pages to print
   1.2. The user can view a preview of the pages before printing
   1.3. The user can change the margins, paper size(e.g., letter, A4) and orientation on the page

2. Spell Checking
   2.1. The user can check for spelling mistakes. The system can operate in one of two modes as selected by the users.
   2.1.1. Mode 1 (Manual): The user will activate the spell checker and it will move the user to the next misspelled word.
   2.1.2. Mode 2 (Automatic): As the user types, the spell checker will flag misspelled words so the user immediately see the misspelling.
   2.2. The user can add words to the dictionary
   2.3. The user can mark words as not misspelled but not add them to the dictionary

FIGURE 4-2
Sample Requirements Definition

a computer system should operate. Many systems failed because they did not adequately address the true business needs of the users. Gradually, the presumption changed so that the users, as the business experts, were seen as being the best position to define how a computer system should operate. However, many systems failed to deliver performance benefits because users simply automated an existing inefficient system, and they failed to incorporate new opportunities offered by technology.

A good analogy is building a house or an apartment. We have all lived in a house or apartment, and most of us have some understanding of what we would like to see in one. However, if we were asked to design one from scratch, it would be a challenge because we lack appropriate design skills and technical engineering skills. Likewise, an architect acting alone would probably miss some of our unique requirements.

Therefore, the most effective approach is to have both businesspeople and analysts working together to determine business requirements. Sometimes, however, users don't know exactly what they want, and analysts need to help them discover their needs. Three kinds of techniques have become popular to help analysts do this: business process automation (BPA), business process improvement (BPI), and business process reengineering (BPR). They are tools that analysts can use when they need to guide the users in explaining what is wanted from a system.

The three kinds of techniques work similarly. They help users critically examine the current state of systems and processes (the as-is system), identify exactly what needs to change, and develop a concept for a new system (the to-be system). A different amount of change is associated with each technique; BPA creates a small amount of change, BPI creates a moderate amount of change, and BPR creates significant change that affects much of the organization. All three will be described in greater detail later in the chapter.

Although BPA, BPI, and BPR enable the analyst to help users create a vision for the new system, they are not sufficient for extracting information about the detailed business requirements that are needed to build it. Therefore, analysts use a portfolio of requirement-gathering techniques to acquire information from users. The analyst has many gathering techniques from which to choose: interviews, questionnaires, observation, joint application development (JAD), and document analysis. The information gathered using these techniques is critically analyzed and used to craft the requirements definition report. The final section of this chapter describes each of the requirements-gathering techniques in greater depth.

Creating the Requirements Definition

Creating the requirements definition is an iterative and ongoing process whereby the analyst collects information with requirements-gathering techniques (e.g., interviews, document analysis), critically analyzes the information to identify appropriate business requirements for the system, and adds the requirements to the requirements definition report. The requirements definition is kept up to date so that the project team and business users can refer to it and get a clear understanding of the new system.

To create the requirements definition, the project team first determines the kinds of functional and nonfunctional requirements that they will collect about the system (of course, these may change over time). These become the main sections of the document. Next, the analysts use a variety of requirement-gathering techniques (e.g., interviews, observation) to collect information, and they list the business requirements that were identified from that information. Finally, the analysts work with the entire project team and the business users to verify, change, and complete the list and to help prioritize the importance of the requirements that were identified.

This process continues throughout the analysis phase, and the requirements definition evolves over time as new requirements are identified and as the project moves into later phases of the SDLC. Beware: the evolution of the requirements definition must be carefully managed. The project team cannot keep adding to the requirements definition, or the system will keep growing and growing and never get finished. Instead, the project team carefully identifies requirements and evaluates which ones fit within the scope of the system. When a requirement reflects a real business need but is not within the scope of the current system or current release, it is added on a list of future requirements, or given a low priority. The management of requirements (and system scope) is one of the hardest parts of managing a project!

REQUIREMENTS ANALYSIS TECHNIQUES

Before the project team can determine what requirements are appropriate for a given system, they need to have a clear vision of the kind of system that will be created, and the level of change that it will bring to the organization. The basic process of *analysis* is divided into three steps: understanding the as-is system, identifying improvements, and developing requirements for the to-be system.

Sometimes the first step (i.e., understanding the as-is system) is skipped or done in a cursory manner. This happens when no current system exists, if the existing system and processes are irrelevant to the future system, or if the project team is using a RAD or Agile development methodology in which the as-is system is not emphasized. Structured design methods such as waterfall and parallel development (see Chapter 1) typically spend significant time understanding the as-is system and identifying improvements before moving to capture requirements for the to-be system. However, newer RAD and Agile methodologies, such as phased development, prototyping, throwaway prototyping, and extreme programming (see Chapter 1) focus almost exclusively on improvements and the to-be system requirements, and they spend little time investigating the current as-is system.

Three requirements analysis techniques—business process automation, business process improvement, or business process reengineering—help the analyst lead users through the three (or two) analysis steps so that the vision of the system can be developed. We should note that requirements analysis techniques and requirements-gathering techniques go hand in hand. Analysts need to use requirements-gathering techniques to collect information; requirements analysis techniques drive the kind of information that is gathered and how it is ultimately analyzed. Although we focus now on the analysis techniques and then discuss requirements gathering at the end of the chapter, they happen concurrently and are complementary activities.

The choice of analysis technique used is based on the amount of change the system is meant to create in the organization. BPA is based on small change that improves process efficiency, BPI creates process improvements that lead to better effectiveness, and BPR revamps the way things work so that the organization is transformed on some level.

To move the users "from here to there," an analyst needs strong *critical thinking skills*. Critical thinking is the ability to recognize strengths and weaknesses and recast an idea in an improved form, and they are needed to really understand issues and develop new business processes. These skills are needed to throughly examine the results of requirements gathering, to identify business requirements, and to translate those requirements into a concept for the new system.

Business Process Automation

Business process automation (BPA) means leaving the basic way in which the organization operates unchanged, and using computer technology to do some of the work. BPA can make the organization more efficient but has the least impact on the business. BPA projects spend a significant time understanding the current as-is system before moving on to improvements and to-be system requirements. Problem analysis and root cause analysis are two popular BPA techniques.

Problem Analysis   The most straightforward (and probably the most commonly used) requirements analysis technique is *problem analysis*. Problem analysis means asking the users and managers to identify problems with the as-is system and to describe how to solve them in the to-be system. Most users have a very good idea of the changes they would like to see, and most will be quite vocal about suggesting them. Most changes tend to solve problems rather than capitalize on opportunities, but this is possible, too. Improvements from problem analysis tend to be small and incremental (e.g., provide more space in which to type the customer's address; provide a new report that currently does not exist).

This type of improvement often is very effective at improving a system's efficiency or ease of use. However, it often provides only minor improvements in business value—the new system is better than the old, but it may be hard to identify significant monetary benefits from the new system.

Root Cause Analysis   The ideas produced by problem analysis tend to be *solutions* to problems. All solutions make assumptions about the nature of the problem, assumptions that may or may not be valid. In our experience, users (and most people in general) tend to jump quickly to solutions without fully considering the nature of the problem. Sometimes the solutions are appropriate, but many times they address a *symptom* of the problem, not the true problem or *root cause* itself.[2]

For example, suppose you notice that a lightbulb is burned out above your front door. You buy a new bulb, get out a ladder, and replace the bulb. A month later, you see that the same bulb is burnt out, so you buy a new bulb, haul out the ladder, and replace it again. This repeats itself several times. At this point, you have two choices. You can buy a large package of lightbulbs and a fancy lightbulb changer on a long pole so you don't need the haul the ladder out each time (thus saving a lot of trips to the store for new bulbs and a lot of effort in working with the ladder). Or you can fix the light fixture that is causing the light to burn out in the first place. Buying the bulb changer is treating the symptom (the burnt-out bulb), while fixing the fixture is treating the root cause.

In the business world, the challenge lies in identifying the root cause—few problems are as simple as the lightbulb problem. The solutions that users propose (or systems that analysts think of) may either address symptoms or root causes, but without a careful analysis, it is difficult to tell which one. And finding out that you've just spent a million dollars on a new lightbulb changer is a horrible feeling!

Root cause analysis therefore focuses on problems, not solutions. The analyst starts by having the users generate a list of problems with the current system, and then prioritize the problems in order of importance. Then starting with the most important, the users and/or the analysts generate all the possible root causes for the problems. Each possible root cause is investigated (starting with the most likely or easiest to check) until the true root cause(s) are identified. If any possible root causes are identified for several problems, those should be investigated first, because there is a good chance they are the real root causes influencing the symptom problems.

In our lightbulb example, there are several possible root causes. A decision tree (Chapter 6) sometimes helps with the analysis. As Figure 4-3 shows, there are many possible root causes, so buying a new fixture may or may not address the true root cause. In fact, buying a lightbulb changer may actually address the root cause. The key point in root cause analysis is to always challenge the obvious.

Business Process Improvement

Business process improvement (BPI) means making moderate changes to the way in which the organization operates to take advantage of new opportunities offered by technology or to copy what competitors are doing. BPI can improve efficiency

---

[2] Two good books that discuss the problems in finding the root causes to problems are: E.M. Goldratt, and J. Cox, *The Goal*, Croton-on-Hudson, NY: North River Press, 1986, and E.M. Goldratt, *The Haystack Syndrome*, Croton-on-Hudson, NY: North River Press, 1990.

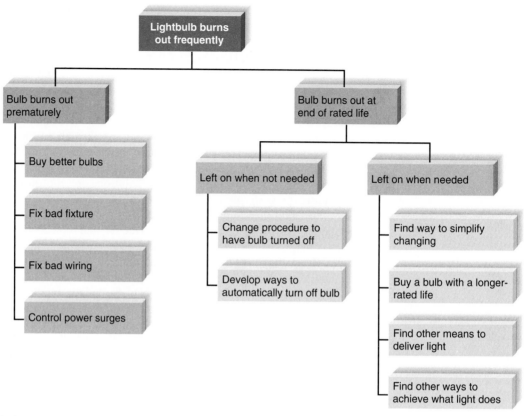

FIGURE 4-3
Root Cause Analysis for the Example of the Burned-Out Lightbulb

(i.e., doing things right) and improve effectiveness (i.e., doing the right things). BPI projects also spend time understanding the as-is system, but much less time than BPA projects; their primary focus is on improving business processes, so time is spent on the as-is only to help with the improvement analyses and the to-be system requirements. Duration analysis, activity-based costing, and information benchmarking are three popular BPI activities.

Duration Analysis   *Duration analysis* requires a detailed examination of the amount of time it takes to perform each process in the current as-is system. The analysts begin by determining the total amount of time it takes, on average, to perform a set of business processes for a typical input. They then time each of the individual steps (or subprocesses) in the business process. The time to complete the basic steps are then totaled and compared to the total for the overall process. When there is a significant difference between the two—and in our experiences the total time often can be 10 or even 100 times longer than the sum of the parts—this indicates that this part of the process is badly in need of a major overhaul.

For example, suppose that the analysts are working on a home mortgage system and discover that on average, it takes 30 days for the bank to approve a mortgage. They then look at each of the basic steps in the process (e.g., data entry, credit check, title search, appraisal, etc.) and find that the total amount of time actually

CONCEPTS 4-B WHEN OPTIMAL ISN'T

IN ACTION

In 1984, I developed an information system to help schedule production orders in paper mills. Paper is made in huge rolls that are six to ten feet wide. Customer orders (e.g., newspaper, computer paper) are cut from these large rolls. The goal of production scheduling is to decide which orders to combine on one roll to reduce the amount of paper wasted because no matter how you place the orders on the rolls, you almost always end up throwing away the last few inches—customer orders seldom add up to exactly the same width as the roll itself (imagine trying to cut several rolls of toilet paper from a paper towel roll).

The scheduling system was designed to run on an IBM PC, which in those days was not powerful enough to use advanced mathematical techniques like linear programming to calculate which orders to combine on which roll. Instead, we designed the system to use the rules of thumb that the production schedulers themselves had developed over many years. After several months of fine-tuning, the system worked very well. The schedulers were happy and the amount of waste decreased.

By 1986, PCs had increased in power, so we revised the system to use the advanced linear programming techniques to provide better schedules and reduce the waste even more. In our tests, the new system reduced waste by a few percentage points over the original system, so it was installed. However, after several months, there were big problems; the schedulers were unhappy and the amount of waste actually grew.

It turned out that although the new system really did produce better schedules with less waste, the real problem was *not* figuring how to place the orders on the rolls to reduce waste. Instead, the real problem was finding orders for the next week that could be produced with the current batch. The old system combined orders in a way in which the schedulers found it easy to find "matching" future orders. The new system made odd combinations with which the schedulers had a hard time working. The old system was eventually re-installed. *Alan Dennis*

QUESTION
How could the problem with the new system have been foreseen?

spent on each mortgage is about 8 hours. This is a strong indication that the overall process is badly broken, because it takes 30 days to perform 1 day's work.

These problems likely occur because the process is badly fragmented. Many different people must perform different activities before the process finishes. In the mortgage example, the application probably sits on many peoples' desks for long periods of time before it is processed. Processes in which many different people work on small parts of the inputs are prime candidates for *process integration* or *parallelization*. Process integration means changing the fundamental process so that fewer people work on the input, which often requires changing the processes and retraining staff to perform a wider range of duties. Process parallelization means changing the process so that all the individual steps are performed at the same time. For example in the mortgage application example, there is probably no reason that the credit check cannot be performed at the same time as the appraisal and title check.

Activity-Based Costing *Activity-based costing* is a similar analysis that examines the cost of each major process or step in a business process rather than the time taken.[3] The analysts identify the costs associated with each of the basic functional

---

[3] Many books have been written on activity based costing. Useful ones include K.B. Burk and D. W. Webster, *Activity Based Costing,* Fairfax, VA: American Management Systems, 1994; and D. T. Hicks, *Activity-Based Costing: Making It Work for Small and Mid-Sized Companies,* New York: John Wiley, 1998. The two books by Eli Goldratt mentioned previously (*The Goal* and *The Haystack Syndrome*) also offer unique insights into costing.

CONCEPTS

IN ACTION

4-C DURATION ANALYSIS

A group of executives from a Fortune 500 company used Duration Analysis to discuss their procurement process. Using a huge wall of Velcro and a handful of placards, a facilitator proceeded to map out the company's process for procuring a $50 software upgrade. Having quantified the time it took to complete each step, she then assigned costs based on the salaries of the employees involved. The 15-minute exercise left the group stunned. Their procurement process had gotten so convoluted that it took 18 days, countless hours of paperwork and nearly $22,000 in people time to get the product ordered, received, and up and running on the requester's desktop.

Source: "For Good Measure," *CIO Magazine*, March 1 1999, by Debby Young.

steps or processes, identify the most costly processes, and focus their improvement efforts on them.

Assigning costs is conceptually simple. You just examine the direct cost of labor and materials for each input. Materials costs are easily assigned in a manufacturing process, while labor costs are usually calculated based on the amount of time spent on the input and the hourly cost of the staff. However, as you may recall from a managerial accounting course, there are indirect costs such as rent, depreciation, and so on that also can be included in activity costs.

Informal Benchmarking   *Benchmarking* refers to studying how other organizations perform a business process in order to learn how your organization can do something better. Benchmarking helps the organization by introducing ideas that employees may never have considered, but have the potential to add value.

Informal benchmarking is fairly common for "customer-facing" business processes (i.e., those processes that interact with the customer). With informal benchmarking, the managers and analysts think about other organizations, or visit them as customers to watch how the business process is performed. In many cases, the business studied may be a known leader in the industry or simply a related firm. For example, suppose the team is developing a Web site for a car dealer. The project sponsor, key managers, and key team members would likely visit the Web sites of competitors, as well as those of others in the car industry (e.g., manufacturers, accessories suppliers) and those in other industries that have won awards for their Web sites.

Business Process Reengineering

Business process reengineering (BPR) means changing the fundamental way in which the organization operates—"obliterating" the current way of doing business and making major changes to take advantage of new ideas and new technology. BPR projects spend little time understanding the as-is, because their goal is to focus on new ideas and new ways of doing business. Outcome analysis, technology analysis, and activity elimination are three popular BPR activities.

Outcome Analysis   *Outcome analysis* focuses on understanding the fundamental outcomes that provide value to customers. While these outcomes sound as though they should be obvious, they often aren't. For example, suppose you are an insurance company, and one of your customers has just had a car accident. What is the fundamental outcome from the *customer's* perspective? Traditionally, insurance companies have answered this question by assuming the customer wants to receive the insurance payment quickly. To the customer, however, the payment is only a *means* to the real outcome: a repaired car. The insurance company might benefit by extending their view of the business process past its traditional boundaries to include not paying for repairs, but performing the repairs or contracting with an authorized body shop to do them.

With this approach, the system analysts encourage the managers and project sponsor to pretend they are customers and to think carefully about what the organization's products and services enable the customers to do—and what they *could* enable the customer to do.

Technology Analysis   Many major changes in business over the past decade have been enabled by new technologies. *Technology analysis* therefore starts by having the analysts and managers develop a list of important and interesting technologies. Then the group systematically identifies how each and every technology could be applied to the business process and identifies how the business would benefit.

For example, one useful technology might be the Internet. Saturn, the car manufacturer, took this idea and developed an Extranet application for its suppliers. Rather than ordering parts for its cars, Saturn makes its production schedule available electronically to its suppliers, who ship the parts Saturn needs so that they arrive at the plant just in time. This saves Saturn significant costs because it eliminates the need for people to monitor the production schedule and issue purchase orders.

Activity Elimination   *Activity elimination* is exactly what it sounds like. The analysts and managers work together to identify how the organization could eliminate each and every activity in the business process, how the function could operate without it, and what effects are likely to occur. Initially, managers are reluctant to conclude that processes can be eliminated, but this is a "force-fit" exercise in that they must eliminate each activity. In some cases the results are silly, but nonetheless, participants must address each and every activity in the business process.

For example, in the home mortgage approval process discussed earlier, the managers and analysts would start by eliminating the first activity, entering the data into the mortgage company's computer. This leads to two obvious possibilities (1) eliminate the use of a computer system; or (2) make someone else do the data entry (e.g., the customer over the Web). They would then eliminate the next activity, the credit check. Silly right? After all, making sure the applicant has good credit is critical in issuing a loan. Not really. The real answer depends upon how many times the credit check identifies bad applications. If all or almost all applicants have good credit and are seldom turned down by a credit check, then the cost of the credit check may not be worth the cost of the few bad loans it prevents. Eliminating it may actually result in lower costs even with the cost of bad loans.

Selecting the Appropriate Technique

Each of the techniques discussed in this chapter has its own strengths and weaknesses (see Figure 4-4). No one technique is inherently better than the others, and in practice most projects use a combination of techniques.

4-2 IBM Credit

IBM Credit was a wholly owned subsidiary of IBM responsible for financing mainframe computers sold by IBM. While some customers bought mainframes outright, or obtained financing from other sources, financing computers provided significant additional profit.

When and IBM sales representative made a sale, he or she would immediately call IBM Credit to obtain a financing quote. The call was received by a credit officer who would record the information on a request form. The form would then be sent to the credit department to check the customer's credit status. This information would be recorded on the form, which was then sent to the business practices department who would write a contract (sometimes reflecting changes requested by the customer). The form and the contract would then go to the pricing department, which used the credit information to establish an interest rate and recorded it on the Form. The Form and contract was then sent to the clerical group, where an administrator would prepare a cover letter quoting the interest rate and send the letter and contract via Federal Express to the customer.

The problem at IBM Credit was a major one. Getting a financing quote took anywhere for four to eight days (six days on average), giving the customer time to rethink the order or find financing elsewhere. While the quote was being prepared, sales representatives would often call to find out where the quote was in the process, so they could tell the customer when to expect it. However, no one at IBM Credit could answer the question because the paper forms could be in any department and it was impossible to locate one without physically walking through the departments and going through the piles of forms on everyone's desk.

IBM Credit examined the process and changed it so that each credit request was logged into a computer system so that each department could record an application's status as they completed it and sent it to the next department. In this way, sales representatives could call the credit office and quickly learn the status of each application. IBM used some sophisticated management science queuing theory analysis to balance workloads and staff across the different departments so none would be overloaded. They also introduced performance standards for each department (e.g., the pricing decision had to be completed within one day after that department received an application).

However, process times got worse, even though each department was achieving almost 100 percent compliance on its performance goals. After some investigation, managers found that when people got busy, they conveniently found errors that forced them to return credit requests to the previous department for correction, thereby removing it from their time measurements.

QUESTIONS:

What techniques can you use to identify improvements?
   Choose one technique and apply it to this situation—what improvements did you identify?

Source: *Reengineering the Corporation*, New York: Harper Business, 1993, by M. Hammer and J. Champy.

Potential Business Value   The *potential business value* varies with analysis strategy. While BPA has the potential to improve the business, most of the benefits from BPA are tactical and small in nature. Since BPA does not seek to change the

	Business Process Automation	Business Process Improvement	Business Process Reengineering
Potential business value	Low–moderate	Moderate	High
Project cost	Low	Low–moderate	High
Breadth of analysis	Narrow	Narrow–moderate	Very broad
Risk	Low–moderate	Low–moderate	Very high

FIGURE 4-4
Characteristics of Analysis Strategies

Suppose you are the analyst charged with developing a new Web site for a local car dealer who wants to be very innovative and try new things. What analysis techniques would you recommend? Why?

business processes, it can only improve their efficiency. BPI usually offers moderate potential benefits, depending upon the scope of the project, because it seeks to change the business in some way. It can increase both efficiency and effectiveness. BPR creates large *potential* benefits because it seeks to radically improve the nature of the business.

Project Cost   *Project cost* is always important. In general, BPA requires the lowest cost because it has the narrowest focus and seeks to make the fewest number of changes. BPI can be moderately expensive, depending upon the scope of the project. BPR is usually expensive, both because of the amount of time required of senior managers and the amount of redesign to business processes.

Breadth of Analysis   *Breadth of analysis* refers to the scope of analysis, or whether analysis includes business processes within a single business function, processes that cross the organization, or processes that interact with those in customer or supplier organizations. BPR takes a broad perspective, often spanning several major business processes, even across multiple organizations. BPI has a much narrower scope that usually includes one or several business functions. BPA typically examines a single process.

Risk   One final issue is *risk* of failure, which is the likelihood of failure due to poor design, unmet needs, or too much change for the organization to handle. BPA and BPI have low to moderate risk because the to-be system is fairly well defined and understood, and its potential impact on the business can be assessed before it is implemented. BPR projects, on the other hand, are less predictable. BPR is extremely risky and not something to be undertaken unless the organization and its senior leadership are committed to making significant changes. Mike Hammer, the father of BPR, estimates that 70% of BPR projects fail.

REQUIREMENTS-GATHERING TECHNIQUES

An analyst is very much like a detective (and business users sometimes are like elusive suspects). He or she knows that there is a problem to be solved and therefore must look for clues that uncover the solution. Unfortunately, the clues are not always obvious (and often missed), so the analyst needs to notice details, talk with witnesses, and follow leads just as Sherlock Holmes would have done. The best analysts will thoroughly gather requirements using a variety of techniques and make sure that the current business processes and the needs for the new system are well understood before moving into design. You don't want to discover later that you have key requirements wrong—surprises like this late in the SDLC can cause all kinds of problems.

This section of the chapter will describe five techniques that can be used to gather the business requirements for the proposed system: interviews, joint application development, questionnaires, document analysis, and observation.

Interviews

The *interview* is the most commonly used requirements-gathering technique. After all, it is natural—usually if you need to know something, you ask someone. In general, interviews are conducted one-on-one (one interviewer and one interviewee), but sometimes, due to time constraints, several people are interviewed at the same time. There are five basic steps to the interview process: selecting interviewees, designing interview questions, preparing for the interview, conducting the interview, and postinterview follow-up[4].

Selecting Interviewees The first step to interviewing is to create an *interview schedule* that lists all of the people who will be interviewed, when, and for what purpose (see Figure 4-5). The schedule can be an informal list that is used to help set up meeting times, or a formal list that is incorporated into the workplan. The people who appear on the interview schedule are selected based on the analyst's information needs. The project sponsor, key business users, and other members of the project team can help the analyst determine who in the organization can best provide important information about requirements. These people are listed on the interview schedule in the order in which they should be interviewed.

People at different levels of the organization will have different perspectives on the system, so it is important to include both managers who manage the processes and staff who actually perform the processes to gain both high-level and low-level perspectives on an issue. Also, the kinds of interview subjects that you need may change over time. For example, at the start of the project, the analyst has a limited understanding of the as-is business process. It is common to begin by

Name	Position	Purpose of Interview	Meeting
Andria McClellan	Director, Accounting	Strategic vision for new accounting system	Mon, March 1 8:00–10:00 AM
Jennifer Draper	Manager, Accounts Receivable	Current problems with accounts receivable process; future goals	Mon, March 1 2:00–3:15 PM
Mark Goodin	Manager, Accounts Payable	Current problems with accounts payable process; future goals	Mon, March 1 4:00–5:15 PM
Anne Asher	Supervisor, Data Entry	Accounts receivable and payable processes	Wed, March 3 10:00–11:00 AM
Fernando Merce	Data Entry Clerk	Accounts receivable and payable processes	Wed, March 3 1:00–3:00 PM

FIGURE 4-5
Sample Interview Schedule

[4] A good book on interviewing is that by Brian James, *The Systems Analysis Interview,* Manchester: NCC Blackwell, 1989.

CONCEPTS   4-D SELECTING THE WRONG PEOPLE

IN ACTION

In 1990, I led a consulting team for a major development project for the U.S. Army. The goal was to replace eight existing systems used on virtually every Army base across the United States. The as-is process and data models for these systems had been built, and our job was to identify improvement opportunities and develop to-be process models for each of the eight systems.

For the first system, we selected a group of mid-level managers (captains and majors) recommended by their commanders as being the experts in the system under construction. These individuals were the first and second line managers of the business function. The individuals were expert at managing the process, but did not know the exact details of how the process worked. The resulting to-be process model was very general and non-specific. *Alan Dennis*

QUESTION

Suppose you were in charge of the project. Create an interview schedule the remaining seven projects.

interviewing one or two senior managers to get a strategic view, and then move to mid-level managers who can provide broad, overarching information about the business process and the expected role of the system being developed. Once the analyst has a good understanding of the "big picture," lower-level managers and staff members can fill in the exact details of how the process works. Like most other things about systems analysis, this is an iterative process—starting with senior managers, moving to mid-level managers, then staff members, back to mid-level managers, and so on, depending upon what information is needed along the way.

It is quite common for the list of interviewees to grow, often by 50% to 75%. As you interview people, you likely will identify more information that is needed and additional people who can provide the information.

Designing Interview Questions   There are three types of interview questions: closed-ended questions, open-ended questions, and probing questions. *Closed-ended questions* are those that require a specific answer. You can think of them as being similar to multiple choice or arithmetic questions on an exam (see Figure 4-6). Closed-ended questions are used when the analyst is looking for specific, precise

Types of Questions	Examples
Closed-Ended Questions	• How many telephone orders are received per day?
	• How do customers place orders?
	• What information is missing from the monthly sales report?
Open-Ended Questions	• What do you think about the current system?
	• What are some of the problems you face on a daily basis?
	• What are some of the improvements you would like to see in a new system?
Probing Questions	• Why?
	• Can you give me an example?
	• Can you explain that in a bit more detail?

FIGURE 4-6
Three Types of Questions

information (e.g., how many credit card requests are received per day). In general, precise questions are best. For example, rather than asking "Do you handle a lot of requests?" it is better to ask "How many requests do you process per day?"

Closed-ended questions enable analysts to control the interview and obtain the information they need. However, these types of questions don't uncover *why* the answer is the way it is, nor do they uncover information that the interviewer does not think to ask ahead of time.

Open-ended questions are those that leave room for elaboration on the part of the interviewee. They are similar in many ways to essay questions that you might find on an exam (see Figure 4-6 for examples). Open-ended questions are designed to gather rich information and give the interviewee more control over the information that is revealed during the interview. Sometimes the information that the interviewee chooses to discuss uncovers information that is just as important as the answer (e.g., if the interviewee talks only about other departments when asked for problems, it may suggest that he or she is reluctant to admit his or her own problems).

The third type of question is the *probing question*. Probing questions follow-up on what has just been discussed in order to learn more, and they often are used when the interviewer is unclear about an interviewee's answer. They encourage the interviewee to expand on or to confirm information from a previous response, and they are a signal that the interviewer is listening and interested in the topic under discussion. Many beginning analysts are reluctant to use probing questions because they are afraid that the interviewee might be offended at being challenged or because they believe it shows that they didn't understand what the interviewee said. When done politely, probing questions can be a powerful tool in requirements gathering.

In general, you should not ask questions about information that is readily available from other sources. For example, rather than asking what information is used to perform to a task, it is simpler to show the interviewee a form or report (see document analysis later) and ask what information on it is used. This helps focus the interviewee on the task, and saves time, because he or she does not need to describe the information detail—he or she just needs to point it out on the form or report.

No question type is better than another, and usually a combination of questions is used during an interview. At the initial stage of an IS development project, the as-is process can be unclear, so the interview process begins with *unstructured interviews*, interviews that seek a broad and roughly defined set of information. In this case, the interviewer has a general sense of the information needed, but few close-ended questions to ask. These are the most challenging interviews to conduct because they require the interviewer to ask open-ended questions and probe for important information "on the fly."

As the project progresses, the analyst comes to understand the business process much better, and he or she needs very specific information about how business processes are performed (e.g., exactly how a customer credit card is approved). At this time, the analyst conducts *structured interviews*, in which specific sets of questions are developed prior to the interviews. There usually are more close-ended questions in a structured interview than in the unstructured approach.

No matter what kind of interview is being conducted, interview questions must be organized into a logical sequence, so that the interview flows well. For example, when trying to gather information about the current business process, it can be useful to move in logical order through the process or from the most important issues to the least important.

There are two fundamental approaches to organizing the interview questions: top-down or bottom-up; see Figure 4-7. With the *top-down interview*, the interviewer starts with broad, general issues and gradually works towards more specific ones. With the *bottom-up interview*, the interviewer starts with very specific questions and moves to broad questions. In practice, analysts mix the two approaches, starting with broad general issues, moving to specific questions, and then back to general issues.

The top-down approach is an appropriate strategy for most interviews (it is certainly the most common approach). The top-down approach enables the interviewee to become accustomed to the topic before he or she needs to provide specifics. It also enables the interviewer to understand the issues before moving to the details because the interviewer may not have sufficient information at the start of the interview to ask very specific questions. Perhaps most importantly, the top-down approach enables the interviewee to raise a set of "big picture" issues before becoming enmeshed in details, so the interviewer is less likely to miss important issues.

One case in which the bottom-up strategy may be preferred is when the analyst already has gathered a lot of information about issues and just needs to fill in some holes with details. Or, bottom-up may be appropriate if lower-level staff members are threatened or unable to answer high-level questions. For example, "How can we improve customer service?" may be too broad a question for a customer service clerk, whereas a specific question is readily answerable (e.g., "How can we speed up customer returns?"). In any event, all interviews should begin with non-controversial questions first, and then gradually move into more contentious issues after the interviewer has developed some rapport with the interviewee.

Preparing for the Interview   It is important to prepare for the interview in the same way that you would prepare to give a presentation. You should have a general interview plan that lists the questions that you will ask in the appropriate order; anticipates possible answers and provides how you will follow up with them; and identifies segues between related topics. Confirm the areas in which the interviewee has knowledge so you do not ask questions that he or she cannot

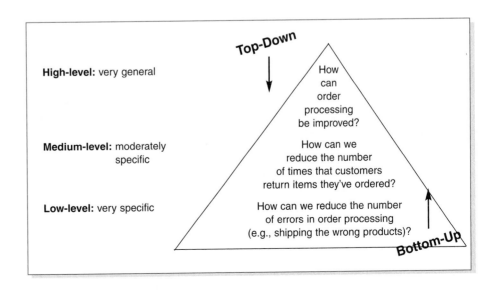

FIGURE 4-7
Top-Down and Bottom-Up Questioning
Strategies

answer. Review the topic areas, the questions, and the interview plan, and clearly decide which have the greatest priority in case you run out of time.

In general, structured interviews with closed-ended questions take more time to prepare than unstructured interviews. So, some beginning analysts prefer unstructured interviews, thinking that they can "wing it." This is very dangerous and often counterproductive, because any information not gathered in the first interview would require follow-up efforts, and most users do not like to be interviewed repeatedly about the same issues.

Be sure to prepare the interviewee as well. When you schedule the interview, inform the interviewee of the reason for the interview and the areas you will be discussing far enough in advance so that he or she has time to think about the issues and organize his or her thoughts. This is particularly important when you are an outsider to the organization, and for lower-level employees who often are not asked for their opinions and who may be uncertain about why you are interviewing them.

Conducting the Interview   When you start the interview, the first goal is to build rapport with the interviewee, so that he or she trusts you and is willing to tell you the whole truth, not just give the answers that he or she thinks you want. You should appear to be professional and an unbiased, independent seeker of information. The interview should start with an explanation of why you are there and why you have chosen to interview the person, and then move into your planned interview questions.

PRACTICAL	4-1 Developing Interpersonal Skills
TIP	

Interpersonal skills are those skills than enable you to develop rapport with others, and they are very important for interviewing. They help you to communicate with others effectively. Some people develop good interpersonal skills at an early age; they simply seem to know how to communicate and interact with others. Other people are less "lucky" and need to work hard to develop their skills.

Interpersonal skills, like most skills, can be learned. Here are some tips:

- **Don't worry, be happy.** Happy people radiate confidence and project their feelings on others. Try interviewing someone while smiling and then interviewing someone else while frowning and see what happens!
- **Pay attention.** Pay attention to what the other person is saying (which is harder than you might think). See how many times you catch yourself with your mind on something other than the conversation at hand.
- **Summarize key points.** At the end of each major theme or idea that someone explains, you should repeat the key points back to the speaker (e.g., "Let me make sure I understand. The key issues are..."). This demonstrates that you consider the information important—and also forces you to pay attention (you can't repeat what you didn't hear).
- **Be succinct.** When you speak, be succinct. The goal in interviewing (and in much of life) is to learn, not to impress. The more you speak, the less time you give to others.
- **Be honest.** Answer all questions truthfully, and if you don't know the answer, say so.
- **Watch body language (yours and theirs).** The way a person sits or stands conveys much information. In general, a person who is interested in what you are saying sits or leans forward, makes eye contact, and often touches his or her face. A person leaning away from you or with an arm over the back of a chair is disinterested. Crossed arms indicate defensiveness or uncertainty, while "steepling" (sitting with hands raised in front of the body with fingertips touching) indicates a feeling of superiority.

It is critical to carefully record all the information that the interviewee provides. In our experience, the best approach is to take careful notes—write down *everything* the interviewee says, even if it does not appear immediately relevant. Don't be afraid to ask the person to slow down or to pause while you write, because this is a clear indication that the interviewee's information is important to you. One potentially controversial issue is whether or not to tape-record the interview. Recording ensures that you do not miss important points, but it can be intimidating for the interviewee. Most organizations have policies or generally accepted practices about the recording of interviews, so find out what they are before you start an interview. If you are worried about missing information and cannot tape the interview, then bring along a second person to take detailed notes.

As the interview progresses, it is important that you understand the issues that are discussed. If you do not understand something, be sure to ask. Don't be afraid to ask "dumb questions" because the only thing worse than appearing "dumb" is to be "dumb" by not understanding something. If you don't understand something during the interview, you certainly won't understand it afterward. Try to recognize and define jargon, and be sure to clarify jargon you do not understand. One good strategy to increase your understanding during an interview is to periodically summarize the key points that the interviewee is communicating. This avoids misunderstandings and also demonstrates that you are listening.

Finally, be sure to separate facts from opinion. The interviewee may say, for example, "we process too many credit card requests." This is an opinion, and it is useful to follow this up with a probing question requesting support for the statement (e.g., "Oh, how many do you process in a day?"). It is helpful to check the facts because any differences between the facts and the interviewee's opinions can point out key areas for improvement. Suppose the interviewee complains about a high or increasing number of errors, but the logs show that errors have been decreasing. This suggests that errors are viewed as a very important problem that should be addressed by the new system, even if they are declining.

As the interview draws to a close, be sure to give the interviewee time to ask questions or provide information that he or she thinks is important but was not part of your interview plan. In most cases, the interviewee will have no additional concerns or information, but in some cases this will lead to unanticipated, but important information. Likewise, it can be useful to ask the interviewee if there are other people who should be interviewed. Make sure that the interview ends on time (if necessary, omit some topics or plan to schedule another interview).

As a last step in the interview, briefly explain what will happen next (see the next section). You don't want to prematurely promise certain features in the new system or a specific delivery date, but you do want to reassure the interviewee that his or her time was well spent and very helpful to the project.

Post interview Follow-up   After the interview is over, the analyst needs to prepare an *interview report* that describes the information from the interview (Figure 4-8). The report contains *interview notes*, information that was collected over the course of the interview and is summarized in a useful format. In general, the interview report should be written within 48 hours of the interview, because the longer you wait, the more likely you are to forget information.

Often, the interview report is sent to the interviewee with a request to read it and inform the analyst of clarifications or updates. Make sure the interviewee is convinced that you genuinely want his or her corrections to the report. Usually there are few changes, but the need for any significant changes suggests that a second

Interviewing is not as simple as it first appears. Select two people from class to go to the front of the room to demonstrate an interview. (This also can be done in groups.) Have one person be the interviewer, and the other the interviewee. The interviewer should conduct a 5-minute interview regarding the school course registration system. Gather information about the existing system and how the system can be improved. If there is time, repeat with another pair.

QUESTIONS:
1. Describe the body language of the interview pair.
2. What kind of interview was conducted?
3. What kinds of questions were asked?
4. What was done well? How could the interview be improved?

interview will be required. Never distribute someone's information without prior approval.

Joint Application Development(JAD)

Joint application development (or *JAD* as it is more commonly known) is an information gathering technique that allows the project team, users, and management to work together to identify requirements for the system. IBM developed the JAD tech-

---

Interview Notes Approved By: Linda Estey

Person Interviewed: Linda Estey,
   Director, Human Resources

Interviewer: Barbara Wixom

Purpose of Interview:

- Understand reports produced for Human Resources by the current system
- Determine information requirements for future system

Summary of Interview:
- Sample reports of all current HR reports are attached to this report. The information that is not used and missing information are noted on the reports.
- Two biggest problems with the current system are:
   1. the data is too old (the HR Department needs information within 2 days of month end; currently information is provided to them after a 3-week delay)
   2. the data is of poor quality (often reports must be reconciled with departmental HR database)
- The most common data errors found in the current system include incorrect job level information and missing salary information.

Open Items:
- Get current employee roster report from Mary Skudrna (extension 4355)
- Verify calculations used to determine vacation time with Mary Skudrna.
- Schedule interview with Jim Wack (extension 2337) regarding the reasons for data quality problems.

Detailed Notes: See attached transcript.

FIGURE 4-8
Interview Report

nique in the late 1970s, and it is often the most useful method for collecting information from users.[5] Capers Jones claims that JAD can reduce scope creep by 50%, and it avoids the requirements for a system from being too specific or too vague, both of which cause trouble during later stages of the SDLC.[6] JAD is a structured process in which 10 to 20 users meet together under the direction of *facilitator* skilled in JAD techniques. The facilitator is a person who sets the meeting agenda and guides the discussion, but does not join in the discussion as a participant. He or she does not provide ideas or opinions on the topics under discussion to remain neutral during the session. The facilitator must be an expert in both group process techniques and systems analysis and design techniques. One or two *scribes* assist the facilitator by recording notes, making copies, and so on. Often the scribes will use computers and CASE tools to record information as the JAD session proceeds.

The JAD group meets for several hours, several days, or several weeks until all of the issues have been discussed and the needed information is collected. Most JAD sessions take place in a specially prepared meeting room, away from the participants' offices so that they are not interrupted. The meeting room is usually arranged in a U shape so that all participants can easily see each other (see Figure 4-9). At the front of the room (the open part of the"U"), there is a whiteboard, flip chart and/or overhead projector for use by the facilitator who leads the discussion.

One problem with JAD is that it suffers from the traditional problems associated with groups; sometimes people are reluctant to challenge the opinions of others (particularly their boss), a few people often dominate the discussion, and not everyone participates. In a 15-member group, for example, if everyone participates equally, then each person can talk for only 4 minutes each hour and must listen for the remaining 56 minutes—not a very efficient way to collect information.

A new form of JAD called *electronic JAD* or *e-JAD* attempts to overcome these problems by using groupware. In an e-JAD meeting room, each participant uses special software on a networked computer to send anonymous ideas and opinions to everyone else. In this way, all participants can contribute at the same time, without fear of reprisal from people with differing opinions. Initial research suggests that e-JAD can reduce the time required to run JAD sessions by 50% to 80%.[7]

Selecting Participants   Selecting JAD participants is done in the same basic way as selecting interview participants. Participants are selected based on the information they can contribute, to provide a broad mix of organizational levels, and to build political support for the new system. The need for all JAD participants to be away from their office at the same time can be a major problem. The office may need to be closed or run with a "skeleton" staff until the JAD sessions are complete.

Ideally, the participants who are released from regular duties to attend the JAD sessions should be the very best people in that business unit. However, without strong management support, JAD sessions can fail because those selected to

[5] More information on JAD can be found in J. Wood and D. Silver, *Joint Application Development,* New York: John Wiley & Sons, 1989; and Alan Cline, "Joint Application Development for Requirements Collection and Management, http://www.carolla.com/wp-jad.htm.

[6] See Kevin Strehlo, "Catching up with the Jones and 'Requirement' Creep," *InfoWorld,* July 29, 1996, and Kevin Strehlo, "The Makings of a Happy Customer: Specifying Project X" *Infoworld.* Nov 11, 1996.

[7] For more information on e-JAD, see A. R. Dennis, G. S. Hayes, and R. M. Daniels, "Business Process Modeling with Groupware," *Journal of Management Information Systems* 15(4), 1999, 115–142.

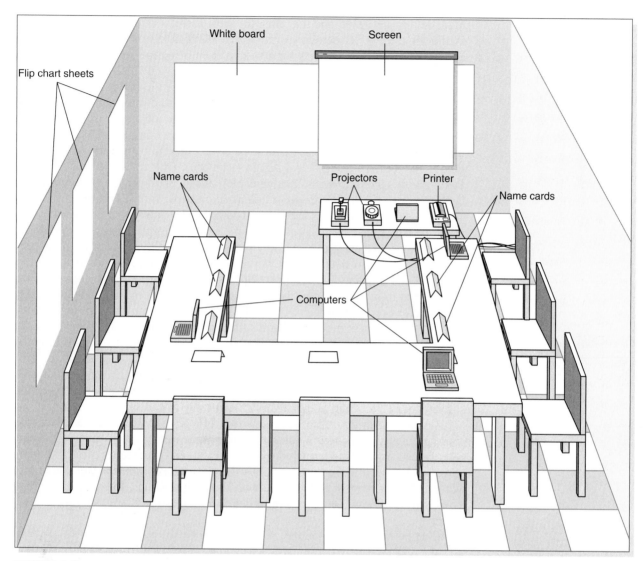

FIGURE 4-9
Joint Application Design Meeting Room

attend the JAD session are people who are less likely to be missed (i.e., the least competent people).

The facilitator should be someone who is an expert in JAD or e-JAD techniques and ideally someone who has experience with the business under discussion. In many cases, the JAD facilitator is a consultant external to the organization because the organization may not have a regular day-to-day need for JAD or e-JAD expertise. Developing and maintaining this expertise in-house can be expensive.

Designing the JAD Session. JAD sessions can run from as little as a half day to several weeks, depending upon the size and scope of the project. In our experience, most JAD sessions tend to last 5 to 10 days spread over a 3-week period. Most e-JAD sessions tend to last 1 to 4 days in a 1-week period. JAD and e-JAD sessions usually go beyond the collection of information and move into analysis. For example, the users and the analysts collectively can create analysis deliverables, such as use case, process models, or the requirements definition.

As with interviewing, success depends upon a careful plan. JAD sessions usually are designed and structured using the same principles as interviews. Most JAD sessions are designed to collect specific information from users, and this requires the development of a set of questions prior to the meeting. A difference between JAD and interviewing is that all JAD sessions are structured—they *must* be carefully planned. In general, closed-ended questions are seldom used because they do not spark the open and frank discussion that is typical of JAD. In our experience, it is better to proceed top-down in JAD sessions when gathering information. Typically 30 minutes is allocated to each separate agenda item, and frequent breaks are scheduled throughout the day because participants tire easily.

Preparing for the JAD Session   As with interviewing, it is important to prepare the analysts and participants for the JAD session. Because the sessions can go beyond the depth of a typical interview and are usually conducted off-site, participants can be more concerned about how to prepare. It is important that the participants understand what is expected of them. If the goal of the JAD session, for example, is to develop an understanding of the current system, then participants can bring procedure manuals and documents with them. If the goal is to identify improvements for a system, then they can think about how they would improve the system prior to the JAD Session.

Conducting the JAD Session   Most JAD sessions try to follow a formal agenda, and most have formal *ground rules* that define appropriate behavior. Common ground rules include following the schedule, respecting others, opinions, accepting disagreement, and ensuring that only one person talks a time.

The role of the JAD facilitator can be challenging. Many participants come to the JAD session with strong feelings about the system to be discussed. Channeling these feelings so that the session moves forward in a positive direction and getting participants to recognize and accept—but not necessarily agree on—opinions and situations different from their own requires significant expertise in systems analysis and design, JAD, and interpersonal skills. Few systems analysts attempt to facilitate JAD sessions without being trained in JAD techniques, and most apprentice with a skilled JAD facilitator before they attempt to lead their first session.

The JAD facilitator performs three key functions. First, he or she ensures that the group sticks to the agenda. The only reason to digress from the agenda is when it becomes clear to the facilitator, project leader, and project sponsor that the JAD session has produced some new information that is unexpected and requires the JAD session (and perhaps the project) to move in a new direction. When participants attempt to divert the discussion away from the agenda, the facilitator must be firm but polite in leading discussion back to the agenda and getting the group back on track.

Second, the facilitator must help the group understand the technical terms and jargon that surround the system development process, and help the participants understand the specific analysis techniques used. Participants are experts in their area, their part of the business, but they are not experts in systems analysis. The facilitator must therefore minimize the learning required and teach participants how to effectively provide the right information.

Third, the facilitator records the group's input on a public display area, which can be a whiteboard, flip chart, or computer display. He or she structures the information that the group provides and helps the group recognize key issues and important solutions. Under no circumstance should the facilitator insert his or her opinions into the discussion. The facilitator *must* remain neutral at all times and simply help the group through the process. The moment the facilitator offers an opinion on an issue, the group will no longer see him or her as a neutral party, but rather as someone who could be attempting to sway the group into some predetermined solution.

However, this does not mean that the facilitator should not try to help the group resolve issues. For example, if two items appear to be the same to the facilitator, the facilitator should not say, "I think these may be similar." Instead, the facilitator should ask, "Are these similar?" If the group decides they are, the facilitator can combine them and move on. However, if the group decides they are not similar (despite what the facilitator believes), the facilitator should accept the decision and move on. The group is *always* right, and the facilitator has no opinion.

Post JAD Follow-up   As with interviews, a JAD *postsession report* is prepared and circulated among session attendees. The postsession report is essentially the same as the interview report in Figure 4-8. Since the JAD sessions are longer and provide more information, it usually takes a week or two after the JAD session before the report is complete.

Questionnaires

A questionnaire is a set of written questions for obtaining information from individuals. Questionnaires often are used when there is a large number of people from whom information and opinions are needed. In our experience, questionnaires are commonly used for systems intended for use outside of the organization (e.g., by customers or vendors) or for systems with business users spread across many geographic locations. Most people automatically think of paper when they think of questionnaires, but today more questionnaires are being distributed in electronic form, either via e-mail or on the Web. Electronic distribution can save a significant amount of money compared to distributing paper questionnaires.

Selecting Participants   As with interviews and JAD sessions, the first step is to select the individuals to whom the questionnaire will be sent. However, it is not usual to select every person who could provide useful information. The standard approach is to select a *sample*, or subset, of people who are representative of the

---

YOUR TURN

4-5 JAD Practice

Organize yourselves into groups of 4 to 7 people, and pick one person in each group to be the JAD facilitator. Using a blackboard, whiteboard, or flip chart, gather information about how the group performs some process (e.g., working on a class assignment, making a sandwich, paying bills, getting to class). How did the JAD session go? Based on your experience, what are pros and cons of using JAD in a real organization?

PRACTICAL 4-2 MANAGING PROBLEMS IN JAD SESSIONS

TIP

I have run more than a hundred JAD sessions and have learned several standard "facilitator tricks." Here are some common problems and some ways to deal with them.

- **Reducing domination.** The facilitator should ensure that no one person dominates the group discussion. The only way to deal with someone who dominates is head on. During a break, approach the person, thank him or her for their insightful comments, and ask them to help you make sure that others also participate.
- **Encouraging noncontributors.** Drawing out people who have participated very little is challenging because you want to bring them into the conversation so that they will contribute again. The best approach is to ask a direct factual question that you are *certain* they can answer. And it helps to ask the question using some repitition to give them time to think. For example "Pat, I know you've worked shipping orders a long time. You've probably been in the Shipping Department longer than anyone else. Could you help us understand exactly what happens when an order is received in Shipping?"
- **Side discussions.** Sometimes participants engage in side conversations and fail to pay attention to the group. The easiest solution is simply to walk close to the people and continue to facilitate right in front of them. Few people will continue a side conversion when you are two feet from them and the entire group's attention is on you and them.
- **Agenda merry-go-round.** The merry-go-round occurs when a group member keeps returning to the same issue every few minutes and won't let go. One solution is to let the person have five minutes to ramble on about the issue while you carefully write down every point on a flip chart or computer file. This flip chart or file is then posted conspicuously on the wall. When the person brings up the issue again, you interrupt them,

walk to the paper and ask them what to add. If they mention something already on the list, you quickly interrupt, point out that it is there, and ask what other information to add. Don't let them repeat the same point, but write any new information.

- **Violent agreement.** Some of the worst disagreements occur when participants really agree on the issues but don't realize that they agree because they are using different terms. An example is arguing whether a glass is half empty or half full; they agree on the facts, but can't agree on the words. In this case, the facilitator has to translate the terms into different words and find common ground so the parties recognize that they really agree.
- **Unresolved conflict.** In some cases, participants don't agree and can't understand how to determine what alternatives are better. You can help by structuring the issue. Ask for criteria by which the group will identify a good alternative (e.g., "Suppose this idea really did improve customer service. How would I recognize the improved customer service?"). Then once you have a list of criteria, ask the group to assess the alternatives using them.
- **True conflict.** Sometimes, despite every attempt, participants just can't agree on an issue. The solution is to postpone the discussion and move on. Document the issue as an "open issue" and list it prominently on a flip chart. Have the group return to the issue hours later. Often the issue will resolve itself by then and you haven't wasted time on it. If the issue cannot be resolved later, move it to the list of issues to be decided by the project sponsor or some other more senior member of management.
- **Use humor.** Humor is one of the most power tools a facilitator has and thus must be used judiciously. The best JAD humor is always in context; never tell jokes but take the opportunity to find the humor in the situation. *Alan Dennis*

entire group. Sampling guidelines are discussed in most statistics books, and most business schools include courses that cover the topic, so we will not discuss it here. The important point in selecting a sample, however, is to realize that not everyone who receives a questionnaire will actually complete it. On average, only 30 to 50% of paper and e-mail questionnaires are returned. Response rates for Web-based questionnaires tend to be significantly lower (often only 5 to 30%).

Designing the Questionnaire   Developing good questions is critical for questionnaires because the information on a questionnaire cannot be immediately clarified for a confused respondent. Questions on questionnaires must be very clearly written and leave little room for misunderstanding, so closed-ended questions tend to be most commonly used. Questions must clearly enable the analyst to separate facts from opinions. Opinion questions often ask the respondent the extent to which they agree or disagree (e.g., "Are network problems common?"), while factual questions seek more precise values (e.g., "How often does a network problem occur: once an hour, once a day, once a week?"). See Figure 4-10 for guidelines on questionnaire design.

Perhaps the most obvious issue—but one that is sometimes overlooked—is to have a clear understanding of how the information collected from the questionnaire will be analyzed and used. You must address this issue before you distribute the questionnaire, because it is too late afterward.

Questions should be relatively consistent in style, so that the respondent does not have to read instructions for each question before answering it. It is generally good practice to group related questions together to make them simpler to answer. Some experts suggest that questionnaires should start with questions important to respondents, so that the questionnaire immediately grabs their interest and induces them to answer it. Perhaps the most important step is to have several colleagues review the questionnaire and then pretest it with a few people drawn from the groups to whom it will be sent. It is surprising how often seemingly simple questions can be misunderstood.

Administering the questionnaire   The key issue in administering the questionnaire is getting participants to complete the questionnaire and send it back. Dozens of marketing research books have been written about ways to improve response rates. Commonly used techniques include: clearly explaining why the questionnaire is being conducted and why the respondent has been selected; stating a date by which the questionnaire is to be returned; offering an inducement to complete the questionnaire (e.g., a free pen); and offering to supply a summary of the questionnaire responses. Systems analysts have additional techniques to improve responses rates inside the organization, such as personally handing out the questionnaire and personally contacting those who have not returned them after a week or two, as well as requesting the respondents' supervisors to administer the questionnaires in a group meeting.

Questionnaire Follow-up   It is helpful to process the returned questionnaires and develop a questionnaire report soon after the questionnaire deadline. This ensures

---

- Begin with nonthreatening and interesting questions.
- Group items into logically coherent sections.
- Do not put important items at the very end of the questionnaire.
- Do not crowd a page with too many items.
- Avoid abbreviations.
- Avoid biased or suggestive items or terms.
- Number questions to avoid confusion.
- Pretest the questionnaire to identify confusing questions.
- Provide anonymity to respondents.

FIGURE 4-10
Good Questionnaire Design

4-6 Questionnaire Practice

Organize yourselves into small groups. Have each person develop a short questionnaire to collect information about the frequency in which group members perform some process (e.g., working on a class assignment, making a sandwich, paying bills, getting to class), how long it takes them, how they feel about the process, and opportunities for improving the process.

Once everyone has completed his or her questionnaire, ask each member to pass it to the right, and then complete his or her neighbor's questionnaire. Pass the questionnaire back to the creator when it is completed.

QUESTIONS:
1. How did the questionnaire you completed differ from the one you created?
2. What are the strengths of each questionnaire?
3. How would you analyze the survey results if you had received 50 responses?
4. What would you change about the questionnaire that you developed?

that the analysis process proceeds in a timely fashion and that respondents who requested copies of the results receive them promptly.

Document Analysis

Project teams often use *document analysis* to understand the as-is system. Under ideal circumstances, the project team that developed the existing system will have produced documentation, which was then updated by all subsequent projects. In this case, the project team can start by reviewing the documentation and examining the system itself.

Unfortunately, most systems are not well documented because project teams fail to document their projects along the way, and when the projects are over—there is no time to go back and document. Therefore, there may not be much technical documentation about the current systems available, or it may not contain updated information about recent system changes. However, there are many helpful documents that do exist in the organization: paper reports, memorandums, policy manuals, user training manuals, organization charts, and forms.

But these documents (forms, reports, policy manuals, organization charts) only tell part of the story. They represent the *formal system* that the organization uses. Quite often, the "real" or *informal system* differs from the formal one, and these differences, particularly large ones, give strong indications of what needs to be changed. For example, forms or reports that are never used likely should be eliminated. Likewise, boxes or questions on forms that are never filled in (or are used for other purposes) should be rethought. See Figure 4-11 for an example of how a document can be interpreted.

The most powerful indication that the system needs to be changed is when users create their own forms or add additional information to existing ones. Such changes clearly demonstrate the need for improvements to existing systems. Thus, it is useful to review both blank and completed forms to identify these deviations. Likewise, when users access multiple reports to satisfy their information needs, it is a clear sign that new information or new information formats are needed.

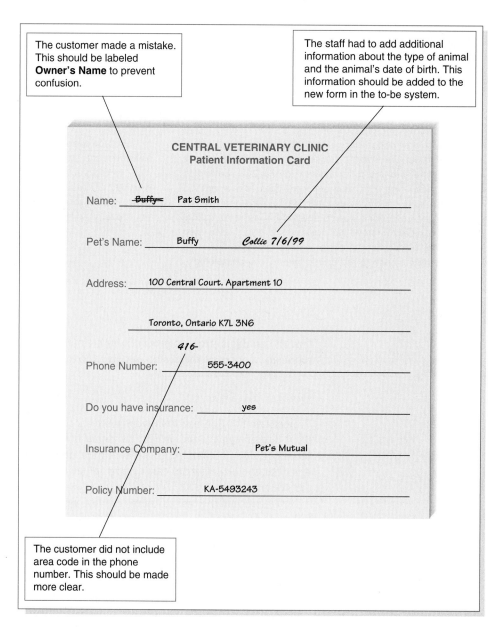

FIGURE 4-11
Performing a Document Analysis

Observation

Observation, the act of watching processes being performed, is a powerful tool for gathering information about the as-is system because it enables the analyst to see the reality of a situation, rather than listening to others describe it in interviews or JAD sessions. Several research studies have shown that many managers really do not remember how they work and how they allocate their time. (Quick, how many hours did you spend last week on each of your courses?) Observation is a good way to check the validity of information gathered from indirect sources such as interviews and questionnaires.

In many ways, the analyst becomes an anthropologist as he or she walks through the organization and observes the business system as it functions. The goal

4-E PUBLIX CREDIT CARD FORMS

At my neighborhood Publix grocery store, the cashiers always hand write the total amount of the charge on every credit card charge form, even though it is printed on the form. Why? Because the "back office" staff people who reconcile the cash in the cash drawers with the amount sold at the end of each shift find it hard to read the small print on the credit card forms. Writing in large print makes it easier for them to add the values up. However, cashiers sometimes make mistakes and write the wrong amount on the forms, which causes problems. *Barbara Wixom*

QUESTIONS:
1. What does the credit card charge form indicate about the existing system?
2. How can you make improvements with a new system?

is to keep a low profile, to not interrupt those working, and to not influence those being observed. Nonetheless, it is important to understand that what analysts observe may not be the normal day-to-day routine because people tend to be extremely careful in their behavior when they are being watched. Even though normal practice may be to break formal organizational rules, the observer is unlikely to see this. (Remember how you drove the last time a police car followed you?) Thus, what you see may *not* be what you get.

Observation is often used to supplement interview information. The location of a person's office and its furnishings gives clues as to their power and influence in the organization, and can be used to support or refute information given in an interview. For example, an analyst might become skeptical of someone who claims to use the existing computer system extensively if the computer is never turned on while the analyst visits. In most cases, observation will support the information that users provide in interviews. When it does not, it is an important signal that extra care must be taken in analyzing the business system.

Selecting the Appropriate Techniques

Each of the requirements-gathering techniques just discussed has strengths and weaknesses. No one technique is always better than the others, and in practice most projects use a combination of techniques. Thus, it is important to understand the

4-7 Observation Practice

Visit the library at your college or university and observe how the book check-out process occurs. First watch several students checking books out, and then check one out yourself. Prepares a brief summary report of your observations.

When you return to class, share your observations with others. You may notice that not all the reports present the same information. Why? How would the information be different had you used the interview or JAD technique?

strengths and weaknesses of each technique and when to use each (see Figure 4-12). One issue not discussed is that of the analysts' experience. In general, document analysis and observation require the least amount of training, while JAD sessions are the most challenging.

Type of Information   The first characteristic is type of information. Some techniques are more suited for use at different stages of the analysis process, whether understanding the as-is system, identifying improvements, or developing the to-be system. Interviews and JAD are commonly used in all three stages. In contrast, document analysis and observation usually are most helpful for understanding the as-is, although occasionally they provide information about current problems that need to be improved. Questionnaires are often used to gather information about the as-is system, as well as general information about improvements.

Depth of information   The depth of information refers to how rich and detailed the information is that the technique usually produces, and the extent to which the technique is useful at obtaining not only facts and opinions, but also an understanding of *why* those facts and opinions exist. Interviews and JAD sessions are very useful at providing a good depth of rich and detailed information and helping the analyst to understand the reasons behind them. At the other extreme, document analysis and observation are useful for obtaining facts, but little beyond that. Questionnaires can provide a medium depth of information, soliciting both facts and opinions with little understanding of why.

Breadth of Information   Breadth of information refers to the range of information and information sources that can be easily collected using that technique. Questionnaires and document analysis both are easily capable of soliciting a wide range of information from a large number of information sources. In contrast, interviews and observation require the analyst to visit each information source individually and, therefore, take more time. JAD sessions are in the middle because many information sources are brought together at the same time.

Integration of Information   One of the most challenging aspects of requirements gathering is the integration of information from different sources. Simply put, different people can provide conflicting information. Combining this information

	Interviews	Joint Application Design	Questionnaires	Document Analysis	Observation
Type of information	As-is, improvements, to-be	As-is, improvements, to-be	As-is, improvements	As-is	As-is
Depth of information	High	High	Medium	Low	Low
Breadth of information	Low	Medium	High	High	Low
Integration of information	Low	High	Low	Low	Low
User involvement	Medium	High	Low	Low	Low
Cost	Medium	Low–Medium	Low	Low	Low–Medium

FIGURE 4-12
Table of Information-Gathering Techniques

and attempting to resolve differences in opinions or facts is usually very time consuming because it means contacting each information source in turn, explaining the discrepancy, and attempting to refine the information. In many cases, the individual wrongly perceives that the analyst is challenging his or her information, when in fact it is another user in the organization. This can make the user defensive and make it hard to resolve the differences.

All techniques suffer integration problems to some degree, but JAD sessions are designed to improve integration because all information is integrated when it is collected, not afterward. If two users provide conflicting information, the conflict becomes immediately obvious, as does the source of the conflict. The immediate integration of information is the single most important benefit of JAD that distinguishes it from other techniques, and this is why most organizations use JAD for important projects.

User Involvement   User involvement refers to the amount of time and energy the intended users of the new system must devote to the analysis process. It is generally agreed that as users become more involved in the analysis process, the greater the chance of success. However, user involvement can have a significant cost, and not all users are willing to contribute valuable time and energy. Questionnaires, document analysis, and observation place the least burden on users, while JAD sessions require the greatest effort.

Cost   Cost is always an important consideration. In general, questionnaires, document analysis, and observation are low cost techniques (although observation can be quite time consuming). The low cost does not imply that they are more or less effective than the other techniques. We regard interviews and JAD sessions as having moderate costs. In general, JAD sessions are much more expensive initially, because they require many users to be absent from their offices for significant periods of time, and they often involve highly paid consultants. However, JAD sessions significantly reduce the time spent in information integration and thus cost less in the long term.

Combining Techniques   In practice, requirements gathering combines a series of different techniques. Most analysts start by using interviews with senior manager(s) to gain an understanding of the project and the "big picture" issues. From this, the scope becomes clear as to whether large or small changes are anticipated. These are often followed with analysis of documents and policies to gain some understanding of the as-is system. Usually interviews come next to gather the rest of the information needed for the as-is.

In our experience, identifying improvements is most commonly done using JAD sessions because the JAD session enables the users and key stakeholders to work together through an analysis technique and come to a shared understanding of the possibilities for the to-be system. Occasionally, these JAD sessions are followed by questionnaires sent to a much wider set of users or potential users to see whether the opinions of those who participated in the JAD sessions are widely shared.

Developing the concept for the to-be system is often done through interviews with senior managers, followed by JAD sessions with users of all levels to make sure the key needs of the new system are well understood.

APPLYING THE CONCEPTS AT CD SELECTIONS

Once the CD Selections approval committee approved the system proposal and feasibility analysis, the project team began performing analysis activities. These included gathering requirements using a variety of techniques, and analyzing the requirements that were gathered. An Internet marketing and sales consultant, Chris Campbell, was hired to advise Alec, Margaret, and the project team during the analysis phase. Some highlights of the project team's activities are presented next.

Requirements Analysis Techniques

Margaret suggested that the project team conduct several JAD sessions with store managers, marketing analysts, and Web-savvy members of the IT staff. Together, the groups could work through some BPI techniques and brainstorm how improvements could be made to the current order process using a new Web-based system.

Alec facilitated three JAD sessions that were conducted over the course of a week. Alec's past facilitation experience helped the eight-person meetings run smoothly and stay on track. First, Alec used technology analysis and suggested several important Web technologies that could be used for the system. The JAD session generated ideas about how CD Selections could apply each of the technologies to the Internet order project. Alec had the group categorize the ideas into three sets: "definite" ideas that would have a good probability of providing business value; "possible" ideas that might add business value; and "unlikely" ideas.

Next Alec applied informal benchmarking by introducing the Web sites of several leading retailers and pointing out the features that they offered online. He selected some sites based on their success with Internet sales, and others based on their similarity to the vision for CD Selections' new system. The group discussed the features that were common across most retailers versus unique functionality, and they created a list of suggested business requirements for the project team.

Requirements-Gathering Techniques

Alec believed that it would be important to understand the order processes and systems that already existed in the organization because they would have to be closely integrated with the Web order system. Three requirements-gathering techniques proved to be helpful in understanding the current systems and processes—document analysis, interviews, and observation.

First, the project team collected existing reports (e.g., order forms, screenshots of the online order screens) and system documentation (data models, process models) that shed light on the as-is system. They were able to gather a good amount of information about the brick-and-mortar order processes and systems in this way. When questions arose, they conducted short interviews with the person who provided the documentation for clarification.

Next, Alec interviewed the senior analysts for the order and inventory systems to get a better understanding of how those systems worked. He asked if they had any ideas for the new system, as well as any integration issues that would need to be addressed. Alex also interviewed a contact from the ISP and the IT person who supported CD Selections' current Web site—both provided information about the existing communications infrastructure at CD Selections and its Web capabilities. Finally, Alex spent a half day visiting two of the retail stores and observing exactly how the order and hold processes worked in the brick-and-mortar facilities.

Requirements Definition

Throughout all of these activities, the project team collected information and tried to identify the business requirements for the system from the information. As the project progressed, requirements were added to the requirements definition and grouped by requirement type. When questions arose, they worked with Margaret, Chris, and Alec to confirm that requirements were in scope. The requirements that fell outside of the scope of the current system were typed into a separate document that would be saved for future use.

At the end of the analysis phase, the requirements definition was distributed to Margaret, two marketing employees who would work with the system on the business side, and several retail store managers. This group then met for a two-day JAD session to clarify, finalize, and prioritize business requirements and to create use cases (Chapter 5) to show how the system would be used.

The project team also spent time creating process models (Chapter 6) and data models (Chapter 7) that depicted the data and processes in the future system. Members of marketing and IT reviewed the documents during interviews with the project team. Figure 4-13 shows a portion of the final requirements definition.

System Proposal

Alec reviewed the requirements definition and the other deliverables that the project team created during the analysis phase. Given Margaret's desire to have the system operating before next year's Christmas season, Alec decided to timebox the project, and he determined what functionality could be included in the system by that schedule deadline (see Chapter 3). He suggested that the project team develop the system in three versions rather than attempting to develop a complete system that provided all the features initially (Phased Development, see Chapter 2). The first version, to be operational well before the holidays, would implement a "basic" system that would have the "standard" order features of other Internet retailers. The second version, planned for late spring or early summer, would have several features unique to CD Selections. The third version would add more "advanced" features, such as the ability to listen to a sample of music over the Internet, to find similar CDs, and to write reviews.

Alec revised the workplan accordingly, and he worked with Margaret and the folks in Marketing to review the feasibility analysis and update it where appropriate. All of the deliverables from the project were then combined into a system proposal and submitted to the approval committee. Figure 4-14 shows the outline of the CD Selections system proposal. Margaret and Alec met with the committee and presented the highlights of what was learned during the analysis phase and the final concept of the new system. Based on the proposal and presentation, the approval committee decided that they would continue to fund the Internet order system.

Nonfunctional Requirements

1. Operational Requirements

   1.1 The Internet system will draw information from the main CD information database, which contains basic information about CDs (e.g., title, artist, ID number, price, quantity in inventory). The Internet order system will not write information to the main CD information database.

   1.2 The Internet system will store orders for new CDs in the special order system and will rely on the special order system to complete the special orders generated.

   1.3 A new module for the in-store system will be written to manage the "holds" generated by the Internet system. The requirements for this new module will be documented as part of the Internet system because they are necessary for the Internet system to function.

2. Performance Requirements

   No special requirements performance requirements are anticipated

3. Security Requirements

   No special security requirements are anticipated

4. Cultural and Political Requirements

   No special cultural and political requirements are anticipated

Functional Requirements

1. Place Requests for CDs

   1.1 Customers will access the Internet system to look for CDs of interest. Some customers will search for specific CDs or CDs by specific artists, while other customers want to browse for interesting CDs in certain categories (e.g., rock, jazz, classical).

   1.2 When the customer has found a CD he or she wants, the customer will check to see which store(s) have the CD in stock. They will use zip code to find stores close to their location.

   1.3 Customers can immediately place a hold on any CD in stock at any of the stores and then come into the store and pick it up (see requirement 3 below)

   1.4 If the CD is not available in the customer's preferred store, the customer can request that the CD be special ordered to that store for later pickup. The customer will be notified by e-mail when the requested CD arrives at the requested store; the CD will be placed on hold (which will expire after 7 days). This process will work similarly to the current special order system.

2. CD Marketing

   2.1 The Internet system provides an additional opportunity to market CDs to current and new customers. The system will provide a database of marketing materials about selected CDs that will help Web users learn more about them (e.g., music reviews, links to Web sites, artist information, and sample sound clips). When information about a CD that has additional marketing information is displayed, a link will be provided to the additional information.

   2.2 Marketing materials will be supplied primarily by vendors and record labels so that we can better promote their CDs. The Marketing Department will determine what marketing materials will be placed in the system and will be responsible for adding, changing, and deleting the materials.

3. Process In-store Holds

   3.1 When a CD is available in a store, the system will send a hold request to the in-store system at the selected store.

   3.2 The in-store system will alert the store staff (through an audible alarm and a pop-up message).

   3.3 Staff will print a label for the requested CD(s), pull them from the shelves, attach the label, and place them on the special order shelf. Just like the special orders, the hold items will be held for 7 days.

   3.4 Once the hold has been placed on the shelf, the staff will enter a hold confirmation, and the system will do an inventory adjustment to the main inventory database so that no other holds are accepted for the item.

FIGURE 4-13
CD Selections Requirements Definition

---

1. Table of Contents

2. Executive Summary

A summary of all the essential information in the proposal so a busy executive can read it quickly and decide what parts of the plan to read in more depth.

3. System Request

The revised system request form (see Chapter 2).

4. Workplan

The original workplan, revised after having completed the analysis phase (see Chapter 3).

5. Feasibility Analysis

A revised feasibility analysis, using the information from the analysis phase (see Chapter 2).

6. Requirements Definition

A list of the functional and nonfunctional business requirements for the system (this chapter).

7. Use Cases

A set of use cases that illustrate the basic processes that the system needs to support (see Chapter 5).

8. Process Model

A set of process models and descriptions for the to-be system (see Chapter 6). This may include process models of the current as-is system that will be replaced.

9. Data Model

A set of data models and descriptions for the to-be system (see Chapter 7). This may include data models of the as-is system that will be replaced.

Appendices

These contain additional material relevant to the proposal, often used to support the recommended system. This might include results of a questionnaire survey or interviews, industry reports and statistics, etc.

FIGURE 4-14
Outline of the CD Selections System Proposal

SUMMARY

The Analysis Phase

The analysis phase is the second phase of the SDLC, and it focuses on capturing the business requirements for the system. The analysis phase identifies the "what" of the system, and it leads directly into the design phase, during which the "how" of the system is determined. Many deliverables are created during the analysis phase, including the requirements definition, use cases, process models, and a data model. At the end of the analysis phase, all of these deliverables, along with revised planning and project management deliverables, are combined into a system proposal and submitted to the approval committee for a decision regarding whether or not to move ahead with the project.

Requirements Determination

Requirements determination is the part the analysis phase whereby the project team turns the very high-level explanation of the business requirements stated in the system request into a more precise list of requirements. A requirement is simply a statement of what the system must do or what characteristic it needs to have. Business requirements describe the "what" of the systems, and system requirements describe "how" the system will be implemented. A functional requirement relates directly to a process the system has to perform or information it needs to contain. Nonfunctional requirements refer to behavioral properties that the system must have, such as performance and usability. All of the functional and nonfunctional business requirements that fit within the scope of the system are written in the requirements definition, which is used to create other analysis deliverables and leads to the initial design for the new system.

Requirements Analysis Techniques

The basic process of analysis is divided into three steps: understanding the as-is system, identifying improvements, and developing requirements for the to-be system. Three requirements analysis techniques—business process automation, business process improvement, or business process reengineering—help the analyst lead users through the three (or two) analysis steps so that the vision of the system can be developed. Business process automation (BPA) means leaving the basic way in which the organization operates unchanged, and using computer technology to do some of the work. Problem analysis and root cause analysis are two popular BPA techniques. Business process improvement (BPI) means making moderate changes to the way in which the organization operates to take advantage of new opportunities offered by technology or to copy what competitors are doing. Duration analysis, activity-based costing, and information benchmarking are three popular BPI activities. Business process reengineering (BPR) means changing the fundamental way in which the organization operates. Outcome analysis, technology analysis, and activity elimination are three popular BPR activities.

Requirements-Gathering Techniques

Five techniques can be used to gather the business requirements for the proposed system: interviews, joint application development, questionnaires, document analysis, and observation. Interviews involve meeting one or more people and asking them questions. There are five basic steps to the interview process: selecting interviewees, designing interview questions, preparing for the interview, conducting the interview, and postinterview follow-up. Joint application development (JAD) allows the project team, users, and management to work together to identify requirements for the system. Electronic JAD attempts to overcome common problems associated with groups by using groupware. A questionnaire is a set of written questions for obtaining information from individuals. Questionnaires often are used when there is a large number of people from whom information and opinions are needed. Document analysis entails reviewing the documentation and examining the system itself. It can provide insights into the formal and informal system. Observation, the act of watching processes being performed, is a powerful tool for gathering information about the as-is system because it enables the analyst to see the reality of a situation firsthand.

KEY TERMS

Activity elimination
Activity-based costing
Analysis
Analysis plan
As-is system
Benchmarking
Bottom-up interview
Breadth of analysis
Business process automation (BPA)
Business process improvement (BPI)
Business process reengineering (BPR)
Business requirement
Closed-ended question
Critical thinking skills
Document analysis
Duration analysis
Electronic JAD (e-JAD)
Facilitator
Formal system

Functional requirement
Ground rule
Informal benchmarking
Informal system
Interpersonal skill
Interview
Interview notes
Interview report
Interview schedule
Joint application development (JAD)
Nonfunctional requirements
Observation
Open-ended question
Outcome analysis
Parallelization
Process integration
Postsession report
Potential business value
Probing question

Problem analysis
Project cost
Questionnaire
Requirement
Requirements definition
Requirements determination
Risk
Root cause
Root cause analysis
Sample
Scribe
Structured interview
System proposal
System requirements
Technology analysis
To-be system
Top-down interview
Unstructured interview
Walk-through

QUESTIONS

1. What are the key deliverables that are created during the analysis phase? What is the final deliverable from the analysis phase, and what does it contain?
2. Explain the difference between an as-is system and a to-be system.
3. What is the purpose of the requirements definition?
4. What are the three basic steps of the analysis process? Which step is sometimes skipped or done in a cursory fashion? Why?
5. Compare and contrast the business goals of BPA, BPI, and BPR.
6. Compare and contrast problem analysis and root cause analysis. Under what conditions would you use problem analysis? Under what conditions would you use root cause analysis?
7. Compare and contrast duration analysis and activity-based costing.
8. Assuming time and money were not important concerns, would BPR projects benefit from additional time spent understanding the as-is system? Why or why not?

9. What are the important factors in selecting an appropriate analysis strategy?
10. Describe the five major steps in conducting interviews.
11. Explain the difference between a closed-ended question, an open-ended question, and a probing question. When would you use each?
12. Explain the differences between unstructured interviews and structured interviews. When would you use each approach?
13. Explain the difference between a top-down and bottom-up interview approach. When would you use each approach?
14. How are participants selected for interviews and JAD sessions?
15. How can you differentiate between facts and opinions? Why can both be useful?
16. Describe the five major steps in conducting JAD sessions.
17. How does a JAD facilitator differ from a scribe?
18. What are the three primary things that a facilitator does in conducting the JAD session?

19. What is e-JAD, and why might a company be interested in using it?
20. How does designing questions for questionnaires differ from designing questions for interviews or JAD sessions?
21. What are typical response rates for questionnaires and how can you improve them?
22. Describe document analysis.
23. How does the formal system differ from the informal system? How does document analysis help you understand both?
24. What are the key aspects of using observation in the information-gathering process?
25. Explain factors that can be used to select information-gathering techniques.

EXERCISES

A. Review the Amazon.com Web site. Develop the requirements definition for the site. Create a list of functional business requirements that the system meets. What different kinds of nonfunctional business requirements does the system meet? Provide examples for each kind.

B. Pretend that you are going to build a new system that automates or improves the interview process for the Career Services Department of your school. Develop a requirements definition for the new system. Include both functional and nonfunctional system requirements. Pretend you will release the system in three different versions. Prioritize the requirements accordingly.

C. Describe in very general terms the as-is business process for registering for classes at your university. What BPA technique would you use to identify improvements? With whom would you use the BPA technique? What requirements-gathering technique would help you apply the BPA technique? List some example improvements that would you expect to find.

D. Describe in very general terms the as-is business process for registering for classes at your university. What BPI technique would you use to identify improvements? With whom would you use the BPI technique? What requirements-gathering technique would help you apply the BPI technique? List some example improvements that you would expect to find.

E. Describe in very general terms the as-is business process for registering for classes at your university. What BPR technique would you use to identify improvements? With whom would you use the BPR technique? What requirements-gathering technique would help you apply the BPR technique? List some example improvements that would you expect to find.

F. Suppose your university is having a dramatic increase in enrollment and is having difficulty finding enough seats in courses for students so they can take courses required for graduation. Perform a technology analysis to identify new ways help students complete their studies and graduate.

G. Suppose you are the analyst charged with developing a new system for the university bookstore with which students can order books online and have them delivered to their dorms and off-campus housing. What requirements-gathering techniques will you use? Describe in detail how you would apply the techniques.

H. Suppose you are the analyst charged with developing a new system to help senior managers make better strategic decisions. What requirements-gathering techniques will you use? Describe in detail how you would apply the techniques.

I. Find a partner and interview each other about what tasks you/they did in the last job held (full-time, part-time, past or current). If you haven't worked before, then assume your job is being a student. Before you do this, develop a brief interview plan. After your partner interviews you, identify the type of interview, interview approach, and types of questions used.

J. Find a group of students and run a 60-minute JAD session on improving alumni relations at your university. Develop a brief JAD plan, select two techniques that will help identify improvements, and then develop an agenda. Conduct the session using the agenda, and write your postsession report.

K. Find a questionnaire on the Web that has been created to capture customer information. Describe the purpose of the survey, the way questions are worded, and how the questions have been organized. How can it be improved? How will the responses be analyzed?

L. Develop a questionnaire that will help gather information regarding processes at a popular restaurant, or the college cafeteria (e.g., ordering, customer service). Give the questionnaire to 10 to 15 students,

analyze the responses, and write a brief report that describes the results.

M. Contact the Career Services Department at your university and find all the pertinent documents designed to help students find permanent and/or part-time jobs. Analyze the documents and write a brief report.

MINICASES

1. The State Firefighter's Association has a membership of 15,000. The purpose of the organization is to provide some financial support to the families of deceased member firefighters and to organize a conference each year bringing together firefighters from all over the state. Annually members are billed dues and calls. "Calls" are additional funds required to take care of payments made to the families of deceased members. The bookkeeping work for the association is handled by the elected treasurer, Bob Smith, although it is widely known that his wife, Laura, does all of the work. Bob runs unopposed each year at the election, since no one wants to take over the tedious and time-consuming job of tracking memberships. Bob is paid a stipend of $8,000 per year, but his wife spends well over 20 hours per week on the job. The organization however, is not happy with their performance.

A computer system is used to track the billing and receipt of funds. This system was developed in 1984 by a Computer Science student and his father. The system is a DOS-based system written using dBase 3. The most immediate problem facing the treasurer and his wife is the fact that the software package no longer exists, and there is no one around who knows how to maintain the system. One query in particular takes 17 hours to run. Over the years, they have just avoided running this query, although the information in it would be quite useful. Questions from members concerning their statements cannot be easily answered. Usually Bob or Laura just jot down the inquiry and return a call with the answer. Sometimes it takes 3 to 5 hours to find the information needed to answer the question. Often, they have to perform calculations manually since the system was not programmed to handle certain types of queries. When member information is entered into the system, each field is presented one at a time. This makes it very difficult to return to a field

and correct a value that was entered. Sometimes a new member is entered but disappears from the records. The report of membership used in the conference materials does not alphabetize members by city. Only cities are listed in the correct order.

What requirements analysis strategy or strategies would you recommend for this situation? Explain your answer.

2. Brian Callahan, IS Project Manager, is just about ready to depart for an urgent meeting called by Joe Campbell, Manager of Manufacturing Operations. A major BPI project sponsored by Joe recently cleared the approval hurdle, and Brian helped bring the project through project initiation. Now that the approval committee has given the go-ahead, Brian has been working on the project's analysis plan.

One evening, while playing golf with a friend who works in the Manufacturing Operations Department, Brian learned that Joe wants to push the project's time frame up from Brian's original estimate of 13 months. Brian's friend overheard Joe say, "I can't see why that IS project team needs to spend all that tune 'analyzing' things. They've got two weeks scheduled just to look at the existing system! That seems like a real waste. I want that team to get going on building my system."

Because Brian has a little inside knowledge about Joe's agenda for this meeting, he has been considering how to handle Joe. What do you suggest Brian tell Joe?

3. Barry has recently been assigned to a project team that will be developing a new retail store management system for a chain of submarine sandwich shops. Barry has several years of experience in programming but has not done much analysis in his career. He was a little nervous about the new work he would be doing, but was confident he could handle any assignment he was given.

One of Barry's first assignments was to visit one of the submarine sandwich shops and prepare an observation report on how the store operates. Barry planned to arrive at the store around noon, but he chose a store in an area of town he was unfamiliar with, and due to traffic delays and difficulty in finding the store, he did not arrive until 1:30. The store manager was not expecting him and refused to let a stranger behind the counter until Barry had him contact the project sponsor (the Director of Store Management) back at company headquarters to verify who he was and what his purpose was.

After finally securing permission to observe, Barry stationed himself prominently in the work area behind the counter so that he could see everything. The staff had to maneuver around him as they went about their tasks, and there were only minor occasional collisions. Barry noticed that the store staff seemed to be going about their work very slowly and deliberately, but he supposed that was because the store wasn't very busy. At first, Barry questioned each worker about what he or she was doing, but the store manager eventually asked him not to interrupt their work so much—he was interfering with their service to the customers.

By 3:30, Barry was a little bored. He decided to leave, figuring he could get back to the office and prepare his report before 5:00 that day. He was sure his team leader would be pleased with his quick completion of his assignment. As he drove, he reflected, "There really won't be much to say in this report. All they do is take the order, make the sandwich, collect the payment, and hand over the order. It's really simple!" Barry's confidence in his analytical skills soared as he anticipated his team leader's praise.

Back at the store, the store manager shook his head, commenting to his staff, "He comes here at the slowest time of day on the slowest day of the week. He never even looked at all the work I was doing in the back room while he was here—summarizing yesterday's sales, checking inventory on hand, making up resupply orders for the weekend … plus he never even considered our store opening and closing procedures. I hate to think that the new store management system is going to be built by someone like that. I'd better contact Chuck (the Director of Store Management) and let him know what went on here today." Evaluate Barry's conduct of the observation assignment.

4. Anne has been given the task of conducting a survey of sales clerks who will be using a new order entry system being developed for a household products catalog company. The goal of the survey is to identify the clerks' opinions on the strengths and weaknesses of the current system. There are about 50 clerks who work in three different cities, so a survey seemed like an ideal way of gathering the needed information from the clerks.

Anne developed the questionnaire carefully and pretested it on several sales supervisors who were available at corporate headquarters. After revising it based on their suggestions, she sent a paper version of the questionnaire to each clerk, asking that it be returned within 1 week. After one week, she had only three completed questionnaires returned. After another week, Anne received just two more completed questionnaires. Feeling somewhat desperate, Anne then sent out an e-mail version of the questionnaire, again to all the clerks, asking them to respond to the questionnaire by e-mail as soon as possible. She received two e-mail questionnaires and three messages from clerks who had completed the paper version expressing annoyance at being bothered with the same questionnaire a second time. At this point, Anne has just a 14% response rate, which she is sure will not please her team leader. What suggestions do you have that could have improved Anne's response rate to the questionnaire?

PLANNING

ANALYSIS

☑ **Apply Requirements Analysis Techniques (Business Process Automation, Business Process Improvement, or Business Process Reengineering)**

☑ **Use Requirements Gathering Techniques (Interview, JAD Session, Questionnaire, Document Analysis, or Observation)**

☐ Develop Use Cases

☐ Develop Data Flow Diagrams

☐ Develop Entity Relationship Model

☐ Normalize Entity Relationship Model

TASK CHECKLIST

PLANNING   ANALYSIS   DESIGN

CHAPTER 5

USE-CASE
ANALYSIS

Use cases describe in more detail the key elements of the requirements definition. They explain the process by which the system will meet the functional requirements defined in the previous chapter. The use cases are then used to build a process model, which defines the business processes in a more formal manner.

OBJECTIVES

- Understand the role of use cases.
- Understand the process used to create use cases.
- Be able to create use cases.

CHAPTER OUTLINE

IMPLEMENTATION

INTRODUCTION

The previous chapter discussed the process of requirements determination that results in the requirements definition. The requirements definition defines what the system is to do. In this chapter, we discuss how these requirements are further refined into a set of use cases that provide more detail on the process by which the system is to meet these requirements and the data the system needs to capture and store. Once the use cases have been developed, the next steps are to use the requirements definition and the use cases to create even more detailed description of the processes and data in the form of a process model and a data model for the new system.

Use cases are a relatively new technique. For many years, systems analysts simply sat down with users and began drawing process and data models. However, users often found it difficult to learn the process and data modeling languages used by the analysts. In recent years, many organizations have begun using the use case approach in which the analysts first work with the users to create simple text descriptions of complex processes, and then later use these to build formal models. The approach is the same whether the project team is defining the as-is model or the to-be model, but obviously the focus is different; the as-is model focuses on current business processes, whereas the to-be model focuses on desired business processes.

This two-step approach for complex processes (use cases first and process and data models second) has come from two different parts of the systems analysis and design community, and thus there are two different views on how best to create the use case. Organizations using structured design techniques have begun to use what they call business scenarios to describe processes, while organizations using object-oriented techniques (see Chapter 15) have begun to use what they call use cases. At present, there are no formal standards for either business scenarios or use cases, so we have tried to incorporate what we believe are the best elements of both approaches. We have adopted the term *use case* rather than *business scenario* because use case is gradually becoming more popular.[1]

In this chapter, we first explain how to read use cases and describe their basic syntax. Then we describe the process used to build use cases.

USE CASES

A *use case* is a set of activities that produce some output result. Each use case describes how the system reacts to an *event* that *triggers* the system. For example, in a library system, a trigger event might be someone borrowing a book, someone returning a book, or a book becoming overdue. With this type of *event-driven modeling*, everything in the system can be thought as a response to some trigger event. When there are no events, the system is at rest, patiently waiting for the next event

---

[1] As you will see in Chapter 15, object-oriented techniques take the text-based use cases we describe in this chapter and create use case diagrams before moving to modeling structure and behavior (similar to the data and process models we describe in the next chapters). In the non-object-oriented world, the use case diagrams are not created so we omit them in this chapter; they are described in Chapter 15. We focus only on the text descriptions of the use cases. For a more detailed description of business scenarios, see Karen McGraw and Karen Harbison, *User-Centered Requirements: The Scenario-Based Engineering Process,* Mahwah, NJ: Lawrence Erlbaum Associates, 1997. For a more detailed description of use cases, see I. Jacobson, M. Christerson, P. Jonsson, and G. Overgaard, *Object-Oriented Software Engineering: A Use-Case Driven Approach,* Reading, MA: Addison-Wesley, 1992.

to trigger it. When a trigger event occurs, the system (and the people using it) responds, performs the actions defined in the use case, and then returns to the waiting state.

In some situations, the process may be "small," such as the actions that are performed when a book is borrowed in the previous example. In more complex systems (such as the CD Selections example in this book), a use case may require several distinct activities, some of which are performed each time the use case is activated, and some of which are performed only occasionally (e.g., consider the return of a library book, which very rarely will be returned with damage that needs to be repaired). Simple use cases may have only one path through them, while complex use cases may have several possible paths.

We only create use cases when they are likely to help us better understand the situation and help make the following design steps simpler. For very simple processes that are well explained in the requirements definition, we often do not bother to create a use case, but simply use the information in requirements definition itself to build the process and data models.

It is also important to create use cases when we are reengineering processes, or the new system is likely to significantly change the way people work.

When creating use cases, the project team must work closely with the users to gather the information needed. This is often done using requirements-gathering techniques such as interviews, JAD sessions, and observation. Gathering the information needed for use cases is a relatively straightforward process—but one that takes considerable practice. Users work closely with the project team to create the use cases and in some cases, the users themselves actually write the use cases.

Elements of a Use Case

A use case contains a fairly complete description of all the activities that occur in response to a trigger event. Figure 5-1 shows a sample use case for the appointments system in a doctor's office. While there are numerous pieces of information in the use case, the information is organized into three main parts: basic information, inputs and outputs, and details.

Basic Information   Each use case has a name, number, and brief description. The name should be as simple yet descriptive as possible. The number is simply a sequential number so that it is easy to refer to use cases (e.g., use case 3). The description is also very short and provides a bit more information about what the use case is all about.

Another element of basic information is the trigger for the use case—the event that causes the use case to begin. A trigger can be an *external trigger*, such as a customer placing an order or the fire alarm ringing, or it can be *temporal trigger*, such as a book being overdue at the library or being time to pay the rent.

It is helpful to have a consistent *viewpoint* when writing use cases. For example, the use case in Figure 5-1 is written from the viewpoint of the doctor and his or her staff, not of the patient. All the steps and descriptions are written as activities that the staff performs.

Inputs and Outputs   The second major part of a use case is the set of major *inputs* and *outputs*. Each of the major inputs and outputs to the use case are described, along with their source or destination. These are all possible inputs and outputs,

Scenario Name: Patient makes, cancels, or changes an appointment   ID number: 1

Short description: This describes how we make a new appointment as well as changing or canceling an appointment

Trigger: Patient calls and asks for an appointment or asks to cancel an existing appointment

Type: (External) Temporal

Major Inputs

Description	Source
Patient name	Patient
Desired appointment	Patient
Appointment to change/cancel	Patient
Patient information	Patient's DB
Available appointments	Appointment's DB

Major Outputs

Description	Destination
Appointment	Appointment
Appointment	Patient
Possible appointments	Patient

Major Steps Performed

Information for Steps

Patient name

1. If this is a change or cancellation, then find current appointment in the Appointment File and cancel it.

Appointment to change/cancel

Revised appointment

Patient name

2. Check to make sure the patient is a current patient and has no unpaid bills. If this is a new patient, perform the "Add New Patient" use case before continuing. If the patient has unpaid bills, then transfer the call to the business office.

Patient information

3. Find the available times for an appointment and select ones to propose to the patient (some appointment times will be too short or too long for the patient's problem). This may be repeated several times until a good time is found.

Available appointments

Possible appointments

Patient information

4. The patient picks an appointment time to be scheduled.

Desired appointment

Appointment

FIGURE 5-1
Sample Use Case

not just those that always or usually are part of the use case. The goal is to include every one, but it is common for users and analysts to miss inputs and outputs when they first define use cases. This is not a major problem because the process of building use cases is one of gradual refinement: as they work through the parts of the use case, they often return to previous parts to correct them.

Details   The third major part of a use case are the detailed individual steps within the use case and the inputs and outputs they use. These steps are the activities that are performed during the use case, such as taking the patient's name and address, checking to see if the appointment is available, and so on. The steps are listed in the order in which they are performed and any conditional steps are clearly noted (e.g., what steps are performed if the patient's appointment is available versus what steps are performed if the appointment is not available).

Building Use Cases

Use cases can be used for both the as-is and to-be systems; as-is use cases focus on the current system, whereas to-be use cases focus on the desired new system. When used for the to-be system, it is fairly common for the use cases to identify additional requirements that were not completely specified in the Requirements Definition. This, in fact, is one of the reasons why use cases are important, and why after the use cases have been built, that analysts often return to the Requirements Definition and revise it, based on their improved understanding of the system.

The most common ways to gather information for the use cases are through the same requirement determination techniques discussed in the last chapters, especially interviews and JAD sessions. Observation also is sometimes used for as-is use cases. Regardless of whether interviews or JAD sessions are used, research shows that some ways to gather the information for use cases are better than others. The most effective process has four steps[2] (see Figure 5-2). These four steps are performed in order, but of course the analyst often cycles among them in an *iterative* fashion as he or she moves from use case to use case.

Identify the Major Use Cases   The first step is to identify the use cases and fill in the top parts of use-case form (basic information and inputs and outputs). The goal is to develop a set of major use cases, with the major information about each, rather than jumping into one use case and describing it completely. This prevents the users and analysts from forgetting key use cases and helps the users explain the overall set of business processes they are responsible for. It also helps users understand how to describe the use cases and reduces the chance of overlap between use cases. In this step, the analysts and users identify a set of major use cases that could benefit from additional definition beyond the requirements definition.

This step starts with the requirements definition. Much of the information needed to identify the use cases should be contained in the requirements definition, because the use cases are just a more detailed explanation of the requirements. The

[2] The approach in this section is based on the work of George Marakas and Joyce Elam, "Semantic Structuring in Analyst Acquisition and Representation of Facts in Requirements Analysis," *Information Systems Research,* 1998, 9(1), 37–63 as well as our own: Alan Dennis, Glenda Hayes, and Robert Daniels, "Business Process Modeling with Group Support Systems," *Journal of Management Information Systems,* 1999, 15(4), 115–142.

Step	Activities	Typical Questions Asked[a]
1. Identify the use cases.	Start a use case report form for each use case by filling in the name, description, trigger, and the easily identified major inputs and outputs. If there are more than nine use cases, group them into packages.	Ask *who*, *what*, *when*, and *where* about the use cases (or tasks) and their inputs and outputs (e.g., forms and reports). What are the major tasks that are performed? What triggers this task? What tells you to perform this task? What information/forms/reports do you need to perform this task? Who gives you these information/forms/reports? What information/forms/report does this produce and where do they go?
2. Identify the major steps within each use case.	For each use case, fill in the major steps needed to process the inputs and produce the outputs.	Ask *how* about each use case. How do you produce this report? How do you change the information on the report? How do you process forms? What do tools do you use to do this step (e.g., on paper, by e-mail, by phone)?
3. Identify elements within steps.	For each step, identify its triggers and its inputs and outputs.	Ask *how* about each step. How does the person know when to perform this step? What forms/reports/data does this step produce? What forms/reports/data does this step need? What happens when this form/report/data is not available?
4. Confirm the use case.	For each use case, validate that it is correct and complete.	Ask the user to execute the process using the written steps in the use case—that is, have the user role-play the use case.

[a] We have used the typical questions for the as-is model (e.g., "What are the..."). These same questions can be used for the to-be model, but they would be phrased in the future tense (e.g., "What should be the...").

FIGURE 5-2
Steps for Writing for Use Case Reports

users and analysts work together to identify the use cases from the requirements definition, often realizing the additional requirements will need to be added. The users and analysts first focus on just the basic information in the use case: they give names to the use cases, provide short descriptions, and identify the triggers.

Identifying use cases is an iterative process, with users often changing their minds about what is a use case and what it includes. It is very easy to get trapped in the details at this point, so you need to remember that the goal at this step is to just identify the major use cases. For example, in the doctor's office example in Figure 5-1, we defined one use case as "patient makes, cancels, or changes an appointment." This use case included appointments for both new patients and existing patients, as well as when the patient changes or cancels the appointment. We could have defined each of these activities (makes an appointment, changes an appointment, or cancels an appointment) as a separate use case, but this would have created a much larger set of much smaller use cases. The trick is to select the right size so that you end up with the major use cases that need additional explanation beyond the requirements definition. Remember that a use case is a set of end-to-end activities that starts with a trigger event and continues through many possible paths until some output has been produced, and the system is again at rest.

If the project team discovers more than eight or nine major use cases, this suggests that the system is complex (or that the use cases are not defined at the right level of detail). If there really are more than eight or nine major use cases, the use cases are grouped together into *packages* of related use cases. For example, in a doctor's office system you might group together all the patient-oriented use cases into one package (e.g., maintaining patient records, maintaining appointments) and all the accounting use cases into another package (e.g., customer billing, insurance billing). These packages are then treated as the major processes for the top level of the process model with the use cases appearing on lower levels, or are treated as separate systems and modeled as separate systems (process modeling will be described in the next chapter).

Once the use cases have been identified the users and analysts complete the second principle part of the use case: inputs and outputs (also called *data flows*), the major pieces of information that the use case needs or produces. Some use cases require physical inputs or produce physical outputs (e.g., a repair car use case would have a broken car as an input and a repaired car as an output), but physical inputs are not included because the use case is used to build an information system and such physical items do not go into the system, although information about them does.

At this point, the analysts are not concerned with defining all the inputs and outputs—just the major ones that come to mind quickly. In later steps, they will return to this list to ensure that every single input and output is identified. However, it is important to understand and define acronyms and jargon so that the project team and others from outside the user group can clearly understand the use case. Typical questions asked by the analysts in this step are given in Figure 5-2.

Identify the Major Steps for Each Use Case   At this point, the use cases and major inputs and outputs have been defined. In short, you have filled in the top two parts of the use case (basic information and inputs and outputs). The next step is to complete the third part: the detailed information. The users and analysts go back through the use cases to fill in the three to nine major *steps* within each use case. The steps focus on what the business process or system does to complete the use case, as opposed to what actions the users or other external entities do. In general, the steps should be listed in the order in which they are performed, from first to last, but there also may be steps that are performed only occasionally, have no formal sequence in which they are done, or loop back and forth. The order of steps implies a sequence but does not require it. It is fine to list steps that have no sequence in any order you like, but if there is a sequence, you should list the steps in that way.

Each step should be about the same size as the others. For example, if we were writing steps for preparing a meal, steps such as "take fork out of drawer" and "put fork on table" are much smaller than "prepare cake using mix." If you end up with more than nine steps or steps that vary greatly in size, you must go back and adjust the steps. Recognizing the size of the steps takes practice but will become natural in time.

One good approach to produce the steps for a use case is to have the users visualize themselves actually performing the use case and to write down the steps as if they were writing a recipe for a cookbook. In most cases, the users will be able to quickly define what they do in as-is use cases. Defining the steps for to-be use cases may take a bit more coaching. In our experience, the descriptions of the steps change greatly as the users work through a use case. Our advice is to use blackboard or whiteboard that can be easily erased (or paper with pencil) to develop the list of steps. Once the set of steps is fairly well defined, only then do you write it on the use case form.

Occasionally, a use case is so simple that further refinement is not needed. The analyst simply wrties a brief description and does not bother to develop the steps within the use case. The information at the top of the use case form is sufficient, because the use case need not be explained in more detail. Some of the use cases used for the exercises at the end of this chapter are simple enough that they do not need information beyond what is at the top of the use case form.

Identify Elements within Steps At this point, the steps have been described, but not the elements that further define and link the steps. In other words, the use case form in Figure 5-1 is complete except for the last column ("Information for Steps"), which is blank and has no arrows drawn between steps. The next step is to delve more deeply into the steps within the use case to understand and describe their inputs and outputs. Each step should have at least one input and at least one output.

The goal at this point is to identify the major inputs and outputs for each step. One could identify the inputs and outputs in great detail, but this would make it difficult to list them concisely at the top of the form. The solution is to identify details within the description of the steps, but to provide only general categories at the top of the use-case form. For example, if a step needs the patient name, address, and phone number, we might note these in the step description but list only "patient information" as the major input at the top of the form. In Figure 5-1, for example, we list "appointment" at the top of the form but mention "date and time" in step 3.

The users and analysts now return to the steps in the use case and begin linking the steps together. Typically, this mean asking what inputs (e.g., information, forms, reports) are used by each step and what outputs they produce. These are written in the last column on the use case form with an arrow pointing into or out of a step (see Figure 5-1). Sometimes forms, reports, and information will flow from one step to the next to the next; these are shown using arrows from step to step.

It is common at this point for users to discover that there are major inputs and outputs that they forgot to list during their first time through the use case. Sometimes users realize they have forgotten entire steps in the description. These previously omitted inputs, outputs, and steps are simply added to the use case. Our experience has shown that users can forget to include seldom-used activities that occur in special cases (e.g., when data is not available or when something unexpected occurs), so it is useful to question the steps carefully to make sure that no steps have been omitted.

Confirm the Use Case The final step is for the users to confirm that the use case is correct as written, which means reviewing the use case with the users to make sure each step and input and output are correct. The most powerful approach is to ask the user to *role-play*, or execute the use case using the written steps in the use case. The analyst will hand the user pieces of paper labeled as the major inputs to the use case and have the user follow the written steps like a recipe to make sure that those steps and inputs really can produce the outputs defined for the use case.

APPLYING THE CONCEPTS AT CD SELECTIONS

Identifying the Major Use Cases

The first step in creating the use cases is to identify the major use cases, based on the requirements definition, which was developed in the last chapter and shown in Figure

5-1 Campus Housing

Create a set of use-case reports for the following high-level requirements in a housing system run by the Campus Housing Service. The Campus Housing Service helps students find apartments. Owners of apartments fill in information forms about the rental units they have available (e.g., location, number of bedrooms, monthly rent), which are entered into a database. Students can search through this database via the Web to find apartments that meet their needs (e.g., a two-bed-room apartment for $400 or less per month within 1/2 mile of campus). They then contact the apartment owners directly to see the apartment and possibly rent it. Apartment owners call the service to delete their listing when they have rented their apartment(s).

In building the major use cases, follow the four-step process: identify the use cases, identify the steps within them, identify the elements within the steps, and confirm the use cases.

4-13. Take a minute and carefully read the requirements definition. Identify the major use cases that you think need additional definition before you continue reading.

It is important that you think about the use cases before you read what we have to say about them, so please, if you haven't tried to do this, take 5 minutes now and do it. We'll wait.

5-A BUILDING A BAD SYSTEM?

Several years ago, a well-known national real estate company built a computer-based system to help its real estate agents sell houses more quickly. The system, which worked in many ways like an early version of realtor.com, enabled its agents to search the database of houses for sale to find houses matching the buyer's criteria using a much easier interface than the traditional system. The system also enabled the agent to show the buyer a virtual tour of selected houses listed by the company itself. It was believed that by more quickly finding a small set of houses more closely matching the buyer's desires, and by providing a virtual tour, the buyers (and the agent) would waste less time looking at unappealing houses. This would result in happier buyers and in agents that were able to close sales more quickly, leading to more sales for the company and higher commissions for the agent.

The system was designed with input from agents from around country and was launched with great hoopla. The initial training of agents met with a surge of interest and satisfaction among the agents, and the project team received many congratulations.

Six months later, satisfaction with the system had dropped dramatically, absenteeism had increased by 300%, and agents were quitting in record numbers; turnover among agents had risen by 500%, and in exit interviews, many agents mentioned the system as the primary reason for leaving. The company responded by eliminating the system—with great embarrassment.

One of agents' key skills was the ability to find houses that match buyer's needs. The system destroyed the value of this skill by providing a system that could enable less skilled agents to perform almost as well as highly skilled ones. Worse still—from the viewpoint of the agent—the buyer could interact directly with the system, thus bypassing the "expertise" of the agent.

QUESTIONS
1. How were the problems with the system missed?
2. How might these problems have been foreseen and possibly avoided?
3. In perfect hindsight, given the widespread availability of such systems on the Internet today, what should the company have done?

Source: "The Hidden Minefields in Sales Force Automation Technologies," *Journal of Marketing*, July 2002, by C. Speier and V. Venkatesh.

The information in the functional requirements definition sometimes just flows into the use cases, but usually it requires some thought. After you read the requirements definition, you probably identified one or two major use cases needing additional definition: place requests for CDs and CD marketing. As you will shortly see, there is another use case.

If we adopt the viewpoint of CD Selections, then the first use case is better named the *take requests use case*, which includes requirements 1.1, 1.2, 1.3, and 1.4. You might have considered each of the requirements in section 1 as separate use cases (e.g., 1.1, 1.2, 1.3, etc.). Each of these requirements is not a separate and distinct use case; they are all—or most as we will shortly see—part of the same overall process. While the customer could end their interactions with the system after requirement 1.1 (e.g., after searching and not finding anything of interest), requirement 1.1 is just the first step in the series of steps that are needed to make the use case run to its conclusion in requirements 1.3 or 1.4. Even though requirements 1.3 and 1.4 are mutually exclusive for any one CD, they are treated together as part of the same use case. Therefore, these requirements are all part of the same use case.

The first requirement in CD marketing (2.1) is quite different from the second requirement. The first requirement defines what marketing information will be displayed and when it will be displayed. The display of this information occurs during the first use case, when the customer is looking for CDs. Thus it best fits in the place request use case.

The second requirement (2.2) is separate from the customer use case. This defines the way in which the marketing materials will be created, changed, and deleted from the system and who is responsible—the Marketing Department. It is clearly a separate use case.

You might have considered adding new marketing materials to the database, changing them, and deleting old materials (the subparts of requirement 2.2) as three separate use cases, which in fact they could be. If you think about it, you should see that in addition to these three, we also need to have use cases for finding information about the marketing materials (e.g., when the manager forgets what CDs have marketing materials) and for printing reports about the marketing materials. However, our goal at this point is to develop a set of only the major use cases that need extra definition. Therefore, we will put these activities together as part of one overall larger use case called maintain marketing materials.

These five activities (creating, changing, deleting, finding, and printing) are the standard processes usually required for every database. You should see this same pattern repeating itself over and over again in most information systems.

The project team at CD Selections identified these same two use cases, called *take requests for CDs* and *maintain marketing materials*. They then needed to gather additional information to define them in more detail. This was done on the basis of the results of the earlier analyses described in Chapter 4, as well as through a series of JAD meetings with the project sponsor and the key Marketing Department managers and staff who would ultimately operate the system. Although the marketing staff was involved with all JAD sessions, other users were brought in for some of the sessions (e.g., the managers responsible for the special order system, and two store managers) to make sure no important elements were overlooked.

For each use case, they then identified its trigger and the major inputs and outputs. The first use case, take requests for CDs has an external trigger: the customer's actions. It has many inputs and outputs, and it is often easiest to start by thinking about the outputs first. One major output is a special order stored in the special order database for CDs that are not in stock when the user requests them. The special orders flow nicely into the special order system, which is well suited to managing them; as far as the Internet system is concerned, all we have to do is create the special order and then just rely on the special order system to take care of it. In this day and age, when most organizations have well-developed information systems, this ability to integrate with an existing system and reuse it for a new purpose not originally intended—sometimes with changes as we shall see—is quite common.

A second major output is the "hold" placed on a CD that is already in stock at a specific store. At this point, the project team, with the help of the store managers in the user team, recognized a major requirement not explicitly stated in the requirements definition: a way to capture and manage the "holds." No process existed in the stores to take CDs off the shelf and place them in the special order section. Thus an important outcome of developing the use cases was the recognition of the impact on the in-store processes and the in-store information system. Changing the in-store system was not planned as part of the Internet system, but was critical to its success; this was a major change in scope. Alec, the project leader, made a note of the need to revise the in-store system to integrate with the Internet system. A separate use case for handling the in-store "holds" was developed (and is explained shortly).

The take requests use case has many more inputs. One input is the search information the customer provides when looking for the CD (e.g., album name). Another input is the set of CDs the customer actually selects for purchase, because not all CDs searched for will be ordered. A third input is the customer's information (e.g., name, e-mail address, phone number).

The second use case, maintain marketing materials, has a similar structure. It is triggered by the receipt of marketing materials from vendors. The major input is the marketing material and the major output is the same information, reformatted and stored in the marketing material data store.

The third use case was the use case for processing the in-store holds. CD Selections recently completed an upgrade of its in-store information system so that all stores are linked together over the Internet via an always-on DSL or cable modem connection. This enables the in-store system to continuously communicate with CD Selections' main computer systems, such as the inventory system. It also means that the in-store system can send and receive information in real time to and from other systems, such as the Internet system. In this case, the take requests use case will generate a message that is sent to the manage holds use case that serves as the trigger. The input is the hold request. One output is a hold label that identifies the customer and the products to put on hold in the special order shelf.

Note that at this point, only the top half of each use case has been completed. Take a moment to review the use cases (see Figure 5-3) and make sure you understand them. You will shortly discover (if you haven't already) that the inputs and outputs in the use cases in Figure 5-3 are incomplete; the users have overlooked several important inputs and outputs. This is not a problem, because this is typical of the way most use cases evolve.

Identifying the Major Steps for Each Use Case

The next step is to define the major steps within each use case. The goal at this point is to describe how the use case operates. In this example, we will focus on the most

Use case name: Take requests for CDs   ID number: __1__

Short description: This describes how customers can search the Web site and place requests to hold CD's in stock or place special orders

Trigger: Customer searches Web and places request to hold a CD or to special order it

Type: (External) Temporal

Major Inputs:

Description	Source	Major Outputs: Description	Destination
Search request	Customer	Special order	Special order DBs
CDs selected for request	Customer	Hold for in-stock CD	In-store hold DB
Customer information	Customer		
Marketing materials	Marketing DB		

Major Steps

Use case name: Maintain marketing materials   ID number: __2__

Short description: This adds, deletes, and modifies the additional marketing material from vendors (e,g, reviews, musics clips)

Trigger: Materials from vendors, distributors, wholesalers, record companies, and articles in trade magazines

Type: (External) Temporal

Major Inputs:

Description	Source	Major Outputs: Description	Destination
Marketing materials	Vendor	Marketing materials	Marketing DB

---

use case name: Process in-store holds   ID number: __3__

Short description: This alerts the store staff to pull a requested CD from the shelves and place it in the special order section

Trigger: Hold request from take request use case

Type: (External) Temporal

Major Inputs:

Description	Source	Major Outputs: Description	Destination
Major Steps			
Hold request	Take request use case	Hold label	Store staff

Major Steps Performed   Information for Steps

FIGURE 5-3
Initial Use Cases for CD Selections

complex use case, take requests for CD. The best way to begin to understand how the customer works through this use case is to visualize yourself placing a CD request over the Web and to think about how other electronic commerce Web sites work. The techniques of visualizing yourself interacting with the process and of thinking about how other systems work (informal benchmarking) are important techniques that help analysts and users understand how processes work and how to write the use case. Both techniques (*visualization* and informal benchmarking) are commonly used in practice.

After you connect to the Web site, you probably begin searching, perhaps for a specific CD, perhaps for a category of music, but in any event, you enter some information for your search. The Web site then presents a list of CDs matching your request along with some basic information about the CDs (e.g., artist, title, price). If one of the CDs is of interest to you, you might seek more information about it, such as the list of songs, liner notes, and reviews.

Once you, the customer, find a CD you like, you then need to see which store, if any, has it in stock. The easiest way to find a store near you is probably based on zip code. After you enter a zip code, the system will show the list of stores in the zip code (and nearby zip codes) and the availability of the CD in that store. If the CD is not available in the store, you could special order it to the store.

Once you add a CD to your request (or choose not to) you could continue looking for more CDs. Or, you could "check out" by presenting your order with information on the CDs you want and giving additional information, such as your name and e-mail address.

When you write the use case, your focus should be on the steps that the business process or system performs to execute the use case, rather than on the actions the users perform. One might argue that the first step is to present the customer with the home page or a form to fill in to search for an album. This is correct, but this type of step is usually very small compared to other steps that follow. It is analogous to making the first step "hand the user a piece of paper." Usually we avoid documenting such small steps at this point in the process.[3]

The first major step performed by the system is to respond to the customer's search inquiry, which might include a search for a specific album name or albums by a specific artist. Alternatively, it might be the customer's wanting to see all the classical or alternative music CDs in stock; or it might be a request to see the list of special deals or CDs on sale. In any event, the system finds all the CDs matching the request and shows a list of CDs in response.

The user will view this response and perhaps will decide to seek more information about one or more CDs. He or she will click on it, and the system will provide additional information. Perhaps the user will also want to see any extra marketing material that is a available.

The user will then select the CD to request and perhaps continue with a new search. The user may later make changes to the list of CDs selected, either by removing some or by changing the number of CDs requested.

At some point, the user will check out by verifying the CDs he or she has selected and providing information about himself or herself (e.g., name, e-mail address, phone number).

Figure 5-4 shows the use case at this point. Note that the steps have been added to the form, but nothing else has changed.

---

[3] They are important and will be documented when we design the system at a much lower level of detail. At this point, though, we are only looking for the three to nine major steps.

Use case name: **Take requests for CDs**   ID number: ___1___

Short description: **This describes how customers can search the Web site and place requests to hold CDs in stock or place special orders.**

Trigger: **Customer searches Web and places requests to hold a CD or to special order it**

Type: (**External**) Temporal

Major Inputs		Major Outputs	
Description	Source	Description	Destination
Search request	Customer	Special order	Special order DB
CDs selected to request	Customer	Hold for CDs in stock	Hold DB
Customer information	Customer		
Marketing materials	Marketing DB		

Major Steps Performed   Information for Steps

1. Find CDs matching customer's request, whether it is a search by author, title, etc., a search by category (e.g., jazz, classical), or a request for "sale" items.

2. Provide information about one CD. This starts with some basic information but may also include extra marketing material such as reviews and music clips.

3. Find stores close to the customer and display the availability of the CD in those stores' inventory.

4. Customer selects the CD at a store to hold or special order.

5. Customer "checks out" confirming the CDs the user has selected, calculating the total amount, and accepting user's name and contact information.

6. Place hold(s) for CD(s) in stock.

7 Place special order for CDs not in stock.

FIGURE 5-4
Take Requests for CDs Use Case after Step 3

Identifying the Elements within Steps

The next step is to add more detail to the steps by identifying their inputs and outputs. This means identifying what inputs (e.g., information, forms, reports) are used by each step and what outputs each step produces. As we noted earlier, it is common for users to discover that there are major inputs and outputs that they forgot to list on their first time through the use case—and CD Selections is no exception.

The first step (find matching CDs) has at least one input and one output. The input is the search information from the user. We could list every type of search information (e.g., artist name, album name, type of music, sale items), but this would make for a long list. Instead, we can bundle these inputs together under the general name of "CD information request." The output from this step is the list of CDs matching the search request, which is shown to the customer. We forgot to list this output in the list of major outputs at the top of the use case, so we go back and add it (Figure 5-5).

Take a moment to read down through the rest of inputs and outputs added to each step in Figure 5-5. Most should be fairly obvious and straightforward, and you should also note that they have been added at the top of the form under the major inputs and outputs. The only unusual one may be step 4, which has CD(s) selected to request as an input, as an output, and as information passed to the check-out step. If you visualize this step, you will be able to easily see why it is an input: the user clicks on some CDs to indicate that he or she wants to purchase them. Why are they an output? Because the user will want to the Web site to show him or her what CD(s) he or she has selected for purchase as a confirmation. Likewise, they are an output of this step into the checkout step because this step (step 3) is where the user makes the selection decisions and these selections are needed in step 4 to calculate the cost and to put them on hold or place a special order. Likewise, the customer information entered in step 5 is needed to produce the order in step 5. This may be a bit hard to identify, but it should become clear if you visualize yourself trying to write an order to put in the file; the order needs the names of the CDs, information about the customer, and the cost.

You may also note that sometimes customers will move from step 4 back to step 1 to find more CDs. This is not shown in the use case, because use cases do not show the flow of control through the system or process.

Confirming the Use Case

Once all the use cases had been defined, the final step in the JAD session was to confirm that they were accurate. The project team had the users role-play the use cases. A few minor problems were discovered and easily fixed. The final use cases are shown in Figure 5-6.

YOUR TURN

5-2 CD SELECTIONS' INTERNET SYSTEM

Complete the use cases for the maintain marketing materials use case and the process in-store holds use case by adding the steps and the inputs and outputs for each step.

Use case name: Take requests for CDs			ID number: ___1___

Short description: This describes how customers can search the Web site and place requests to hold CDs in stock or place special orders

Trigger: Customer searches Web and places request to hold a CD or to special order it

Type: (External) Temporal

Major Inputs:

Description	Source
Search request	Customer
CDs selected to request	Customer
Customer information	Customer
Marketing materials	Marketing DB
CD information request	Customer
CD inventory	Inventory DB

Major Outputs:

Description	Destination
Special order	Special order system
Hold for in-stock CD	In-store hold CD
CDs imatching search request	Customer
CDs requested	Customer
CD information	Customer
Marketing material	Customer

Major Steps Performed

1. Find CDs matching customers request, whether it is a search by author, title, etc., a search by category (e.g., jazz, classical), or a request for "sale" items.

2. Provide information about one CD. This starts with some basic information but may also include extra marketing material such as reviews and music clips.

3. Find stores close to the customer and display the availability of the CD in those stores inventory

4. Customer selects the CD at a store to hold or special order

   CD select to request

5. Customer "checks-out" confirming the CDs the user has selected, calculating the total amount, and accepting user's name and contact information.

   CD select to request

6. Place a hold for CDs in stock

   CD select to request

7. Place a special order for CDs not in stock

Information for Steps

Search request
CDs matching search request

CD information request
CD information
Marketing materials

Zip code
CD availability by store

CD selected to request
CD selected to request

CD selected to request

Customer Information

Hold for in-stock CD

Special order

FIGURE 5-5
Take Requests for CD Use Case after Step 4

Use case name: Take requests for CDs   ID number: _1_

Short description: This describes how customers can search the Web site and place requests to hold CDs in stock or place special orders

Trigger: Customer searches Web and places request to hold a CD or to special order it

Type: (External) Temporal

Major Inputs:

Description	Source
Search request	Customer
CDs selected for request	Customer
Customer information	Customer
Marketing materials	Marketing DB
CD information request	Customer
CD inventory	Inventory DB

Major Outputs:

Description	Destination
Special Order	Special order DBs
Hold for in-stock CD	In-store hold DB
CDs matching search request	Customer
CDs requested	Customer
CD information	Customer
Marketing materials	Customer

Major Steps

Use case name: Maintain marketing materials   ID number: _2_

Short description: This adds, deletes, and modifies the additional marketing material from vendors (e.g, reviews, musics clips)

Trigger: Materials from vendors, distributors, wholesalers, record companies, and articles in trade magazines

Type: (External) Temporal

Major Inputs:

Description	Source
Marketing materials	Vendor
Marketing materials	Marketing manager
CD information	CD DB
Vendor information	Vendor

Major Outputs:

Description	Destination
Marketing materials	Marketing DB
Marketing material report	Marketing manager

---

use case name: Process in-store holds   ID number: _3_

Short description: This alerts the store staff to pull a request CD from the shelves and place it in the special order section

Major Steps

Trigger: Hold request from take request use case

Type: (External) Temporal

Major Inputs:

Description	Source
Hold request	In-store hold DB
Hold confirmation	In-store staff

Major Outputs:

Description	Destination
Hold Label	In-store staff
Hold request alert	In-store staff
Hold confirmation	In-store hold DB
Inventory adjustment	Inventory DB

Major Steps Performed   Information for Steps

FIGURE 5-6

Final Use Cases for CD Selections

NonFunctional Requirements

1. Look and Feel

2. Usability

3. Performance

4. Operational

   4.1 The Internet order system will draw information from the main CD information database, which contains basic information about CDs (e.g., title, artist, ID number, price, quantity in inventory). The Internet order system will not write information to the main CD information database.

   4.2 The Internet order system will store orders for new CDs in the special order system, and will rely on the special order system to complete the special orders generated.

   4.3 A new module for the In-store system will be written to manage the "holds" generated by the Internet system. The requirements for this new module will be documented as part of the Internet system because they are necessary for the Internet system to function.

5. Maintainability and Portability

6. Security

7. Cultural and Political

8. Legal

FIGURE 5-7

Revised Nonfunctional Requirements Definition for CD Selections' Internet System

---

Functional Requirements

1. Place Requests for CDs

   1.1 Customers will access the Internet system to look for CDs of interest. Some customers will search for specific CDs or CDs by specific artists, while other customers want to browse for interesting CDs in certain categories (e.g., rock, jazz, classical).

   1.2 When the customer has found a CD he or she wants, the customer will check to see which store(s) have the CD in stock. They will use zip code to find stores close to their location.

   1.3 Customers can immediately place a hold on any CD in stock at any of the stores and then come into the store and pick it up (see requirement 3).

   1.4 If the CD is not available in the customer's preferred store, the customer can request that the CD be special ordered to that store for later pickup. The customer will be notified by e-mail when the requested CD arrives at the requested store; the CD will be placed on hold (which will expire after 7 days). This process will work similarly to the current special order system.

2. CD Marketing

   2.1 The Internet system provides an additional opportunity to market CDs to current and new customers. The system will provide a database of marketing materials about selected CDs that will help Web users learn more about them (e.g., music reviews, links to Web sites, artist information, and sample sound clips). When information about a CD that has additional marketing information is displayed, a link will be provided to the additional information.

   2.2 Marketing materials will be supplied primarily by vendors and record labels so that we can better promote their CDs. The Marketing Department will determine what marketing materials will be placed in the system and will be responsible for adding, changing, and deleting the materials.

3. Process In-store Holds

   3.1 When a CD is available in a store, the system will send a hold request to the in-store store system at the selected store.

   3.2 The in-store system will alert the store staff (through an audible alarm and a pop-up message).

   3.3 Staff will print a label for the requested CD(s), pull them from the shelves, attach the label, and place them on the special order shelf. Just like the special orders, the hold items will be held for 7 days.

   3.4 Once the hold has been placed on the shelf, the staff will enter a hold confirmation and the system will do an inventory adjustment to the main inventory database so that no other holds are accepted for the item.

FIGURE 5-8

Revised Functional Requirements Definition for CD Selections' Internet System

Revising the Requirements Definition

We should also pause at this point to discuss how the results of the use case models feed back into the requirements definition. In this case, the project team discovered that they would need to write a new "Manage Holds" module for the in-store system. This module would need to operate as part of an existing system, running in a different place from the rest of the system: while most of the Internet system will run over the Internet, this module would be running on the desktop of the in-store computers. The nonfunctional requirements were revised to better document this; see Figure 5-7. The functional requirements were also expanded to better explain this process; see Figure 5-8.

SUMMARY

Use Cases
A use case contains all the information needed to build one part of a process model, but expresses it in a less formal way that is usually simpler for users to understand. A use case has a name, number, brief description, trigger(s), major inputs and outputs, and a list of the major steps required to perform it.

Creating Use Cases
The first step in writing use cases is to develop a fairly complete set of the major steps, with basic information about each. The second step is to go back through the steps to fill-in the major steps required to produce them. The third step is to identify each step's inputs and outputs, which often leads to the discovery of additional inputs and outputs included at the top of the use case. The final step is to have the users confirm that the use case is correct as written, which often has them role-playing the steps.

KEY TERMS

Business scenario	Iteration	Use case
Data flow	Output	Use-case package
Event	Role-play	Viewpoint
Event-driven modeling	Step	Visualization
External trigger	Temporal trigger	
Input	Trigger	

QUESTIONS

1. How is creating use cases related to requirements determination and process modelling?
2. What are the major elements of a use case?
3. Describe how to create use cases.
4. What is the difference between a temporal trigger and an external trigger? Provide two examples for each.
5. Why do we strive to develop use cases for only the major use cases?

6. What happens if you have a large number of use cases? Discuss two ways to handle this situation.
7. Why is iteration important in creating use cases?
8. What is the viewpoint of a use case, and why is it important?

9. Is it important to write use cases for all requirements? Why or why not?
10. How can role-play help analysts build use cases?

EXERCISES

A. Create a set of use cases for the process of buying glasses from the viewpoint of the patient, but do not bother to identify the steps within each use case (just complete the information at the "top" of the use-case form). The first step is to see an eye doctor who will give you a prescription. Once you have a prescription, you go to a glasses store, where you select your frames and place the order for your glasses. Once the glasses have been made, you return to the store for a fitting and pay for the glasses.

B. Create a set of use cases for the following dentist office system but do not bother to identify the steps within each use case (just complete the information at the "top" of the use-case form). Whenever new patients are seen for the first time, they complete a patient information form that asks their name, address, phone number, and brief medical history, which are stored in the patient information file. When a patient calls to schedule a new appointment or change an existing appointment, the receptionist checks the appointment file for an available time. Once a good time is found for the patient, the appointment is scheduled. If the patient is a new patient, an incomplete entry is made in the patient file; the full information will be collected when they arrive for their appointment. Because appointments are often made so far in advance, the receptionist usually mails a reminder postcard to each patient 2 weeks before the appointment.

C. Complete the use cases for the dentist office system in exercise B by identifying the steps and the data flows within the use cases.

D. Create a set of use cases for an online university registration system. The system should enable the staff of each academic department to examine the courses offered by their department, add and remove courses, and change the information about them (e.g., the maximum number of students permitted). It should permit students to examine currently avail-able courses, add and drop courses to and from their schedules, and examine the courses for which they are enrolled. Department staff should be able to print a variety of reports about the courses and the students enrolled in them. The system should ensure that no student takes too many courses and that students who have any unpaid fees are not permitted to register (assume that a fees data store is maintained by the university's financial office which the registration system accesses but does not change).

E. Create a set of use cases for the following system. A Real Estate Inc. (AREI) sells houses. People who want to sell their houses sign a contract with AREI and provide information on their house. This information is kept in a database by AREI and a subset of this information is sent to the city-wide multiple listing service used by all real estate agents. AREI works with two types of potential buyers. Some buyers have an interest in one specific house. In this case, AREI prints information from its database, which the real estate agent uses to help show the house to the buyer (a process beyond the scope of the system to be modeled). Other buyers seek AREI's advice in finding a house that meets their needs. In this case, the buyer completes a buyer information form that is entered into a buyer database, and AREI real estate agents use its information to search AREI's database and the multiple listing service for houses that meet their needs. The results of these searches are printed and used to help the real estate agent show houses to the buyer.

F. Create a set of use cases for the following system. A Video Store (AVS) runs a series of fairly standard video stores. Before a video can be put on the shelf, it must be catalogued and entered into the video database. Every customer must have a valid AVS customer card in order to rent a video. Customers rent videos for 3 days at a time. Every time a customer rents a video, the system must ensure that they do not have any overdue videos. If so, the overdue videos must be returned and an overdue fee paid before customer can rent more videos. Likewise, if the customer has returned overdue videos, but has not paid the overdue fee, the fee must be paid before new videos can be rented. Every morning, the store manager prints a report that lists overdue videos; if a video is 2 or more days overdue, the manager calls the customer to remind them to return the video. If a video is returned in damaged condition, the manager removes it from the video database and may sometimes charge the customer.

G. Create a set of use cases for the following health club membership system. When members join the health club, they pay a fee for a certain length of time. Most memberships are for 1 year, but memberships as short as 2 months are available. Throughout the year, the health club offers a variety of discounts on their regular membership prices (e.g., two memberships for the price of one for Valentine's Day). It is common for members to pay different amounts for the same length of membership. The club wants to mail out reminder letters to members asking them to renew their memberships 1 month before their memberships expire. Some members have become angry when asked to renew at a much higher rate than their original membership contract, so the club wants to track the price paid so that the manager can override the regular prices with special prices when members are asked to renew. The system must track these new prices so that renewals can be processed accurately. One of the problems in the health club industry is the high turnover rate of members. While some members remain active for many years, about half of the members do not renew their memberships. This is a major problem, because the health club spends a lot in advertising to attract each new member. The manager wants the system to track each time a member comes into the club. The system will then identify the heavy users and generate a report so the manager can ask them to renew their memberships early, perhaps offering them a reduced rate for early renewal. Likewise, the system should identify members who have not visited the club in more than a month, so the manager can call them and attempt to reinterest them in the club.

H. Create a set of use cases for the following system. Picnics R Us (PRU) is a small catering firm with five employees. During a typical summer weekend, PRU caters 15 picnics with 20 to 50 people each.

The business has grown rapidly over the past year, and the owner wants to install a new computer system for managing the ordering and buying process. PRU has a set of 10 standard menus. When potential customers call, the receptionist describes the menus to them. If the customer decides to book a picnic, the receptionist records the customer information (e.g., name, address, phone number, etc.) and the information about the picnic (e.g., place, data, time, which one of the standard menus, total price) on a contract. The customer is then faxed a copy of the contract and must sign and return it along with a deposit (often a credit card or by check) before the picnic is officially booked. The remaining money is collected when the picnic is delivered. Sometimes, the customer wants something special (e.g., birthday cake). In this case, the receptionist takes the information and gives it to the owner who determines the cost; the receptionist then calls the customer back with the price information. Sometimes the customer accepts the price; other times, the customer requests some changes which have to go back to the owner for a new cost estimate. Each week, the owner looks through the picnics scheduled for that weekend and orders the supplies (e.g., plates) and food (e.g., bread, chicken) needed to make them. The owner would like to use the system for marketing as well. It should be able to track how customers learned about PRU, and identify repeat customers, so that PRU can mail special offers to them. The owner also wants to track the picnics on which PRU sent a contract, but the customer never signed the contract and actually booked a picnic.

I. Create a set of use cases for the following system. Of-the-Month Club (OTMC) is an innovative young firm that sells memberships to people who have an interest in certain products. People pay membership fees for 1 year and each month receive a product by mail. For example, OTMC has a coffee-of-the-month club that sends members one pound of special coffee each month. OTMC currently has six memberships (coffee, wine, beer, cigars, flowers, and computer games) each of which costs a different amount. Customers usually belong to just one, but some belong to two or more. When people join OTMC, the telephone operator records the name, mailing address, phone number, e-mail address, credit card information, start date, and membership service(s) (e.g., coffee). Some customers request a double or triple membership (e.g., two pounds of coffee, three cases of beer). The computer game membership operates a bit differently from the others. In this case, the member must also select the type of game (action, arcade, fantasy/science-fiction, educational, etc.) and age level. OTMC is planning to greatly expand the number of memberships it offers (e.g., video games, movies, toys, cheese, fruit, vegetables) so the system needs to accommodate this future expansion. OTMC is also planning to offer 3-month and 6-month memberships.

J. Create a set of use cases for a university library borrowing system (do not worry about catalogue searching, etc.). The system will record the books owned by the library and will record who has borrowed what books. Before someone can borrow a book, they must show a valid ID card that is checked to ensure that it is still valid against the student database maintained by the Registrar's Office (for student borrowers), the faculty/staff database maintained by the Personnel Office (for faculty/staff borrowers), or against the library's own guest database (for individuals issued a "guest" card by the library). The system must also check to ensure the borrower does not have any overdue books or unpaid fines before he or she can borrow another book. Every Monday, the library prints and mails postcards to those people with overdue books. If a book is overdue by more than 2 weeks, a fine will be imposed, and a librarian will telephone the borrower to remind him or her to return the book(s). Sometimes books are lost or are returned in damaged condition. The manager must then remove them from the database and will sometimes impose a fine on the borrower.

MINICASES

1. Williams Specialty Company is a small printing and engraving organization. When Pat Williams, the owner, brought computers into the business office 5 years ago, the business was very small and very simple. Pat was able to utilize an inexpensive PC-based accounting system to handle the basic information processing needs of the firm. As time has gone on, however, the business has grown and the work being perfomed has become significantly more complex. The simple accounting software still in use is no longer adequate to keep track of many of the company's sophisticated deals and arrangements with its customers.

Pat has a staff of four people in the business office who are familiar with the intricacies of the company's record-keeping requirements. Pat recently met with her staff to discuss her plan to hire as IS consulting firm to evaluate their information system needs and recommend a strategy for upgrading their computer system.

The staff is excited about the prospect of a new system, since the current system causes them much aggravation. No one on the staff has ever done anything like this before, however, and they are a little wary of the consultants who will be conducting the project.

Assume that you are a systems analyst on the consulting team assigned to the Williams Specialty Co. engagement. At your first meeting with the Williams staff, you want to be sure that they understand the work that your team will be performing, and how they will participate in that work.

   a. Explain in clear, nontechnical terms, the goals of the analysis phase of the project.
   b. Explain in clear, nontechnical terms, how use cases will be used by the project team. Explain what these models are, what they represent in the system, and how they will be used by the team.

ANALYSIS

☑ **Apply Requirements Analysis Techniques (Business Process Automation, Business Process Improvement, or Business Process Reengineering)**

☑ **Use Requirements Gathering Techniques (Interview, JAD Session, Questionnaire, Document Analysis, or Observation)**

☑ **Develop Use Cases**

☐ **Develop Data Flow Diagrams**

☐ Develop Entity Relationship Model

☐ Normalize Entity Relationship Model

T A S K CHECKLIST

PLANNING   ANALYSIS   DESIGN

CHAPTER 6

PROCESS

MODELING

A process model describes business processes—the activities that people do—and can be used to describe both the as-is system and the to-be system being developed. This chapter describes data flow diagramming, one of the most commonly used process modeling techniques.

OBJECTIVES

- Understand the rules and style guidelines for data flow diagrams.
- Understand the process used to create data flow diagrams.
- Be able to create data flow diagrams.

CHAPTER OUTLINE

IMPLEMENTATION

INTRODUCTION

The previous chapters discussed requirements activities, such as interviewing and JAD, and how to transform those requirements into more detailed use cases. In this chapter, we discuss how the requirements definition and use cases are further refined into a proces model. A *process model* is a formal way of representing how a business system operates. It illustrates the processes or activities that are performed and how data moves among them. A process model can be used to document the current system (i.e., as-is system) or the new system being developed (i.e., to-be system), whether computerized or not.

There are many different process modeling techniques in use today. In this chapter, we focus on one of the most commonly used techniques[1]: data flow diagramming. Data flow diagramming is a technique that diagrams the business processes and the data that passes among them. In this chapter, we first describe the basic syntax rules and illustrate how they can be used to draw simple one-page data flow diagrams (DFDs). Then we describe how to create more complex multipage diagrams.

Although the name *data flow diagram* implies a focus on data, this is not the case. The focus is mainly on the processes or activities that are performed. Data modeling, discussed in the next chapter, presents how the data created and used by processes are organized. Process modeling—and creating DFDs in particular—is one of the most important skills needed by systems analysts.

In this chapter, we focus on *logical process models,* which are models that describe processes without suggesting how they are conducted. When reading a logical process model, you will not be able to tell if a process is computerized or manual, if a piece of information is collected by paper form or via the Web, or if information is placed in a filing cabinet or a large database. These physical details are defined during the design phase when these logical models are refined into *physical models*, which provide information that is needed to ultimately build the system (see Chapter 8). By focusing on logical processes first, analysts can focus on how the business should run without being distracted with implementation details.

In this chapter, we first explain how to read DFDs and describe their basic syntax. Then we describe the process used to build DFDs that draws information from the use cases and from additional requirements information gathered from the users.

DATA FLOW DIAGRAMS

Reading Data Flow Diagrams

Figure 6-1 shows one part of a DFD for a doctor's office. By examining this DFD, an analyst can understand the process by which the doctor's office makes appointments. Take a moment and examine the diagram before reading the next paragraph. How much do you understand?

---

[1] Another commonly used process modeling technique is IDEF0. IDEF0 is used extensively throughout the U.S. federal government. For more information about IDEF0, see FIPS 183: *Integration Definition for Function Modeling (IDEF0),* Federal Information Processing Standards Publications, Washington, D.C.: U.S. Department of Commerce, 1993.

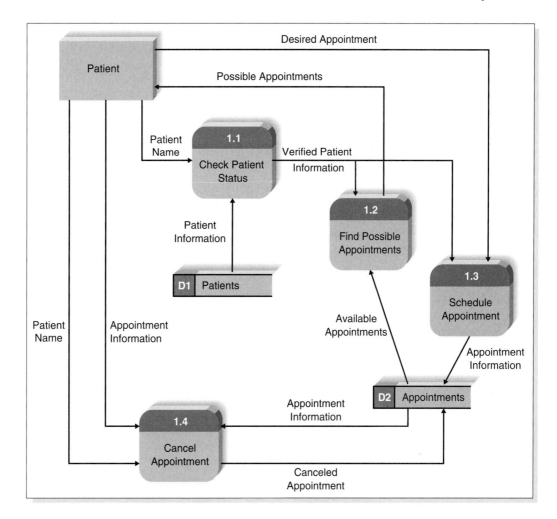

FIGURE 6-1
One Part of a DFD for a Doctor's Office

Perhaps you recognized this as the use case in the last chapter (Figure 5-1), which described how patients made, canceled, and changed appointments. Most people start reading in the upper left corner of the DFD, so this is where most analysts try to make the DFD start, although this is not always possible. The first item in the upper left corner of Figure 6-1 is the "Patient" external entity, which is a rectangle that represents individual patients who visit the doctor's office. It has four arrows pointing away from itself to rounded rectangles, and one arrow coming into it. The arrows represent data flows, and they show that four pieces of data (e.g., patient name) are sent as outputs provided by the external entity and used by the rounded rectangles, which indicate processes (e.g., "Check Patient Status")—actions that are performed. As you follow the arrow from the "Patient" external entity to the "Check Patient Status" process, imagine yourself as a receptionist sitting at a computer asking a patient for his or her name and entering it in the computer to find the appropriate record.

The "Check Patient Status" process has three arrows, or data flows, pointing in or out. Data flows going into the process are inputs used by the process, and data flows leaving the process are outputs changed or created by the process. Sometimes information is not provided by an external entity; instead, it comes from a data store, which holds information. Notice the long open-ended rectangle labeled "Patients." This is a data store that holds patient information; notice that it provides patient information to the "Check Patient Status" process, as shown by the arrow from the data store to the process. Examine the DFD and see how much you can understand about it from reading the data flows going into and out of processes and external entities.

Hopefully, you understood that the "Check Patient Status" process uses the patient name given by the patient to retrieve information from the "Patients" data store. From there, the information can be used to find possible appointments. The "Schedule Appointment" process uses the set of available appointments from the appointments data store to enable the staff to propose possible appointments to the patient (not all available appointments may be suitable for the patient because some patients may require longer or shorter appointments depending upon their health problems). The patient information and the desired appointment is then used to schedule an appointment, which is stored in the appointment data store.

We have one process for "Cancel Appointment" but none for "Change Appointment." If you think carefully, you can see that changing an appointment is the same as canceling an appointment and making a new appointment.

This diagram is an example of a single-page DFD. In practice, most DFDs are much more complex than this. Most systems have so many processes that if we attempted to draw one DFD showing all processes, it would not fit on one page, so instead they are depicted with a series of DFDs (each drawn on a separate page) to represent the entire system.

Elements of Data Flow Diagrams

Now that you have had a glimpse of a DFD, we will present the language of DFDs, which includes a set of symbols and syntax rules. There are four symbols in the DFD language (processes, data flows, data stores, and external entities), each of which is represented by a different graphic symbol. There are two commonly used styles of symbols, one set developed by Chris Gane and Trish Sarson and the other by Tom DeMarco and Ed Yourdan[2] (Figure 6-2). Neither is better than the other; some organizations use the Gane and Sarson style of symbols and others use the DeMarco/Yourdan style. We will use the Gane and Sarson style in this book.

Process A *process* is an activity or a function that is performed for some specific business reason. Processes can be manual or computerized, and every process has a name that starts with a verb and ends with a noun (e.g., "Find Patient," "Update Patient"). Names should be short, yet contain enough information so that the reader can easily understand exactly what they do. In general, each process per-

---

[2] See: Chris Gane and Trish Sarson, *Structured Systems Analysis: Tools and Techniques,* Englewood Cliffs, NJ: Prentice Hall, 1979; Tom DeMarco, *Structured Analysis and System Specification,* Englewood Cliffs, NJ: Prentice Hall, 1979; and E. Yourdan and Larry L. Constantine, *Structured Design: Fundamentals of a Discipline of Computer Program and Systems Design,* Englewood Cliffs, NJ: Prentice Hall, 1979.

Data Flow Diagram Element	Typical Computer-Aided Software Engineering Fields	Gane and Sarson Symbol	DeMarco and Yourdan Symbol

Every *process* has
  A number
  A name (verb phase)
  A description
  One or more output data flows
  Usually one or more input data flows
	Label (name)
	Type (process)
	Description
	(what is it)
	Process number
	Process description
	(Structured English)
	Notes
		[symbol: rounded box "1 Name"]	[symbol: oval "Name"]

Every *data flow* has
  A name (a noun)
  A description
  One or more connections to a process
	Label (name)
	Type (flow)
	Description
	Alias (another name)
	Composition
	(description of data elements)
	Notes
		Name →	Name →

Every *data store* has
  A number
  A name (a noun)
  A description
  One or more input data flows
  Usually one or more output data flows
	Label (name)
	Type (store)
	Description
	Alias (another name)
	Composition
	(description of data elements)
	Notes
		[symbol: D1 Name]	[symbol: D1 Name]

Every *external entity* has
  A name (a noun)
  A description
	Label (name)
	Type (entity)
	Description
	Alias (another name)
	Entity description
	Notes
		[symbol: box "Name"]	[symbol: box "Name"]

FIGURE 6-2
Data Flow Diagram Elements

forms only one activity, so most system analysts avoid using the word and in-process names because it suggests that the process performs several activities.

Figure 6-2 shows the basic elements of a process and how they are usually named in CASE tools. Every process has a unique identification number, a name, and a description, all of which are noted in the CASE repository. Descriptions clearly and precisely describe the steps and details of the processes; ultimately, they are used to guide the programmers who need to computerize the processes (or the writers of policy manuals for noncomputerized processes). The process descriptions become more detailed as information is learned about the process through the analysis phase. Many process descriptions are written as simple text statements about what happens. More complex processes use more formal techniques such as structured English, decision tables, or decision trees which are discussed in a later section.

Data Flow A *data flow* is a single piece of data (e.g., patient name) (sometimes called a data element), or a logical collection of several pieces of information (e.g., patient information). Every data flow has a descriptive name that is a noun, and a description. Typically, the description of a data flow will list exactly what data elements the flow contains. For example, the patient information data flow can list the patient name, address, and phone number as its data elements.

Data flows are the glue that holds the processes together. One end of every data flow will always come from or go to a process, with the arrow showing the direction into or out of the process. Data flows show what inputs go into each process and what outputs each process produces. Every process must create at least one output data flow because if there is no output, the process does not do anything. Likewise, each process usually has at least one input data flow because it is difficult if not impossible to produce an output with no input.

Data Store   A *data store* is a collection of data that is stored in some way (which is determined later when creating the physical model). As with processes, every data store has a descriptive name (a noun), an identification number, and a description. Data stores form the starting point for the data model (discussed in the next chapter) and are the principal link between the process model and the data model.

Data flows coming out of a data store indicate that information is retrieved from the data store, and data flows going into a data store indicate that information is added to the data store or that information in the data store is changed. Whenever a process updates a data store (e.g., by retrieving a record from a data store, changing it, and storing it back), we document both the data coming from the data store and the data written back into the data store.

All data stores must have at least one input data flow (or else they never contain any data), unless they are created and maintained by another information system or on another page of the DFD. Likewise, they usually have at least one output data flow. (Why store data if you never use it?) In cases in which the same process both stores data and retrieves data from a data store, there is a temptation to draw one data flow with an arrow on both ends. However, for clarity, it is better practice to draw two separate data flows.

External Entity   An *external entity* is a person, organization, or system that is external to the system but interacts with it (e.g., patient, doctor, government organization, accounting system). Every external entity has a name and a description. The key point to remember about an external entity is that it is external to the system but may or may not be part of the organization.

A common mistake is to include people who are part of the system as external entities. The people who execute a process are part of the process and are not external to the system (e.g., data-entry clerks, order takers). The person who performs a process is often described in the process description but never on the DFD itself. However, people who use the information from the system to perform other processes or who decide what information goes into the system are documented as external entities (e.g., managers, staff).

Using Data Flow Diagrams to Define Business Processes

Most business processes are too complex to be explained in one DFD. Most process models are therefore composed of a set of DFDs. The first DFD provides a summary of the overall system, with additional DFDs providing more and more detail about each part of the overall business process. Thus, one important principle in process modeling with DFDs is the decomposition of the business process into a series of DFDs, each representing a lower level of detail. Figure 6-3 shows how one business process can be decomposed into several levels of DFDs.

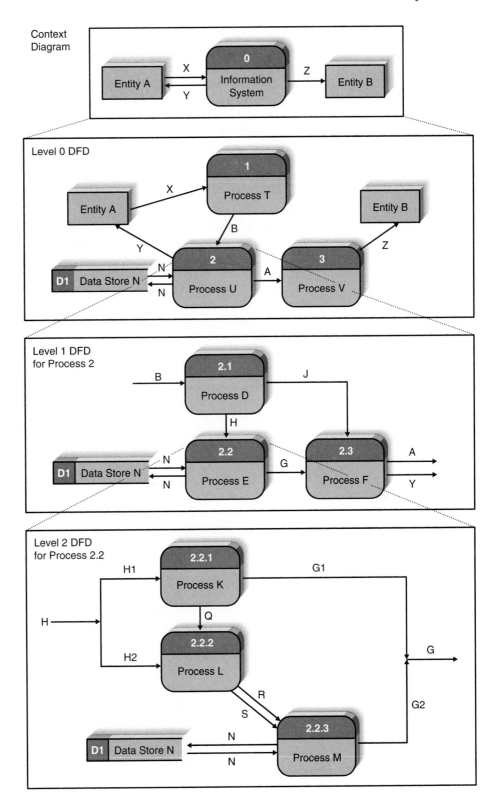

FIGURE 6-3
Relationships among Levels of Data Flow Diagrams (DFDs)

Context Diagram The first DFD in every business process, whether a manual system or a computerized system, is the *context diagram* (see Figure 6-3). As the name suggests, the context diagram shows the context into which the business process fits. All process models have one context diagram.

The context diagram shows the overall business process as just one process (i.e., the system itself) and shows the data flows to and from external entities. Data stores are not usually included on the context diagram, unless they are "owned" by systems or processes other than the one being documented. For example, an information system used by the university library that records who has borrowed books would likely check the registrar's student information database to see if a student is currently registered at the university. In this context diagram, the registrar's student information data store likely would be shown on the context diagram because it is external to the library system, but used by it. Many organizations would show this not as a data store but as an external entity called "Library Information System."

Level 0 Diagram The next DFD is called the *level 0 diagram* or *level 0 DFD* (see Figure 6-3). The level 0 diagram shows all the processes at the first level of numbering (i.e., processes numbered 1 through 9), the data stores, external entities, and data flows among them. The purpose of the level 0 DFD is to show all the major high-level processes of the system and how they are interrelated. All process models have one and only one level 0 DFD.

A second key principle in creating sets of DFDs is *balancing*. Balancing means ensuring that all information presented in a DFD at one level is accurately represented in the next level DFD. This doesn't mean that the information is identical, but that it is shown appropriately. There is a subtle difference in meaning between these two words that will become apparent shortly, but for the moment, let's compare the context diagram and the level 0 DFD in Figure 6-3 to see how the two are balanced. In this case, we see that the external entities (A, B) are identical between the two diagrams and the data flows to and from the external entities in the context diagram (X, Y, Z) also appear in the level 0 DFD. The level 0 DFD replaces the context diagram's single process (always numbered 0) with three processes (1, 2, 3), adds a data store (D1), and includes two additional data flows that were not in the context diagram (data flow B from process 1 to process 2; data flow A from process 2 to process 3).

These three processes and two data flows are contained within process 0. They were not shown on the context diagram because they are the internal components of process 0. The context diagram deliberately hides some of the system's complexity to make it easier for the reader to understand. Only after the reader understands the context diagram does the analyst "open up" process 0 to show its internal operations by decomposing the context diagram into the level 0 DFD, which shows more detail about the processes and data flows inside the system.

Level 1 Diagrams In the same way that the context diagram deliberately hides some of the system's complexity, so, too, does the level 0 DFD. The level 0 DFD shows only how the major high-level processes in the system interact. Each process on the level 0 DFD can be decomposed into a more explicit DFD, called a *level 1 diagram,* or *level 1 DFD* which shows how it operates in greater detail. The DFD illustrated in Figure 6-1 is a level 1 DFD.

In general, all process models have as many level 1 diagrams as there are processes on the level 0 diagram; every process in the level 0 DFD would be *decomposed* into its own level 1 DFD, so the level 0 DFD in Figure 6-3 would have three level 1 DFDs (one for process 1, one for process 2, one for process 3). For simplicity, we have chosen to show only one level 1 DFD in this figure, the DFD for process 2. The processes in level 1 DFDs are numbered on the basis of the process being decomposed. In this example, we are decomposing process 2, so the processes in this level 1 DFD are numbered 2.1, 2.2, and 2.3.

Processes 2.1, 2.2, and 2.3 are the *children* of process 2, and process 2 is the *parent* of processes 2.1, 2.2, and 2.3. These three children processes wholly and completely make up process 2. The set of children and the parent are identical; they are simply different ways of looking at the same thing. When a parent process is decomposed into children, its children must completely perform all of its functions, in the same way that cutting up a pie produces a set of slices that wholly and completely make up the pie. Even though the slices may not be the same size, the set of slices is identical to the entire pie; nothing is omitted by slicing the pie.

Once again, it is very important to ensure that the level 0 and level 1 DFDs are balanced. The level 0 DFD shows that process 2 accesses data store D1, has one input data flow (B), and has two output data flows (A and Y). A check of the level 1 DFD shows the same data store and data flows. Once again, we see that three new data flows have been added (G,H,J) at this level. These data flows are contained within process 2 and therefore are not documented in the level 0 DFD. Only when we decompose or open up process 2 via the level 1 DFD do we see that they exist.

The level 1 DFD shows more precisely which process uses the input data flow B (process 2.1) and which produces the output data flows A and Y (process 2.3). Note however, that the level 1 DFD does not show where these data flows come from or go to. To find the source of data flow B, for example, we have to move up to the level 0 DFD, which shows data flow B coming from external entity B. Likewise, if we follow the data flow from A up to the level 0 DFD, we see it goes to process 3, but we still do not know exactly which process within process 3 uses it (e.g., process 3.1, 3.2). To determine the exact source, we would have to examine the level 1 DFD for process 3.

This example shows one downside to the decomposition of DFDs across multiple pages. To find the exact source and destination of data flows, one often must follow the data flow across several DFDs on different pages. Several alternatives to this approach to decomposing DFDs have been proposed, but none are as commonly used as the "traditional" approach. The most common alternative is to show the source and destination of data flows to and from external entities (as well as data stores) at the lower level DFDs. Since most data flows are to or from data stores and external entities, rather than processes on other DFD pages, this can significantly simplify the reading of multiple page DFDs. We believe this to be a better approach, so when we teach our courses, we show external entities on all DFDs, including level 1 DFDs and below.

Level 2 Diagrams   The bottom of Figure 6-3 shows the next level of decomposition, a *level 2 diagram* or *level 2 DFD* for process 2.2. This DFD shows that process 2.2 is decomposed into three processes (2.2.1, 2.2.2, and 2.2.3). The level 1 diagram for process 2.2 shows interactions with data store D1, which we see in the level 2 DFD as occurring in process 2.2.3. Likewise, the level 2 DFD for 2.2 shows one input data flow (H) and one output data flow (G), which we also see on the level 2 diagram, along with several new data flows (Q,R,S,H1,H2,G1,G2). The two DFDs are therefore balanced.

It is sometimes difficult to remember which DFD level is which. It may help to remember that the level numbers refer to the number of decimal points in the process numbers on the DFD. A level 0 DFD has process numbers with no decimal points (e.g., 1, 2), whereas a level 1 DFD has process numbers with one decimal point (e.g., 2.3, 5.1), a level 2 DFD has numbers with two decimal points (e.g., 1.2.5, 3.3.2), and so on.

Data Flow Splits and Joins   Data flows H1 and H2 in the level 2 DFD in Figure 6-3 illustrate a *split* of data flow H in which it is broken into its component parts. Some data flows are actually made up of many different data elements, such as patient name, patient address, appointment time, and doctor. The split in this figure could be used to document that part of the data (e.g., patient name and patient address) is used by process 2.2.1 and another part (e.g., doctor and appointment time) is used by process 2.2.3. Unlike the decomposition of processes, there is no requirement that the data flows in the split are mutually exclusive or include all of the parent data flow. So, for example, patient name and patient address could split off to one process, while patient name could split off to another. Or, as in Figure 6-1 the same information could be used in two or more processes, and the split is the simplest way to draw the data flow.

The reason for the split is because in the higher-level DFDs we did not care about the exact details of what components of data flow were used where. It was sufficient to simply say "H" (or "Patient Information") at the higher level. As we move to lower levels, however, we need to become more precise about the data flows in the same way we become more precise about the processes.

Data flow G in this same DFD illustrates a *join*. In this case, separate parts of data flow G (G1 and G2) join together to form the data flow. In our doctor's office example, G1 might be the patient information (name, address) and G2 might be the information about the appointment (e.g., appointment time, doctor). These two data flows are produced by different processes at the lowest-level DFD but are shown as one "Appointment Information" data flow in higher-level DFDs.

Alternative Data Flows   Suppose that a process produces two different data flows under different circumstances. For example, a quality-control process could produce a quality-approved widget or a defective widget, or our credit card authorization request could produce an "approved" or "rejected" result. How do we show these alternative paths in the DFD? The answer is that we show both data flows and use the process description to explain that they are alternatives. Nothing on the DFD itself shows that the data flows are mutually exclusive. For example, process 2.1 on the level 1 DFD produces two output data flows (H, J). Without reading the text description of process 2.1, we do not know if these are produced simultaneously or whether they are mutually exclusive.

Process Descriptions

The purpose of the process descriptions are to explain what the process does and provide additional information that the DFD does not provide. As we move through the SDLC, we gradually move from the general text descriptions of requirements into more and more precise descriptions that are eventually translated into very precise program languages. In most case, a process is straightforward enough that the requirements definition, a use case, and a DFD with a simple text description provides sufficient detail to support the activities in the Design Phase. Sometimes, however, the process is sufficiently complex that it can benefit from a more detailed process description that explains the logic that occurs inside the process. Three techniques are commonly used to describe more complex processing logic: structured English, decision trees, and decision tables. Very complex processes may use a combination of structured English and either decision trees or decision tables.

Structured English *Structured English* is simply a formal way of writing instructions that describe the steps of a process. Because structured English is the first step toward the program (or policy manual) that will ultimately define how the process is performed, it looks much like a simple programming language. Structured English uses short sentences that clearly describe exactly what work is performed on what data. There are many versions of structured English because there are no formal standards; each organization has its own type of structured English.

Figure 6-4 shows some examples of commonly used structured English statements. *Action statements* are simple statements that perform some action. *If statements* control actions that are performed under different conditions, while *for statements* (or *while statements*) perform some actions until a condition is reached. Finally, a *case statement* is an advanced form of an if statement that has several mutually exclusive branches.

For example, the use case for making appointments discussed in Figure 5-1 that was used to create the DFD shown in Figure 6-1 discusses what to do in the "Check Patient Status" process if the patient is a new patient or if the patient has unpaid bills. The DFD by itself does not explain this. Therefore, the process description for process 1.1 in the DFD must provide the additional information needed. Figure 6-5 shows how this might be expressed in structured English.

Common Statements	Example
Action statement	Profits = revenues − expenses
	Generate inventory report
	Add product record to product data store
If statement	*If* customer not in customer data store
	Then add customer record to customer data store
	Else
	Add current sale to customer's total sales
	Update customer record in customer data store
For statement	*For* all customers in customer data store, *do*
	Generate a new line in the customer report
	Add customer's total sales to report total
Case statement	*Case*
	If income < 10,000, marginal tax rate = 10%
	If income < 20,000, marginal tax rate = 20%
	If income < 30,000, marginal tax rate = 31%
	If income < 40,000, marginal tax rate = 35%
	Else marginal tax rate = 38%
	Endcase

FIGURE 6-4
Structured English

1.1 Check Patient Status

Process Description

The purpose of this process is to ensure that we do not make appointments for patients unless their status indicates that we can. The process is as follows:

IF patient is a new patient

THEN Perform Add New Patient Process

ELSE

    IF patient has unpaid bills

    THEN transfer the patient to the business office

FIGURE 6-5
An Example of Structured English

Decision Trees *Decision trees* are a graphical way of describing the logic contained in the structured English if statements. Decision trees are only helpful when there is a large number of nested if statements; that is, if statements within if statements within if statements. Figure 6-6 shows a simple decision tree for the first step in the make appointment process. Each circle is a decision point that presents a question (an If statement). The decision point has a set of outcomes that branch off in different directions that lead to other decision points. In Figure 6-6, we have a series of yes/no questions that have two branches each, but we can build decision trees that have more than two branches at each decision point by using questions other than simple yes/no questions (e.g., different income amounts that lead to different tax rates), in which case the decision points are similar to case statements in structured English.

Decision trees are visual. Some people find it easier to use decision trees while others prefer structured English statements. Generally speaking, most people find decision trees simpler than structured English if we have three to four decision points in a process. If we have more than four decision points, decision trees can become difficult to draw on a single sheet of paper, so for more complex logic, we use decision tables.

Decision Tables *Decision tables* are used for very complex processes that have a multiple decision rules. Figure 6-7a shows a complete decision table for the same process. The top left part of the decision table shows the conditions to be tested

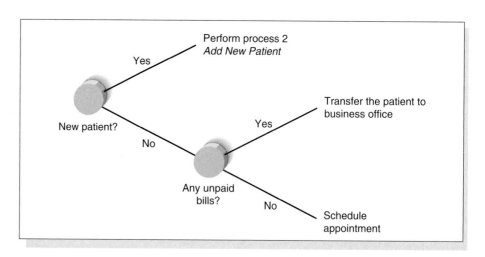

FIGURE 6-6
A Decision Tree for Checking the Patient's Status

Conditions	Rule 1	Rule 2	Rule 3	Rule 4
C1: New patient	yes	no	yes	no
C2: Unpaid bills	no	no	yes	yes
Actions				
A1: Perform process 2 *Add New Patient*	X			
A2: Schedule appointment		X		
A3: Refer patient to Business Office				X

(a)

Conditions	Rule 1	Rule 2	Rule 3
C1: New patient	yes	no	no
C2: Unpaid bills		no	no
Actions			
A1: Perform process 2 *Add New Patient*	X		
A2: Schedule appointment		X	
A3: Transfer patient to Business Office			X

(b)

FIGURE 6-7
(a) Complete Decision Table; (b) Reduced Decision Table

(new patient, and unpaid bills), while the bottom left shows the actions (perform process 2, schedule appointment, and refer patient to Business Office). The conditions and actions are linked by the rules. The top right part of the complete decision table lists the rules and all possible values that can occur when the conditions are tested (yes or no). The bottom right part of the table shows which actions are performed for each rule (an "X" means the action is performed, while a blank indicates that it is not performed). Rule 1, for example, states that when condition 1 is a "yes" (a new patient) and condition 2 is a "no" (no unpaid bills) then action A1 is performed (perform process 2).

A close look at the complete decision table will show that rule 3 is never used, for the simple reason that we can never have a new patient with unpaid bills because we haven't see the patient before. The second condition is not needed in order to decide whether to perform action A1. Figure 6-7b shows the reduced decision table that rearranges the rules to condense the table and explicitly show that condition C2 is not tested when condition C1 is a yes (as indicated by the blank beside C2 in the Rule 1 column).

Most people find decision tables are a bit more difficult to understand than decision trees or structured English, but they are a much more concise way of explaining a process's logic when the logic is very complex. Decision tables are best used when there are five or more separate conditions that need to be considered.

CREATING DATA FLOW DIAGRAMS

Data flow diagrams start with the information in the use cases and the requirements definition. Although the use cases are created by the users and project team working together, the DFDs are usually created by the project team and then reviewed by the users. Generally speaking, the set of DFDs that make up the process model simply integrate the individual use cases (and add in any processes in the requirements definition not selected as use cases). The project team takes the use cases and rewrites them as DFDs. However, because DFDs have formal rules about symbols and syntax that use cases do not, the project team sometimes has to revise some of the information in the use cases to make them conform to the DFD rules. The most common types of changes are to the names of the use cases that become processes and the inputs and outputs that become data flows. The second most common type of change is to combine several small inputs and outputs in the use cases into larger data flows in the DFDs (e.g., combining three separate inputs, such as "customer name," "customer address," and "customer phone number," into one data flow, such as "customer information").

Project teams usually use process modeling tools or CASE tools to draw process models. Simple tools such as Visio, are just fancy drawing tools that work like PowerPoint, in that they do not understand the syntax or the meaning of DFD elements. Other process modeling tools such as BPWin understand the DFD and can perform simple syntax checking to make sure that the DFD is at least somewhat correct. A full CASE tool, such as Visible Analyst Workbench, provides many capabilities in addition to process modeling (e.g., data modeling as discussed in the next chapter). CASE tools tend to be complex and while they are valuable for large and complex projects, they often cost more than they add for simple projects. Figure 6-8 shows a sample screen from the Visible Analyst CASE tool.

Building a process model that has many levels of DFDs usually entails five steps. First, the team builds the context diagram that shows all the external entities and the data flows into and out of the system from them. Second, the team creates a DFD fragment for each use case that shows how the use case exchanges data flows with the external entities and data stores. Third, these DFD fragments are organized into a level 0 DFD. Fourth, the team develops level 1 DFDs, based on the steps within each use case, to better explain how they operate. In some cases, these level 1 DFDs are further decomposed into level 2 DFDs, level 3 DFDs, level 4 DFDs, and so on. Fifth, the team validates the set of DFDs to make sure they are complete and correct.

Creating the Context Diagram

The context diagram defines how the business process or computer system interacts with its environment, primarily the external entities. To create the context diagram, you simply draw one process symbol for the business process or system being modeled (numbered 0 and named for the process). You read through the use cases and add the inputs and outputs listed at the top of the form, as well as their sources and destinations. Usually, all the inputs and outputs will come from or go to external entities such as a person, organization, or other information system, but sometimes they will connect directly to data stores in an external system. Such an external data store can be listed as the data store it is, or simply as an external entity that is the name of the system that owns the data store. None of the data stores inside the process/system that are created by the process or system itself are included in the context diagram because they are "inside" the system. Because there are sometimes so many inputs and outputs, we often combine several small data flows into larger data flows.

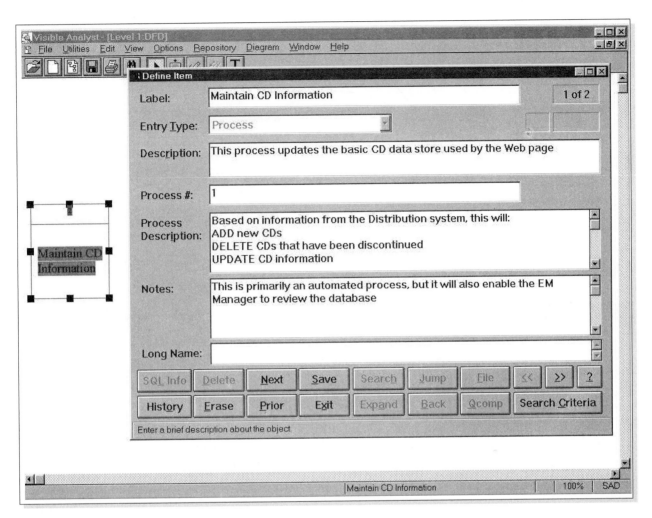

FIGURE 6-8
Entering Data Flow Diagram Processes in a Computer-Aided Software Engineering Tool

Figure 6-9 shows a context diagram that describes a patient information system, from which the level 1 DFD for the appointment process in Figure 6-1 was taken. Take a moment to compare these two figures. You should see that both diagrams have the same external entity: the patient. All of the inputs and outputs on the level 1 DFD are also present in the context diagram, all although "Patient Name" in the level 1 DFD has been stated as the higher level data flow of "Patient Information," which would include patient name as well as host of other information such as medical history and insurance information. Several other external entities are shown (e.g., doctor, insurance company) as are other data flows (e.g., bill, payment information) because other use cases have them.

Creating Data Flow Diagram Fragments

A *DFD fragment* is one part of a DFD that will eventually be combined with other DFD fragments to form a DFD diagram. In this step, each use case is converted into

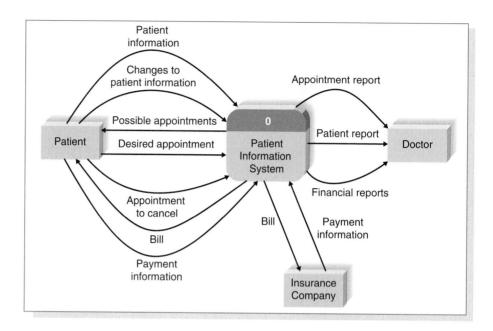

FIGURE 6-9
Context Diagram for a Doctor's Office
Patient Information System

one DFD fragment. You start by taking each use case and drawing a DFD fragment using the information given on the top of the use case: the name, ID number, and major inputs and outputs. The information about the major steps that make up each use case is ignored at this point (it will be used in a later step). Figure 6-10 shows a use case and the DFD fragment that was created from it.

Once again, some subtle but important changes are often made in converting the use case into a DFD. The two most common changes are modifications to the process names and the addition of data flows. There were no formal rules for use case names, but there are formal rules for naming processes on the DFD. All process names must be a verb phase—they must start with a verb and include a noun (see Figure 6-2). Not all of our use case names are structured in this way, so we sometimes need to change them. It is also important to have a consistent *viewpoint* when naming processes. For example, the DFD in Figure 6-10 is written from the viewpoint of the doctor's office, not of the patient. All the process names and descriptions are written as activities that the staff performs. It is traditional to design the processes from the viewpoint of the organization running the system, so this sometimes requires some additional changes in name.

The second common change is the addition of data flows. Use cases are written to describe how the system interacts with the user. Typically they do not describe how the system obtains data, so the use case often omits data flows read from a data store. When creating DFD fragments, it is important to make sure that any information given to the user is obtained from a data store. The easiest way to do this is to first create the DFD fragment using the major inputs and outputs listed at the top of the use case and then verify that all outputs have sufficient inputs to create them.

There are no formal rules covering the *layout* of processes, data flows, data stores, and external entities within a DFD. They can be placed anywhere you like on the page; however, because we in Western cultures tend to read top to bottom and left to right, most systems analysts try to put the process in the middle of the

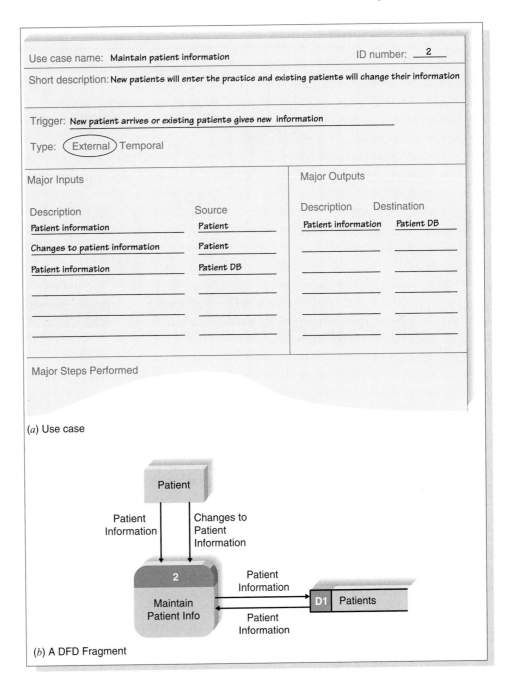

Use case name: Maintain patient information   ID number: __2__

Short description: New patients will enter the practice and existing patients will change their information

Trigger: New patient arrives or existing patients gives new information

Type: (External) Temporal

Major Inputs

Description	Source
Patient information	Patient
Changes to patient information	Patient
Patient information	Patient DB

Major Outputs

Description	Destination
Patient information	Patient DB

Major Steps Performed

(a) Use case

(b) A DFD Fragment

FIGURE 6-10
A DFD For a Doctor's Office

DFD fragment, with the major inputs starting from the left side or top entering the process and outputs leaving from the right or the bottom. Data stores are often written below the process.

Take a moment and draw a DFD fragment for the use case in Figure 5-1 in the previous chapter. Figure 6-11 shows one possible way in which the fragment could be drawn. (Don't look at it until you've drawn your own). There are many other good ways to draw this, so yours might be different.

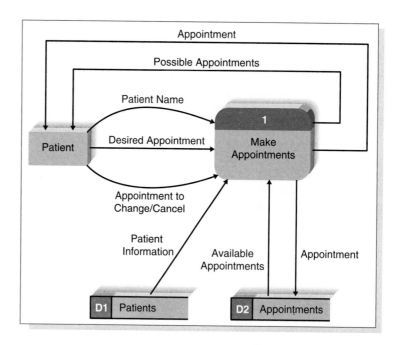

FIGURE 6-11
A DFD Fragment for a Doctor's Office

Creating the Level 0 Data Flow Diagram

Once you have the set of DFD fragments (one for each of the major use cases), you simply combine them into one DFD drawing that becomes the level 0 DFD. As mentioned earlier, there are no format layout rules for DFDs. However, most systems analysts try to put the process that is first chronologically in the upper left corner of the diagram and work their way from top to bottom, left to right (e.g., Figure 6-1). Generally speaking, most analysts try to reduce the number of times that data flow lines cross or to ensure that when they do cross, they cross at right angles so there is less confusion (many give one line a little "hump" to imply that one data flow jumps over the other without touching it). Minimizing the number of data flows that cross is challenging.

Iteration is the cornerstone of good DFD design. Even experienced analysts seldom draw a DFD perfectly the first time. In most cases, they draw it once to understand the pattern of processes, data flows, data stores, and external entities and then draw it a second time on a fresh sheet of paper (or in fresh file) to make it easier to understand and to reduce the number of data flows that cross. Often, a DFD is drawn many times before it is finished.

Figure 6-12 combines the DFD fragments in Figures 6-9 and 6-10 with DFD fragments from two other use cases (process 3, perform billing, and process 4, prepare management reports). Take a moment to examine Figure 6-12 and find the DFD fragments from Figures 6-10 and 6-11 contained within it.

Creating Level 1 Data Flow Diagrams (and Below)

The team now begins to create lower-level DFDs for each process in the level 0 DFD that needs a level 1 DFD. Each one of the use cases is turned into its own DFD. The process for creating the level 1 DFDs is to take the steps as written on the use cases and convert them into a DFD in much the same way as for the level 0

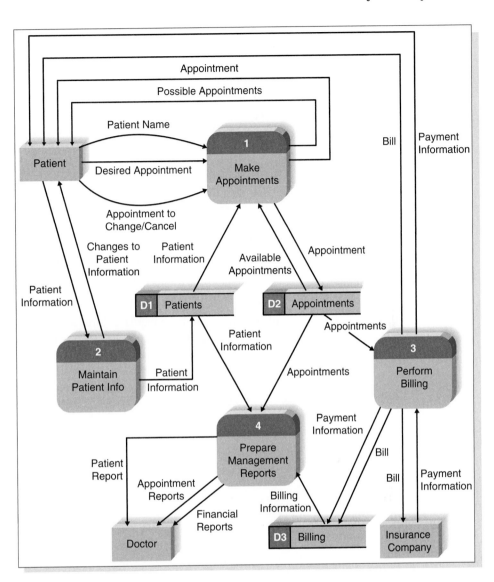

FIGURE 6-12
A Level 0 DFD for a Doctor's Office System

DFD. Usually, each step in the use case becomes a process on the level 1 DFD, with the inputs and outputs becoming the input and output data flows, although once again, sometimes subtle changes are required to go from the informal descriptions in the use case to the more formal process model, such as adding input data flows that were not included in the use case. And because the analysts are now starting to think more deeply about how the processes will be supported by an information system, they sometimes slightly change the use case steps to make the process easier to use.

With the traditional approach to creating DFDs, no source and destination is given on the level 1 DFD (and lower) for the inputs that come from and go to external entities (or other processes outside of this process). But, the source and destination of data flows for data stores and data flows that go to processes within this DFD, are included (i.e., from one step to another in the same use case, such "Patient Information" from step 1.1 to step 1.2 in Figure 6-1). The reason for this is because the information is redundant; you can see the destination of data flows by reading the level 0 DFD.

The problem with this approach is that in order to really understand the level 1 DFD, you must refer back to the level 0 DFD. For small systems that only have one or two level 1 DFDs, this is not a major problem. But, for large systems that have many levels of DFDs, the problem grows; in order to understand the destination of a data flow on a level 3 DFD, you have to read the level 2 DFD, the level 1 DFD, and the level 0 DFD—and if the destination is to another activity, then you have to trace down in the lower-level DFDs in the other process.

We believe that including external entities in level 1 and lower DFDs dramatically simplifies the readability of DFDs, with very little downside. In our work in several dozen projects with the U.S. Department of Defense, several other federal agencies, and the military of two other countries, we came to understand the value of this approach and converted them to our viewpoint. We acknowledge that this is nontraditional, but encourage you to be nontraditional, too. Because DFDs are not standardized, each organization uses them slightly differently. So, the ultimate decision of whether or not to include external entities on level 1 DFDs is yours—or your instructor's! In this book, we will include them.

Ideally, we try to keep the data stores in the same general position on the page in the level 1 DFD as they were in the level 0 DFD, but this is not always possible. We try to draw input data flows arriving from the left edge of the page and output data flows leaving from the right edge. For example, see the level 1 DFD in Figure 6-3.

One important issue is how to draw small data flows that were *bundled* into larger data flows at the level 0 DFD. For example, the level 0 might show an input data flow of "customer information," whereas the individual steps in the use case (that become the processes on the level 1 DFD) use parts of the data flow (e.g., "customer name" in one place and "credit card" in another). This is done using splits (for input data flows) and joins (for output data flows). The level 1 DFD in Figure 6-3 shows "H" as an input to process 2.2, whereas the level 2 DFD shows "H" entering the DFD but being split into two parts ("H1," used by process 2.2.1, and "H2," used by process 2.2.2). This same figure also shows joins: "G" at the level 1 DFD being a product of a join of "G1" and "G2" in the level 2 DFD.

One of the most challenging design questions is knowing when to decompose a level 1 DFD into lower levels. The decomposition of DFDs can be taken to almost any level, so for example, we could decompose process 1.2 on the level 1 DFD into processes 1.2.1, 1.2.2, 1.2.3, and so on in the level 2 DFD. This can be repeated to any level of detail, so one could have level 4 or even level 5 DFDs.

There is no simple answer to the "ideal" level of decomposition, because it depends on the complexity of the system or business process being modeled. In general, you decompose a process into a lower-level DFD whenever that process is sufficiently complex that additional decomposition can help explain the process. Most experts believe that there should be at least three processes and no more than seven to nine processes on every DFD, so if you begin to decompose a process and end up with only two processes on the lower-level DFD, you probably don't need to decompose it. There seems little point in decomposing a process and creating another lower-level DFD for only two processes; you are better off simply showing two processes on the original higher level DFD. Likewise, a DFD with more than nine processes becomes difficult for users to read and understand because it is very complex and crowded. Some of these processes should be combined and explained on a lower-level DFD.

CONCEPTS 6-A U.S. ARMY AND MARINE CORPS BATTLEFIELD LOGISTICS

IN ACTION

Shortly after the Gulf War in 1991 (Desert Storm), the U.S. Department of Defense realized that there were significant problems in its battlefield logistics systems that provided supplies to the troops at the division level and below. During the Gulf War, it had proved difficult for army and marine units fighting together to share supplies back and forth because their logistics computer systems would not easily communicate. The goal of the new system was to combine the army and marine corps logistics systems into one system to enable units to share supplies under battlefield conditions.

The army and marines built separate as-is process models of their existing logistics systems that had 165 processes for the army system and 76 processes for the marines. Both process models were developed over a 3-month time period and cost several million dollars to build, even though they were not intended to be comprehensive.

I helped them develop a model for the new integrated battlefield logistics system that would be used by both services (i.e., the to-be model). The initial process model contained 1,500 processes and went down to level 6 DFDs in many places. It took 3,300 pages to print. They realized that this model was too large to be useful. The project leader decided that level 4 DFDs was as far as the model would go, with additional information contained in the process descriptions. This reduced the model to 375 processes (800 pages) and made it far more useful. *Alan Dennis*

QUESTIONS:
1. What are the advantages and disadvantages to setting a limit for the maximum depth for a DFD?
2. Is a level 4 DFD an appropriate limit?

The process model is more likely to be drawn to the lowest level of detail for a to-be model if a structured design development process is used (i.e., not rapid application development [RAD]; see Chapter 1) or if the system will be built by an external contractor. Without the complete level of detail, it may be hard to specify in a contract exactly what the system should do. If a RAD approach, which involves a lot of interaction with the users and quite often prototypes, is being used, we would be less likely to go to as low a level of detail, because the design will evolve through interaction with the users. In our experience, most systems go to only level 2 at most.

There is no requirement that all parts of the system must be decomposed to the same level of DFDs. Some parts of the system may be very complex and require many levels, whereas other parts of the system may be simpler and require fewer.

Validating the Data Flow Diagrams

Once you have created a set of DFDs, it is important to check them for quality. Figure 6-13 provides a quick checklist for identifying the most common errors. There are two fundamentally different types of problems that can occur in DFDs: *syntax errors* and *semantics errors*. Syntax refers to structure of the DFDs and whether the DFDs follow the rules of the DFD language. Syntax errors can be thought of as grammatical errors made by the analyst when he or she creates the DFD. Semantics refers to the meaning of the DFDs and whether they accurately describe the business process being modeled. Semantics errors can be thought of as misunderstandings by the analyst in collecting, analyzing, and reporting information about the system.

Syntax	
Within DFD	
Process	• Every process has a unique name that is an action-oriented verb phase, a number, and a description
	• Every process has at least one input data flow
	• Every process has at least one output data flow
	• Output data flows usually have different names than input data flows (a common exception is that data read from a data store often has the same name as data written to it)
	• There are between 3 and 7 processes per DFD
Data Flow	• Every data flow has a unique name that is a noun, and a description
	• Every data flow connects to at least one process
	• Data flows only in one direction (no two-headed arrows)
	• A minimum number of data flow lines cross
Data Store	• Every data flow has a unique name that is a noun, and a description
	• Every data store has at least one input data flow (which means to add new data or change existing data in the data store) on some page of the DFD
	• Every data store has at least one output data flow (which means to read data from the data store) on some page of the DFD
External Entity	• Every external entity has a unique name that is a noun, and a description
	• Every external entity has at least one input or output data flow
Across DFDs	
Context diagram	• Every set of DFDs must have one context diagram
Viewpoint	• There is a consistent viewpoint for the entire set of DFDs.
Decomposition	• Every process is wholly and completely described by the processes on its children DFDs
Balance	• Every data flow, data store, and external entity on a higher level DFD is shown on the lower-level DFD that decomposes it
	• No data stores or data flows appear on lower-level DFDs that do not appear on their parent DFD
Semantics	
Appropriate Representation	• User validation
	• Role-play processes
Consistent Decomposition	• Examine lowest-level DFDs
Consistent Terminology	• Examine names carefully

FIGURE 6-13
Data Flow Diagram Quality Checklist

In general, syntax errors are easier to find and fix than are semantics errors, because there are clear rules that can be used to identify them (e.g., a process must have a name). Most CASE tools have syntax checkers that will detect errors within one page of a DFD in much the same way that word processors have spelling checkers and grammar checkers. Finding syntax errors that span several pages of a DFD (e.g., from a level 1 to a level 2 DFD) is slightly more challenging, particularly for consistent viewpoint, decomposition, and balance. Some CASE tools can detect balance errors, but that is about all. In most cases, analysts must carefully and painstakingly review every process, external entity, data flow, and data store on all DFDs by hand to make sure that they have a consistent viewpoint and that the decomposition and balance are appropriate.

In our experience, the most common syntax error that novice analysts make in creating DFDs is violating the Law of Conservation of Data.[3] The first part of the Law states that:

1. Data at rest stays at rest until moved by a process.

In other words, data cannot move without a process. Data cannot go to or come from a data store or an external entity without having a process to push it or pull it.

The second part of the Law states that:

2. Processes cannot consume or create data.

In other words, data only enters or leaves the system by way of the external entities. A process cannot destroy input data; all processes must have outputs. Drawing a process without an output is sometimes called a "black hole" error. Likewise, a process cannot create new data; it can transform data from one form to another, but it cannot produce output data without inputs. Drawing a process without an input is sometimes called a "miracle" error (because output data miraculously appears). There is one exception to the part of the Law requiring inputs, but it is so rare that most analysts never encounter it.[4] Figure 6-14 shows some common syntax errors.

Generally speaking, semantics errors cause the most problems in system development. Semantics errors are much harder to find and fix because doing so requires a good understanding of the business process. And even then, what may be identified as an error may actually be a misunderstanding by the person reviewing the model. There are three useful checks to help ensure that models are semantically correct (see Figure 6-13).

The first check to ensure the model is an appropriate representation is to ask the users to validate the model in a walk-through (i.e., the model is presented to the users and they examine it for accuracy). A more powerful technique is for the users to role-play the process from the DFDs in the same way in which they role-played the use case. The users pretend to execute the process exactly as it is described in the DFDs. They start at the first process and attempt to perform it using only the inputs specified and producing only the outputs specified. Then they move to the second process, and so on.

One of the most subtle forms of semantics error occurs when a process creates an output but has insufficient inputs to create it. For example, in order to create water (H_2O), we need to have both hydrogen (H) and oxygen (O) present. The same is true of computer systems, in that the outputs of a process can only be combinations and transformations of its inputs. Suppose for example, we want to record an order; we need the customer name and mailing address, and the quantities and prices for the CDs the customer is ordering. We need information from the customer data store (e.g., address) and information from the CD data store (e.g., price). We cannot draw a process that produces an output order data flow without inputs from these two data stores. Role-playing with strict adherence to the inputs and outputs in a model is one of the best ways to catch this type of error.

A second semantics error check is to ensure consistent decomposition, which can be tested by examining the lowest-level processes in the DFDs. In most cir-

[3] This law was developed by Prof. Dale Goodhue at the University of Georgia.

[4] The exception is a temporal process that issues a trigger output based on an internal time clock. Whenever some predetermined time period elapses, the process produces an output. The time-keeping process has no inputs because the clock is internal to the process.

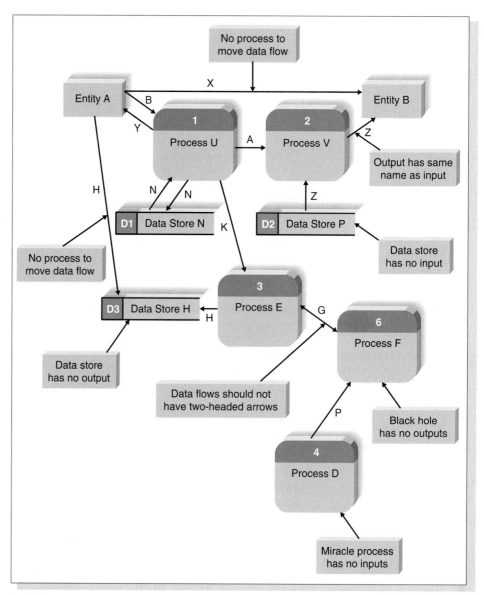

FIGURE 6-14
Some Common Errors

cumstances, all processes should be decomposed to the same level of detail—which is not the same as saying the same number of levels. For example, suppose we were modeling the process of driving to work in the morning. One level of detail would be to say the following: (1) enter car; (2) start car; (3) drive away. Another level of detail would be to say the following: (1) unlock car; (2) sit in car; (3) buckle seat belt; and so on. Still another level would be to say the following: (1) remove key from pocket; (2) insert key in door lock; (3) turn key; and so on. None of these is inherently better than another, but barring unusual circumstances, it is usually best to ensure that all processes at the very bottom of the model provide the same consistent level of detail.

Likewise, it is important to ensure that the terminology is consistent throughout the model. The same item may have different names in different parts of the organization, so one person's "sales order" may be another person's "customer order." Likewise, the same term may have different meanings; for example, "ship date" may mean one thing to the sales representative taking the order (e.g., promised date) and something else to the warehouse (e.g., the actual date shipped). Resolving these differences before the model is finalized is important in ensuring that everyone who reads the model or who uses the information system built from the model has a shared understanding.

APPLYING THE CONCEPTS AT CD SELECTIONS

Creating the Context Diagram

The project team began by creating the context diagram. They read through the top part of the three major use cases in Figure 5-6 to find the major inputs and inputs. The first use case (Take Requests for CDs) lists four inputs from the customer and two from databases. The databases are internal to the system, so they are not documented on the context diagram. This use case also has four outputs to the customer, one output to the special order database, and one to the in-store hold database. The special order database is external to the Internet system, so it is documented in the context diagram but as a system.

The in-store database is a bit more complex. It is not strictly internal to the Internet system because it will be part of the in-store system, but it is part of this system development project, so it could be consider within the scope of the "system," interpreted broadly. Because the use case for managing the in-store system will be part of the Internet system, and because the in-store hold database is only used by Internet system, the project team decided that the in-store hold database would be considered internal to the Internet system. Figure 6-15 shows the incompletes context diagram after considering only this first use case.

The team moved on to the next two use cases. Take a moment and complete the context diagram in Figure 6-15 by adding the information from the remaining two use cases. Figure 6-16 shows the final context diagram developed by the project teams.

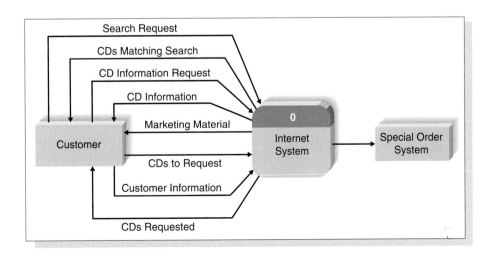

FIGURE 6-15

Incomplete Context Diagram for CD Selections Internet System

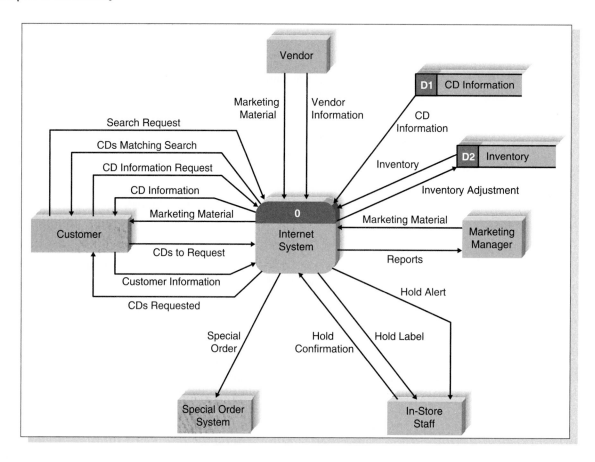

FIGURE 6-16
Context Diagram for CD Selections Internet Sales System

This context diagram was very busy, especially around the customer external entity. Some analysts might at this point try to bundle together some of these data flows into higher-level data flows. For example, the two output data flows that are responses to these requests (i.e., CDs matching search, CD information, and marketing materials) might be bundled into one data flow called "search response." Using bundles makes the context diagram simpler. However, it then requires the level 0 DFD to use these same bundles and ultimately will require some level 1 DFD or lower-level DFD to use splits and joins, as these bundles are shown as inputs to the DFD but the specific elements within them are used by the processes on the DFD. Such a use of splits and joins will increase complexity at the lower levels. This is a trade-off with no clear right or wrong answer. We have chosen not to bundle these data flows on the context diagram to avoid the complexity of splits and joins later on.

Creating Data Flow Diagram Fragments

The next step was to create one DFD fragment for each use case. This was done by drawing the process in middle of the page, making sure that process number and name were appropriate and connecting all the input and output data flows to it. Unlike the context diagram, the DFD fragment includes data flows to external entities and to both internal and external data stores.

After using the use case to draw the DFD fragment, the team realized that one input was missing. The process had an output data flow called CD information sent to the customer external entity. However, there was no input data flow providing the information. Therefore, the team added it to the DFD (and the use case).

Figure 6-17 shows the DFD fragment for the first use case. If you compare Figure 6-17 to Figure 6-16 you will see that they look almost the same, for the simple reason that the first step in creating the context diagram is the same as creating a DFD fragment, except that the DFD fragment includes internal data stores.

The DFD fragments for the other two use case are equally straightforward. Take a moment and draw them before you look at Figure 6-18. There are many good ways to draw both.

Creating the Level 0 Data Flow Diagram

The next step is to create the level 0 DFD by integrating the DFD fragments, which was anticlimactic. The team simply took the DFD fragments and drew them together on one piece of paper. Although it sometimes is challenging to arrange all the DFD fragments on one piece of paper, it was primarily a mechanical exercise (Figure 6-19). They tried to position the external entities in similar places on this DFD as they were on the context diagram in Figure 6-16 because it makes it simpler understand the two diagrams.

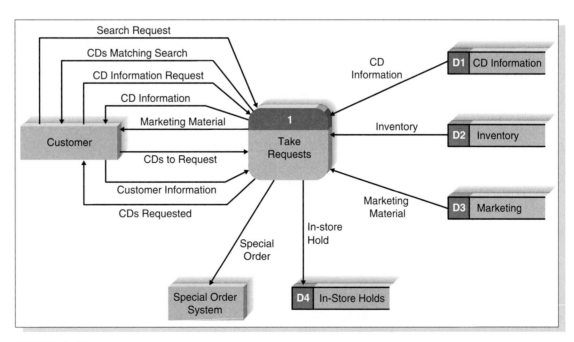

FIGURE 6-17
DFD Fragments for CD Selections

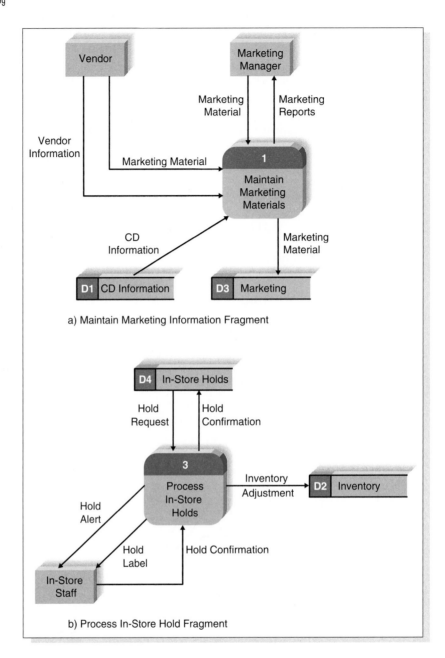

a) Maintain Marketing Information Fragment

b) Process In-Store Hold Fragment

FIGURE 6-18
The Other Two DFD Fragments for CD
Selections Internet System

Creating Level 1 Data Flow Diagrams (and Below)

The next step was to create the level 1 DFDs for those processes that could benefit
from them. The analysts started with the first use case (take requests) and started to
draw a DFD for the individual steps it contained. The first three steps in the use case
were straightforward, but as is common, the team had to choose names and numbers for the processes and to add input data flows from data stores not present in the
use case.

The last four steps in the use case required some rethinking. Take a moment
and reread the use case in Figure 5-5 in the last chapter. The last four steps include

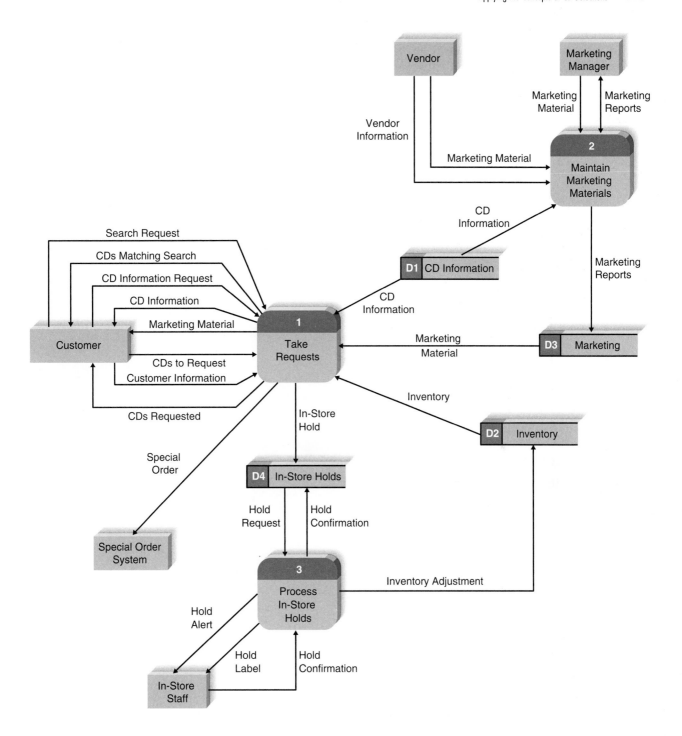

FIGURE 6-19
Level 0 DFD for CD Selections Internet System

selecting one or more CDs, checking out by giving customer information, and actually placing the special orders and in-store holds. In thinking about how this was done at other Web sites, the team decided to use a shopping cart approach. The customer would place the selected CDs in a digital shopping cart, and perhaps continue by finding other CDs, before checking out. Placing CDs in the shopping cart would be one process, and the check-out step would be a second. The check-out step would also include sending the special order(s) and in-store hold(s) rather than placing those in a separate process. Their version of the level 1 DFD for the first use case is shown in Figure 6-20.

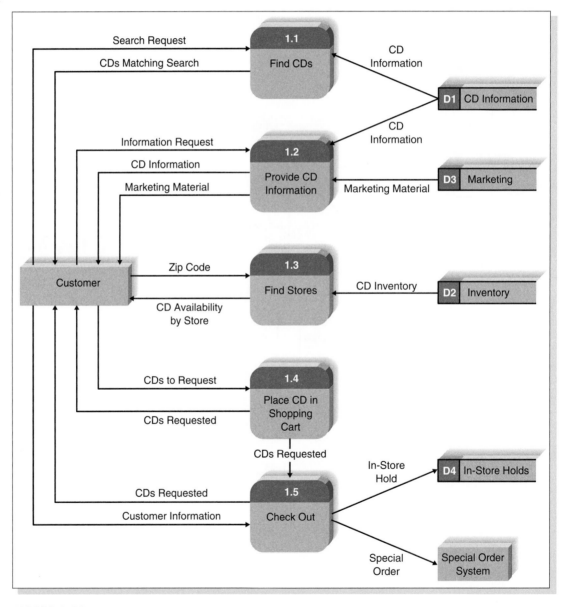

FIGURE 6-20
Level 1 DFD for CD Selections Process 1, Take Requests

6-1 CAMPUS HOUSING

Draw a context diagram, a level 0 DFD, and a set of level 1 DFDs (where needed) for the campus housing use cases that you developed for the Your Turn 5–1 box in Chapter 5.

If you take a close look at process 1.4 in Figure 6-20, you will see no explicit mention of the shopping cart as a data store—yet obviously we do need to store what CDs have been selected somewhere. It would be quite appropriate to include another data store called D5: Shopping Cart that receives an output from process 1.4 and sends input to process 1.5. This data store is a temporary data store that contains data only until the customer completes the use case by placing an order, or by abandoning the transaction. Once the customer checks out, it is deleted. By tradition, we do not include temporary data stores that do not survive a transaction on DFDs. But again, to do so is not wrong, just not common. Instead, we show process 1.4 passing data to process 1.5 via a data flow.

Although it would have been possible to decompose several of the processes on this level 1 DFD into more detail on a level 2 DFD, the project team decided not to. Instead, they made sure that the description for each of these processes was very detailed in what was expected to occur. This detail would be critical in developing the data models and designing the user interface and programs during the design phase.

Validating the Data Flow Diagrams

The final set of DFDs was validated by the project team and then by the users in a final JAD meeting. A few minor changes were identified. Once the system was defined on paper, the project team entered it into a CASE tool.

SUMMARY

Data Flow Diagram Syntax

Four symbols are used on data flow diagrams (processes, data flows, data stores, and external entities). A process is an activity that does something. Each process has a name (a verb phrase), a description, and a number that shows where it is in relation to other processes and its children processes. Every process must have at least one output and usually has at least one input. A data flow is a piece of data or an object and has a name (a noun) and a description and either starts or ends at a process (or both). A data store is a manual or computer file, and it has a number, a name (a noun), and at least one input data flow and one output data flow (unless the data store is created by a process external to the data flow diagram [DFD]). An external entity is a person or organization outside the scope of the system and has a name (a noun) and a description. Every set of DFDs starts with a context diagram and a level 0 DFD and has numerous level 1 DFDs, level 2 DFDs, and so on. Every element on the higher-level DFDs (i.e., data flows, data stores, and external entities) must appear on lower-level DFDs or else they are not balanced. Data flows can split and join at any level in a set of DFDs, but this is most common at lower levels.

Creating Data Flow Diagrams

The DFDs are created using the use cases. First, the team builds the context diagram that shows all the external entities and the data flows into and out the system from them. Second, the team creates DFD fragments for each use case that show how the use case exchanges data flows with the external entities and data stores. Third, these DFD fragments are organized into a level 0 DFD. Fourth, the team develops level 1 DFDs on the basis of the steps within each use case to better explain how they operate. Fifth, the team validates the set of DFDs to make sure they are complete and correct and contain no syntax or semantics errors. Analysts seldom create DFDs perfectly the first time, so iteration is important in ensuring that both single-page and multipage DFDs are clear and easy to read.

KEY TERMS

Action statement	Decomposition	Logical process model
Balancing	DFD fragment	Parent
Bundle	External entity	Physical model
Case statement	For statement	Process model
Children	If statement	Process
Context diagram	Iteration	Semantics error
Data flow	Join	Split
Data flow diagram (DFD)	Layout	Structured English
Data store	Level 0 DFD	Syntax error
Decision table	Level 1 DFD	Viewpoint
Decision tree	Level 2 DFD	While statement

QUESTIONS

1. How is data flow diagramming related to process modeling?
2. Explain the following terms. Use layperson's language as though you were describing them to a user: process, data flow, data store, external entity.
3. Every process must have several things. What are they and why?
4. Every data flow must be connected to at least one _____. Why?
5. What is a split? A join?
6. What are five common processes associated with data stores?
7. What is the parent process for process 3.2.1? 4.3.2.3? 1.2?
8. Describe how to create DFDs.

9. What are some guidelines for laying out a single-page DFD?
10. Why is iteration important in creating DFDs?
11. What is the viewpoint of a DFD, and why is it important?
12. What is decomposition?
13. What is balancing?
14. How are mutually exclusive data flows (i.e., alternate routes through a process) included in DFDs?
15. What are some guidelines for designing a set of DFDs?
16. How can an analyst assess model quality?
17. Explain the difference between a syntax error and a semantics error. Which is usually the most difficult to find and fix? Why?

18. Until recently, most analysts did not bother to develop use cases when working with users. During interviews or JAD sessions they usually began directly with DFDs. Why do you think the trend today is to start with use cases?

19. How can you make a use case easier to understand? (It might help if you first think about how to make one difficult to understand).

20. How can you make a DFD easier to understand? (It might help if you first think about how to make one difficult to understand.)

21. Suppose your goal is to create a set of DFDs. How would you begin an interview? How would you begin a JAD session?

22. What do you think are three common mistakes novice analysts make in creating DFDs?

EXERCISES

A. Draw a level 0 data flow diagram (DFD) for the process of buying glasses in Exercise A, Chapter 5.

B. Draw a level 0 data flow diagram (DFD) for the dentist office system in exercise B, Chapter 5.

C. Draw a level 0 data flow diagram (DFD) for the university system in exercise D, Chapter 5.

D. Draw a level 0 data flow diagram (DFD) for the real estate system in exercise E, Chapter 5.

E. Draw a level 0 data flow diagram (DFD) for the video store system in exercise F, Chapter 5.

F. Draw a level 0 data flow diagram (DFD) for the health club system in exercise G, Chapter 5.

G. Draw a level 0 data flow diagram (DFD) for the Picnics R Us system in exercise H, Chapter 5.

H. Draw a level 0 data flow diagram (DFD) for the Of-the-Month Club system in exercise I, Chapter 5.

I. Draw a level 0 data flow diagram (DFD) for the university library system in exercise J, Chapter 5.

MINICASES

1. Professional and Scientific Staff Management (PSSM) is a unique type of temporary staffing agency. Many organizations today hire highly skilled, technical employees on a short-term, temporary basis, to assist with special projects or to provide a needed technical skill. PSSM negotiates contracts with its client companies in which it agrees to provide temporary staff in specific job categories for a specified cost. For example, PSSM has a contract with an oil and gas exploration company in which it agrees to supply geologists with at least a master's degree for $5,000 per week. PSSM has contracts with a wide range of companies and can place almost any type of professional or scientific staff members, from computer programmers to geologists to astrophysicists.

When a PSSM client company determines that it will need a temporary professional or scientific employee, it issues a staffing request against the contract it had previously negotiated with PSSM. When a staffing request is received by PSSM's contract manager, the contract number referenced on the staffing request is entered into the contract database. Using information from the database, the contract manager reviews the terms and conditions of the contract and determines whether the staffing request is valid. The staffing request is valid if the contract has not expired, the type of professional or scientific employee requested is listed on the original contract, and the requested fee falls within the negotiated fee range. If the staffing request is not valid, the contract manager sends the staffing request back to the client with a letter stating why the staffing request cannot be filled, and a copy of the letter is filed. If the staffing request is valid, the contract manager enters the staffing request into the staffing request database as an outstanding staffing request. The staffing request is then sent to the PSSM placement department.

In the Placement Department, the type of staff member, experience, and qualifications requested on the staffing request are checked against the database of available professional and scientific staff. If a qualified individual is found, he or she is marked "reserved" in the staff database. If a qualified individual cannot be found in the database, or is not immediately available, the Placement Department creates a memo that explains the inability to meet the staffing request and attaches it to the staffing request. All staffing requests are then sent to the Arrangements Department.

In the Arrangement Department the prospective temporary employee is contacted and asked to agree to the placement. After the placement details have been worked out and agreed to, the staff member is marked "placed" in the staff database. A copy of the staffing request and a bill for the placement fee is sent to the client. Finally, the staffing request, the "unable to fill" memo (if any), and a copy of the placement fee bill is sent to the contract manager. If the staffing request was filled, the contract manager closes the open staffing request in the staffing request database. If the staffing request could not be filled the client is notified. The

staffing request, placement fee bill, and "unable to fill" memo are then filed in the contract office.

a. Develop a use case for each of the major processes just described.
b. Create the context diagram for the system just described.
c. Create the DFD fragments for each of the four use cases outlined in part a, and then combine them into the level 0 DFD.
d. Create a level 1 DFD for the most complicated use case.

ANALYSIS

☑ **Apply Requirements Analysis Technique (Business
Process Automation, Business Process
Improvement, or Business Process Reengineering)**

☑ **Use Requirements-Gathering Techniques (Interview,
JAD Session, Questionnaire, Document Analysis,
or Observation)**

☑ **Develop Use Cases**

☑ **Develop Data Flow Diagrams**

☐ **Develop Entity Relationship Model**

☐ **Normalize Entity Relationship Model**

TASK CHECKLIST

PLANNING   ANALYSIS   DESIGN

CHAPTER 7

DATA

MODELING

A data model describes the data that flows through the business processes in an organization. During the analysis phase, the data model presents the logical organization of data without indicating how the data are stored, created, or manipulated so that analysts can focus on the business without being distracted by technical details. Later, during the design phase, the data model is changed to reflect exactly how the data will be stored in databases and files. This chapter describes entity relationship diagramming, one of the most common data modeling techniques used in industry.

OBJECTIVES

- Understand the rules and style guidelines for creating entity relationship diagrams.
- Be able to create an entity relationship diagram.
- Become familiar with the data dictionary and metadata.
- Become familiar with the process of normalization.
- Understand how to balance between entity relationship diagrams and data flow diagrams.

CHAPTER OUTLINE

IMPLEMENTATION

INTRODUCTION

During the analysis phase, analysts create process models to represent how the business system will operate. At the same time, analysts need to understand the information that is used and created by the business system (e.g., customer information, order information). In this chapter, we discuss how the data that flow through the processes are organized and presented.

A *data model* is a formal way of representing the data that are used and created by a business system; it illustrates people, places, or things about which information is captured and how they are related to each other. The data model is drawn using an iterative process in which the model becomes more detailed and less conceptual over time. During analysis, analysts draw a logical data model, which shows the logical organization of data without indicating how data are stored, created, or manipulated. Because this model is free of any implementation or technical details, the analysts can focus more easily on matching the diagram to the real business requirements of the system.

In the design phase, analysts draw a *physical data model* to reflect how the data will physically be stored in databases and files. At this point, the analysts investigate ways to store the data efficiently and to make the data easy to retrieve. The physical data model (Chapter 8) and performance tuning (Chapter 9) are discussed later in the textbook.

Project teams usually use CASE tools to draw data models. Some of the CASE tools are data modeling packages, such as ER*win* by Platinum Technology that help analysts create and maintain logical and physical data models; they have a wide array of capabilities to aid modelers, and they can automatically generate many different kinds of databases from the models that are created. Other CASE tools (e.g., Oracle Designer) come bundled with database management systems (e.g., Oracle), and they are particularly good for modeling databases that will be built in their companion database products. A final option is to use a full-service CASE tool, such as Visible Analyst Workbench in which data modeling is one of many capabilities, and the tool can be used with many different databases. These tools have decent data modeling tools although they do not have as much functionality and support for data modeling as the two other kinds of CASE software. A benefit of the full-service CASE tool is that it well integrates the data model information with other relevant parts of the project.

In this chapter, we focus on creating a logical data model. Although there are several ways to model data, we will present one of the most commonly used techniques: entity relationship diagramming, a graphic drawing technique developed by Peter Chen[1] that shows all the data components of a business system. We will first describe how to create an entity relationship diagram (ERD) and discuss some style guidelines. Then, we will present a technique called normalization that helps analysts validate the data models that they draw. The chapter ends with a discussion of how data models balance, or interrelate, with the process models that you learned about in Chapter 6.

[1] P. Chen, "The Entity-Relationship Model—Toward a Unified View of Data," *ACM Transactions on Database Systems,* 1976, 1:9–36.

THE ENTITY RELATIONSHIP DIAGRAM

An *entity relationship diagram (ERD)* is a picture that shows the information that is created, stored, and used by a business system. An analyst can read an ERD to find out the individual pieces of information in a system and how they are organized and related to each other. On an ERD, similar kinds of information are listed together and placed inside boxes called entities. Lines are drawn between entities to represent relationships among the data, and special symbols are added to the diagram to communicate high-level business rules that need to be supported by the system. The ERD implies no order, although entities that are related to each other are usually placed close together.

For example, consider the doctor's office system that was described in Chapter 6. Although we understand how the system will work from studying the data flow diagram (DFD), we have very little detailed understanding about the information itself that flows through the system. What exactly is "patient information"? What pieces of information are captured when making an "appointment"? How is a patient related to the appointment that he or she makes?

Reading an Entity Relationship Diagram

The analyst can answer these questions and many more by examining the ERD that is presented in Figure 7-1. First, the analyst knows that the data to support the appointment system can be organized into six main categories—patient, appointment, doctor, bill, payment, and insurance company—and that patient information includes the patient's ID number, last name, first name, address, phone number, and birthdate. The analyst understands what information is used to uniquely identify a patient, an appointment, a doctor, and so on by looking for the information in the top box of each category. For example, the patient ID number is used to identify a particular patient.

The lines connecting the six categories of information communicate the relationships that the categories share. By reading the relationship lines, the analyst understands that a doctor can be scheduled for an appointment that is scheduled by a patient.

The ERD also communicates high-level *business rules*. Business rules are constraints or guidelines that are followed during the operation of the system; they are rules like "only one person can be seen by a doctor at a time" or "payments must be made in the form of cash or check" or "a bill can be paid by insurance, a patient, or a combination of both." Over the course of a workday, people are constantly applying business rules to do their jobs, and they know what the rules are because they either have been taught them (the office manager will know from experience that credit card payments are not accepted) or they can look them up somewhere. If a situation arises and people don't know how to handle it, they may have to refer to a policy guide or written procedure to determine what the business rules are.

On a data model, business rules are communicated by the kinds of relationships that the entities share. From the ERD, for example, we know that a patient can schedule many appointments from the round dot placed on the line closest to appointment, and we know that a person does not have to be insured to become a patient because of the diamond on the line nearest the insurance company. A doctor can be scheduled for many appointments, and each appointment generates one

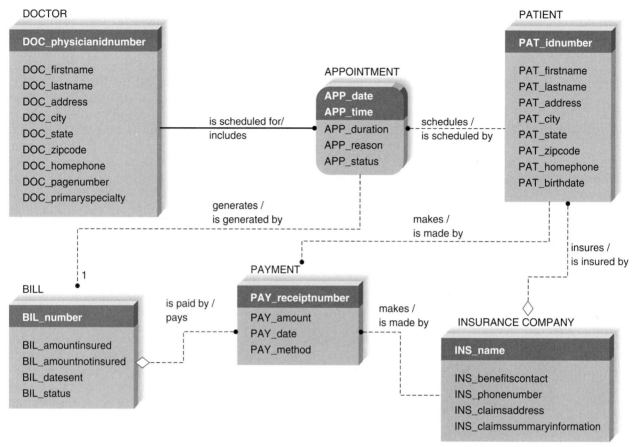

FIGURE 7-1
Example Entity Relationship Diagram

bill (communicated by the number one on the line closest to bill). Both the patient and insurance company make payments, and this rule makes sense given that a bill can have many payments associated with it. Ultimately, the new system should support the business rules we just described, and it should ensure that users don't violate the rules when performing the processes of the system.

Now that you've seen an ERD, let's step back and learn the ERD basics. In the following sections, we will first describe the syntax of the ERD using the diagram in Figure 7-1. Then we will teach you how to create an ERD using an example from CD Selections.

Elements of an Entity Relationship Diagram

There are three basic elements in the data modeling language (entities, attributes, and relationships), each of which is represented by a different graphic symbol. There are many different sets of symbols that can be used on an ERD. No one set of symbols dominates industry use, and none is necessarily better than another. We will use *IDEF1X* in this book. Figure 7-2 summarizes the three basic elements of ERDs and the symbols we will use.

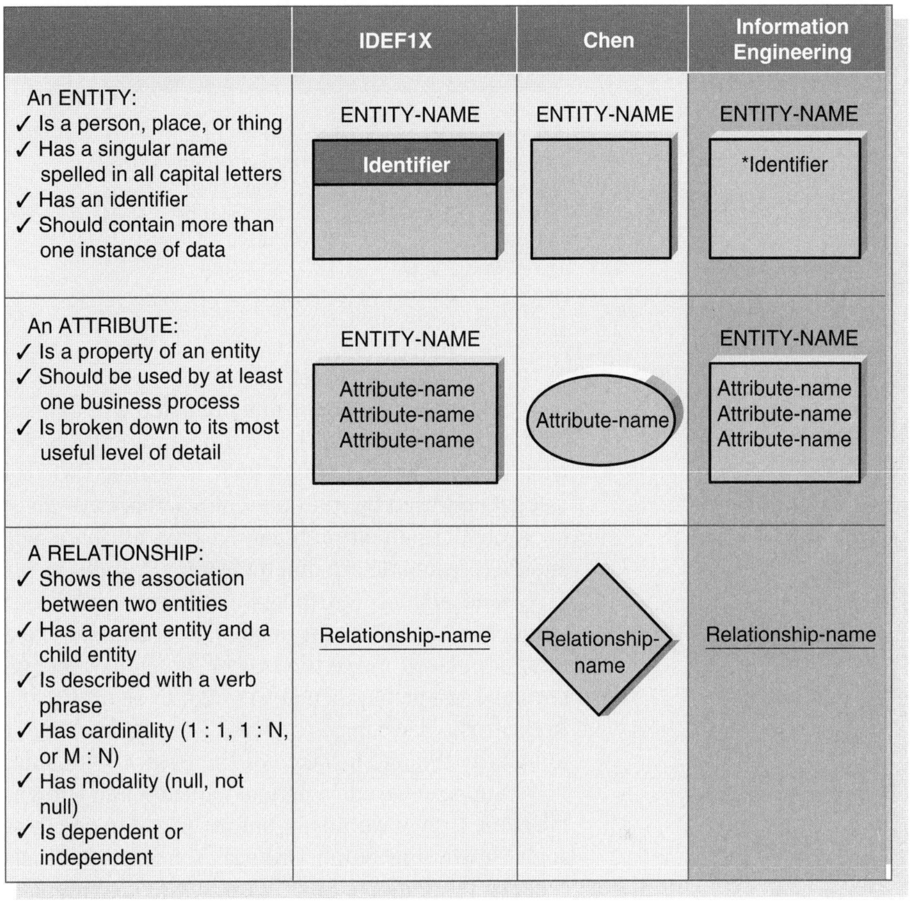

	IDEF1X	Chen	Information Engineering
An ENTITY: ✓ Is a person, place, or thing ✓ Has a singular name spelled in all capital letters ✓ Has an identifier ✓ Should contain more than one instance of data	ENTITY-NAME Identifier	ENTITY-NAME	ENTITY-NAME *Identifier
An ATTRIBUTE: ✓ Is a property of an entity ✓ Should be used by at least one business process ✓ Is broken down to its most useful level of detail	ENTITY-NAME Attribute-name Attribute-name Attribute-name	Attribute-name	ENTITY-NAME Attribute-name Attribute-name Attribute-name
A RELATIONSHIP: ✓ Shows the association between two entities ✓ Has a parent entity and a child entity ✓ Is described with a verb phrase ✓ Has cardinality (1 : 1, 1 : N, or M : N) ✓ Has modality (null, not null) ✓ Is dependent or independent	Relationship-name	Relationship-name	Relationship-name

FIGURE 7-2
Data Modeling Symbol Sets

Entity   The *entity* is the basic building block for a data model. It is a person, place, event, or thing about which data is collected—for example, an employee, an order, or a product. An entity is depicted by a rectangle, and it is described by a singular noun spelled in capital letters. All entities have a name, a short description that explains what they are, and an *identifier* that is the way to locate information in the entity (which is discussed later). In Figure 7-1, the entities are patient, doctor, appointment, payment, insurance company, and bill.

Entities represent something for which there exists multiple *instances*, or occurrences. For example, John Smith and Susan Jones could be instances of the entity patient (Figure 7-3). We would expect the patient entity to stand for all of the people who have scheduled an appointment, and each of them would be an instance in the patient entity.

Attribute   An *attribute* is some type of information that is captured about an entity. For example, date of birth, home address, and last name are all attributes of a patient. It is easy to come up with hundreds of attributes for an entity (e.g., a patient has an eye color, a favorite hobby, a religious affiliation), but only those that will actually be used by a business process should be included in the model.

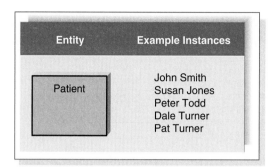

FIGURE 7-3
Entities and Instances

Attributes are nouns that are listed within an entity. Usually, some form of the entity name is appended to the beginning of each attribute to make it clear as to what entity it belongs (e.g., PAT_lastname, PAT_address). Without doing this, you can get confused by multiple entities that have the same attributes—for example, a doctor and a patient both can have an attribute called "lastname." DOC_lastname and PAT_lastname are much clearer attributes to place on the data model.

One or more attributes can serve as the identifier, the attribute(s) that can uniquely identify one instance of an entity, and the attributes that serve as the identifier are placed in the top box of the entity rectangle. If there are no patients with the same last name, then last name can be used as the identifier of the patient entity. In this case, if we need to locate John Brown, the name Brown would be sufficient to identify the one instance of the Brown last name.

Suppose we add a patient named Sarah Brown. Now we have a problem: using the name Brown would not uniquely lead to one instance—it would lead to two (i.e., John Brown and Sarah Brown). You have three choices at this point, and all are acceptable solutions. First, you can use a combination of multiple fields to serve as the identifier (last name and first name). This is called a *concatenated identifier* because several fields are combined, or concatenated, to uniquely identify an instance. Second, you can find a field that is unique for each instance, like the patient ID number. Third, you can wait to assign an identifier (like a randomly generated number that the system will create) until the design phase of the SDLC (Figure 7-4). Many data modelers don't believe that randomly generated identifiers belong on a logical data model because they do not logically exist in the business process.

Relationship *Relationships* are associations between entities, and they are lines that connect the entities together. Every relationship has a *parent entity* and a *child entity*, the parent being the first entity in the relationship, and the child being the second.

Relationships should be clearly labeled with active verbs so that the connections between entities can be understood. If one verb is given to each relationship, it is read in two directions. For example, we could write the verb *schedules* alongside the relationship for the *patient* and *appointment* entities, and this would be read as "a patient schedules an appointment" and "an appointment is scheduled by a patient." In Figure 7.1, we have included words for both directions of the relationship line; the top words are read from parent to child, and the bottom words are read from child to parent. Notice that the *doctor* entity is the parent entity in the *doctor – appointment* relationship.

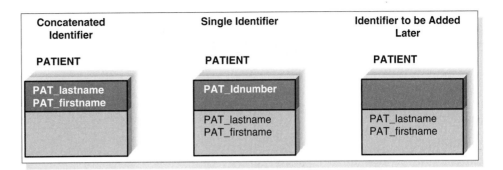

FIGURE 7-4
Choices for Identifiers

Cardinality   Relationships have two properties. First, a relationship has cardinality, which is ratio of parent instances to child instances. To determine the cardinality for a relationship, we ask ourselves: "How many instances of one entity are associated with an instance of the other?" (Remember that an instance is one occurrence of an entity, such as patient John Smith, Dr. Dave Brousseau, or Aetna Insurance Company.) For example, how many appointments can a patient schedule? How many patients are associated with an appointment? How many insurance companies can a patient have? The cardinality for binary relationships (i.e., relationships between two entities) are 1 : 1, 1 : N, or M : N, and we will discuss each in turn.

The 1 : 1 *relationship* (read as "one to one"), means that one instance of the parent entity is associated with one instance of the child entity. Notice that there is a 1 : 1 relationship between appointment and bill in Figure 7-1. Each appointment (the parent entity) always results in a single bill that is generated, and a bill (the child entity) is associated with only one appointment. The model contains a solid dot and a number "1" on the relationship line nearest the child entity to designate the 1 : 1 relationship.

More often, relationships are 1 : N relationships (read as "one to many"). In this kind of relationship, a single instance in a parent entity is associated with many instances of a child entity; however, the child-entity instance is only related to one instance in the parent. For example, a patient (parent entity) can schedule many appointments (child entity), but a particular appointment is scheduled by only one patient, suggesting a 1 : N relationship between patient and appointment. Another 1 : N relationship is between *insurance company* and *payment* (an insurance company can make many payments, but a specific payment is only associated with one insurance company). A character resembling a solid dot is placed closest to *appointment* to show the "many" end of the relationship. The parent entity is always on the "1" side of the 1 : N relationship. Can you identify several other 1 : N relationships in Figure 7-1? Identify the parent and child entities for each relationship.

A third kind of relationship is the M : N relationship (read as "many to many"). In this case, many instances of a parent entity can be related to many instances of a child entity. There are no M : N relationships in Figure 7-1, but take a look at Figure 7-5, which provides two M : N relationship examples. The first example demonstrates that one patient (parent entity) can display many different symptoms at a time (e.g., sore throat, fever, runny nose), and a symptom (child entity) can be associated with many different patients. The second M : N in Figure 7-5 communicates that a doctor can have many different specialties (e.g., infant contagious disease, cardiology, neurology), and a specialty can be associated with many different doctors. M : N relationships are depicted on an ERD by having solid dots at both ends of the relationship line.

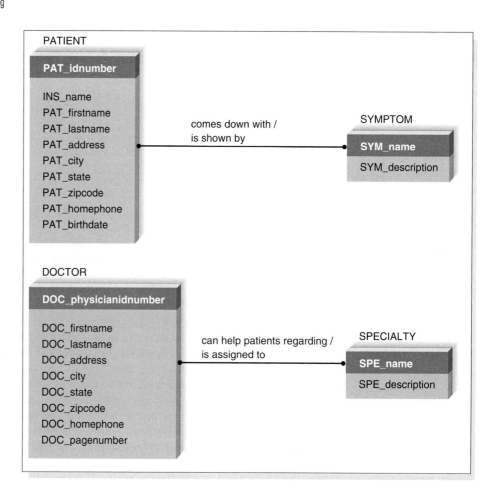

FIGURE 7-5
M : N Relationships

Modality Second, relationships have a *modality* of null or not null, which refers to whether or not an instance of a child entity can exist without a related instance in the parent entity. Basically, the modality of a relationship indicates whether the child-entity instance must participate in the relationship. It forces you to ask questions like: Can you have an appointment without a doctor? Can you have a bill without an appointment? Modality is depicted by placing a diamond on the relationship line next to the parent entity if nulls are allowed.

In the first two questions we asked, the answer is no—of course you need a doctor to have an appointment, and of course you can't send a bill without an associated appointment. Thus, the relationships are left alone, and the analyst understands that the modality is "not null" for both of these relationships. Notice, however, that a diamond has been placed on the relationship line next to insurance company. This means that we can have a patient in our system who does not have an insurance company, and the modality is "null."

The Data Dictionary and Metadata

As we described earlier, a CASE tool is used to help build ERDs. Every CASE tool has something called a *data dictionary*, which quite literally is where the analyst goes to define or look up information about the entities, attributes, and relationships

7-1 UNDERSTANDING THE ELEMENTS OF AN ERD

A wealthy businessman owns a large number of paintings that he loans to museums all over the world. He is interested in setting up a system that records what he loans to whom so that he doesn't lose track of his investments. He would like to keep information about the paintings that he owns as well as the artists who painted them. He also wants to track the various museums that reserve his art along with the actual reservations. Obviously artists are associated with paintings, paintings are associated with reservations, and reservations are associated with museums.

QUESTIONS:
1. Draw the four entities that belong on this data model.

2. Provide some basic attributes for each entity, and select an identifier, if possible.
3. Draw the appropriate relationships between the entities and label them.
4. What is the cardinality for each relationship? Depict this on your drawing.
5. What is the modality for each relationship? Depict this on your drawing.
6. List two business rules that are communicated by your ERD.

on the ERD. Figures 7-6, 7-7, and 7-8 illustrate common data dictionary entries for an entity, an attribute, and a relationship; notice the kinds of information the data dictionary captures about each element.

The information you see in the data dictionary is called *metadata*, which, quite simply, is data about data. Metadata is anything that describes an entity, attribute, or relationship, such as entity names, attribute descriptions, and relationship cardinality, and it is captured to help designers better understand the system that they are building and to help users better understand the system that they will use. Figure 7-9 lists typical metadata that are found in the data dictionary. Notice that the metadata can describe an ERD element (like entity name) and also information that is helpful to the project team (like the user contact, the analyst contact, and special notes).

Metadata is stored in the data dictionary so that it can be shared and accessed by developers and users throughout the SDLC. The data dictionary allows you to record the standard pieces of information about your elements in one place, and it makes that information accessible to many parts of a project. For example, the data elements in a data model also appear on the process models as data stores and data flows, and on the user interface as fields on an input screen. When you make a change in the data dictionary, the change ripples to the relevant parts of the project that are affected.

When metadata is complete, clear, and shareable, the information can be used to integrate the different pieces of the analysis phase and ultimately lead to a much better design. It becomes much more detailed as the project evolves though the SDLC.

CREATING AN ENTITY RELATIONSHIP DIAGRAM

Drawing an ERD is an iterative process of trial and revision. It usually takes considerable practice. ERDs can become quite complex—in fact, there are systems that

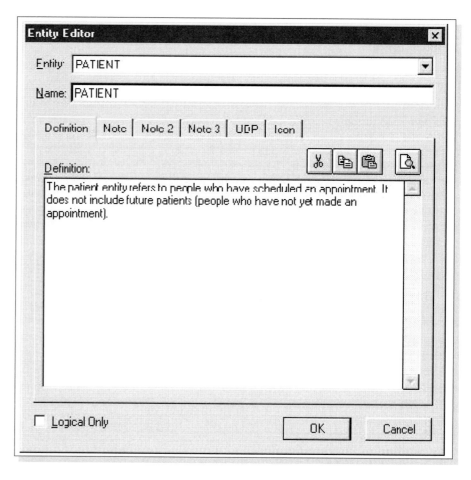

FIGURE 7-6
Data Dictionary Entry for the Patient Entity (Shown Using ER*win*)

have ERDs containing hundreds or thousands of entities. The basic steps in build-ing an ERD are these: (1) identify the entities, (2) add the appropriate attributes to each entity, and then (3) draw relationships among entities to show how they are associated with one another. First, we will describe the three steps in creating ERDs using the data model example from Figure 7-1. Then, we will present an ERD for CD Selections.

YOUR TURN

7-2 EVALUATE YOUR CASE TOOL

Examine the CASE tool that you will be using for your project, or find a CASE tool on the Web that you are interested in learning about. What kind of metadata does its data dictionary capture? Does the CASE tool integrate data model information with other parts of a project? How?

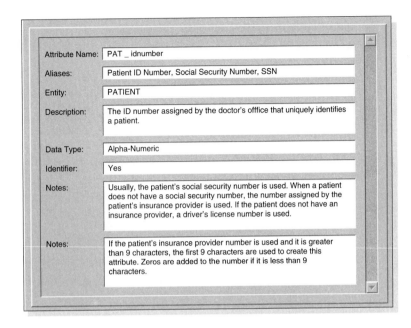

FIGURE 7-7
CASE Entry for Patient ID Number

Building Entity Relationship Diagrams

Step 1: Identify the Entities   As we explained, the most popular way to start an ERD is to first identify the entities for the model and their attributes. The entities should represent the major categories of information that you need to store in your system. If the process models (e.g., DFDs) have been prepared, the easiest way to start is with them: the data stores on the DFDs, the external entities, and the data flows indicate the kinds of information that are captured and flow through the system. If you begin your data model using a use case, look at the major inputs to the use case, the major outputs, and the information used for the use-case steps.

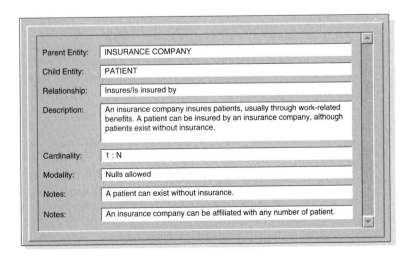

FIGURE 7-8
CASE Entry for the Insurance Company and Patient Relationship

ERD Element	Kinds of Metadata	Example
Entity	Name	Doctor
	Definition	A doctor is any medical professional who has been hired to see patients during scheduled appointments in the doctor's office.
	Special notes	A nurse practitioner is considered a doctor within this system.
	User contact	Virginia Baker is the office manager (x4335), and she can provide information about the doctor entity.
	Analyst contact	Barbara Wixom is the analyst assigned to this entity.
Attribute	Name	DOC_physicianidnumber
	Definition	A physicianidnumber is a number assigned to a doctor by the medical industry. The number was put in place to track physicians and the prescription they make.
	Alias	Physician identification number, PID
	Sample values	334997300
	Acceptable values	Any 9-digit number
	Format	9-digits, no spaces or dashes
	Type	Numeric
	Special Notes	If a nurse practitioner does not have a PID, then his or her social security number will be used.
Relationship	Verb Phrase	Schedules
	Parent Entity	Doctor
	Child Entity	Appointment
	Definition	One doctor is scheduled to participate in an appointment, and he or she can be scheduled for many appointments over time. A single appointment can only have one doctor affiliated with it.
	Cardinality	1 : N
	Modality	Not null
	Special Notes	Investigate how walk-in patients are handled! Are they placed on the schedule like regularly scheduled appointments?

FIGURE 7-9
Types of Metadata Captured by the Data Dictionary

Examine the process model (Figure 6.12) and the use cases (Figure 5.1) for the doctor's office system. As you look at the doctor's office system's level 0 DFD, you see that there are three data stores: patient, appointment, and billing. Each of these unique types of data probably will be represented using entities on our data model.

As a next step, you should examine the external entities and ask, "Will the system need to capture information about any of these entities?" We are already capturing patient information, but will we also need doctor and insurance company data? The answer is yes—we need information about doctors and insurance companies. The *make appointments* process will certainly need to include what doctor is affiliated with the appointment, and the *perform billing* process likely cannot be performed without information about the patient's insurer.

We should note that if the doctor's office only contained one doctor, then there would be no need for a doctor entity. A rule of thumb is to include only entities that contain more than one instance. We will assume that this doctor's office has multiple doctors who can see patients, so we include doctor on the ERD.

Finally, it is good to look at the data flows and see if there is information that flows through the system that has not yet been captured by the ERD. Notice that there is information called a bill that flows from the *perform billing* process. This seems logically different from the other entities. Remember that creating ERDs is an iterative process, so we can go ahead and add an entity called *bill* for now to store bill information—we always can make changes down the road if this turns out to be a bad assumption.

Step 2: Add Attributes and Assign Identifiers   The information that describes each entity becomes its attributes. It is likely that you identified a few attributes if you read the doctor's office system use cases and paid attention to the information flows on its DFDs. For example, a patient has a name and information, and an appointment has a date. Unfortunately, much of the information from the process models and use cases is at too high a level to identify the exact attributes that should exist in each of our six entities.

On a real project, there are a number of places where you can go to figure out what attributes belong in your entity. For one, you can check in the CASE tool— often an analyst will describe a process model data flow in detail when he or she enters the data flow into the CASE repository. For example, an analyst may create an entry for the patient information data flow like the one shown in Figure 7-10 that lists nine data elements that comprise patient information. The elements of the data flow should be added to the ERD as attributes in your entities. A second approach is to check the requirements definition. Often there is a section under functional requirements called data requirements. This section describes the data needs for the system that were identified while gathering requirements. A final approach to identifying attributes is to use requirements-gathering techniques. The most effective techniques would be interviews (e.g., asking people who create and use reports about their data needs) or document analysis (e.g., examining existing reports or input screens).

Once the attributes are identified, one or more of them will become the entity's identifier. The identifier must be an attribute(s) that is able to uniquely identify a single instance of the entity. Look at Figure 7-1 and notice the identifiers that were selected for each entity. Five of the entities had some attribute that could identify an instance by itself (e.g., physician ID number, bill number, payment receipt number, insurance company name, and patient ID).

Step 3: Identify Relationships   The last step in creating ERDs is to determine how the entities are related to each other. Lines are drawn between entities that have relationships, and each relationship is labeled and assigned both a cardinality and a modality. The easiest approach is to begin with one entity and determine all the entities with which it shares relationships. For example, there likely there are relationships between a patient and appointment, payment, and insurance com-

FIGURE 7-10
Elements of the Patient Information Data Flow

Data flow name:	Patient information
Data elements:	identification number + first name + last name + address + city + state + zip code + home phone number + birth date

pany. We know that a patient schedules appointments, makes payments, and is insured by an insurance company. The same exercise is done with each entity until all of the relationships have been identified. Each relationship should be labeled appropriately as it is added.

When you find a relationship to include on the model, you need to determine its cardinality and modality. For cardinality, ask how many instances of each entity participate in the relationship. You know that an insurance company can have many patients, but that a patient is affiliated with only one insurer. This suggests that there is a 1: N relationship in which the insurance company is the parent entity (the "1") and the patient is the child entity (the "many"). Next we examine the relationship's modality. Can a patient exist without an insurance company? We already determined that the answer is "yes," so the modality for the relationship is "null." This exercise is done for every relationship. Notice the cardinality and modality for each relationship in Figure 7-1.

Again, remember that data modeling is an iterative process. Often the assumptions you make and the decisions you make change as you learn more about the business requirements and as changes are made to the use cases and process models. But, you have to start somewhere—so do the best you can with the three steps we just described and keep iterating until you have a model that works. Later in this chapter, we will show you a few ways to validate the ERDs that you draw.

Applying the Concepts at CD Selections

Let's go through one more example of creating a data model using the context of CD Selections. For now, review the example use case that was presented in Figure 5-6 and the final level 0 process model presented in Figure 6-19.

Identify the Entities When you examine the CD Selections level 0 DFD, you see that there are four data stores: CD information, inventory, marketing, and in-store holds. Each of these unique types of data likely will be represented using entities on a data model.

As a next step, you should examine the external entities and ask, "Will the system need to capture information about any of these entities?" You may be tempted to include marketing manager and in-store staff, but there really is no need to track information about either of these in our system. Later, we may want to track system users, passwords, and data access privileges, but this information has to do with the *use* of the new system and would not be added until the physical data model (which is created in the design phase).

Customer and vendor are different matters. We need to capture information about both of these external entities. First consider customer. The *take-request* process requires that a customer provide a name and some basic contact information that is used for holding and special ordering CDs. A customer entity is needed on the data model to store this information. Likewise, the *maintain marketing materials* process accepts basic vendor information when a vendor sends in marketing material. This information should be stored in a vendor entity.

The last external entity, special order system, would not be added to the data model because we don't need to store or capture any information about it. Additionally, the special order system entity refers to only one system. A rule of thumb is to exclude entities on your data model that have only one instance. If, however, our Internet order system interfaced with multiple systems and we tracked information about each of them for some business reason, then it would be wise to create an entity called *system* to represent the multiple systems involved (the special order system would be one instance of this entity).

It is good practice to also look at the data flows on your process model and make sure that all of the information that flows through the system has been covered by your ERD. Unlike the doctor's office example (which found an additional entity by looking at the data flows), it appears that the main entities for CD Selections have been identified after examining the data stores and external entities.

To recap, by examining our process models, we have determined that our data model will contain six entities: CD, inventory, marketing material, in-store hold, customer, and vendor. See Figure 7-11 for the beginning of our data model.

Identify the Attributes   The next step is to select which attributes should be used to describe each entity. It is likely that you identified a handful of attributes if you read the CD Selections use cases and examined the DFDs. For example, a CD has an artist, title, category, and "sale" status and some attributes of customer are name and contact information, which likely includes address, phone number, and e-mail address.

As a second example, the inventory external entity suggests that a data store is needed to hold inventory information, and the take-request use case provides some indication of the kind of information that it will have. The use case says that the new system will find a store close to the customer and display the availability of the CD in that store. To do this, we need to capture the CD availability, or inventory information, such as the retail store that has the CD in stock. We should also include the store's zip code because the system will find the nearest store to the customer based on this information.

See if you can think about possible attributes to include in some of the other entities like Marketing Material and CD. Remember that on a real project, you would have the data dictionary, the requirements definition, and a variety of requirements-gathering techniques to use as sources for identifying attributes.

Figure 7-12 shows the data model with some of the attributes that likely would be found on the CD Selections data model. The model shows that we would capture the type of marketing material (e.g., video clip, audio clip), a short description, the e-mail address of the person who provided the material, and the actual contents in the marketing material entity. The CD entity would have the following attributes: SKU, title, artist, category, and sale status (i.e., whether or not the CD is on sale).

One or more of the attributes will be used to uniquely identify an instance of each entity. E-mail address may be a good choice for the customer entity in case

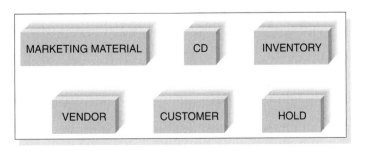

FIGURE 7-11
Entities for CD Selections ERD

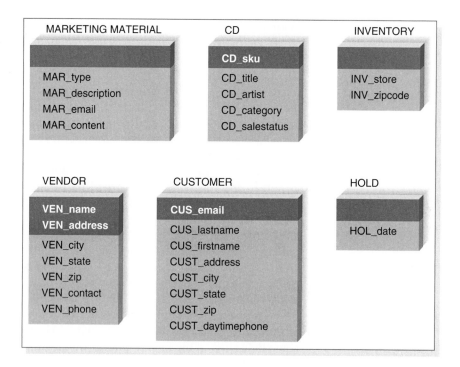

two customers have the same name. You know that every CD has a unique SKU (stock keeping unit), so this can be used as the identifier in the CD entity. In case there are two vendors with the same name, we may want to use a concatenated identifier that combines vendor name and address for the vendor entity. Notice that there do not seem to be obvious identifiers to use for the marketing material, inventory, and in-store hold entities. We will leave the identifier off for the time being until we learn about some special types of entities later in this chapter.

Identify the Relationships   The last step in creating ERDs is to determine how the entities are related to each other. Lines are drawn between entities that have relationships, and each relationship is labeled and assigned both a cardinality and a modality. For example, there likely is some kind of relationship between a CD and marketing material used to promote it. When you probe into the cardinality of the relationship, you discover that a CD can have many different kinds of marketing material (e.g., audio clip, video clip, jacket cover photo), but a specific piece of marketing material is only associated with one CD. This suggests that there is a 1 : N relationship between the CD and marketing material entities.

Next we examine the modality. Can marketing material exist without a CD? Well, maybe. There is a chance that a vendor will send us information before a CD is available for sale. In this case, we may want to store the marketing material until we are ready to use it. Therefore, we would make the modality of the relationship "null." Look at Figure 7-13. Notice the labels, cardinality, and modality that have been included on the diagram.

Advanced Syntax   Now that we have created a data model according to the basic syntax that was presented earlier, we can move to some more advanced concepts. There are three special types of entities that need to be explained and added to the diagram to make it clearer.

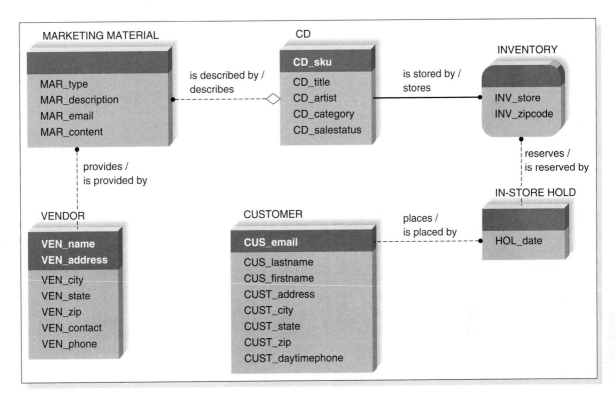

FIGURE 7-13
Relationships for CD Selections ERD

7-A THE USER'S ROLE IN DATA MODELING

I have two very different stories regarding data models. First, when I worked with First American Corporation, the head of Marketing kept a data model for the marketing systems hanging on a wall in her office. I thought this was a little unusual for a high-level executive, but she explained to me that data was critical for most of the initiatives that she puts in place. Before she can approve a marketing campaign or new strategy, she likes to confirm that the data exists in the systems and that it's accessible to her analysts. She has become very good at understanding ERDs over the years because they had been such an important communications tool for her to use with her own people and with IT.

On a very different note, here is a story I received from a friend of mine who heads up an IT department:

"We were working on a business critical, time dependent development effort, and VERY senior management decided that the way to ensure success was to have the various teams do technical design walkthroughs to senior management on a weekly basis. My team was responsible for the data architecture and database design. How could senior management, none of whom

probably had ever designed an Oracle architecture, evaluate the soundness of our work?

So, I had my staff prepare the following for the one (and only) design walkthrough our group was asked to do. First, we merged several existing data models and then duplicated each one … that is, every entity and relationship printed twice (imitating, if asked, the redundant architecture). Then we intricately color coded the model and printed the model out on a plotter and printed one copy of every inch of model documentation we had. On the day of the review, I simply *wheeled* in the documentation and stretched the plotted model across the executive boardroom table. 'Any questions,' I asked? 'Very impressive,' they replied. That was it! My designs were never questioned again." *Barbara Wixom*

QUESTIONS:
1. Based on these two stories, what do you think is the user's role in data modeling?
2. When is it appropriate to involve users in the ERD creation process?
3. How can users help analysts create better ERDs?

7-3 Campus Housing System

Consider the following system that was described in Chapter 5. Use the use cases and process models that you created in Chapters 5 and 6 to help you answer the questions below.

The Campus Housing Service helps students find apartments. Owners of apartments fill in information forms about the rental units they have available (e.g., location, number of bedrooms, monthly rent). Students who register with the service can search the rental information to find apartments that meet their needs (e.g., a two-bedroom apartment for $400 or less per month within 1/2 mile of campus). They then contact the apart-ment owners directly to see the apartment and possibly rent it. Apartment owners call the service to delete their listing when they have rented their apartment(s).

Questions:
1. What entities would you include on a data model?
2. What attributes would you list for each entity? Select an identifier for each entity, if possible.
3. What relationships exist between the entities that you identified? Label the relationships appropriately, and denote the cardinality and modality of reach relationship.

Independent Entity   The first type of entity is an independent entity, an entity that can exist without the help of another entity, such as patient, doctor, and insurance company. These three entities all have identifiers that were created using their own attributes (i.e., patient ID number, physician ID number, and insurer name). In other words, attributes from other entities were not needed to uniquely identify an instance. For example, using the patient ID number is sufficient for uniquely identifying a patient in the system shown in Figure 7-1. Information from the insurance company, appointment, or payment are not needed to identify a patient, even though these entities share relationships with the patient entity. Independent entities are depicted using rectangles with straight corners.

When relationships have an independent child entity, they are called *non-identifying relationships*, and they are denoted using a dotted relationship line. This name originated because the attributes from parent entity were not used as part of the child's identifier. The nonidentifying relationships in Figure 7-1 include the relationships between appointment and bill, appointment and patient, patient and payment, bill and payment, insurance company and payment, and insurance company and patient.

Dependent Entity   You may have guessed that there are situations when a child entity does rely on attributes from the parent entity to uniquely identify an instance.

7-4 Independent Entities

Locate the independent entities on Figure 7-13. How do you know which of the entities are independent? Locate the nonidentifying relationships. How did you find them? Can you create a rule that describes the association between independent entities and nonidentifying relationships?

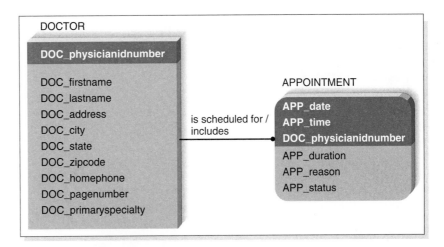

FIGURE 7-14
Dependent Entity Examples

In these cases, the child entity is called a *dependent entity* (or associative entity), and it needs external attributes to use for identification purposes.

Look at Figure 7-14, which shows the *appointment* entity that originally appeared in Figure 7-1. Alone, an appointment cannot be identified using the appointment's date and time because likely there are many appointments that are scheduled concurrently. (Several doctors may be seeing patients at 10 AM on March 26.) In this case, you actually would need to add the identifier from the parent entity (doctor) to help identify a unique appointment. Appointment is considered a dependent entity, and the fact that it is a dependent entity is depicted using a rectangle with rounded corners.

When relationships have a dependent child entity, they are called *identifying relationships*, and they are denoted using a solid relationship line. This name originated because attributes from parent entity were needed as part of the child's identifier. The only identifying relationship in Figure 7-1 includes the doctor and appointment relationship.

Intersection Entity   A third kind of entity is called an *intersection entity*, and it exists on the basis of a relationship between two other entities. Typically, intersection entities are created to store information about two entities that share a M : N relationship. Think back to the M : N relationship between patient and symptom that was described in Figure 7-5. Currently, one instance of a symptom (e.g., sore throat, fever, runny nose) could occur with many patients, and a patient could have many symptoms at one time. A difficulty arises if we want to capture the date upon which a particular patient demonstrates one or more symptoms. For

YOUR	**7-5 DEPENDENT ENTITIES**
TURN	

The CD Selections ERD has been updated in Figure 7-15. Which entities are dependent entities on this version? Locate the identifying relationships. How did you find them? Can you create a rule that describes the association between dependent entities and identifying relationships?

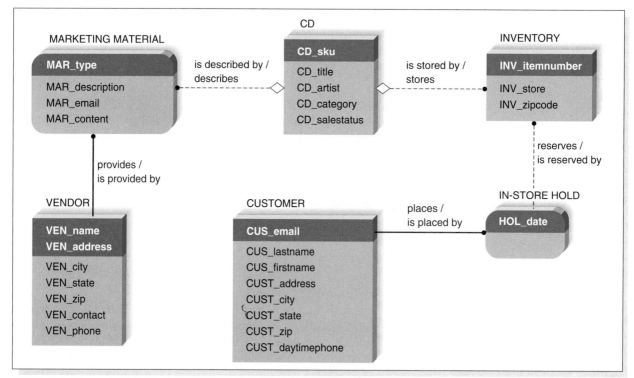

FIGURE 7-15
Dependent Entities for CD Selections

example, we can't place a date in the patient entity because a patient gets sick many times. Alternatively, the date doesn't belong in the symptom entity because it contains symptoms that are shared by all kinds of patients. In this case, an intersection entity is needed to capture attributes about the relationship itself—attributes that describe when a particular patient actually has a particular symptom.

There are three steps to adding an intersection entity, and this process usually is called "resolving a M : N relationship" because it eliminates the M : N relationship and problems associated with it from the data model. Step 1: remove the M : N relationship line between the two entities and add a third entity in between the two existing ones. Step 2: add two 1 : N relationships to the model. The two original entities should serve as the parent entities for each 1 : N, and the new intersection entity becomes the child entity in both relationships. Step 3: Name the intersection entity. Many times intersection entities are named using a concatenation of the two entities that created it (e.g., patient symptom), making its purpose clear, or the entity can be given another appropriate name (e.g., illness). Figure 7-16 shows the M : N relationship and how it was resolved using an intersection entity.

Are intersection entities dependent or independent? Actually, it depends. Sometimes an intersection entity has a logical identifier that can uniquely identify instances within it. For example, an intersection entity between student and course (a student can take many courses and a course can be taken by many students) may be called *transcript*. If transcripts have unique transcript numbers, then the entity would be considered dependent. In contrast, the illness intersection entity in Figure 7-16 requires the identifiers from both symptom and patient for an instance to be uniquely identified. Thus, illness is a dependent entity.

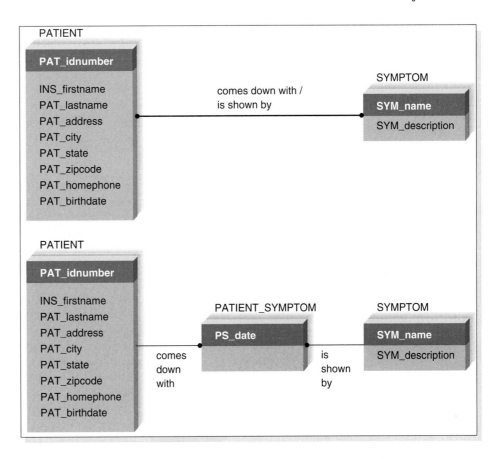

FIGURE 7-16
Resolving an M : N Relationship

VALIDATING AN ERD

As you probably guessed from the previous section, creating ERDs is pretty tough. It takes a lot of experience to draw ERDs well, and there are not many black-and-white rules to help guide you. Luckily, there are some general design guidelines that you can keep in mind as you build ERDs, and once the ERDs are drawn, you can use a technique called normalization to validate that your models are well formed. Another technique is to check your ERD against your process models to make sure that both models balance with each other.

<table>
<tr><td>**YOUR**</td><td>**7-6 INTERSECTION ENTITIES**</td></tr>
<tr><td>TURN</td><td></td></tr>
</table>

Resolve the M : N relationship between the doctor and specialty that is shown in Figure 7-5. What kinds of information could you capture about this relationship? What would the new ERD look like? Would the intersection entity be considered dependent or indepdent?

Can you think of other kinds of M : N relationships that exist in the real world? How would you resolve these M : N relationships if you were to include them on an ERD?

Design Guidelines

Design guidelines are not rules that must be followed; rather, they are "best practices" that often lead to better-quality diagrams. For example, labels and naming conventions are important for creating clear ERDs. Names should not be ambiguous (e.g., name, number); instead, they should clearly communicate what the model component represents. These names should be consistent across the model and reflect the terminology used by the business. If CD Selections refers to people who order products as customers, the data model should include an entity called customer, not client or stakeholder.

There are no rules covering the layout of ERD components. They can be placed anywhere you like on the page, although most systems analysts try to put the entities together that are related to each other. If the model becomes too complex or busy (some companies have hundreds of entities on a data model), the model can be broken down into *subject areas*. Each subject area would contain related entities and relationships, and the analyst can work with one group of entities at a time to make the modeling process less confusing.

In general, data modeling can be quite tricky, mainly because the data model is heavily based on interpretation; therefore, when business rules change, likely the relationships or other data model components will have to be altered. *Assumptions* are an important part of data modeling. For example, we assumed that only one doctor can be scheduled for an appointment. What if a patient needs to see two different specialists at once? The data model would have to change to allow one appointment to have multiple doctors.

Therefore, when you data model, don't panic or become overwhelmed by details. Rather, you should slowly add components to the diagram knowing that they will be changed and rearranged many times. Make assumptions along the way and then confirm these assumptions with the business users. Work iteratively and constantly challenge the data model with business rules and exceptions to see if the diagram is communicating the business system appropriately. Figure 7-17 summarizes the guidelines presented in this chapter to help you evaluate your data model.

Normalization

Once you have created your ERD, there is a technique called *normalization* that can help analysts validate the models that they have drawn. It is a process whereby a series of rules are applied to a logical data model or a file to determine how well formed it is (Figure 7-18). Normalization rules help analysts identify entities that are not represented correctly in a logical data model, or entities that can be broken out from a file. We describe here three normalization rules that are applied regularly in practice.

First Normal Form A logical data model is in *first normal form (1NF)* if it does not contain attributes that have repeating values for a single instance of an entity. Often this problem is called *repeating attributes*, or *repeating groups*. Every attribute in an entity should have only one value per instance for the model to "pass" 1NF.

Let's pretend that the CD Selections project team was given the layout for the special order file that is used by the existing special order system. The team is anxious to incorporate the data from this file into their own system, and they decide to

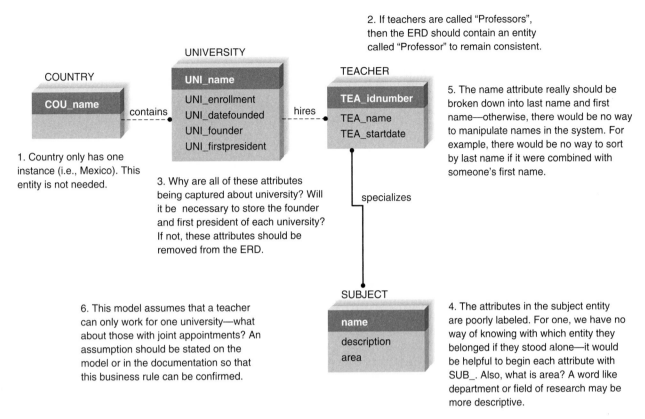

2. If teachers are called "Professors",
then the ERD should contain an entity
called "Professor" to remain consistent.

5. The name attribute really should be
broken down into last name and first
name—otherwise, there would be no way
to manipulate names in the system. For
example, there would be no way to sort
by last name if it were combined with
someone's first name.

1. Country only has one
instance (i.e., Mexico). This
entity is not needed.

3. Why are all of these attributes
being captured about university? Will
it be necessary to store the founder
and first president of each university?
If not, these attributes should be
removed from the ERD.

6. This model assumes that a teacher
can only work for one university—what
about those with joint appointments? An
assumption should be stated on the
model or in the documentation so that
this business rule can be confirmed.

4. The attributes in the subject entity
are poorly labeled. For one, we have no
way of knowing with which entity they
belonged if they stood alone—it would
be helpful to begin each attribute with
SUB_. Also, what is area? A word like
department or field of research may be
more descriptive.

FIGURE 7-17
A Few Data Modeling Guidelines

put the file into third normal form to make the information easier to understand and
ultimately easier for them to add to the data model for the new Internet order system. See Figure 7-19 for the file layout that the project team received.

If you examine the file carefully, you should notice that there are two cases in
which multiple values are captured for the same attribute for a single special order
instance. The first violation is the customer's book preferences, which are the kinds
of books that the customer enjoys to read (e.g., horror, nonfiction, sci-fi). The fact
that the attribute is plural leads us to believe that many different preferences are
captured for each instance of a special order, and preference is a repeating attribute.
This needs to be resolved by creating a new entity that contains preference information, and a relationship is added between special order and preference. See Figure 7-20 for the data model in 1NF. The new relationship is M : N because special
order can be associated with many preferences, and a preference can be found on
many special orders.

Additionally, notice that the special order file in Figure 7-19 captures information about many books for a single special order—another violation of 1NF. This
time, an entire group of attributes repeats (in this case up to three times) information
about books each time an order is placed. This is called a repeating group, and it can
be removed by creating a *book* entity and placing all of the book attributes into it.
The relationship between the special order and book entities also is a M : N; a book
can be found on many special orders, and a special order can include many books.

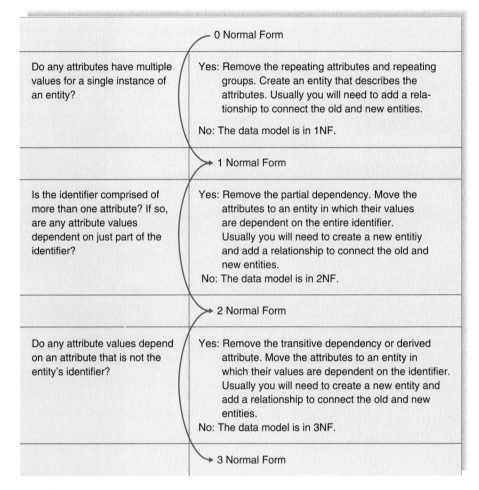

FIGURE 7-18
Normalization Steps

Second Normal Form *Second normal form (2NF)* requires first that the data model is in 1NF and second that the data model leads to entities containing attributes that are *dependent* on the whole identifier. This means that the value of all attributes that serve as identifier can determine the value for all of the other attributes for an instance in an entity. Sometimes nonidentifier attributes are only dependent on part of the identifier (i.e., *partial dependency*), and these attributes belong in another entity.

Figure 7-21 shows the special order data model placed in 2NF. Notice that originally, the *special order* entity had three attributes that were used as identifiers: special order date, customer last name, and customer first name. The problem was that some of the attributes were dependent on the customer last name and first name, but had no dependency on special order date. These attributes were all of the attributes that describe a customer: phone, address, and birth date. To resolve this problem, a new entity called *customer* was created, and the customer attributes were moved into the new entity. A 1 : N relationship exists between customer and spe-

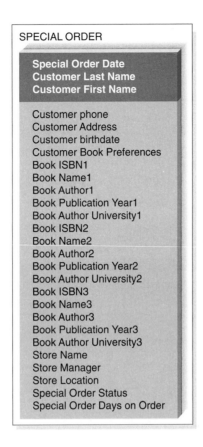

FIGURE 7-19
Normal Form

cial order because a customer can place many special orders, but a special order is only associated with one customer.

Remember that the customer last name and first name are still used in the *special order* entity—we know this because of the identifying 1 : N relationship between *customer* and *special order*. The identifying relationship implies that the customer

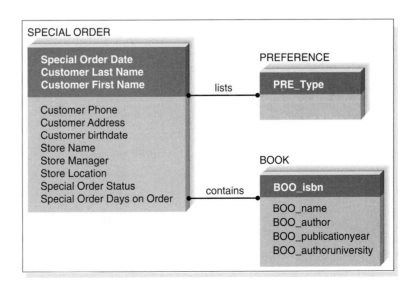

FIGURE 7-20
First Normal Form

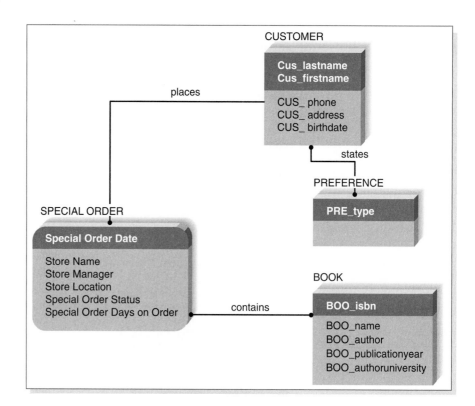

FIGURE 7-21
Second Normal Form

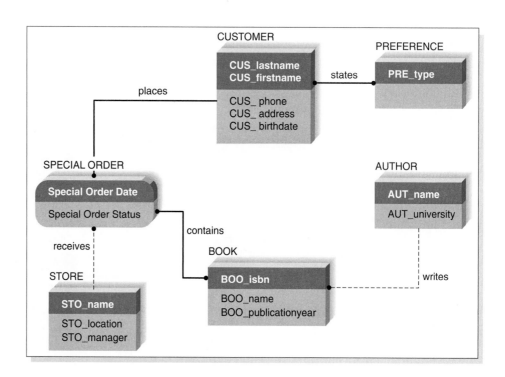

FIGURE 7-22
Third Normal Form

identifier (i.e., last name and first name) are used in special order as a part of its identifier.

Notice that we moved the relationship with reference to the new customer entity. Logically, a preference should be associated with a customer, not a particular special order.

Third Normal Form   *Third normal form (3NF)* occurs when a model is in both 1NF and 2NF and when in the resulting entities none of the attributes are dependent on a nonidentifier attribute (i.e., *transitive dependency*). Violations of 3NF can be found in both the *book* and *special order* entities in Figure 7-21.

The problem with the book entity is that author university is dependent on the author, not the ISBN. In other words, by knowing a book's ISBN, you do not automatically know the author's current university—but you would be able to find that value using the author name, a nonidentifier attribute. Therefore, we create a separate entity called *author* and move the author attributes to the new entity. The 1 : N relationship assumes that an author can write many books, but our system only captures information about one author (the first author) for each book.

In a similar vein, notice how store manager and location depend on store name, which is not an identifier for special order. This transitive dependency can be resolved by creating a *store* entity with store attributes.

Third normal form also addresses problems caused by *derived*, or *calculated*, *attributes*. The values of derived attributes don't need to be stored in a database (and thus, can be eliminated from the data model) because they can be derived from other attributes. For example, a system would never capture someone's age if it also contains birthdate—age can be derived by knowing the birthdate and comparing that date to today. In Figure 7-21 the number of days since a special order had been placed is removed from the data model. There is no need to capture this attribute since we can determine it using the attribute that captures the value for the date an order was placed. Figure 7-22 shows the final data model in 3NF.

YOUR

TURN

7-7 NORMALIZING A STUDENT ACTIVITY FILE

Pretend that you have been asked to build a system that tracks student involvement in activities around campus. You have been given a file with information that needs to be imported into the system, and the file contains the following fields:

- Student Social Security number (identifier)
- Activity 1 code (identifier)
- Activity 1 description
- Activity 1 start date
- Activity 1 years with activity
- Activity 2 code
- Activity 2 description
- Activity 2 start date
- Activity 3 years with activity

- Activity 3 code
- Activity 3 description
- Activity 3 start date
- Activity 3 years with activity
- Student last name
- Student first name
- Student birthdate
- Student age
- Student advisor name
- Student advisor phone

Normalize the file. Show how the logical data model would change as you move from 1NF to 2NF to 3NF.

7-8 BOAT CHARTER COMPANY

A charter company owns boats that are used to chart trips to islands. The company has created a computer system to track the boats it owns, including each boat's ID number, name, and seating capacity. The company also tracks information about the various islands, such as their names and populations. Every time a boat is chartered, it is important to know the date that the trip is to take place and the number of people on the trip. The company also keeps information about each captain, such as Social Security number, name, birthdate, and contact information for next of kin. Boats travel to only one island per visit.

QUESTIONS:
1. Create a data model. Include entities, attributes, identifiers, and relationships.
2. Which entities are dependent? Which are independent?
3. Use the steps of normalization to put your data model in 3NF. Describe how you know that it is in 3NF.

Balancing Entity Relationship Diagrams with Data Flow Diagrams

All the analysis activities of the systems analyst are interrelated. For example, the requirements analysis techniques are used to determine how to draw both the process models and data models, and the CASE repository is used to collect information that is stored and updated throughout the entire analysis phase. Now we will see how the process models and data models are interrelated.

Although the process model focuses on the processes of the business system, it contains two data components—the data flow (which is composed of data elements) and the data store. The purposes of these are to illustrate what data are used and created by the processes and where those data are kept. These components of the DFD need to *balance* with the ERD. In other words, the DFD data components need to correspond with the ERD's data stores (i.e., entities) and the data elements that comprise the data flows (i.e., attributes) depicted on the data model.

Many CASE tools offer the feature of identifying problems with balance among DFDs and ERDs; however, it is a good idea to understand how to identify problems on your own. For example, examine the process model that was created in Chapter 6 (Figure 6-19). Notice that two of the entities on the data model, vendor and customer, do not exist on the process model as data stores (only as external entities). Because we will need to collect and use both customer and vendor information (and therefore will need places to store that information), we will need to add customer and vendor data stores on the process model. The vendor store will be added to the maintain marketing material process because that process is the one in which the vendor information is manipulated, and the customer data store will be used by both *take requests* and *process in-store holds*.

Why don't you try to see how well balanced the models are for the doctor's office system. Take a look at the DFD in Figure 6-11 and the ERD in Figure 7-1. Can you identify any data stores that are missing from the DFD? (Hint: what entities on the data model do not have a corresponding data store?)

We hope you noticed that doctor, payment, and insurance company do not appear on the process model as data stores, and they should be added to the DFD. Where would you put them?

Similarly, the bits of information that are contained in the data flows (these are usually defined in the CASE entry for the data flow) should match up to the attributes found in entities in the data models. For example, if the customer information data flow that goes from the *customer* entity to the *take-requests* process were defined as having customer name, e-mail address, and home address, then each of these pieces of information should be recorded as attributes in the *customer* entity on the data model. Take a moment now to determine the contents of the search request data flow. Given what you know about the kind of information belonging in a search request, do the contents balance with the data model?

In general, balance occurs when all the data stores can be equated to entities and when all entities are referred to by data stores. Likewise, all data elements should be captured by data attributes on the data model. If the DFD and ERD are not balanced, then information critical to the business process will be missing, or the system will contain unnecessary data.

SUMMARY

Basic Entity Relationship Diagram Syntax

The entity relationship diagram (ERD) is the most common technique for drawing a data model, a formal way of representing the data that are used and created by a business system. There are three basic elements in the data modeling language, each of which is represented by a different graphic symbol. The entity is the basic building block for a data model. It is a person, place, or thing about which data is collected. An attribute is some type of information that is captured about an entity. The attribute that can uniquely identify one instance of an entity is called the identifier. The third data model component is the relationship, which conveys the associations between entities. Relationships have cardinality (the ratio of parent instances to child instances) and modality (a parent needs to exist if a child exists). Information about all of the components is captured by meta-data in the data dictionary.

Creating an Entity Relationship Diagram

The basic steps in building an ERD are: (1) identify the entities, (2) add the appropriate attributes to each entity, and then (3) draw relationships among entities to show how they are associated with one another. There are three special types of entities that ERDs contain. Most entities are independent, because one (or more) attribute can be used to uniquely identify an instance. Entities that rely on attributes from other entities to identify an instance are dependent. An intersection entity is placed between two entities to capture information about their relationship. In general, data models are based on interpretation; therefore, it is important to clearly state assumptions that reflect business rules.

Validating an Entity Relationship Diagram

Normalization is the process whereby a series of rules are applied to the logical data model to determine how well formed it is. A logical data model is in first normal form (1NF) if it does not lead to repeating attributes, which are attributes that capture multiple values for a single instance. Second normal form (2NF) requires that all entities are in 1NF and lead to attributes whose values are dependent on the whole identifier (i.e., no partial dependency). Third normal form (3NF) occurs when a model is in both 1NF and 2NF and none of the resulting attributes are dependent on nonidentifier attributes (i.e., no transitive dependency). With each violation, additional entities should be created to remove the repeating attributes or improper dependencies from the existing entities. Finally, ERDs should be balanced with the data flow diagrams (DFDs)—which were presented in Chapter 6—by making sure that data model entities and attributes correspond to data stores and data flows on the process model.

KEY TERMS

1 : 1 relationship	Entity	Normalization
1 : N relationship	Entity relationship diagram (ERD)	Parent entity
Assumption	First normal form (1NF)	Partial dependency
Attribute	IDEF1X	Physical data model
Balance	Identifier	Relationship
Business rule	Identifying relationship	Repeating attributes
Cardinality	Independent entity	Repeating groups
Child entity	Instance	Second normal form (2NF)
Concatenated identifier	Intersection entity	Subject area
Data dictionary	Logical data model	Third normal form (3NF)
Data model	M : N relationship	Transcript
Dependent	Metadata	Transitive dependency
Dependent entity	Modality	
Derived attribute	Nonidentifying relationship	

QUESTIONS

1. If you must select an identifier for an employee entity, what three options do you have? What are the pros and cons of each choice?
2. Why do identifiers need to contain unique values?
3. Describe to a businessperson the cardinality and modality of a relationship between two entities.
4. Describe the metadata that can be collected about an entity, an attribute, and a relationship.
5. Why is metadata important?
6. What is the difference between a dependent and independent entity? How is each depicted on a data model? What kind of relationships are associated with these entities?
7. What is a business rule? Provide three examples of business rules. How would the business rules be incorporated into an ERD?

8. What is an intersection entity? When would it be a dependent entity? When would it be an independent entity?
9. What would need to occur for a data flow diagram (DFD) and an entity relationship diagram (ERD) to be balanced?
10. Why is it important to balance DFDs and ERDs?
11. How can you make an ERD easier to understand? (It might help if you first think about how to make one difficult to understand.)
12. What do you think are three common mistakes novice analysts make in creating ERDs?
13. What is the purpose of normalization?
14. How does a model meet the requirements of third normal form?

EXERCISES

A. Draw data models for the following entities:
- Movie (title, producer, length, director, genre)
- Ticket (price, adult or child, showtime, movie)
- Patron (name, adult or child, age)

B. Draw a data model for the following entities, considering the entities as representing a system for a patient billing system and including only the attributes that would be appropriate for this context:
- Patient (age, name, hobbies, blood type, occupation, insurance carrier, address, phone)
- Insurance carrier (name, number of patients on plan, address, contact name, phone)
- Doctor (specialty, provider identification number, golf handicap, age, phone, name)

C. Draw the following relationships. Would the relationships be identifying or nonidentifying? Why?
- A patient must be assigned to only one doctor, and a doctor can have many patients.
- An employee has one phone extension, and a unique phone extension is assigned to an employee.
- A movie theater shows many different movies, and the same movie can be shown at different movie theaters around town.

D. Draw an entity relationship diagram (ERD) for the following situations:
1. Whenever new patients are seen for the first time, they complete a patient information form that asks their name, address, phone number, and insurance carrier, all of which is stored in the patient information file. Patients can be signed up with only one carrier, but they must be signed up to be seen by the doctor. Each time a patient visits the doctor, an insurance claim is sent to the carrier for payment. The claim must contain information about the visit, such as the date, purpose, and cost. It would be possible for a patient to submit two claims on the same day.
2. The state of Georgia is interested in designing a database that will track its researchers. Information of interest includes researcher name, title, position; university name, location, enrollment; and research interests. Each researcher is associated with only one institution, and each researcher has several research interests.
3. A department store has a bridal registry. This registry keeps information about the customer (usually the bride), the products that the store carries, and the products for which each cus-

tomer registers. Customers typically register for a large number of products, and many customers register for the same products.
4. Jim Smith's dealership sells Fords, Hondas, and Toyotas. The dealership keeps information about each car manufacturer with whom it deals so that employees can get in touch with manufacturers easily. The dealership also keeps information about the models of cars that it carries from each manufacturer. It keeps such information as list price, the price the dealership paid to obtain the model, and the model name and series (e.g., Honda Civic LX). The dealership also keeps information about all sales that it has made (for instance, employees will record the buyer's name, the car the buyer bought, and the amount the buyer paid for the car). To allow employees to contact the buyers in the future, contact information is also kept (e.g., address, phone number).

E. Examine the data models that you created for Exercise D. How would the respective models change (if at all) on the basis of these corresponding new assumptions?
- Two patients have the same first and last names.
- Researchers can be associated with more than one institution.
- The store would like to keep track of purchased items.
- Many buyers have purchased multiple cars from Jim over time because he is such a good dealer.

F. Visit a Web site that allows customers to order a product over the Web (e.g., Amazon.com). Create a data model that the site needs to support its business process. Include entities to show what types of information the site needs. Include attributes to represent the type of information the site uses and creates. Finally, draw relationships, making assumptions about how the entities are related.

G. Create metadata entries for the following data model components and, if possible, input the entries into a computer-aided software engineering (CASE) tool of your choosing:
- Entity—product
- Attribute—product number
- Attribute—product type
- Relationship—company makes many products, and any one product is made by only one company

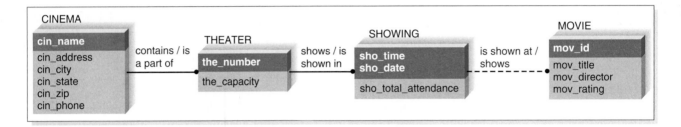

H. Describe the assumptions that are implied from the data model shown above.

I. Create a data model for one of the processes in the end-of-chapter Exercises for Chapter 6. Explain how you would balance the data model and process model.

J. Apply the steps of normalization to validate the models you drew in Exercise D.

K. You have been given a file that contains the following fields relating to CD information. Using the steps of normalization, create a logical data model that represents this file in third normal form. The fields include the following:
- Musical group name
- Musicians in group
- Date group was formed
- Group's agent
- CD title 1
- CD title 2
- CD title 3
- CD 1 length
- CD 2 length
- CD 3 length

The assumptions are as follows:
- Musicians in group contains a list of the members of the people in the musical group.
- Musical groups can have more than one CD, so both group name and CD title are needed to uniquely identify a particular CD.

MINICASES

1. West Star Marinas is a chain of 12 marinas that offer lakeside service to boaters; service and repair of boats, motors, and marine equipment; and sales of boats, motors, and other marine accessories. The systems development project team at West Star Marinas has been hard at work on a project that eventually will link all the marina's facilities into one unified, networked system.

The project team has developed a logical process model of the current system. This model has been carefully checked for syntax errors. Last week, the team invited a number of system users to role-play the various data flow diagrams, and the diagrams were refined to the users' satisfaction. Right now, the project manager feels confident that the as-is system has been adequately represented in the process model.

The Director of Operations for West Star is the sponsor of this project. He sat in on the role-playing of the process model and was very pleased by the thorough job the team had done in developing the model. He made it clear to you, the project manager, that he was anxious to see your team begin work on the process model for the to-be system. He was a little skeptical that it was necessary for your team to spend any time modeling the current system in the first place but grudgingly admitted that the team really seemed to understand the business after going through that work.

The methodology you are following, however, specifies that the team should now turn its attention to developing the logical data model for the as-is system. When you stated this to the project sponsor, he seemed confused and a little irritated. "You are going to spend even more time looking at the current system? I thought you were done with that! Why is this necessary? I want to see some progress on the way things will work in the future!"

a. What is your response to the Director of Operations?

b. Why do we perform data modeling?

c. Is there any benefit to developing a data model of the current system at all?

d. How does the process model help us develop the data model?

2. Holiday Travel Vehicles sells new recreational vehicles and travel trailers. When new vehicles arrive at Holiday Travel Vehicles, a new vehicle record is created. Included in the new vehicle record is a vehicle serial number, name, model, year, manufacturer, and base cost.

When a customer arrives at Holiday Travel Vehicles, he or she works with a salesperson to negotiate a vehicle purchase. When a purchase has been agreed to, a sales invoice is completed by the salesperson. The invoice summarizes the purchase, including full customer information, information on the trade-in vehicle (if any), the trade-in allowance, and information on the purchased vehicle. If the customer requests dealer-installed options, they will be listed on the invoice as well. The invoice also summarizes the final negotiated price, plus any applicable taxes and license fees. The transaction concludes with a customer signature on the sales invoice.

a. Identify the data entities described in this scenario (you should find six). Customers are assigned a customer ID when they make their first purchase from Holiday Travel Vehicles. Name, address, and phone number are recorded for the customer. The trade-in vehicle is described by a serial number, make, model, and year. Dealer installed options are described by an option code, description, and price.

b. Develop a list of attributes for each of the entities.

Each invoice will list just one customer. A person does not become a customer until he or she purchase a vehicle. Over time, a customer may purchase a number of vehicles from Holiday Travel Vehicles.

Every invoice must be filled out by only one salesperson. A new salesperson may not have sold any vehicles, but experienced salespeople have probably sold many vehicles.

Each invoice only lists one new vehicle. If a new vehicle in inventory has not been sold, there will be no invoice for it. Once the vehicle sells, there will be just one invoice for it. A customer may decide to have no options added to the vehicle, or may choose to add many options. An option may be listed on no invoices, or it may be listed on many invoices.

A customer may trade in no more than one vehicle on a purchase of a new vehicle. The trade-in vehicle may be sold to another customer, who later trades it in on another Holiday Travel vehicle.

c. Based on these business rules in force at Holiday Travel Vehicles, draw an ERD and document the relationships with the appropriate cardinality and modality.

3. The system development team at the Wilcon Company is working on developing a new customer order entry system. In the process of designing the new system, the team has identified the following data entity attributes:

Inventory Order
Order Number (identifier)
Order Date
Customer Name
Street Address
City
State
Zip
Customer Type
Initials
District Number
Region Number
1 to 22 occurrences of:
    Item Name
    Quantity Ordered
    Item Unit
    Quantity Shipped
    Item Out
    Quantity Received

a. State the rule that is applied to place an entity in first normal form. Revise this data model so that it is in first normal form.

b. State the rule that is applied to place an entity into second normal form. Revise the data model (if necessary) to place it in second normal form.

c. State the rule that is applied to place an entity into third normal form. Revise the data model to place it in third normal form.

d. What other guidelines and rules can you follow to validate that your data model is in good form?

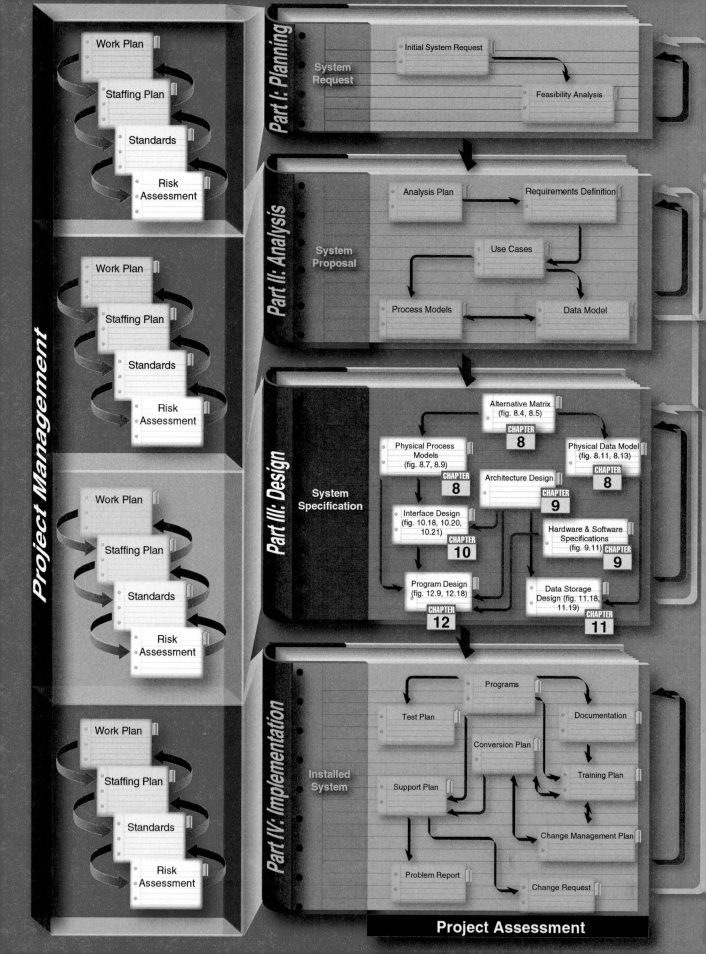

PART THREE
DESIGN
PHASE

PROJECT BINDER

The Design Phase
decides how the system will operate.
This collection of deliverables
is the system specification
that is handed to the program team
for implementation.

At the end of the Design Phase,
the feasibility analysis and project plan
are reexamined and revised,
and another decision is made by the project
sponsor and approval committee
about whether to terminate the project
or continue.

System Design — CHAPTER 8

Architecture Design — CHAPTER 9

User Interface Design — CHAPTER 10

Data Storage Design — CHAPTER 11

Program Design — CHAPTER 12

Design Plan

Physical Process Model

Physical Data Model

Architecture Design

Hardware/ Software Specification

Interface Design

Data Storage Design

Program Design

PLANNING

ANALYSIS

DESIGN

- [ ] **Select Design**
- [ ] **Develop Physical Data Flow Diagrams**
- [ ] **Develop Physical Entity Relationship Diagrams**
- [ ] Design Architecture
- [ ] Select Hardware and Software
- [ ] Develop Use Scenarios
- [ ] Design Interface Structure
- [ ] Design Interface Standards
- [ ] Design Interface Prototype
- [ ] Evaluate User Interface
- [ ] Design User Interface
- [ ] Select Data Storage Format
- [ ] Denormalize Entity Relationship Diagram
- [ ] Performance Tune Data Storage
- [ ] Size Data Storage
- [ ] Develop Program Structure Chart
- [ ] Develop Program Specification

TASK CHECKLIST

PLANNING   ANALYSIS   DESIGN

CHAPTER 8

SYSTEM
DESIGN

T he design phase of the SDLC uses the requirements that were gathered during analysis to create a blueprint for the future system. A successful design builds on what was learned in earlier phases and leads to a smooth implementation by creating a clear, accurate plan of what needs to be done. This chapter describes the initial transition from analysis to design and presents three ways to approach the design for the new system.

OBJECTIVES

- Understand the initial transition from analysis to design.
- Understand how to create a system specification.
- Be familiar with the custom, packaged, and outsource design alternatives.
- Be able to create an alternative matrix.
- Be able to create physical data flow diagrams and entity relationship diagrams.
- Be able to create a CRUD matrix.

CHAPTER OUTLINE

IMPLEMENTATION

INTRODUCTION

The purpose of the analysis phase is to figure out what the business needs. The purpose of the *design phase* is to decide how to build it. During the initial part of design, the project team converts the business requirements for the system into *system requirements* that describe the technical details for building the system. Unlike business requirements, which are listed in the requirements definition and communicated through use cases and *logical* process and data models, system requirements are communicated using a collection of design documents and *physical* process and data models. Together, the design documents and physical models comprise the blueprint for the new system.

The design phase has a number of steps that lead to the system blueprint (see Figure 8-1). An important initial part of the design phase is the examination of several design strategies to decide which will be used to build the system. Systems can be built from scratch, purchased and customized, or outsourced to others, and the project team needs to investigate the viability of each alternative. The decision to make, to buy, or to outsource influences the design tasks that are performed throughout the rest of the phase.

Next, the project team converts the logical diagrams that show the functional business requirements into physical diagrams that explain how to build the system. The logical DFDs and ERDs are converted into physical DFDs and ERDs, and CASE repository entries are expanded to include much more detailed information about how components of the diagrams map to specific technology. To ensure that the DFDs and ERDs balance properly, analysts create a CRUD matrix to show the way in which data is used by the system's processes. In some situations, the logical use cases may also need to be revised, but this is less common. The transition from logical to physical models and the CRUD matrix will be covered in detail at the end of this chapter.

Steps in the Design Phase	Deliverables	Chapter
1. Present design alternatives (make, buy, or outsource)	Alternative matrix	8
2. Convert logical process and data models to physical process and data models	Physical process model	8
	Physical data model	8
	Updated CASE repository with more detailed information	8
	CRUD matrix	8
3. Design the architecture for the system	Architecture design	9
4. Make hardware and software selections	Hardware and software specification	9
5. Design the inputs and outputs for the system	Interface design	10
6. Design the way in which data will be stored	Data storage design	11
7. Design the programs that will enable the processes of the system	Program design	12
8. Create the final deliverable for the design phase	System specification: all of the above deliverables combined and presented to the approval committee	8

FIGURE 8-1
The Design Phase

The project team also carefully considers the nonfunctional business requirements that were identified during analysis. The nonfunctional business requirements influence the system requirements that drive the design of the system's architecture. Major considerations of the "how" of a system are operational, performance, security, cultural, and political in nature. For example, the project team needs to plan for the new system's performance: how fast the system will operate, what its capacity should be, and its availability and reliability. The team needs to create a secure system by specifying access restrictions and by identifying the need for encryption, authentication, and virus control. The nonfunctional requirements are converted into system requirements that are described in the *architecture design* document (Chapter 9).

At the same time, architecture decisions are made regarding the hardware and software that will be purchased to support the new system (Chapter 9). These decisions are documented in the *hardware and software specification,* which is a document that describes what hardware and software are needed to support the new application. The actual acquisition of hardware and software is the responsibility of the Purchasing Department or the area in the organization that handles capital procurement; however, the project team uses the hardware and software specification to communicate the hardware and software needs to the appropriate people.

The next steps in design include such activities as designing the system inputs and system outputs, which involve the ways that the user interacts with the system. Chapter 10 describes these activities in detail, along with techniques, such as storyboarding and prototyping, that help the project team design a system that meets the needs of its users and is satisfying to use. Design decisions made regarding the interface are communicated using the design document called the interface design.

Finally, Chapters 11 and 12 present programming and data storage activities that are used to map out the nuts and bolts of the system. Such items as program specifications, pseudocode, and file and data design provide the final design details in preparation for the implementation phase and ensure that programmers have sufficient information to build the right system efficiently. The design decisions about programming and data storage are written up in the program design and the data storage design documents, respectively.

The many steps of the design phase are highly interrelated and, as with the steps in the analysis phase, analysts often go back and forth among them. For example, prototyping in the interface design step often uncovers additional information that is needed in the system, leading to a revision of the physical DFDs or ERDs. Alternatively, a system that is being designed for an organization with centralized systems may require substantial hardware and software investments if the project team decides to change a system in which all the processing is distributed.

At the end of the design phase, the project team creates the final deliverable for the phase called the *system specification*. This document contains all of the design documents just described: physical process models, physical data model, architecture report, hardware and software specification, interface design, data storage design, and program design. Collectively, the system specification conveys exactly what system the project team will implement during the implementation phase of the SDLC.

This chapter first examines the three fundamental approaches to developing new systems: make, buy, or outsource. Then it explains the movement from logical to physical DFDs and ERDs and a technique called the CRUD matrix.

PRACTICAL   8-1 Avoiding Classic Design Mistakes

TIP

In Chapters 3 and 4, we discussed several classic mistakes and how to avoid them. Here, we summarize four classic mistakes in the design phase and discuss how to avoid them:

1. **Reducing design time:** If time is short, there is a temptation to reduce the time spent in such "unproductive" activities as design so that the team can jump into "productive" programming. This results in missing important details that have to be investigated later at a much higher time cost (usually at least 10 times longer).

   Solution: If time pressure is intense, use rapid application development (RAD) techniques and timeboxing to eliminate functionality or move it into future versions.

2. **Feature creep:** Even if you are successful at avoiding scope creep, about 25% of system requirements will still change. Changes—big and small—can significantly increase time and cost.

   Solution: Ensure that all changes are vital and that the users are aware of the impact on cost and time. Try to move proposed changes into future versions.

3. **Silver bullet syndrome:** Analysts sometimes believe the marketing claims for some design tools that are said to solve all problems and magically reduce time and costs. No one tool or technique can eliminate overall time or costs by more than 25% (although some can reduce individual steps by this much).

   Solution: If a design tool has claims that appear too good to be true, just say no.

4. **Switching tools in midproject:** Sometimes analysts switch to what appears to be a better tool during design in the hopes of saving time or costs. Usually, any benefits are outweighed by the need to learn the new tool. This also applies to even "minor" upgrades to current tools.

   Solution: Don't switch or upgrade unless there is a compelling need for specific features in the new tool, and then explicitly increase the schedule to include learning time.

Source: Adapted from *Rapid Development*, Redmond, WA: Microsoft Press, 1996, by Steve McConnell.

DESIGN STRATEGIES

Until now, we have assumed that the system will be built and implemented by the project team. However, there are actually three ways to approach the creation of a new system: (1) developing a custom application in-house; (2) buying a packaged system and customizing it; and (3) relying on an external vendor, developer, or service provider to build the system. Each of these choices has its strengths and weaknesses, and each is more appropriate in different situations. The following sections describe each design choice in turn, and then we present criteria you can use to select one of the three approaches for your project.

Custom Development

Many project teams assume that custom development, or building a new system from scratch, is the best way to create a system. For one, teams have complete control over way the system looks and functions. Let's consider the request-taking process for CD Selections. If the company wants a Web request-taking feature that links tightly with its existing special order system, the project may involve a complex, highly specialized program. Alternatively, CD Selections might have a technical environment in which all information systems are built using standard technology and interface designs so that they are consistent and easier to update and support. In both cases, it could be very effective to create a new system from scratch that meets these highly specialized requirements.

Custom development also allows developers to be flexible and creative in the way they solve business problems. CD Selections may envision the Web interface that takes customer requests as an important strategic enabler. The company may want to use the information from the system to better understand its customers who order over the Web, and it may want the flexibility to evolve the system to incorporate technology such as data-mining software and geographic information systems to perform marketing research. A custom application would be easier to change to include components that take advantage of current technologies that can support such strategic efforts.

Building a system in-house also builds technical skills and functional knowledge within the company. As developers work with business users, their understanding of the business grows and they become better able to align information systems with strategies and needs. These same developers climb the technology learning curve so that future projects applying similar technology become much less effortful.

However, custom application development requires a dedicated effort that includes long hours and hard work. Many companies have a development staff that is already overcommitted to filling huge backlogs of systems requests and just does not have time for another project. Also, a variety of skills—technical, interpersonal, functional, project management, modeling—all have to be in place for the project to move ahead smoothly. IS professionals, especially highly skilled individuals, are quite difficult to hire and retain.

CONCEPTS   **8-A BUILDING A CUSTOM SYSTEM—WITH SOME HELP**

IN ACTION

I worked with a large financial institution in the southeast that suffered serious financial losses several years ago. A new chief executive officer was brought in to change the strategy of the organization to being more customer-focused. The new direction was quite innovative, and it was determined that custom systems, including a data warehouse, would have to be built to support the new strategic efforts. The problem was that the company did not have the in-house skills for these kinds of custom projects.

The company now has one of the most successful data warehouse implementations because of its willingness to use outside skills and its focus on project management. To supplement skills within the company, eight sets of external consultants, including hardware vendors, system integrators, and business strategists, were hired to take part and transfer critical skills to internal employees. An in-house project manager coordinated the data warehouse implementation full time, and her primary goals were to clearly set expectations, define responsi-bilities, and communicate the interdependencies that existed among the team members.

This company showed that successful custom development can be achieved even when the company may not start off with the right skills in-house. However, this kind of project is not easy to pull off—it takes a talented project manager to keep the project moving along and to transition the skills to the right people over time.

Barbara Wixom

QUESTIONS:
1. What are the risks in building a custom system without having the right technical skills available within the organization?
2. Why did the company select a project manager from within the organization?
3. Would it have been better to hire an external professional project manager to coordinate the project? Why or why not?

The risks associated with building a system from the ground up can be quite high, and there is no guarantee that the project will succeed. Developers could be pulled away to work on other projects, technical obstacles could cause unexpected delays, and the business users could become impatient with a growing timeline.

Packaged Software

Many business needs are not unique, and because it makes little sense to reinvent the wheel, many organizations buy *packaged software*, a software program that has already been written, rather than developing their own custom solution. In fact, there are thousands of commercially available software programs that have already been written to serve a multitude of purposes. Think about your own need for a word processor—did you ever consider writing your own word processing software? That would be very silly, considering the number of good software packages available for a relatively inexpensive cost.

Likewise, most companies have needs, such as payroll or accounts receivable, that can be met quite well by packaged software. It can be much more efficient to buy programs that have already been created, tested, and proven, and a packaged system can be bought and installed in a relatively short period of time compared with a custom system. Plus, packaged systems incorporate the expertise and experience of the vendor who created the software.

Let's examine the request-taking process once again. It turns out that there are programs available, called shopping cart programs, that allow a company to sell products on the Internet by keeping track of customers' selections, totaling them, and then e-mailing the order to a mailbox. You can easily install one on an existing Web page, and it allows you to take orders. Some shopping cart programs are even available over the Internet for free. Therefore, CD Selections could decide that acquiring a ready-made order-taking application might be much a more efficient way to capture customer requests than developing a program from scratch.

Packaged software can range from small single-function tools, such as the shopping cart program, to huge, all-encompassing systems, such as *enterprise resource planning (ERP)* applications that are installed to automate an entire business. Implementing ERP systems is a popular practice in which large organizations spend millions of dollars installing packages by such companies as SAP, People-Soft, Oracle, and Baan and then change their businesses accordingly. Installing ERP software is much more difficult than installing small application packages because benefits can be harder to realize and problems are much more serious.

One problem is that companies buying packaged systems must accept the functionality that is provided by the system, and rarely is there a perfect fit. If the packaged system is large in scope, its implementation could mean a substantial change in the way the company does business. Letting technology drive the business can be a dangerous way to go.

Most packaged applications allow for customization, or the manipulation of system parameters to change the way certain features work. For example, the package might have a way to accept information about your company or the company logo that would then appear on input screens. An accounting software package could offer a choice of various ways to handle cash flow or inventory control so that it could support the accounting practices in different organizations. If the amount of customization is not enough and the software package has a few features that don't quite work the way the company needs them to work, the project team can create a *workaround*.

A workaround is a custom-built add-on program that interfaces with the packaged application to handle special needs. It can be a nice way to create needed functionality that does not exist in the software package. However, workarounds should be a last resort for several reasons. First, workarounds are not supported by the vendor who supplied the packaged software, so when upgrades are made to the main system, they may make the workaround ineffective. Also, if problems arise, vendors have a tendency to blame the workaround as the culprit and refuse to provide support.

Although choosing a packaged software system is simpler than going with custom development, it also can benefit from following a formal methodology, just as if you were building a custom application.

Systems integration refers to the process of building new systems by combining packaged software, existing legacy systems, and new software written to integrate these. Many consulting firms specialize in systems integration, so it is not uncommon for companies to select the packaged software option and then outsource the integration of a variety of packages to a consulting firm. (Outsourcing is discussed in the next section.)

The key challenge in systems integration is finding ways to integrate the data produced by the different packages and legacy systems. Integration often hinges on taking data produced by one package or system and reformatting it for use in another package/system. The project team starts by examining the data produced by and needed by the different packages/systems and identifying the transformations that must occur to move the data from one to the other. In many cases, this involves fooling the different packages/systems into thinking that the data were produced by an existing program module that the package/system expects to produce the data rather than by the new package/system that is being integrated.

For example, CD Selections needs to integrate the new Internet Order System with existing legacy systems, such as the special order system. The special order system was written to support a different application—retail store special orders, and it currently exchanges data with the system that supports CD Selections' retail stores. The Internet Order System will need to produce data in the same format as this existing system so the special order system will think the data is from the same system as always. Conversely, the project team may need to revise the existing special order system so it can accept data from the Internet Order System in a new format.

Outsourcing

The design choice that requires the least amount of in-house resources is *outsourcing*, which means hiring an external vendor, developer, or service provider to create the system. Outsourcing has become quite popular in recent years, with a market that is expected to exceed $100 billion by the year 2005. Some estimate that as many as 50% of companies with IT budgets over $5 million are currently outsourcing or evaluating the approach. Although these figures include outsourcing for all kinds of systems functions, this section focuses on outsourcing a single development project. There can be great benefit to having others develop your system. They may be more experienced in the technology or have more resources, such as experienced programmers. Many companies embark on outsourcing deals to reduce costs, whereas others see it as an opportunity to add value to the business. For example, instead of creating a program that handles the request-taking process or buying a preexisting package, CD Selections may decide to let a Web service provider provide commercial services for them.

For whatever reason, outsourcing can be a good alternative for a new system; however, it does not come without costs. If you decide to leave the creation of a new system in the hands of someone else, you could compromise confidential information or lose control over future development. In-house professionals are not benefiting from the skills that could be learned from the project; instead, the expertise is transferred to the outside organization. Ultimately, important skills can walk right out the door at the end of the contract.

Most risks can be addressed if you decide to outsource, but two are particularly important. First, assess the requirements for the project thoroughly—you should never outsource what you don't understand. If you have conducted rigorous planning and analysis, then you should be well aware of your needs. Second, carefully choose a vendor, developer, or service with a proven track record with the type of system and technology that your system needs.

There are three primary types of contracts that can be drawn to control the outsourcing deal. A *time and arrangements* deal is very flexible because you agree to pay for whatever time and expenses are needed to get the job done. Of course, this agreement could result in a large bill that exceeds initial estimates. This arrangement works best when you and the outsourcer are unclear about what it is going to take to finish the job.

You will pay no more than expected with a *fixed-price contract* because if the outsourcer exceeds the agreed-on price, he or she will have to absorb the costs. Outsourcers are very careful about defining requirements clearly up front, and there is little flexibility for change.

The type of contract gaining in popularity is the *value-added contract* whereby the outsourcer reaps some percentage of the completed system's benefits. You have very little risk in this case but expect to share the wealth once the system is in place.

Creating fair contracts is an art because you need to carefully balance flexibility with clearly defined terms. Needs often change over time, so you don't want the contract to be so specific and rigid that alterations can't be made. Think about how quickly technology like the World Wide Web changes. It is difficult to foresee how a project may evolve over a long period of time. Short-term contracts leave room for reassessment if needs change or if relationships are not working out the way both parties expected. In all cases, the relationship with the outsourcer should be viewed as a partnership in which both parties benefit and communicate openly.

Managing the outsourcing relationship is a full-time job. Thus, someone needs to be assigned full time to manage the outsourcer, and the level of that person should be appropriate for the size of the job (a multimillion-dollar outsourcing engagement should be handled by a high-level executive). Throughout the relationship, progress should be tracked and measured against predetermined goals. If you do embark on an outsourcing design strategy, be sure to get more information. Many books have been written that provide much more detailed information on the topic.[1] Figure 8-2 summarizes some guidelines for outsourcing.

---

[1] For more information on outsourcing, we recommend M. Lacity and R. Hirschheim, *Information Systems Outsourcing: Myths, Metaphors, and Realities,* New York: John Wiley & Sons, 1993, and L. Willcocks and G. Fitzgerald, *A Business Guide to Outsourcing Information Technology,* London: Business Intelligence, 1994.

- Keep the lines of communication open between you and your outsourcer.
- Define and stabilize requirements before signing a contact.
- View the outsourcing relationship as a partnership.
- Select the vendor, developer, or service provider carefully.
- Assign a person to manage the relationship.
- Don't outsource what you don't understand.
- Emphasize flexible requirements, long-term relationships, and short-term contracts.

FIGURE 8-2
Outsourcing Guidelines

DESIGN STRATEGY

Each of the design strategies just discussed has its strengths and weaknesses, and no one strategy is inherently better than the others. Thus, it is important to understand the strengths and weaknesses of each strategy and when to use each. Figure 8-3 summarizes the characteristics of each strategy.

Business Need

If the business need for the system is common and technical solutions already exist in the marketplace that can meet the business need of the system, it makes little sense to build a custom application. Packaged systems are good alternatives for common business needs. A custom alternative should be explored when the business need is unique or has special requirements. Usually if the business need is not critical to the company, then outsourcing is the best choice—someone outside of the organization can be responsible for the application development.

In-house Experience

If in-house experience exists for all the functional and technical needs of the system, it will be easier to build a custom application than if these skills do not exist.

CONCEPTS
IN ACTION

8-B EDUCATIONAL DEVELOPMENT SERVICES' VALUE-ADDED CONTRACT

Value-added contracts can be quite rare—and very dramatic. They exist when a vendor is paid a percentage of revenue generated by the new system, which reduces the up-front fee, sometimes to zero. The landmark deal of this type was signed three years ago by the City of Chicago and EDS (a large consulting and systems integration firm), which agreed to reengineer the process by which the city collects the fines on 3.6 million parking tickets per year. At the time, because of clogged courts and administrative problems, the city collected on only about 25% of all tickets issued. It had a $60 million backlog of uncollected tickets.

Dallas-based EDS invested an estimated $25 million in consulting and new systems in exchange for the right to up to 26% of the uncollected fines, a base pro-

cessing fee for new tickets, and software rights. To date, EDS has taken in well over $50 million on the deal, analysts say. The deal has come under some fire from various quarters as an example of an organization giving away too much in a risk/reward–sharing deal. City officials, however, counter that the city has pulled in about $45 million in previously uncollected fines and has improved its collection rate to 65% with little up-front investment.

QUESTION:
Do you think the city of Chicago got a good deal from this arrangement? Why or why not?

Source: "Outsourcing? Go out on a Limb Together," *Datamation,* February 1, 1999, 41(2): 58–61, by Jeff Moad.

	When to Use Custom Development	When to Use a Packaged System	When to Use Outsourcing
Business need	The business need is unique	The business need is common	The business need is not core to the business
In-house experience	In-house functional and technical experience exists	In-house functional experience exists	In-house functional or technical experience does not exist
Project skills	There is a desire to build in-house skills	The skills are not strategic	The decision to outsource is a strategic decision
Project management	The project has a highly skilled project manager and a proven methodology	The project has a project manager who can coordinate vendor's efforts	The project has a highly skilled project manager at the level of the organization that matches the scope of the outsourcing deal
Time frame	The time frame is flexible	The time frame is short	The time frame is short or flexible

FIGURE 8-3
Selecting a Design Strategy

A packaged system may be a better alternative for companies that do not have the technical skills to build the desired system. For example, a project team that does not have Web commerce technology skills may want to acquire a Web commerce package that can be installed without many changes. Outsourcing is a good way to bring in outside experience that is missing in-house so that skilled people are in charge of building the system.

Project Skills

The skills that are applied during projects are either technical (e.g., Java, Structured Query Language [SQL]) or functional (e.g., electronic commerce), and different design alternatives are more viable depending on how important the skills are to the company's strategy. For example, if certain functional and technical expertise that relates to Internet sales applications and Web commerce application development is important to the organization because the company expects the Internet to play an important role in sales over time, then it makes sense for the company to develop Web commerce applications in-house, using company employees so that the skills can be developed and improved. On the other hand, some skills, such as network security, may be either beyond the technical expertise of employees or not of interest to the company's strategists—it is just an operational issue that needs to be handled. In this case, packaged systems or outsourcing should be considered so internal employees can focus on other business-critical applications and skills.

Project Management

Custom applications require excellent project management and a proven methodology. There are so many things that can push a project off-track, such as funding obstacles, staffing holdups, and overly demanding business users. Therefore, the project team should choose to develop a custom application only if it is certain that the underlying coordination and control mechanisms will be in place. Packaged and outsourcing alternatives also must be managed; however, they are more shielded from internal obstacles because the external parties have their own objectives and priorities (e.g., it may be easier for an outside contractor to say no to a user than for a person within the company to do so). The latter alternatives typically have their own methodologies, which can benefit companies that do not have an appropriate methodology to use.

Time Frame

When time is a factor, the project team should probably start looking for a system that is already built and tested. In this way, the company will have a good idea of how long the package will take to put in place and what the final result will contain. Of course, this assumes that the package can be installed as-is and does not need many workarounds to integrate it into the existing business processes and technical environment. The time frame for custom applications is hard to pin down, especially when you consider how many projects end up missing important deadlines. If you must choose the custom development alternative and the time frame is very short, consider using techniques like timeboxing to manage this problem. The time to produce a system using outsourcing really depends on the system and the outsourcer's resources. If a service provider has services in place that can be used to support the company's needs, then a business need could be met quickly. Otherwise, an outsourcing solution could take as long as a custom development initiative.

SELECTING A DESIGN STRATEGY

Once the project team has a good understanding of how well each design strategy fits with the project's needs, it must begin to understand exactly how to implement these strategies. For example, what tools and technology would be used if a custom alternative were selected? What vendors make packaged systems that address the project needs? What service providers would be able to build this system if the application were outsourced? This information can be obtained by talking to people working in the IS Department and getting recommendations from business users by contacting other companies with similar needs and investigating the types of systems that they have put in place. Vendors and consultants are usually willing to provide information about various tools and solutions in the form of brochures, product demonstrations, and information seminars.

It is likely that after weighing the specific design options, the project team will identify several ways that the system could be constructed. For example, the project team may find three vendors that make packaged systems that potentially

YOUR	8-1 SELECT A DESIGN STRATEGY
TURN	

Suppose that your university were interested in creating a new course registration system that could support Web-based registration.

QUESTION:
What should the university consider when determining whether to invest in a custom, packaged, or outsourced system solution?

could meet the project's needs; or the team may be debating over whether to develop a system using Visual Basic as a development tool and the database management system from Sybase; or the team may think it worthwhile to outsource the development effort to a consulting firm like Accenture or American Management Systems. Each alternative will have pros and cons associated with it that must be considered, and only one solution can be selected in the end.

Alternative Matrix

An alternative matrix can be used to organize the pros and cons of the design alternatives so that the best solution will be chosen in the end (see Figure 8-4). This matrix is created using the same steps as the feasibility analysis, which was presented in Chapter 2. The only difference is that the alternative matrix combines several feasibility analyses into one matrix so that the alternatives can be easily compared. The alternative matrix is a grid that contains the technical, budget, and organizational feasibilities for each system candidate, pros and cons associated with adopting each solution, and other information that is helpful when making comparisons. Sometimes weights are provided for different parts of the matrix to show when some criteria are more important to the final decision.

To create the alternative matrix, draw a grid with the alternatives across the top and different criteria (e.g., feasibilities, pros, cons, and other miscellaneous criteria) along the side. Next, fill in the grid with detailed descriptions about each alternative. This becomes a useful document for discussion because it clearly presents the alternatives being reviewed and comparable characteristics for each one.

Sometimes weights and scores are added to the alternative matrix to create a scorecard that communicates the project's most important criteria and the alternatives that best address them. A scorecard is built by adding a column labeled "weight" that includes a number depicting how much each criterion matters to the final decision. Typically, analysts take 100 points and spread them out across the

Criteria	Weight	Alternative 1: Custom Application Using VB	1-5*	Alternative 2: Customer Application Using Java	1-5	Alternative 3: Packaged Software-Product ABC	1-5
Criteria 1	25		5		1		3
Criteria 2	20		3		5		3
Criteria 3	10		3		5		3
Criteria 4	10		5		5		3
Criteria 5	10	Supporting	5	Supporting	5	Supporting	3
Criteria 6	5	Information	3	Information	3	Information	3
Criteria 7	5		3		1		3
Criteria 8	5		1		1		3
Criteria 9	5		1		1		3
Criteria 10	5		1		3		3
Total	100		360		320		300

* This denotes how well the aternative meets the criteria. 1 = poor fit; 5 = perfect fit

FIGURE 8-4
Sample Alternative Matrix Using Weights

criteria appropriately. If five criteria were used and all mattered equally, each criterion would receive a weight of 20. However, if cost were the most important criterion for choosing an alternative, it may receive "60" points, and the other four criteria may only get 10 points each.

Then, the analysts add a column called "score" to the matrix that communicates how well each alternative meets the criteria. Usually numbers like 1 to 5 or 1 to 10 are used to rank the appropriateness of the alternatives by the criteria. So, for the cost criterion the least expensive alternative may receive a 5 on a 1 to 5 scale; whereas, a costly alternative would receive a 1. When numbers are used in the alternative matrix, project teams can make decisions quantitatively and make decisions based on hard numbers.

Suppose that your company is thinking about implementing a packaged financial system, like Oracle Financials or Platinum, but there is not enough expertise in-house to be able to create a thorough alternative matrix. This situation is quite common—often the alternatives for a project are unfamiliar to the project team, so outside expertise is needed to provide information about the alternatives' criteria.

One helpful tool is the *request for proposal (RFP)*, a document that solicits proposals to provide the alternative solutions from a vendor, developer, or service provider. Basically, the RFP explains the system that you are trying to build and the criteria that you will use to select a system. Vendors, then, respond by describing what it would mean for them to be a part of the solution. They communicate the time, cost, and exactly how their product or services will address the needs of the project.

There is no formal way to write an RFP, but it should include basic information, like the description of the desired system, any special technical needs or circumstances, evaluation criteria, instructions for how to respond, and the desired schedule. An RFP can be a very large document (i.e., hundreds of pages) because companies try to include as much detail as possible about their needs so that the respondent can be just as detailed in the solution that would be provided. Thus, RFPs typically are used for large projects rather than small ones because they take a lot of time to create, and even more time and effort for vendors, developers, and service providers to develop high-quality responses—only a project with a fairly large price tag would be worth a response.

A less effort-intensive tool is a *request for information (RFI)* that takes the same format as the RFP. The RFI is shorter and contains less detailed information about a company's needs, and it requires general information from respondents that communicates the basic services that they can provide.

The final step, of course, is to decide which solution to design and implement. The approval committee should make the decision after the issues involved with the

---

YOUR	**8-2 ALTERNATIVE MATRIX**
TURN	

Pretend that you have been assigned the task of selecting a CASE tool for your class to use for a semester project. Using the Web or other reference resources, select three CASE tools (e.g., Visible Analyst Workbench, Oracle Designer/2000). Create an alternative matrix that can be used to compare the three software products in the way in which a selection decision can be made. Have a classmate select the "right" tools based on the information in your matrix.

different alternatives are well understood. Remember that the line between the analysis and design is quite fuzzy. Sometimes alternatives are described and selected at the end of analysis, and sometimes it is done at the beginning of design. The bottom line is that at some point before moving into the heart of the design phase, the project team and the approval committee must understand all of the feasible ways in which the system can be created, and they must select the way that makes the most sense for the organization. The design selection that is made will then drive many of the activities in the design phase.

Applying the Concepts at CD Selections

Alec Adams, Senior Systems Analyst and Project Manager for CD Selections' Internet Order System, had three different approaches that he could take with the new system: he could develop the entire system using development resources from CD Selections, he could buy a packaged software program (or a set of different packages and integrate them), or he could hire a consulting firm or service provider to create the system. Immediately, Alec ruled out the third option. Building Internet applications, especially e-commerce systems, was becoming increasingly important to the CD Selections' business strategy. By outsourcing the Internet system, CD Selections would not develop Internet application development skills and business skills within the organization.

Instead, Alec decided that a custom development project using the company's standard Web development tools would be the best choice for CD Selections. In this way, the company would be developing critical technical and business skills in-house, and the project team would be able to have a high level of flexibility and control over the final product. Also, Alec wanted the new Internet system to directly interface with the existing special order system, and there was a chance that a packaged solution would not be able to integrate as well into the CD Selections environment.

There was one part of the project that potentially could be handled using packaged software: the request-taking portion of the application. Alec realized that a multitude of programs have been written and are available (at low prices) to handle customer transactions over the Web. These programs, called shopping cart programs, usually allow customers to select items for an order form, input basic information, and finalize the order transaction. Alec believed that the project team should at least consider some of these packaged alternatives so that less time had to be spent writing a program that handled basic Web tasks and more time could be devoted to innovative marketing ideas and custom interfaces with the special order system.

To help better understand some of the shopping cart programs that were available in the market and how their adoption could benefit the project, Alec created an alternative matrix that compared three different shopping cart programs to one another (Figure 8-5). Although all three alternatives had positive points, Alec saw alternative 2 (WebShop) as the best alternative for handling the shopping cart functionality for the new Internet system. WebShop was written in Java, the tool that CD Selections selected as its standard Web development language; the expense was reasonable, with no hidden or recurring costs; and there was a person in-house who had some positive experience with the program. Alec made a note to look into acquiring WebShop as the shopping cart program for the Internet system.

	Alternative 1: Shop With Me	Alternative 2: WebShop	Alternative 3: Shop-N-Go
Technical feasibility	Developed using C; very little C experience in-house Orders sent to company using e-mail files	Developed using C and Java; would like to develop in-house Java skills Flexible export features for passing order information to other systems	Developed using Java; would like to develop in-house Java skills Orders saved to a number of file formats
Economic feasibility	$150 initial charge	$700 up-front charge, no yearly fees	$200/year
Organizational feasibility	Program used by other retail music companies	Program used by other retail music companies	Brand-new application; few companies have experience with Shop-N-Go to date
Other benefits	Very simple to use	Tom in Information Systems support has had limited but positive experience with this program Easy to customize	
Other limitations			The interface is not easily customized

FIGURE 8-5
Alternative Matrix for Shopping Cart Program

MOVING FROM LOGICAL TO PHYSICAL MODELS

Once the design strategy has been selected, the next step is to move from the logical process and data models to the physical ones and create the CRUD matrix. The project team does this if custom development has been chosen. If the project has been outsourced or packaged software has been selected, then this is done by the consultant or software developer. We will assume for the remainder of the book that custom development has been chosen.

So far, the project has defined the processes supporting the business system by drawing *logical DFDs* and the data that is used by those processes by drawing *logical ERDs*. These models do not contain any indication of how the system will actually be implemented when the information system is built; they simply state what the new system will do. In this way, developers do not get distracted by technical details and are not biased by technical limitations in the initial stages of system development, and business users can better understand diagrams that show recognizable business ideas. These diagrams can be thought of as containing the "business view" of the system.

However, when the system is being designed, *physical process models* and *physical data models* are created to show implementation details and to explain how the final system will work. These details can include references to actual technology, the format of information moving through processes, and the human interaction that is involved. In some cases, most often when packages are used, the use cases may need to be revised as well. The models are to-be models because they describe characteristics of the system that will be created, and the diagrams can be thought of as containing the "systems (or programmer) view" of the system; they communicate the system requirements of the new system.

After the physical models are created, analysts can use a technique called the CRUD matrix to depict exactly how data is being used by the processes in the system. The matrix—which stands for create read update delete—is drawn to make sure that the data stores are associated with the right processes in the correct way. First, we will explain the physical DFD and show an example, followed by the physical ERD and a second example. We will end the chapter with a description of the CRUD matrix.

The Physical Data Flow Diagram

The physical DFD contains the same components as the logical DFD (e.g., data stores, data flows), and the same rules apply (e.g., balancing, decomposition). The basic difference between the two models is that a physical DFD contains additional details that describe how the system will be built. There are five steps to perform to make the transition to the physical DFD (Figure 8-6).

Step 1: Add Implementation References   The first step in creating a physical DFD is to begin with the existing logical DFD and add references to the ways in which the data stores, data flows, and processes will be implemented. Data stores on physical DFDs will refer to files and/or database tables; processes, to programs or human actions; and data flows, to the physical medium for the data, such as paper reports, bar code scanning, input screens, or computer reports. The names for the various components on the physical DFD should contain references to these implementation details. By definition, external entities on the DFD are outside of the scope of the system and therefore remain unchanged in the physical diagram.

Figure 8-7 shows the physical DFD that was drawn to depict the physical details for the original logical DFD from Figure 6-20. Notice how the logical data store called *inventory* that will store data in the inventory table of an Oracle database has been renamed "Oracle: Inventory Table" and the logical data flow order now includes

Step	Explanation
Add implementation references	Using the existing logical DFD, place the way in which the data stores, data flows, and processes will be implemented in parentheses below each component.
Draw a human–machine boundary	Draw a line to separate the automated parts of the system from the manual parts.
Add system-related data stores, data flows, and processes	Add system-related data stores, data flows, and processes to the model (components that have little to do with the business process).
Update the data elements in the data flows	Update the data flows to include system-related data elements.
Update the metadata in the CASE repository	Update the metadata in the CASE repository to include physical characteristics.

CASE = computer-aided software engineering; DFD = data flow diagram.

FIGURE 8-6
Steps to Create the Physical Data Flow Diagram

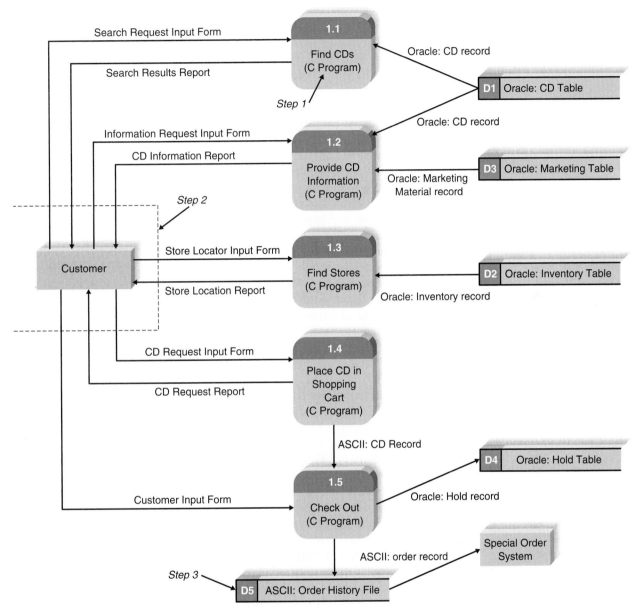

FIGURE 8-7
The Physical Data Flow Diagram (The *How*)

"Oracle: Inventory Record" to show that this information will be in the form of a record from the inventory table. Can you identify other changes that were made to the physical model to communicate how other components will be implemented?

Step 2: Draw a Human–Machine Boundary   The second step is to add a *human–machine boundary*. Physical DFDs also are different from their logical counterparts because they differentiate human and computer interaction using a human–machine boundary, which is a line drawn on the model to separate human action from automated processes. For example, the take request processes (i.e., processes 1.1 – 1.5) require the customer to interact with the Web using an inter-face driven by system programs and processes. The physical model, therefore, contains a line separating the customer from the rest of the process to show exactly what is done by a person as opposed to a "machine" (see Figure 8-7).

Every part of every process in the system may not be automated, so it is up to the project team to determine where to draw a human–machine boundary and how large to draw it. The project team will need to weigh the following criteria when drawing the boundary: cost, efficiency, and integrity. First, a piece of the system should be automated only if the cost of computerizing it is less than doing it man-ually. Next, the system should be more efficient with the mode that is selected. For example, if the project team must decide whether to store a paper copy of a docu-ment in a filing cabinet and have people access it manually or to save an electronic file of information in a central file server that all employees can access, the team likely will find the latter option to be more efficient in terms of letting users access and update the information.

Finally, the team should consider the integrity of the information that is han-dled by the system. It may be cheaper for a clerk to record orders by phone and deliver the order forms to the distribution area; however, errors could be made when the clerk takes the order, and a form could be misplaced en route to distribution. Instead, the project team may be more comfortable with an automated process that accepts a customer's order from the customer directly using a Web form that is then directly transmitted to the distribution system.

Step 3: Add System-Related Data Stores, Data Flows, and Processes

In step 3, you will add to the DFD additional processes, stores, or flows that are specific to the implementation of the system and have little (or nothing) to do with the business process itself. These additions can be due to technical limitations or to the need for audits, controls, or exception handling. Technical limitations occur when tech-nology cannot support the way in which the system is modeled logically. For example, suppose a data store exists on the logical DFD to hold customer infor-mation, but the database technology that will be used to build the system cannot handle the large volume of customers in one table. A physical DFD may need to have two data stores—one for current customers and one for old customers—so that the technology will work properly.

Audits, controls, or exception handling refers to putting checks and balances in place in the system in case something goes wrong. For instance, on rare occa-sions, customers might call and cancel an order that they placed. Instead of just hav-ing the system get rid of the information about that order, a process may be included for control purposes that records the deleted orders along with reasons for the cancellations. Or, consider the take-request processes that we have been work-ing with. Suppose the project team is worried about the transfer of order records from the new Internet system to the special order system. As a precaution, the team can place a data store on the physical DFD that captures information about each batch of records that is sent to the special order system. In this way, if problems were to occur, there would be a history of transactions that could be examined or used for backup purposes (see Figure 8-7).

Step 4: Update the Data Elements in the Data Flows

The fourth step is to update the elements in the data flows. The data flows will appear to be identical in both the logical and physical DFDs, but the physical data flows may contain additional system-related data elements for reasons similar to those described in the previous section. For example, most systems add system-related data elements to data flows that capture when changes were made to information (e.g., a last_update data element) and who made the change (e.g., an updated_by data element). Another physical data element is a system-generated number used to uniquely identify each record in a database. During step 4, the physical data elements are added to the metadata descriptions of the data flows in the CASE repository.

Step 5: Update the Metadata in the Computer-Aided Software Engineering Repository
Finally, the project team needs to make sure that the information about the DFD components in the CASE repository is updated with implementation-specific information. This information can include when batch processes will be run and how often, names of the actual tables or files that are represented by data flows, and the sizes and projected growth rates of the data stores.

Applying the Concepts at CD Selections

To better understand physical DFDs, we will now use an example based on the CD Selections maintain CD marketing material process. Figure 8-8 shows the logical DFD. We will perform each step to create the physical model, using the logical model as our starting point. Before we begin, see if you can identify how the data stores, data flows, and processes will need to change to reflect physical characteristics of the proposed system.

First, we need to identify how the data flows, data stores, and processes will be implemented and add the implementation references on the DFD. From the original business requirements, we know that (1) the marketing material from vendors usually comes as files attached to e-mails, (2) the marketing manager will receive reports via the Web, and (3) vendor information will be received as records from an Oracle table. On the physical DFD, each of the data flows is altered by renaming them appropriately as shown in Figure 8-9.

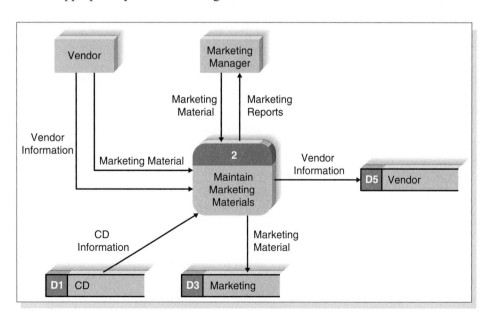

FIGURE 8-8
Logical Model of the Maintain Marketing Process

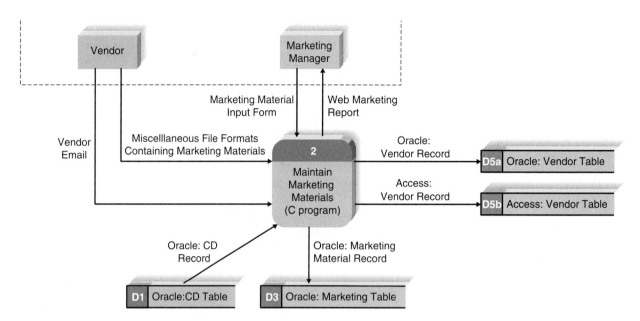

FIGURE 8-9
Physical Model for Maintain Marketing Material Process

Let us assume that the data stores (i.e., vendor, CD, and marketing material) will refer to tables called vendor, CD, and marketing material that are contained in an Oracle database; therefore, all three stores are updated with this information to indicate their physical qualities. Also, the maintain marketing materials process will be written using C, and this information is added to the process model.

As a second step, a dotted line is drawn to represent the human–machine boundary and to communicate how much (and what parts) of the process is automated, and next we add system-related components to the model. Let us assume that vendor information actually will exist in two formats at CD Selections—an Oracle database that supports the order system and a Microsoft Access database used by Marketing for other purposes. For now, we will add two vendor data stores to reflect this situation. See Figure 8-9 for these changes.

Completion of the last two steps, 4 and 5, will not be apparent on the physical DFD. In step 4, we will add system-related data elements to the data flow entries in the CASE repository. For example, we will create a system-related data element called last_update and add it to the data flow that goes from process 2 to the marketing material data store. This field will capture the last time a piece of marketing material was inserted or changed in the system.

Step 5 requires that we add implementation-specific information in the metadata in the CASE repository. This can include such information as the actual field types and sizes of the data elements that will be stored in the marketing material table, or the expected response time for a report to be created for the marketing manager.

The Physical Entity Relationship Diagram

Like the DFD, the ERD contains the same components for both the logical and physical models, including entities, relationships, and attributes. The difference lies

In Chapter 6, you were asked to create a logical data flow diagram (DFD) for the housing system run by the Campus Housing Service ("Your Turn 6-1"). The Campus Housing Service helps students find apartments. Owners of apartments fill in information forms about the rental units they have available (e.g., location, number of bedrooms, monthly rent), and the information is entered into a database. Students can search through this database via the Web to find apartments that meet their needs (e.g., a two-bedroom apartment for $400 or less per month within 1/2 mile of campus). They then contact the apartment owners directly to see the apartment and possibly rent it. Apartment owners call the service to delete their listing when they have rented their apartment(s). Create a physical DFD for the above situation. Compare the diagram that you just drew to the logical diagram that you created in Chapter 6.

in the facts that physical ERDs contain references to exactly how data will be stored in a file or database table and that much more metadata is added to the CASE repository to describe the data model components. The transition from the logical to physical data model is fairly straightforward; see the steps in Figure 8-10.

Step 1: Change Entities to Tables or Files   The first step is to change all the entities in the logical ERD to reflect the files or tables that will be used to store the data. Usually, project teams adhere to strict naming conventions for such things as tables, files, and fields, so the physical ERD would use the names that the real components will have when implemented. Metadata for the tables and files, like the expected size of the table, are added to the CASE repository. See Figure 8-11 for a physical ERD from the doctor's office system that was described in Chapters 6 and 7.

Step 2: Change Attributes to Fields   Second, change the attributes to fields, which are columns in files or tables, and add information like the field's length, data type, default value, and valid value to the CASE repository. There are a number of different data types that fields can have, such as number, decimal, longint, character, and variable character. The analyst inputs the data type along with the size of the field into the CASE tool so that the system can be designed for the

Step	Explanation
Change entities to tables or files	Beginning with the logical entity relationship diagram, change the entities to tables or files and update the metadata.
Change attributes to fields	Convert the attributes to fields and update the metadata.
Add primary keys	Assign primary keys to all entities.
Add foreign keys	Add foreign keys to represent the relationships among entities.
Add system-related components	Add system-related tables and fields.

FIGURE 8-10

Steps to Moving from Logical to Physical Entity Relationship Diagram

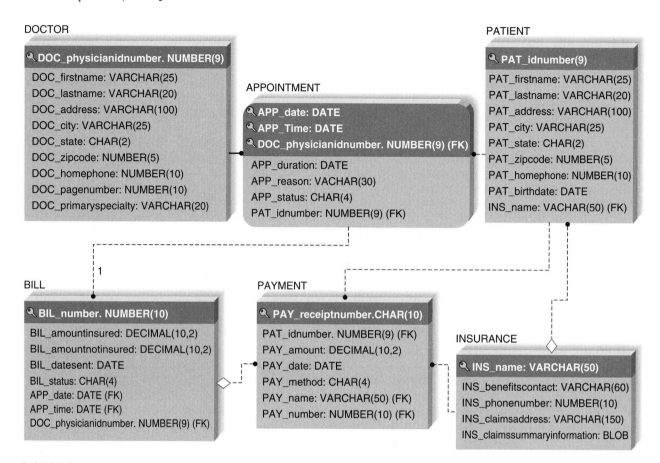

FIGURE 8-11
Steps to Moving from Logical to Physical Entity Relationship Diagram

right kind of information. A *default value* specifies what should be placed in a column if no value is explicitly supplied when a record is inserted into the table. A *valid value* is a fixed list of valid values for a particular column, or an expression to define some form of data validation code for a column or table. Figure 8-12 shows a variety of metadata that describes a field called cust_id.

Inputting complete information regarding the tables and columns into the CASE repository is very important. Many CASE tools will actually generate code to build tables and create files for the new system based upon the information they contain for the physical models. By taking time to describe the physical process and data models in detail, the analyst can save a lot of time when it comes time for implementing the system.

Step 3: Add Primary Keys   As a third step, the attributes that served as identifiers on the logical ERD are converted into *primary keys*, which are fields that contain a unique value for each record in the file or table. For instance, Social Security number would serve as a good primary key for a customer table if every customer record in the table will contain a unique value in the social security number field. Unlike with the logical model, a unique identifier is mandatory for every table placed on the physical ERD; therefore, primary fields must be created for entities

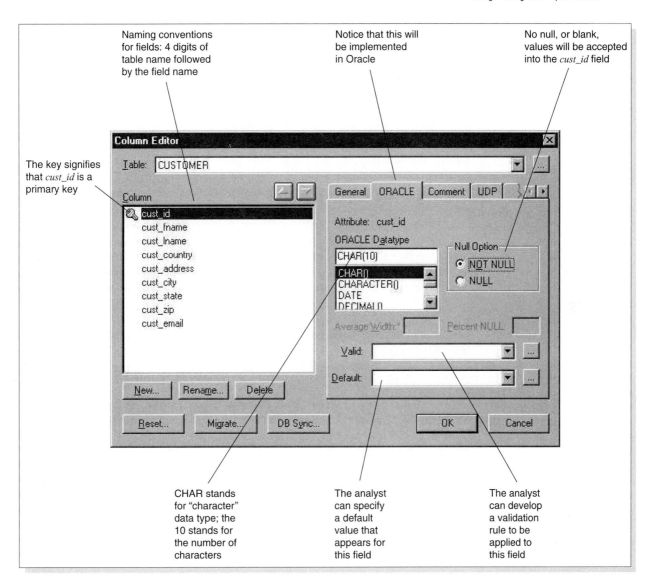

Naming conventions for fields: 4 digits of table name followed by the field name

Notice that this will be implemented in Oracle

No null, or blank, values will be accepted into the *cust_id* field

The key signifies that *cust_id* is a primary key

CHAR stands for "character" data type; the 10 stands for the number of characters

The analyst can specify a default value that appears for this field

The analyst can develop a validation rule to be applied to this field

FIGURE 8-12
Metadata for a cust_id Field

that did not have identifiers previously. If we did not choose an identifier for the customer entity on the logical ERD, we would now create a system-generated field (e.g., cust_id) that could serve as the primary key for the customer table. This field would have no meaning or purpose other than ensuring that each record has a field that contains unique values.

Step 4: Add Foreign Keys   The relationships on the logical ERD show that pairs of entities are associated with each other, and in step 4, the analyst specifies how the associations are going to be maintained from a technical standpoint. In a relational database, for example, an association between two tables is maintained using a technique referred to as a *foreign key*. A foreign key is the primary key field(s) from one table that is repeated in another table to provide a common field between the two tables. The common field contains values that match a record in one table to a record in the other. For example, if we were to create two tables called customer and order that were related to each other, we could include the primary key field from customer (cust_id) in the order table as well. In this way, if we want to find out customer information (e.g., name, address, phone number) when looking at someone's order, we can use the value for cust_id that appears in the order table to go back to the customer table to locate the appropriate information.

Thus, on the physical ERD, the primary key fields in the parent tables (the "1" end of the relationship) are copied and placed as fields in the child tables (the "many" end of the relationship) and designated as foreign keys. The fields will contain values that are common between the two tables. Many times, the CASE tools that are used to draw ERDs will "migrate" foreign keys to the appropriate tables on the model automatically, and the database technology will ensure that the values in the two fields match appropriately. For example, you would not want to have an order table that contains a cust_id value that does not exist in the customer table; this would mean that someone placed an order who is not recognized as a customer. The idea of having values in foreign key fields that match values in primary keys fields is called *referential integrity*, and this concept is described further in Chapter 12.

Step 5: Add System-Related Components   As the fifth and final step, components are added to the physical ERD to reflect special implementation needs, including components that were included on the DFD. We have mentioned balance between DFDs and ERDs in earlier chapters, and this balance must be maintained in the physical models as well. Therefore, implementation-specific data stores and data elements from the physical DFD should be included on the ERD as tables and fields. For example, in Figure 8-7 we added the order history data store to the physical DFD to serve as a control data store for orders that are sent to the special order system. Now we will need to add an order batch history table to the physical ERD model along with its fields and relationships.

Some CASE tools allow you to toggle back and forth between the logical and physical ERDs, and changes that you make to one model (e.g., adding an entity) will be reflected in the other. When you first begin data modeling, you should focus on creating a sound logical model, and then gradually over time begin to add physical details. More detailed information on database and file design is explained in Chapter 11.

Applying the Concepts at CD Selections   Let us now show how to apply some of the concepts that you have learned by creating a physical ERD using the logical ERD that was created in Chapter 7.

When we use the logical model as a starting point, the first step is to rename the entities to match with the tables or files that will be used by the system (Figure 8-13). Outwardly, the data model does not look very different after this step, but notice that the marketing material entity was renamed to represent a table called MKTMAT, and the in-store hold entity becomes the HOLD table. At this time, we will need to include metadata for the tables, such as their estimated size.

Next, the attributes for the entities become fields with such characteristics as data type, length, and valid values, and this is recorded in the CASE repository. For example, cust_state in the customer table will be a text field with a size of two characters, and valid values are the 50 two-letter state abbreviations. If most customers at CD Selections live in the state of California, then it may be worthwhile to make

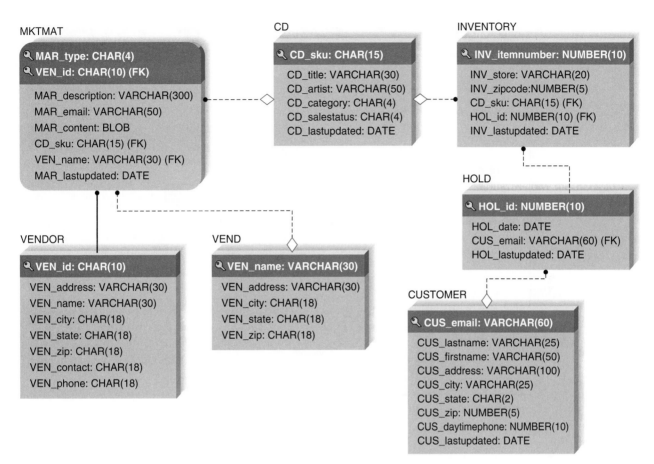

FIGURE 8-13
Physical Entity Relationship Diagram for CD Selections

CA the default value for this field. Figure 8-14 is an example of the CASE repository entry for the cust_state field.

Step 3 suggests that we change the identifiers in the logical ERD to become primary keys (e.g., customer e-mail, CD SKU), and entities without identifiers need to have a primary key created. At this time, we also can decide to use a system-generated primary key if it is more efficient than using logical attributes from the logical model. For example, instead of using a combination of vendor name and address to uniquely identify a vendor, it is easier to add a system-generated field called vendor ID for each record in the vendor table (see Figure 8-13). The same goes for the hold table—a hold ID is easier to use to track records than the combination of customer e-mail and hold date.

The relationships on the logical ERD indicate where foreign key fields need to be placed. For example, customer e-mail is placed as a field in hold to serve as the link between two entities, and hold gets the extra field because it is the child table (it exists at the "many" end of the relationship). Similarly, hold ID is placed in the inventory table.

Finally, system-related components are included within the model. Notice that two vendor tables were added to the model to represent the table for this order

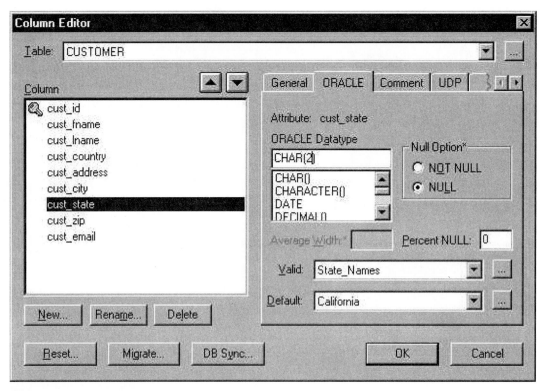

FIGURE 8-14
Computer-Aided Software Engineering Repository Entry for cust_state Field

system and the table maintained by marketing. Also, fields that will capture when a record was last inserted or updated were added to many of the tables.

CRUD Matrix

The physical process and data models must balance with each other just like the logical models balanced (see Chapter 7). There is another way in which analysts can

---

8-4 ISLAND CHARTERS

In Chapter 7, you were asked to create a logical entity relationship diagram (ERD) for a charter company that owns boats that are used to chart trips to the islands ("Your Turn 7-8"). The company has created a computer system to track the boats it owns, including each boat's ID number, name, and seating capacity. The company also tracks information about the various islands, such as name and population. Every time a boat is chartered, it is important to know the data about the trip that takes place and the number of people on the trip. The company also keeps information about each captain, such as Social Security number, name, birthdate, and how to contact next of kin. Boats travel to only one island per visit.

Create a physical ERD for this situation. Compare the diagram that you drew to the logical diagram that you created in Chapter 7.

better understand how process models and data models interrelate—it is called a CRUD matrix.

A *CRUD matrix*—which stands for create read update delete—is a matrix that shows how data are used by the processes within the system. The CRUD matrix can be created in either the analysis or design phase, but in the design phase it helps analysts ensure that all of the data stores used by processes have been created. It also provides important information for the program specifications (Chapter 12) because the matrix shows exactly how data is used and created by the major processes in the system.

To created a CRUD matrix, first draw a table with all of the processes in the system listed across the top and all of the tables (or tables and fields) listed as rows on the left-hand side of the table. Next, complete the table by writing a C, R, U, and/or D in the table to show how the system processes that are listed across the top of the matrix use each field.

The information on the CRUD matrix should match what is presented in the physical process model. If a process does not interact with a data store, then there should be no data flows between them on the process model. If a process reads information from a data store but does not update it, then there should be a data flow coming out of the data store only. When a data store is updated in some way by a process, then data should flow into the data store from the process. Look at Figure 8-15, and notice how the CRUD matrix mirrors what is seen on the process model.

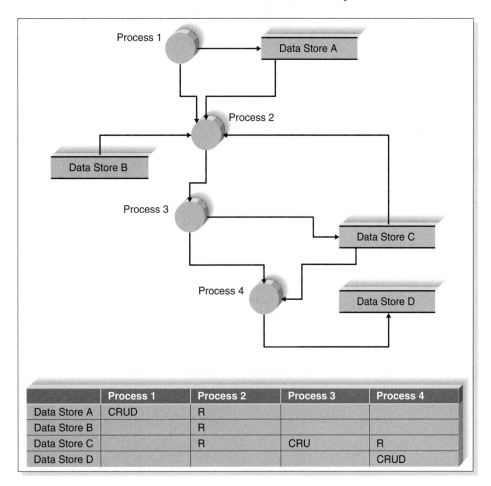

	Process 1	Process 2	Process 3	Process 4
Data Store A	CRUD	R		
Data Store B		R		
Data Store C		R	CRU	R
Data Store D				CRUD

FIGURE 8-15
CRUDE Matrix and Process Model

Figure 8-16 shows an example of a preliminary CRUD matrix that was created for the CD Selections Order System *take-request* processes shown in Figure 8-7. Look at the original process models, and notice how the first three processes are merely reading information from data stores. This is illustrated on the CRUD matrix by placing an "R" in the relevant intersections of the matrix. Can you tell how data is used by the remaining processes?

Once the DFDs and ERDs are converted to physical models that describe exactly how the system will be built and a CRUD matrix is created to show how the processes use data, the project team can focus on the rest of the design phase tasks and techniques that are presented in Chapters 10 through 12. All the deliverables from the design phase will be used to create the system specification, which is used by programmers to build the system during implementation.

	1.1 Find CDs	1.2 Provide CD Information	1.3 Find Stores	1.4 Place CD in Shopping Cart	1.5 Check out
CD table					
cd_sku	R	R			
cd_title	R	R			
cd_artist	R	R			
cd_category	R	R			
cd_salestatus	R	R			
cd_lastupdated					
MKTMAT table					
mar_type		R			
mar_venid					
mar_description					
mar_email					
mar_content		R			
cd_sku		R			
ven_name					
mar_lastupdated					
INVENTORY table					
inv_itemnumber			R		
inv_store			R		
inv_zipcode			R		
cd_sku			R		
hol_id					
invl_lastupdated					
HOLD table					
hol_id					CRUD
hol_date					CRUD
cus_email					CRUD
hol_lastupdated					CRUD

FIGURE 8-16
Preliminary CRUD Matrix for *Take Request* processes

SUMMARY

The Design Phase

The design phase is the phase of the SDLC in which the blueprint for the new system is developed, and it contains many steps that guide the project team through planning exactly how the system needs to be constructed. The requirements that were identified in the analysis phase serve as the primary inputs for design activities. The main deliverable from the design phase is the system specification, which includes the physical process and data models, architecture report, hardware and software specifications, interface design, data storage design, and program design.

Design Strategies

During the design phase, the project team also needs to consider three approaches to creating the new system, including developing a custom application in-house; buying a packaged system and customizing it; and relying on an external vendor, developer, or system provider to build and/or support the system. Custom development allows developers to be flexible and creative in the way they solve business problems, and it builds technical and functional knowledge within the organization. Many companies have a development staff that is already overcommitted to filling huge backloads of systems requests, however, and they just don't have time to devote to a project for which a system is built from scratch. It can be much more efficient to buy programs that have been created, tested, and proven, and a packaged system can be bought and installed in a relatively short period of time, when compared with a custom solution. Workarounds can be used to meet the needs that are not addressed by the packaged application. The third design strategy is to outsource the project and pay an external vendor, developer, or service provider to create the system. It can be a good alternative for how to approach the new system; however, it does not come without costs. If a company does decide to leave the creation of a new system in the hands of someone else, the organization could compromise confidential information or lose control over future development.

Design Strategy

Each of the design strategies just discussed has its strengths and weaknesses, and no one strategy is inherently better than the others. Thus, it is important to consider such issues as the uniqueness of business need for the system, the amount of in-house experience that is available to build the system, and the importance of the project skills to the company. Also, the existence of good project management and the amount of time available to develop the application play a role in the selection process.

Selecting a Design Strategy

Ultimately, the decision must be made regarding the specific type of system that needs to be designed. An alternative matrix can help the design team make this decision by presenting feasibility information for several candidate solutions in a way in which they can be compared easily. The request for proposals and request for information are two ways to gather accurate information regarding the alternatives. At this point, the team decides exactly who will perform each part of the design phase and what packages will be used.

Physical Data Flow Diagrams and Entity Relationship Diagrams

One important aspect of the initial part of design is the movement from logical to physical entity relationship diagrams (ERDs) and data flow diagrams (DFDs), which show implementation details and indicate how the final system will work. Physical DFDs include data stores that refer to files and database tables, programs, or human actions that perform processes, and the physical transfer medium for the data flows. They show the human–machine boundary, which illustrates which parts of the system are automated and which are manual.

Physical ERDs contain references to how data will be stored in a file or database table, and metadata are included to describe the data model components. Both models can reflect design decisions (e.g., creating a temporary storage for data, creating a field to capture when a record was last inserted or changed) that will affect the physical implementation of the system.

A CRUD matrix can be created to show exactly how data is created and used by the processes in the system. Information in the CRUD matrix is helpful to validate the physical process and data models and to provide information for the program design that is described in Chapter 12.

KEY TERMS

Alternative matrix	Logical data flow diagram	Request for information (RFI)
CRUD matrix	Logical entity relationship diagram	Request for proposal (RFP)
Custom development	Outsourcing	System requirement
Design phase	Packaged software	System specification
Enterprise resource planning (ERP)	Physical data model	Systems integration
Fixed-price contract	Physical process model	Time and arrangements deal
Foreign key	Primary key	Value-added contract
Human–machine boundary	Referential integrity	Workaround

QUESTIONS

1. What are the main activities that happen during the design phase of the systems development life cycle (SDLC)?
2. What is the main deliverable from the design phase? What does it include?
3. Compare and contrast analysis and design activities. What is the relationship between things that happen in analysis and things that happen in design?
4. What situations are most appropriate for a custom development design strategy?
5. What are some problems with using a packaged software approach to building a new system? How can these problems be addressed?
6. Why do companies invest in enterprise resource planning (ERP) systems?
7. What are the pros and cons of using a workaround?

8. When is outsourcing considered a good design strategy? When is it not appropriate?
9. What are the differences between the time and arrangements, fixed-price, and value-added contracts for outsourcing?
10. How are the alternative matrix and feasibility analysis related?
11. What is a request for proposals (RFP)? How is this different from a request for information (RFI)?
12. Why do you think most companies initially assume a custom development strategy will be used when first considering a new system? Is this a good assumption?
13. Can we eliminate or reduce the analysis phase when we know packaged software will be used instead of outsourcing or custom development? Explain.

14. Why is systems integration becoming more common?
15. What is the purpose of creating logical models and then physical models?
16. What are the differences between the logical and the physical data flow diagram (DFD)?
17. What are the differences between the logical and the physical entity relationship diagram (ERD)?
18. What is a human–machine boundary?
19. What kind of metadata is included in the computer-aided software engineering (CASE) repository for a physical ERD?
20. Describe the purposes of primary keys and foreign keys.
21. What are some system-related decisions that may result in a change to the physical DFD or ERD?
22. What is a CRUD matrix? How does it relate to process models and data models?

EXERCISES

A. Draw a physical level 0 data flow diagram (DFD) for the following dentist office system, and compare it to the logical model that you created in Chapter 6. Whenever new patients are seen for the first time, they complete a patient information form that asks their name, address, phone number, and brief medical history, all of which are stored in the patient information file. When a patient calls to schedule a new appointment or change an existing appointment, the receptionist checks the appointment file for an available time. Once a good time is found for the patient, the appointment is scheduled. If the patient is a new patient, an incomplete entry is made in the patient file; the full information will be collected when the patient arrives for the appointment. Because appointments are often made so far in advance, the receptionist usually mails a reminder postcard to each patient 2 weeks before his or her appointment.

B. Create a physical level 0 DFD for the following, and compare it to the logical model that you created in Chapter 6. A Real Estate Inc. (AREI) sells houses. People who want to sell their houses sign a contract with AREI and provide information on their house. This information is kept in a database by AREI, and a subset of this information is sent to the citywide multiple listing service used by all real estate agents. AREI works with two types of potential buyers. Some buyers have an interest in one specific house. In this case, AREI prints information from its database, which the real estate agent uses to help show the house to the buyer (a process beyond the scope of the system to be modeled). Other buyers seek AREI's advice in finding a house that meets their needs. In this case, buyers complete a buyer information form and information from it is entered into a buyer database, and AREI real estate agents use its information to search AREI's database and the multiple listing service for houses that meet their needs. The results of these searches are printed and used to help the real estate agents show houses to the buyers.

C. Draw a physical level 0 DFD for the following system and compare it to the logical model that you created in Chapter 6. A Video Store (AVS) runs a series of fairly standard video stores. Before a video can be put on the shelf, it must be catalogued and entered into the video database. To rent a video, every customer must have a valid AVS customer card. Customers rent videos for three days at a time. Every time a customer rents a video, the system must ensure that he or she does not have any overdue videos. If so, the overdue videos must be returned and the customer must pay a fine before renting more videos. Likewise, if the customer has returned overdue videos but has not paid the fine, the fine must be paid before new videos can be rented. Every morning, the store manager prints a report that lists overdue videos; if a video is two or more days overdue, the manager calls the customer to remind him or her to return the video. If a video is returned in damaged condition, the manager removes it from the video database and sometimes charges the customer.

D. How would the following ERD be changed to incorporate the design decision listed next?
   • The analyst wants to keep track of the user ID of anyone who changes a grade for a course.
   • A data store is added on the physical DFD so that information regarding the current semester's courses can be stored temporarily during the add/drop time period before the courses become a part of the student's permanent record.

- The system would like to archive alumni into a table once they graduate so that only active students are stored in the student table.
E. Describe what metadata you would place in the computer-aided software engineering (CASE) repository for the following physical data model components. Describe how these entries would be different if you were working on a logical data model.
    - Entity—product
    - Attribute—product number
    - Attribute—product type
F. Consider the situation in exercise A. Which design strategy would you recommend for the construction of this system? Why?
G. Consider the situation in exercise B. Which design strategy would you recommend for the construction of this system? Why?
H. Consider the situation in exercise C. Which design strategy would you recommend for the construction of this system? Why?
I. Assume that you are leading a project that will implement a new course enrollment system for your uni-

versity. You are thinking about either using a packaged course enrollment application or outsourcing the job to an external consultant. Create a request for proposal (RFD) to which interested vendors and consultants could respond.
J. Assume that you and your friends are starting a small business painting houses in the summertime. You need to buy a software package that handles the financial transactions of the business. Create an alternative matrix that compares three packaged systems (e.g., Quicken, Microsoft Money, Quickbooks). Which alternative appears to be the best choice?
K. Draw a physical process model (just the processes and data stores) for the following CRUD matrix.

Student	Register Student	Schedule Student	Create Transcript	Create Bill
Student Data Store	CRUD	R	R	R
Course Data Store		CRUD	R	
Billing Data Store		CRUD		CRUD
Grade Data Store			CRUD	

MINICASES

1. Susan, president of MOTO, Inc., a human resources management firm, is reflecting on the client management software system her organization purchased 4 years ago. At that time, the firm had just gone through a major growth spurt, and the mixture of automated and manual procedures that had been used to manage client accounts became unwieldy. Susan and Nancy, her IS department head, researched and selected the package that is currently used. Susan had heard about the software at a professional conference she attended, and at least initially, it worked fairly well for the firm. Some of their procedures had to change to fit the package, but they expected that and were prepared for it.

Since that time, MOTO, Inc. continued to grow, not only through an expansion of the client base, but through the acquisition of several smaller employment-related businesses. MOTO, Inc. is a much different business than it was 4 years ago. Along with expanding to offer more diversified human resource management services, the firm's support staff has also expanded. Susan and Nancy are particularly proud of the IS Department they have built up over the years. Using strong ties with a local university, an attractive compensation package, and a good

working environment, the IS Department is well staffed with competent, innovative people, plus a steady stream of college interns keeps the department fresh and lively. One of the IS teams pioneered the use of the Internet to offer MOTO's services to a whole new market segment, an experiment that has proven very successful.

It seems clear that a major change is needed in the client management software, and Susan has already begun to plan financially to undertake such a project. This software is a central part of MOTO's operations, and Susan wants to be sure that a quality system is obtained this time. She knows that the vendor of their current system has made some revisions and additions to its product line. There are also a number of other software vendors who offer products that may be suitable. Some of these vendors did not exist when the purchase was made four years ago. Susan is also considering Nancy's suggestion that the IS Department develop a custom software application.

a. Outline the issues that Susan should consider that would support the development of a custom software application in-house.
b. Outline the issues that Susan should consider which would support the purchase of a software package.

c. Within the context of a systems development project, when should the decision of "make-versus-buy" be made? How should Susan proceed? Explain your answer.

2. Refer to the level 1 DFD you prepared in minicase 1, Chapter 6, part d. Several design decisions have been made for the new system. A Visual Basic program will be written to perform the validation process. Staffing requests will come from the clients on a standard preprinted form (designated Form 367). The contract and staffing request tables will be implemented in Access, and Access will be used to create the electronic version of the staffing request. The denial letters will be prepared with MS Word, printed and mailed to the client on PSSM company stationery, and filed electronically in a specific folder on the file server.

Modify the DFD prepared earlier to reflect these physical implementation details. Also, draw in the human–machine boundary.

PLANNING

ANALYSIS

DESIGN

- ☑ **Select Design**
- ☑ **Develop Physical Data Flow Diagrams**
- ☑ **Develop Physical Entity Relationship Diagrams**
- ☐ **Design Architecture**
- ☐ **Select Hardware and Software**
- ☐ Develop Use Scenarios
- ☐ Design Interface Structure
- ☐ Design Interface Standards
- ☐ Design Interface Prototype
- ☐ Evaluate User Interface
- ☐ Design User Interface
- ☐ Select Data Storage Format
- ☐ Denormalize Entity Relationship Diagram
- ☐ Performance Tune Data Storage
- ☐ Size Data Storage
- ☐ Develop Program Structure Chart
- ☐ Develop Program Specification

TASK CHECKLIST

PLANNING   ANALYSIS   DESIGN

CHAPTER 9

ARCHITECTURE

DESIGN

An important component of the design phase is the architecture design, which describes the system's hardware, software, and network environment. The architecture design flows primarily from the nonfunctional requirements, such as operational, performance, security, cultural and political requirements. The deliverable from architecture design includes the architecture design and the hardware and software specification.

OBJECTIVES

- Understand the fundamental components of an information system.
- Understand server-based, client-based, and client–server architectures.
- Understand how operational, performance, security, cultural and political requirements affect the architecture design.
- Be familiar with how to create an architecture design.
- Be familiar with how to create a hardware and software specification.

CHAPTER OUTLINE

IMPLEMENTATION

INTRODUCTION

In today's environment, most information systems are spread across two or more computers. A Web-based system, for example, will run in the browser on your desktop computer but will interact with the Web server (and possibly other computers) over the Internet. A system that operates completely inside a company's network may have a Visual Basic program installed on your computer but interact with a database server elsewhere on the network. Therefore, an important step of the design phase is the creation of the *architecture design,* the plan for how the system will be distributed across the computers and what hardware and software will be used for each computer (e.g., Windows, Linux).

Most systems are built to use the existing hardware and software in the organization, so often the current architecture and hardware and software infrastructure restricts the choice. Other factors, such as corporate standards, existing site licensing agreements, and product-vendor relationships, also can mandate what architecture, hardware, and software the project team must design. However, many organizations now have a variety of infrastructures available or are openly looking for pilot projects to test new architectures, hardware and software, enabling a project team to select an architecture on the basis of other important factors.

Designing an architecture can be quite difficult, and therefore many organizations hire expert consultants or assign very experienced analysts to the task.[1] In this chapter, we will examine the key factors in architecture design, but it is important to remember that it takes lots of experience to do it well. The nonfunctional requirements developed early in the analysis phase (see Chapter 4) play a key role in architecture design. These requirements are reexamined and refined into more detailed requirements that influence system's architecture. In this chapter, we first explain the how designers think about application architectures and describe the three primary architectures: server-based, client-based, and client–server. Then we examine how the very general nonfunctional requirements from the analysis phase are refined into more specific requirements, and the implications that they have for architecture design. Finally, we consider how the requirements and architecture can be used to develop the hardware and software specifications that define exactly what hardware and other software (e.g., database systems) are needed to support the information system being developed.

ELEMENTS OF AN ARCHITECTURE DESIGN

The objective of architecture design is to determine what parts of the application software will be assigned to what hardware. In this section we first discuss the major *architectural components* of the software to understand how the software can be divided into different parts. Then we briefly discuss the major types of hardware onto which the software can be placed. Although there are numerous ways in which the software components can be placed on the hardware components, there are three principal application architectures in use today: server-based architectures, client-based architectures and client–server architectures. The most common architecture is the client–server architecture, so we focus on it.

---

[1] For more information on architecture design, see the Zachman Institute at www.zifa.com.

Architectural Components

The major architectural components of any system are the software and the hardware. The major software components of the system being developed have to be identified and then allocated to the various hardware components on which the system will operate. Each of these components can be combined in a variety of different ways.

All software systems can be divided into four basic *functions*. The first is *data storage*. Most information systems require data to be stored and retrieved, whether it is a small file, such as a memo produced by a word processor, or a large database, such as one that stores an organization's accounting records. These are the data entities documented in ERDs. The second function is the *data access logic*, the processing required to access data, which often means database queries in Structured Query Language (SQL). The third function is the *application logic*, which is the logic that is documented in the DFDs, use cases, and functional requirements. The fourth function is the *presentation logic*, the display of information to the user and the acceptance of the user's commands (the user interface). These four functions (data storage, data access logic, application logic, and presentation logic) are the basic building blocks of any information system.

The three primary hardware components of a system are *client computers, servers*, and the *network* that connects them. Client computers are the input/output devices employed by the user and are usually desktop or laptop computers, but can also be handheld devices, cell phones, special purpose terminals, and so on. Servers are typically larger computers that are used to store software and hardware that can be accessed by anyone who has permission. Servers can come in several flavors: *mainframes* (very large, powerful computers usually costing millions of dollars), *minicomputers*, (large computers costing hundreds of thousands of dollars), and *microcomputers* (small desktop computers like the one you use to ones costing $50,000 or more). The network that connects the computers can vary in speed from a slow cell phone or modem connection that must be dialed, to medium speed always-on frame relay networks, to fast always-on broadband connections such as cable modem, DSL, or T1 circuits, to high speed always-on Ethernet, T3, or ATM circuits.[2]

Server-Based Architectures

The very first computing architectures were *server based,* with the server (usually a central mainframe computer) performing all four application functions. The clients (usually *terminals*) enabled users to send and receive messages to and from the server computer. The clients merely captured keystrokes and sent them to the server for processing, accepting instructions from the server on what to display (Figure 9-1).

This very simple architecture often works very well. Application software is developed and stored on one computer, and all data are on the same computer. There is one point of control because all messages flow through the one central server. The fundamental problem with server-based networks is that the server must process all messages. As the demands for more and more applications grow, many

---

[2] For more information on networks, see Alan Dennis, *Networking in the Internet Age,* John Wiley & Sons, 2002.

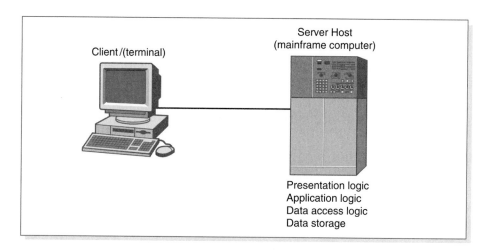

FIGURE 9-1
Server-Based Architecture

server computers become overloaded and unable to quickly process all the users' demands. Response time becomes slower, and network managers are required to spend increasingly more money to upgrade the server computer. Unfortunately, upgrades come in large increments and are expensive (e.g., $500,000); it is difficult to upgrade "a little."

Client-Based Architectures

With *client-based architectures,* the clients are microcomputers on a local area network, and the server computer is a server on the same network. The application software on the client computers is responsible for the presentation logic, the application logic, and the data access logic; the server simply stores the data (Figure 9-2). In very simple one-user systems, the data may remain on the client computer itself, and no server is used.

This simple architecture often works very well. However, as the demands for more and more network applications grow, the network circuits can become overloaded. The fundamental problem in client-based networks is that all data on the server must travel to the client for processing. For example, suppose the user wishes to display a list of all employees with company life insurance. All the data in the database must travel from the server, where the database is stored, over the network to the client, which then examines each record to see if it matches the data requested by the user. This can overload both the network and the power of the client computers.

Client–Server Architectures

Most organizations today are moving to *client–server architectures,* which attempt to balance the processing between the client and the server. In these architectures, the client is responsible for the presentation logic, whereas the server is responsible for the data access logic and data storage. The application logic may either reside on the client, reside on the server, or be split between both (Figure 9-3). When the client has most or all of the application logic (as in Figure 9-3), it is called a *thick client*. If the client contains only the presentation function and most of application function resides on the server, it is called a *thin client*. For example, many Web-

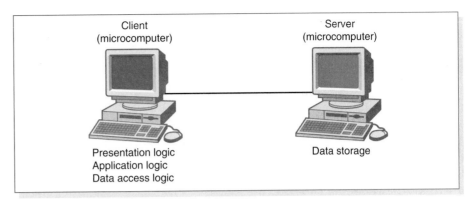

FIGURE 9-2
Client-Based Architecture.

based systems are designed with the Web browser performing presentation and only minimal application logic using such programming languages as Java, while the Web server has the application logic, data access logic, and data storage.

Client–server architectures have four important benefits. First and foremost, they are *scalable*. That means it is easy to increase or decrease the storage and processing capabilities of the servers. If one server becomes overloaded, you simply add another server so that many servers are used to perform the application logic, data access logic, or data storage. The cost to upgrade is much more gradual, and you can upgrade in smaller steps (e.g., $5,000) rather than spending hundreds of thousands to upgrade a mainframe server.

Second, client–server architectures can support many different types of clients and servers. It is possible to connect computers that use different operating systems so that users can choose which type of computer they prefer (e.g., combining both Windows computers and Apple Macintoshes on the same network). You are not locked into one vendor, as is often the case with server-based networks. *Middleware* is a type of system software designed to translate between different

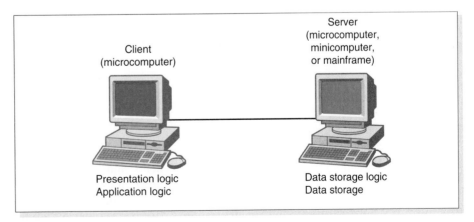

FIGURE 9-3
Client–Server Architecture.   *Source:* Jerry FitzGerald and Alan Dennis, *Business Data Communications and Networking,* 6th ed., p.75. Copyright © 1999 by John Wiley & Sons, Inc. Used with permission.

vendors' software. Middleware is installed on both the client computer and the server computer. The client software communicates with the middleware that can reformat the message into a standard language that can be understood by the middleware that assists the server software.

Third, for thin client–server architectures that use Internet standards, is it simple to clearly separate the presentation logic, the application logic, and the data access logic and design each to be somewhat independent. For example, the presentation logic can be designed in HTML or XML to specify how the page will appear on the screen (e.g., the colors, fonts, order of items, specific words used, command buttons, the type of selection lists and so on; see Chapter 10). Simple program statements are used to link parts of the interface to specific application logic modules that perform various functions. These HTML or XML files defining the interface can be changed without affecting the application logic. Likewise, it is possible to change the application logic without changing the presentation logic or the data, which are stored in databases and accessed using SQL commands.

Finally, because no single server computer supports all the applications, the network is generally more reliable. There is no central point of failure that will halt the entire network if it fails, as there is with server-based architectures. If any one server fails in a client–server environment, the network can continue to function using all the other servers.

Client–server architectures also have some critical limitations, the most important of which is its complexity. All applications in client–server computing have two parts, the software on the client and the software on the server. Writing this software is more complicated than writing the traditional all-in-one software used in server-based architectures. Updating the network with a new version of the software is more complicated, too. In server-based architectures, there is one place in which application software is stored; to update the software, you simply replace it there. With client–server architectures, you must update all clients and all servers.

Much of the debate about server-based versus client–server architectures has centered on cost. One of the great claims of server-based networks in the 1980s was that they provided economies of scale. Manufacturers of big mainframes claimed it was cheaper to provide computer services on one big mainframe than on a set of smaller computers. The microcomputer revolution changed this. Since the 1980s, the cost of microcomputers has continued to drop, whereas their performance has increased significantly. Today, microcomputer hardware is more than *1,000 times cheaper* than mainframe hardware for the same amount of computing power.

Client–Server Tiers

There are many ways in which the application logic can be partitioned between the client and the server. The example in Figure 9-3 is one of the most common. In this case, the server is responsible for the data and the client is responsible for the application and presentation. This is called a *two-tiered architecture* because it uses only two sets of computers, clients, and servers.

A *three-tiered architecture* uses three sets of computers, as shown in Figure 9-4. In this case, the software on the client computer is responsible for presentation logic, an application server(s) is responsible for the application logic, and a separate database server(s) is responsible for the data access logic and data storage.

An *n-tiered architecture* uses more than three sets of computers. In this case, the client is responsible for presentation, a database server(s) is responsible for the

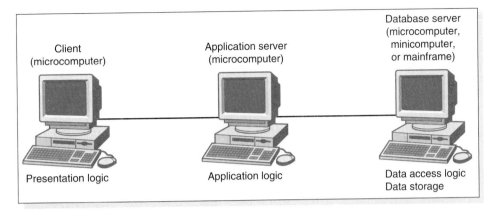

FIGURE 9-4

Three-Tiered Client–Server Architecture   *Source:* Jerry FitzGerald and Alan Dennis, *Business Data Communications and Networking,* 6th ed., p.75. Copyright © 1999 by John Wiley & Sons, Inc. Used with permission.

data access logic and data storage, and the application logic is spread across two or more different sets of servers. Figure 9-5 shows an example of an n-tiered architecture of a software product called Consensus @nyWARE®.[3] Consensus @nyWARE® has four major components. The first is the Web browser on the client computer employed by a user to access the system and enter commands (presentation logic). The second component is a Web server that responds to the user's requests, either by providing HTML pages and graphics (application logic) or by sending the request to the third component (a set of 28 programs written in the C programming language) on another application server that performs various functions (application logic). The fourth component is a database server that stores all the data (data access logic and data storage). Each of these four components is separate, making it easy to spread the different components on different servers and to partition the application logic on two different servers.

The primary advantage of an n-tiered client–server architecture compared with a two-tiered architecture (or a three-tiered with a two-tiered) is that it separates out the processing that occurs to better balance the load on the different servers; it is more scalable. In Figure 9-5, we have three separate servers, a configuration that provides more power than if we had used a two-tiered architecture with only one server. If we discover that the application server is too heavily loaded, we can simply replace it with a more powerful server, or just put in several more application servers. Conversely, if we discover the database server is underused, we could store data from another application on it.

There are two primary disadvantages to an n-tiered architecture compared with a two-tiered architecture (or a three-tiered with a two-tiered). First, the configuration puts a greater load on the network. If you compare Figure 9-3, 9-4, and 9-5, you will see that the n-tiered model requires more communication among the servers; it generates more network traffic, so you need a higher-capacity network.

[3] Consensus @nyWARE® was originally developed by Alan Dennis while at the University of Georgia. It is on the Web at www.softbicycle.com

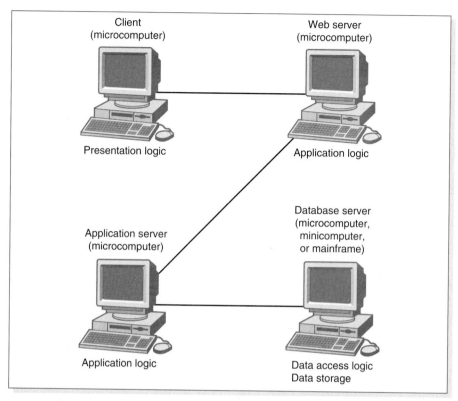

FIGURE 9-5
Four-Tiered Client–Server Architecture   *Source:* Jerry FitzGerald and Alan Dennis, *Business Data Communications and Networking,* 6th ed., p. 76. Copyright © 1999 by John Wiley & Sons, Inc. Used with permission.

Second, it is much more difficult to program and test software in n-tier architectures than in two-tiered architectures because more devices have to communicate to complete a user's transaction.

CREATING AN ARCHITECTURE DESIGN

The architecture design specifies the overall architecture and the placement of software and hardware that will be used. Architecture design is very complex process that is often left to experienced architecture designers and consultants, but we will give a sense of the process here.

Each of the computing architectures discussed earlier has its strengths and weaknesses. Most organizations are trying to move to client–server architectures for cost reasons, so in the event that there is no compelling reason to choose one architecture over another, cost usually suggests client–server.

Creating an architecture design begins with the nonfunctional requirements. The first step is to refine the nonfunctional requirements into more detailed requirements that are then used to help select the architecture to be used (server-based, client-based, or client–server) and what software components will be placed on each device. In a client–server architecture, one also has to decide whether to use a two-

CONCEPTS **9-A A MONSTER CLIENT–SERVER ARCHITECTURE**

IN ACTION

Every spring, Monster.com, one of the largest job sites in the United States with an average of more than 3 million visitors per month, experiences a large increase in traffic. Aaron Braham, vice president of operations, attributes the spike to college students who increase their job search activities as they approach graduation.

Monster.com uses a 3-tier client–server architecture that has 150 Web servers and 30 database servers in its main site in Indianapolis and plans to move that to 400 over the next year by gradually growing the main site, and adding a new site with servers in Maynard, Massachusetts, just in time for the spring rush. The main Web site has a set of load-balancing devices that forward Web requests to the different servers depending upon how busy they are.

Braham says the major challenge is that 90% of the traffic is not simple requests for Web pages, but rather search requests (e.g., what network jobs are available in New Mexico) which require more processing and access to the database servers. Monster.com has more than 350,000 job postings and more than 3 million résumès on file, spread across its database servers. Several copies of each posting and résumès are kept on several database servers to improve access speed and provide redundancy in case a server crashes, so just keeping the database servers in sync so that they contain correct data is a challenge.

Source: "Resume Influx Tests Mettle of Job Sites Scalability," *Internetweek,* May 29, 2000, by Christine Zimmerman.

QUESTIONS:

1. What are the primary two or three nonfunctional requirements that have influenced Monster.com's application architecture?
2. What alternatives do you think Monster.com considered?

tier, three-tier, or n-tier architecture. Then the nonfunctional requirements and the architecture design are used to develop the hardware and software specification.

There are four primary types of nonfunctional requirements that can be important in designing the architecture: operational requirements, performance requirements, security requirements, and cultural/political requirements. We describe each in turn and then explain how they may affect the architecture design.

Operational Requirements

Operational requirements specify the operating environment(s) in which the system must perform and how those may change over time. This usually refers to operating systems, system software, and information systems with which the system must interact, but will on occasion also include the physical environment if the environment is important to the application (e.g., in a noisy factory floor so no audible alerts can be heard). Figure 9-6 summarizes four key operational requirement areas and provides some examples of each.

Technical Environment Requirements *Technical environment requirements* specify the type of hardware and software on which the system will work. These requirements usually focus on the operating system software (e.g., Windows, Linux), database system software (e.g., Oracle), and other system software (e.g., Internet Explorer). Occasionally, there may be specific hardware requirements that impose important limitations on the system, such as the need to operate on a PDA or cell phone with a very small display.

Type of Requirement	Definition	Examples
Technical Environment Requirements	Special hardware, software, and network requirements imposed by business requirements	• The system will work over the Web environment with Internet Explorer. • All office locations will have an always-on network connection to enable real-time database updates. • A version of the system will be provided for customers connecting over the Internet via a small screen PDA.
System Integration Requirements	The extent to which the system will operate with other systems	• The system must be able to import and export Excel spreadsheets. • The system will read and write to the main inventory database in the inventory system.
Portability Requirements	The extent to which the system will need to operate in other environments	• The system must be able to work on future versions of Internet Explorer. • The system may need to operate with handhled devices such as a Palm.
Maintainability Requirements	Expected business changes to which the system should be able to adapt	• The system will be able to support more than one manufacturing plant with 6 months advance notice. • New versions of the system will be released every six months.

FIGURE 9-6
Operational Requirements

System Integration Requirements *System intergration requirements* are those that require the system to operate with other information systems, either inside or outside the company. These typically specify interfaces through which data will be exchanged with other systems.

Portability Requirements Information systems never remain constant. Business needs change and operating technologies change, so the information systems that support them and run on them must change too. *Portability requirements* define how the technical operating environments may evolve over time and how the system must respond (e.g., the system must run on all current and future versions of Windows). Portability requirements also refer to potential changes in business requirements that will drive technical environment changes. For example, in the future, users may want to access a Web site from their cell phones.

Maintainability Requirements *Maintainability requirements* specify the business requirement changes that can be anticipated. Not all changes are predictable, but some are. For example, suppose a small company has only one manufacturing plant but is anticipating the construction of a second plant in the next 5 years. All information systems must be written to make it easy to track each plant separately, whether for personnel, budgeting, or inventory system. The maintainability requirement attempts to anticipate future requirements so that the systems designed today will be easy to maintain if and when those future requirements appear. Maintainability requirements may also define the update cycle for the system, such as the frequency with which new versions will be released.

Performance Requirements

Performance requirements focus on performance issues such as response time, capacity, and reliability. Figure 9-7 summarizes three key performance requirement areas and provides some examples.

Speed Requirements   *Speed requirements* are exactly what they say: how fast should the system must operate. First and foremost, this is the response time of the system: how long does it take the system to respond to a user request. While everyone would prefer low response times with the system responding immediately to each user request, this is not practical. We could design such a system, but it would be expensive. Most users understand that certain parts of a system will respond quickly, while others are slower. Those actions that are performed locally on the user's computer must be almost immediate (e.g., typing, dragging and dropping) while others that require communicating across a network can have higher response times (e.g., a Web request). In general, response times less than 7 seconds are considered acceptable when they require communication over a network.

The second aspect of speed requirements is how long it takes transactions in one part of the system to be reflected in other parts. For example, how soon after an order is placed will the items it contained be shown as no longer available for sale to someone else? If the inventory is not updated immediately, then someone else could place an order for the same item, only to find out later it is out of stock. Or, how soon after an order is placed is it sent to the warehouse to be picked from inventory and shipped? In this case, some time delay might have little impact.

Capacity Requirements   *Capacity requirements* attempt to predict how many users the system will have to support, both in total and simultaneously. Capacity

Type of Requirement	Definition	Examples
Speed Requirements	The time within which the system must perform its functions	• Response time must be less than 7 seconds for any transation over the network • The inventory database must be updated in real time. • Orders will be transmitted to the factory floor every 30 minutes.
Capacity Requirements	The total and peak number of users and the volume of data expected	• There will be a maximum of 100–200 simultaneous users at peak use times. • A typical transaction will require the transmission of 10K of data. • The system will store data on approximately 5,000 customers for a total of about 2 meg of data.
Availability and Reliability Requirements	The extent to which the system will be available to the users and the permissible failure rate due to errors.	• The system should be available 24 × 7, with the exception of scheduled maintenance. • Scheduled maintainance shall not exceed one 6-hour period each month. • The system shall have 99% uptime performance.

FIGURE 9-7
Performance Requirements

requirements are important in understanding the size of the databases, the processing power needed, and so on. The most important requirement is the usually the peak number of simultaneous users, because this has a direct impact on the processing power of the computer(s) needed to support the system.

It is often easier to predict the number of users for internal systems designed to support an organization's own employees than it is to predict the number of users for customer-facing systems, especially those on the Web. How does weather.com estimate the peak number of users who will simultaneously seek weather information? This is as much an art as a science, so often the team provides a range of estimates, with wider ranges used to signal a less accurate estimate.

Availability and Reliability Requirements *Availability and reliability requirements* focus on the extent to which users can assume that the system will be available for them to use. While some systems are just intended to be used during the 40-hour workweek, some systems are designed to be used by people around the world. For such systems, project team members need to consider how the application can be operated, supported, and maintained, *24 × 7* (i.e., 24 hours a day, 7 days a week). This 24 × 7 requirement means that users may need help or have questions at any time, and a support desk that is available 8 hours a day will not be sufficient support. It is also important to consider what reliability is needed in the system. A system that requires high reliability (e.g., a medical device or telephone switch) needs far greater planning and testing than one that does not have such high reliability needs (e.g., personnel system, Web catalog).

It is more difficult to predict the peaks and valleys in usage of the system when the system has a global audience. Typically, applications are backed up on weekends or late evenings when users are no longer accessing the system. Such maintenance activities will have to be rethought if they are global. The development of Web interfaces in particular has escalated the need for 24 × 7 support because by default the Web can be accessed by anyone at any time. For example, the developers of a Web application for U.S. outdoor gear and clothing retailer Orvis were surprised when the first order after going live came from Japan.

CONCEPTS **9-B THE IMPORTANCE OF CAPACITY PLANNING**

IN ACTION

At the end of 1997, Oxford Health Plans posted a $120 million loss to its books. The company's unexpected growth was its undoing because the system, which was originally planned to support the company's 217,000 members, had to meet the needs of a membership that exceeded 1.5 million.

System users found that processing a new-member sign-up took 15 minutes instead of the proposed 6 seconds. Also, the computer problems left Oxford unable to send out bills to many of its customer accounts and rendered it unable to track payments to hundreds of doctors and hospitals. In less than a year, uncollected payments from customers tripled to more than $400 million and the

payments owed to caregivers amounted to more than $650 million. Mistakes in infrastructure planning can cost far more than the cost of hardware, software, and network equipment alone.

Source: "Management: How New Technology was Oxford's Nemesis," *The Wall Street Journal*, December 11, 1997, page A.1, by Ron Winslow and George Anders.

QUESTION:

If you had been in charge of the Oxford project, what things would you have considered when planning the system capacity?

Security Requirements

Security is ability to protect the information system from disruption and data loss, whether caused by an intentional act (e.g., a hacker or a terrorist attack), or a random event (e.g., disk failure, tornado). Security is primarily the responsibility of the operations group—the staff responsible for installing and operating security *controls* such as firewalls, intrusion detection systems, and routine backup and recovery operations. Nonetheless, developers of new systems must ensure that the system's *security requirements* produce reasonable precautions to prevent problems; system developers are responsible for ensuring security within the information systems themselves.

Security is an ever-increasing problem in today's Internet-enabled world. Historically, the greatest security threat has come from inside the organization itself. Ever since the early 1980s when the FBI first began keeping computer crime statistics and security firms began conducting surveys of computer crime, organizational employees have perpetrated the vast majority of computer crimes. For years, 80% of unauthorized break-ins, thefts, and sabotage have been committed by insiders, leaving only 20% to hackers external to the organizations.

In 2001, that changed. Depending upon what survey you read, the percent of incidents attributed to external hackers in 2001 increased to 50% to 70% of all incidents, meaning that the greatest risk facing organizations is now from the outside. While some of this shift may be due to better internal security and better communications with employees to prevent security problems, much of it is simply due to an increase in activity by external hackers.

Developing security requirements usually starts with some assessment of the value of the system and its data. This helps pinpoint extremely important systems so that the operations staff are aware of the risks. Security within systems usually focuses on specifying who can access what data, identifying the need for encryption and authentication, and ensuring the application prevents the spread of viruses; see Figure 9-8.

Type of Requirement	Definition	Examples
System Value Estimates	Estimated business value of the system and its data	• The system is not mission critical but a system outage is estimated to cost $50,000 per hour in lost revenue. • A complete loss of all system data is estimated to cost $20 million.
Access Control Requirements	Limitations on who can access what data	• Only department managers will be able to change inventory items within their own department. • Telephone operators will be able to read and create items in the customer file, but cannot change or delete items.
Encryption and Authentication Requirements	Defines what data will be encrypted where and whether authentication will be needed for user access	• Data will be encrypted from the user's computer to the Web site to provide secure ordering. • Users logging in form outside the office will be required to authenticate.
Virus Control Requirements	Requirements to control the spread of viruses	• All uploaded files will be checked for viruses before being saved in the system.

Figure 9-8
Security Requirements

System Value   The most important computer asset in any organization is not the equipment; it is the organization's data. For example, suppose someone destroyed a mainframe computer worth $10 million. The mainframe could be replaced, simply by buying a new one. It would be expensive, but the problem would be solved in a few weeks. Now suppose someone destroyed all the student records at your university so that no one knew what courses anyone had taken or their grades. The cost would far exceed the cost of replacing a $10 million computer. The lawsuits alone would easily exceed $10 million, and the cost of staff to find and reenter paper records would be enormous and certainly would take more than a few weeks.

In some cases, the information system itself has value that far exceeds the cost of the equipment as well. For example, for an Internet bank that has no brick and mortar branches, the Web site is a *mission critical system*. If the Web site crashes, the bank cannot conduct business with its customers. A mission critical application is an information system that is literally critical to the survival of the organization. It is an application that cannot be permitted to fail, and if it does fail the operations staff drops everything else to fix it. Mission critical applications are usually clearly identified so their importance is not overlooked.

Even temporary disruptions in service can have significant costs. The cost of disruptions to a company's primary Web site or the LANs and backbones that support telephone sales operations are often measured in the millions. Amazon.com, for example, has revenues of more $10 million per hour, so if their Web site were unavailable for hour or even part of an hour it would cost them millions of dollars in lost revenue. Companies that do less e-business or telephone sales have lower costs, but recent surveys suggest losses of $100,000 to $200,000 per hour are not uncommon for major customer-facing information systems.

Access Control Requirements   Some of the data stored in the system needs to be kept confidential; some data need special controls on who is allowed to change or delete it. Personnel records, for example, should only be able to be read by the personnel department and the employee's supervisor; changes should only be permit-

---

9-C POWER OUTAGE COSTS A MILLION DOLLARS

Lithonia Lighting, located just outside of Atlanta, is the world's largest manufacturer of light fixtures with more than $1 billion in annual sales. One afternoon, the power transformer at its corporate headquarters exploded, leaving the entire office complex, including the corporate data center, without power. The data center's backup power system immediately took over and kept critical parts of the data center operational. However, it was insufficient to power all systems, so the system supporting sales for all of Lithonia Lighting's North American agents, dealers, and distributors had to be turned off.

The transformer was quickly replaced and power was restored. However, the 3-hour shutdown of the sales system cost $1 million in potential sales lost. Unfortunately, it is not uncommon for the cost of a disruption to be hundreds or thousands of times the cost of the failed components.

QUESTION:
What would you recommend to avoid similar losses in the future?

ted to be made by the personnel department. *Access control requirements* state who can access what data and what type of access is permitted: whether the individual can create, read, update, and/or delete the data. The requirements reduce the chance that an authorized user of the system can perform unauthorized actions.

Encryption and Authentication Requirements   One of the best ways to prevent unauthorized access to data is *encryption*, which is a means of disguising information by the use of mathematical algorithms (or formulas). Encryption can be used to protect data stored in databases or data that are in transit over a network from a database to a computer. There are two fundamentally different types of encryption: symmetric and asymmetric. A *symmetric encryption algorithm* (such as Data Encryption Standard [DES] or Advanced Encryption Standard [AES] is one in which the key used to encrypt a message is the same as the one used to decrypt it, which means that it is essential to protect the key and that a separate key must be used for each person or organization with whom the system shares information (or else everyone can read all the data).

An *asymmetric encryption algorithm* (such as *public key encryption*) is one in which the key used to encrypt data (called the public key) is different from the one used to decrypt it (called the *private key*). Even if everyone knows the public key, once the data is encrypted, it cannot be decrypted without the private key. Public key encryption greatly reduces the key management problem. Each user has its public key that is used to encrypt messages sent to it. These public keys are widely publicized (e.g., listed in a telephone book-style directory)—that's why they're called "public" keys. The private key, in contrast, is kept secret (that's why it's called "private" key).

Public key encryption also permits *authentication* (or digital signatures). When one user sends a message to another, it is difficult to legally prove who actually sent the message. Legal proof is important in many communications, such as bank transfers and buy/sell orders in currency and stock trading, which normally require legal signatures. Public key encryption algorithms are *invertable*, meaning that text encrypted with either key can be decrypted by the other. Normally, we encrypt with the public key and decrypt with the private key. However, it is possible to do the inverse: encrypt with the private key and decrypt with the public key. Since the private key is secret, only the real user could use it to encrypt a message. Thus a digital signature or authentication sequence is used as a legal signature on many financial transactions. This signature is usually the name of the signing party plus other unique information from the message (e.g., date, time, or dollar amount). This signature and the other information are encrypted by the sender using the private key. The receiver uses the sender's public key to decrypt the signature block and compares the result to the name and other key contents in the rest of the message to ensure a match.

The only problem with this approach lies in ensuring that the person or organization who sent the document with the correct private key is actually the person or organization they claim to be. Anyone can post a public key on the Internet so there is no way of knowing for sure who they actually are. For example, it would be possible for persons other than "Organization A" in this example to claim to be Organization A when in fact they are an imposter.

This is where the Internet's public key infrastructure (PKI) becomes important.[4] The PKI is a set of hardware, software, organizations, and polices designed to

---

[4] For more on the PKI, see www.ietf.org/internet-drafts/draft-ietf-pkix-roadmap-06.txt.

make public key encryption work on the Internet. PKI begins with a *certificate authority (CA)*, which is a trusted organization that can vouch for the authenticity of the person or organization using authentication (e.g., VeriSign). A person wanting to use a CA registers with the CA and must provide some proof of identify. There are several levels of certification, ranging from a simple confirmation from valid e-mail address to a complete police-style background check with an in-person interview. The CA issues a digital certificate that is the requestor's public key encrypted using the CA's private key as proof of identify. This certificate is then attached to the user's e-mail or Web transactions in addition to the authentication information. The receiver then verifies the certificate by decrypting it with the CA's public key—and must also contact the CA to ensure that the user's certificate has not been revoked by the CA.

The encryption and authentication requirements state what encryption and authentication requirements are needed for what data. For example, will sensitive data such as customer credit card numbers be stored in the database in encrypted form, or will encryption be used to take orders over the Internet from the company's Web site? Will users be required to use a digital certificate in addition to a standard password?

Virus Control Requirements   The single most common security problem comes from *viruses*. Recent studies have shown that almost 90% of organizations suffer a virus infection each year. Viruses cause unwanted events—some harmless (such as nuisance messages), some serious (such as the destruction of data). Any time a system permits data to be imported or uploaded from a user's computer, there is the potential for a virus infection. Many systems require that all information systems that permit the import or upload of user files to check those files for viruses before they are stored in the system.

Cultural and Political Requirements

Cultural and political requirements are those that are specific to the specific countries in which the system will be used. In today's global business environment, organizations are expanding their systems to reach users around the world. Although this can make great business sense, its impact on application development should not be underestimated. Yet another important part of the design of the system's architecture is understanding the global cultural and political requirements for the system; see Figure 9-9.

Multilingual Requirements   The first and most obvious difference between applications used in one region and those designed for global use is language. Global applications often have *multilingual requirements*, which means that they have to support users who speak different languages and write using non-English letters (e.g., those with accents, Cyrillic, Japanese). One of the most challenging aspects in designing global systems is getting a good translation of the original language messages into a new language. Words often have similar meanings but can convey subtly different meanings when they are translated, so it is important to use translators skilled in translating technical words.

The other challenge is often screen space. In general, English-language messages usually take 20% to 30% fewer letters than their French or Spanish counterparts. Designing global systems requires allocating more screen space to messages than might be used in the English-language version.

Type of Requirement	Definition	Examples
Multilingual Requirements	The language in which the system will need to operate	• The system will operate in English, French, and Spanish.
Customization Requirements	Specification of what aspects of the system can be changed by local users	• Country managers will be able to define new fields in the product database to capture country-specific information. • Country managers will be able to change the format of the telephone number field in the customer database.
Making Unstated Norms Explicit	Explicitly stating assumptions that differ from country to country	• All date fields will be explicitly identified as using the month-day-year format. • All weight fields will be explicitly identified as being stated in kilograms.
Legal Requirements	The laws and regulations that impose requirements on the system	• Personal information about customers cannot be transferred out of European Union countries into the United States. • It is against U.S. federal law to divulge information on who rented what videotape, so access to a customer's rental history is permitted only to regional managers.

Figure 9-9
Cultural and Political Requirements

Some systems are designed to handle multiple languages on the fly so that users in different countries can use different languages concurrently; that is, the same system supports several different languages simultaneously (a *concurrent multilingual system*). Other systems contain separate parts that are written in each

CONCEPTS

IN ACTION

9-D DEVELOPING MULTILINGUAL SYSTEMS

I've had the opportunity to develop two multilingual systems. The first was a special-purpose decision support system to help schedule orders in paper mills called BCW-Trim. The system was installed by several dozen paper mills in Canada and the United States, and it was designed to work in either English or French. All messages were stored in separate files (one set English, one set French), with the program written to use variables initialized either to the English or French text. The appropriate language files were included when the system was compiled to produce either the French or English version.

The second program was a groupware system called GroupSystems, for which I designed several modules. The system has been translated into dozens of different languages, including French, Spanish, Portuguese, German, Finnish, and Croatian. This system enables the user to switch between languages at will by storing messages in simple text files. This design is far more flexible because each individual installation can revise the messages at will. Without this approach, it is unlikely that there would have been sufficient demand to warrant the development of versions to support less commonly used languages (e.g., Croatian). *Alan Dennis*

QUESTIONS:
1. How would you decide how much to support users who speak languages other than English?
2. Would you create multilingual capabilities for any application that would be available to non-English speaking people? Think about Web sites that exist today.

language and must be reinstalled before a specific language can be used; that is, each language is provided by a different version of the system so that any one installation will use only one language (i.e., a *discrete multilingual system*). Either approach can be effective, but this functionality must be designed into the system well in advance of implementation.

Customization Requirements   For global applications, the project team will need to give some thought to *customization requirements*: how much of the application will be controlled by a central group and how much of the application will be managed locally. For example, some companies allow subsidiaries in some countries to customize the application by omitting or adding certain features. This decision has trade-offs between flexibility and control because customization often makes it more difficult for the project team to create and maintain the application. It also means that training can differ between different parts of the organization, and customization can create problems when staff move from one location to another.

Unstated Norms   Many countries have *unstated norms* that are not shared internationally. It is important for the application designer to make these assumptions explicit because they can lead to confusion otherwise. In the United States, the unstated norm for entering a date is the date format MM/DD/YY; however, in Canada and most European countries, the unstated norm is DD/MM/YY. When you are designing global systems, it is critical to recognize these unstated norms and make them explicit so that users in different countries do not become confused. Currency is the other item sometimes overlooked in system design; global application systems must specify the currency in which information is being entered and reported.

Legal Requirements   *Legal requirements* are those requirements imposed by laws and government regulations. System developers sometimes forget to think about legal regulations but unfortunately, doing so comes at some risk because ignorance of the law is no defense. For example, in 1997, a French court convicted the Georgia Institute of Technology of violating French language law. Georgia Tech operated a small campus in France that offered summer programs for American students. The information on the campus Web server was primarily in English because classes are conducted in English, which violated the law requiring French to be the predominant language on all Internet servers in France. By formally considering legal regulations, they are less likely to be overlooked.

Designing the Architecture

In many cases, the technical environment requirements as driven by the business requirements may simply define the application architecture. In this case, the choice is simple: business requirements dominate other considerations. For example, the business requirements may specify that the system needs to work over the Web using the customer's Web browser. In this case, the application architecture must be thin client–server. Such business requirements are most likely in systems designed to support external customers. Internal systems may also impose business requirements, but usually they are not as restrictive.

In the event that the technical environment requirements do not require the choice of a specific architecture, then the other nonfunctional requirements become important. Even in cases when the business requirements drive the architecture, it is still important to work through and refine the remaining nonfunctional requirements because they are important in later stages of the design and implementation phases. Figure 9-10 summarizes the relationship between requirements and recommended architectures.

Operational Requirements   System integration requirements may lead to one architecture over another, depending upon the architecture and design of the system(s) with which the system needs to integrate. For example, if the system must integrate with a desktop system (e.g., Excel) then this may suggest a thin or thick client–server architecture, while if it must integrate with a server-based system then a server-based architecture may be indicated. Systems that have extensive portability requirements tend to be best suited for a thin client–server architecture because it is simpler to write for Web-based standards (e.g., HTML, XML) that

Requirements	Server-Based	Client-Based	Thin Client-Server	Thick Client-Server
Operational Requirements				
System Integration Requirements	✓		✓	✓
Portability Requirements			✓	
Maintainability Requirements	✓		✓	
Performance Requirements				
Speed Requirements			✓	✓
Capacity Requirements			✓	✓
Availability/Reliability Requirements	✓		✓	✓
Security Requirements				
High System Value	✓		✓	
Access Control Requirements	✓			
Encryption/Authentication Requirements			✓	✓
Virus Control Requirements	✓			
Cultural/Political Requirements				
Multilingual Requirements			✓	
Customization Requirements			✓	
Making Unstated Norms Explicit			✓	
Legal Requirements	✓		✓	✓

Figure 9-10
Nonfunctional Requirements and Their Implications for Architecture Design

extend the reach of the system to other platforms, rather than trying to write and rewrite extensive presentation logic for different platforms in the server-based, client-based, or thick client–server architectures. Systems with extensive maintainability requirements may not be well suited to client-based or thick client–server architectures because of the need to reinstall software on the desktops.

Performance Requirements Generally speaking, information systems that have high performance requirements are best suited to client–server architectures. Client–server architectures are more scalable which mean they respond better to changing capacity needs and thus enable the organization to better tune the hardware to the speed requirements of the system. Client–server architectures that have multiple servers in each tier should be more reliable and have greater availability, because if any one server crashes, requests are simply passed to other servers and users may not even notice (although response time may be worse). In practice, however, reliability and availability depend greatly on the hardware and operating system and Windows-based computers tend to have lower reliability and availability than Linux or mainframe computers.

Security Requirements Generally speaking, server-based architectures tend to be more secure because all software is in one location and because mainframe operating systems are more secure than microcomputer operating systems. For this reason, high-value systems are more likely to be found on mainframe computers, even if the mainframe is used as a server in a client–server architecture. In today's Internet-dominated world, authentication and encryption tools for Internet-based client–server architectures are more advanced than those for mainframe-based server-based architectures. Viruses are potential problems in all architectures because they spread easily on desktop computers. If server-based systems can reduce the functions needed on desktop systems then they may be more secure.

Cultural and Political Requirements As the cultural and political requirements become more important, the ability to separate the presentation logic from the application logic and the data becomes important. Such separation makes it easier to develop the presentation logic in different languages while keeping the application logic and data the same. It also makes it easier to customize the presentation logic for different users and to change it to better meet cultural norms. To the extent that the presentation logic provides access to the application and data, it also makes it easier to implement different versions that enable or disable different features required by laws and regulations in different countries. This separation is the easiest in thin client–server architectures, so systems with many cultural and political requirements often use thin client–server architectures. As with system integration requirements, the impact of legal requirements depends upon the specific nature of the requirements, but in general, client-based systems tend to be less flexible.

HARDWARE AND SOFTWARE SPECIFICATION

The design phase is also the time to begin selecting and acquiring the hardware and software that will be needed for the future system. In many cases, the new system will simply run on the existing equipment in the organization. Other times, however, new hardware and software (usually for servers) must be purchased. The *hard-*

YOUR TURN

9-1 University Course Registration System

Think about the course registration system in your university. First, develop a set of nonfunctional requirements if the system were to be developed today. Consider the operational requirements, perfor- mance requirements, security requirements, and cultural and political requirements. Then create an architecture design to satisfy these requirements.

YOUR TURN

9-2 Global e-Learning System

Many multinational organizations provide global Web-based e-learning courses to their employees. First, develop a set of nonfunctional require- ments for such a system. Consider the operational require- ments, performance requirements, security requirements, and cultural and political requirements. Then create an architecture design to satisfy these requirements.

ware and software specification is a document that describes what hardware and software are needed to support the application. The actual acquisition of hardware and software should be left to the Purchasing Department or the area in the organi- zation that handles capital procurement; however, the project team can use the hard- ware and software specification to communicate the project needs to the appropri- ate people. There are several steps involved in creating the document. Figure 9-11 shows a sample hardware and software specification.

	Standard Client	Standard Web Server	Standard Application Server	Standard Database Server
Operating System	• Windows • Internet Explorer	• Windows	• Windows	• Windows
Special Software	• Active X Components • Adobe Acrobat Reader	• IIS Web Server • Net Components	• C • Net Components	• SQL Server • Net Components
Hardware	• 10-gig disk drive • Pentium • 17-inch Monitor	• 40-gig disk drive • Pentium	• 40-gig disk drive • Pentium	• 200-gig disk drive • RAID • Quad Pentium
Network	• Always-on Broadband preferred • Dial-up at 56Kbps possible with some performance loss	• Dual 100 Mbps Ethernet	• Dual 100 Mbps Ethernet	• Dual 100 Mbps Ethernet

Figure 9-11
Sample Hardware and Software Specification

First, you will need to define the software that will run on each component. This usually starts with the operating system (e.g., Windows, Linux) and includes any special purpose software on the client and servers (e.g., Oracle database). This should consider any additional costs such as technical training, maintenance, extended warranties, and licensing agreements (e.g., a site license for a software package) are hardware- and software-related costs that should be considered during the acquisition process. Again, the needs that you list are influenced by decisions that are made in the other design phase activities.

Next, you must create a list of the hardware that is needed to support the future system. In general, the list can include such things as database servers, network servers, peripheral devices (e.g., printers, scanners), backup devices, storage components, and any other hardware component that is needed to support an application. At this time, you also should note the quantity of each item that will be needed.

Finally, you need to describe, in as much detail as possible, the minimum requirements for each piece of hardware. Many organizations have standard lists of approved hardware and software that must be used, so in many cases, this step simply involves selecting items from the lists. Other times, however, the team is operating in new territory and not constrained by the need to select from an approved list. In these cases, the project team must convey such requirements as the amount of processing capacity, the amount of storage space, and any special features that should be included. This step becomes easier with experience; however, there are some hints that can help you describe hardware needs; see Figure 9-12.

APPLYING THE CONCEPTS AT CD SELECTIONS

Creating an Architecture Design

Alec Adams, Senior Systems Analyst and Project Manager for CD Selections' Internet system, realized that the hardware, software, and networks that would support the new application would need to be integrated into the current infrastructure

---

1. **Functions and Features**   What specific functions and features are needed (e.g., size of monitor, software features)

2. **Performance**   How fast the hardware and software operates (e.g., processor, number of database writes per second)

3. **Legacy Databases and Systems**   How well the hardware and software interacts with legacy systems (e.g., can it write to this database)

4. **Hardware and OS Strategy**   What are the future migration plans (e.g., the goal is to have all of one vendor's equipment)

5. **Cost of Ownership**   What are the costs beyond purchase (e.g., incremental license costs, annual maintenance, training costs, salary costs)

6. **Political Preferences**   People are creatures of habit and are resistant to change, so changes should be minimized

7. **Vendor Performance**   Some vendors have reputations or future prospects that are different from those of a specific hardware or software system they currently sell

---

Figure 9-12
Factors in Hardware and Software Selection

YOUR
TURN

9-3 UNIVERSITY COURSE REGISTRATION SYSTEM

Develop a hardware and software specification for the university course registration system described in "Your Turn 9-1."

YOUR
TURN

9-4 GLOBAL E-LEARNING SYSTEM

Develop a hardware and software specification for the global e-learning system described in "Your Turn 9-2."

at CD Selections. He began using the high-level nonfunctional requirements developed in the analysis phase (see Figure 4-13 in Chapter 4) and conducting a JAD session and a series of interviews with managers in the Marketing Department and three store managers to refine the nonfunctional requirements into more detail. Figure 9-13 shows some of the results. The clear business need for a Web-based architecture required a thin client–server architecture.

CD Selections had a formal architecture group responsible for managing CD Selections architecture and its hardware and software infrastructure. Therefore, Alec set up a meeting with the project team and the architecture group. During the meeting, he confirmed that CD Selections was still moving toward a target client–server architecture, although the central mainframe still existed as the primary server for many server-based applications.

They discussed the Internet system and decided that it should be built using a three-tier thin client–server architecture. Everyone believed that it was hard to know at this point exactly how much traffic this Web site would get and how much power the system would require, but a client–server architecture would allow CD Selections to easily scale-up the system as needed.

By the end of the meeting, it was agreed that a three-tiered client–server architecture was the best configuration for the Internet portion of the Internet system. Customers would use their personal computers running a Web browser as the client. A database server would store the Internet system's databases, whereas an application server would have Web server software and the application software to run the system.

The in-store system was currently built using a two-tier client–server architecture, so the portion of the system responsible for the in-store holds would conform to that architecture. A second two-tier client–server system would enable staff in the Marketing Department to maintain the marketing material information. This system would have an application for the microcomputers of the staff working in the Internet sales group that would communicate directly with the database server and would enable staff to update the information. The database server would have

1. Operational Requirements

Technical Environment

1.1 The system will work over the Web environment with Internet Explorer and real audio.

1.2 Customers will only need IE and RA on their desktops.

System Integration

1.3 The Internet system will read information from the main CD information database, which contains basic information about CDs (e.g., title, artist, ID number, price, quantity in inventory). The Internet system will not write information to the main CD information database.

1.4 The Internet system will transmit orders for new CDs in the special order system and will rely on the special order system to complete the special orders generated.

1.5 The Internet system will read and write to the main inventory database.

1.6 A new module for the in-store system will be written to manage the "holds" generated by the Internet system. The requirements for this new module will be documented as part of the Internet system because they are necessary for the Internet system to function.

Portability

1.7 The system will need to remain current with evolving Web standards especially those pertaining to music formats.

Maintainability

1.8 No special maintainability requirements are anticipated.

2. Performance Requirements

Speed

2.1 Response times must be less than 7 seconds.

2.2 The inventory database must be updated in real time.

2.3 In-store holds must be sent to the store within 5 minutes.

Capacity

2.4 There will be a maximum of 20–50 simultaneous users at peak use times.

2.5 The system will support streaming audio to up to 40 simultaneous users.

2.6 The system will send up to 5K of data to each store daily.

2.7 The in-store hold database will require 10–20K of disk space per store.

Availability and Reliability

2.8 The system should be available 24 × 7.

2.9 The system shall have 99% uptime performance.

3. Security Requirements

System Value

3.1 No special system value requirements are anticipated.

Access Control

3.2 Only store managers will be able to override In-Store Holds.

Encryption/Authentication

3.3 No special encryption/authentication requirements are anticipated.

Virus Control

3.4 No special virus control requirements are anticipated.

4. Cultural and Political Requirements

Multilingual

4.1 No special multilingual requirements are anticipated.

Customization

4.2 No special customization requirements are anticipated.

Unstated Norms

4.3 No special unstated norms requirements are anticipated.

Legal

4.4 No special legal requirements are anticipated.

Figure 9-13
Selected Nonfunctional Requirements for CD Selections

a separate program to enable it to exchange data with CD Selections' special order system, CD inventory, and CD information databases on the company mainframe.

Given that the Web interface could reach a geographically dispersed group, the project team realized that it needed to plan for 24 × 7 system support. He scheduled a meeting to talk with the CD Selections' systems operations group and discussed how they might be able to support the Internet system outside of standard working hours.

Hardware and Software Specification

The architecture group and the Internet project team decided that the only components that needed to be acquired for the project were a database server, a Web server, and five new client computers for the marketing group who will maintain the marketing materials. They developed a hardware and software specification for these components and handed them off to the Purchasing Department to start the acquisition process.

SUMMARY

Application Architectures

All software systems can be divided into four basic functions: data storage, data access logic (e.g., SQL), application logic (which is the logic that is documented in the DFDs, use cases, and functional requirements), and the presentation logic (the user interface). There are three fundamental computing architectures that place these functions on different computers. In server-based architectures, the server performs all the functions. In client-based architectures, the client computers are responsible for presentation logic, application logic, and data access logic, with data stored on a file server. In client–server architectures, the client is responsible for the presentation logic and the server is responsible for the data access logic and data storage. In thin client–server architectures, the server performs the application logic while in thick client–server architectures, the application logic is shared between the servers and clients. In a two-tiered client–server architecture there are two groups of computers: one client and a set of servers. In a three-tiered client–server architecture, there are three groups of computers: a client, a set of application servers, and a set of database servers.

Architecture Design

Creating architecture designs begins with the nonfunctional requirements. Operational requirements specify the operating environment(s) in which the system must perform and how those may change over time (i.e., technical environment, system integration, portability, and maintainability). Performance requirements focus on performance issues such as system speed, capacity, and availability and reliability. Security requirements attempt to protect the information system from disruption and data loss (e.g., system value, access control, encryption and authentication, and virus control). Cultural and political requirements are those that are specific to the countries in which the system will be used (e.g., multilingual, customization, unstated norms, and legal).

Hardware and Software Specification

The hardware and software specification is a document that describes what hardware and software are needed to support the application. When a specification document is created, the hardware that is needed to support the future system is listed and then described in as much detail as possible. Next, the software to run on each hardware component is written down, along with any additional associated costs, such as technical training, maintenance, extended warranties, and licensing agreements. Although the project team may suggest specific products or vendors, ultimately the hardware and software specification is turned over to the people who are in charge of procurement.

KEY TERMS

24 × 7
Access control requirements
Application logic
Architectural component
Architecture design
Asymmetric encryption algorithm
Authentication
Availability and reliability
  requirements
Capacity requirements
Certificate authority (CA)
Client-based architecture
Client computer
Client–server architecture
Concurrent multilingual system
Control
Cultural and political requirements
Customization requirements
Data access logic
Data storage
Discrete multilingual system

Encryption
Encryption and authentication
  requirements
Functions
Hardware and software specification
Legal requirements
Mainframe
Maintainability requirements
Microcomputer
Middleware
Minicomputer
Mission critical system
Multilingual requirements
Network
N-tiered architecture
Operational requirements
Performance requirements
Portability requirements
Presentation logic
Private key
Public key

Public key encryption
Response time
Scalable
Security requirements
Server
Server-based architecture
Speed requirements
Symmetric encryption algorithm
System integration requirements
System value
Technical environment requirements
Terminal
Thick client
Thin client
Three-tiered architecture
Two-tiered architecture
Unstated norms
Virus
Virus control requirements

QUESTIONS

1. What are the four general functions of any information system?
2. What are the three main hardware components of a application architecture?
3. Describe two examples of a server.
4. Compare and contrast server-based architectures, client-based architectures and client–server based architectures.
5. What is the biggest problem with server-based architectures?
6. What is the biggest problems with client-based architectures?
7. Describe the major benefits and limitations of client–server architectures.
8. Describe the differences among two-tiered, three-tiered, and n-tiered client–server architectures.
9. Define scalable. Why is this term important to system developers?
10. Describe the elements in creating an architecture design.
11. Why should the project team consider the existing application architecture in the organization when designing the architecture?

12. Describe the major nonfunctional requirements and how they influence architecture design.
13. Why is it useful to define the nonfunctional requirements in more detail even if the technical environment requirements dictate a specific architecture?
14. What additional hardware- and software-associated costs may need to be included on the hardware and software specification?
15. Who is ultimately in charge of acquiring hardware and software for the project?
16. What do you think are three common mistakes that novice analysts make in architecture design and hardware and software specification?
17. Are some nonfunctional requirements more important than others in influencing the architecture design and hardware and software specification?
18. What do you think are the most important security issues for a system?

EXERCISES

A. Using the Web (or past issues of computer industry magazines, such as *Computerworld),* locate a system that runs in a server-based environment. On the basis of your reading, why do you think the company chose that computing environment?

B. Using the Web (or past issues of computer industry magazines, such as *Computerworld*), locate a system that runs in a client–server environment. On the basis of your reading, why do you think the company chose that computing environment?

C. Using the Web, locate examples of a mainframe component, a minicomputer component, and a microcomputer component. Compare the components in terms of price, speed, available memory, and disk storage. Do you find large differences in prices when the performances of the components are considered?

D. You have been selected to find the best client–server computing architecture for a Web-based order entry system that is being developed for L.L. Bean. Write a short memo that describes to the project manager your reason for selecting an n-tiered architecture over a two-tiered architecture. In the memo, give some idea as to what different components of the architecture you would include.

E. Think about the system that your university currently uses for career services and pretend that you are in charge of replacing the system with a new one. Describe how you would decide on the computing architecture for the new system using the criteria presented in this chapter. What information will you need to find out before you can make an educated comparison of the alternatives?

F. Locate a consumer products company on the Web and read its company description (so that you get a good understanding of the geographic locations of the company). Pretend that the company is about to create a new application to support retail sales over the Web. Create an architecture design that depicts the locations that would include components that support this application.

G. Pretend that your mother is a real estate agent and that she has decided to automate her daily tasks using a laptop computer. Consider her potential hardware and software needs and create a hardware and software specification that describes them. The specification should be developed to help your mother buy her hardware and software on her own.

H. Pretend that the admissions office in your university has a Web-based application so that students can apply for admission online. Recently, there has been a push to admit more international students into the university. What do you recommend that the application include to ensure that it supports this global requirement?

MINICASES

1. The system development project team at Birdie Masters golf schools has been working on defining the architecture design for a new system. The major focus of the project is a networked school location operations system, allowing each school location to easily record and retrieve all school location transaction data. Another system element is the use of the Internet to enable current and prospective students to view class offerings at any of the Birdie Masters' locations, schedule lessons and enroll in classes at any Birdie Masters' location, and maintain a student progress profile—a confidential analysis of the student's golf skill development.

   The project team has been considering the globalization issues that should be factored into the architecture design. The school's plan for expansion into the golf-crazed Japanese market is moving ahead. The first Japanese school location is tentatively planned to open about 6 months after the target completion date for the system project. Therefore, it is important that issues related to the international location be addressed now during Design.

   Prepare a set of nonfunctional requirements, including operational requirements, performance requirements, security requirements, and cultural and political requirements. Much information is incomplete, but do your best.

2. Jerry is the project manager for a team developing a retail store management system for a chain of sporting goods stores. Company headquarters is in Las Vegas, and the chain has 27 locations throughout Nevada, Utah, and Arizona. Several cities have multiple stores. Stores will be linked to one of three regional servers,

and the regional servers will be linked to corporate headquarters in Las Vegas. The regional servers also link to each other. Each retail store will be outfitted with similar configurations of two PC-based point-of-sale terminals networked to a local file server. Jerry has been given the task of developing the architecture design and hardware and software specification a network model that will document the geographic structure of this system. He has not faced a system of this scope before and is a little unsure how to begin. What advice would you give?

3. Java Masters is an employment exchange agency that has offices in Northern California. Java Masters oper-

ates as a broker that links its client companies with independent software experts (commonly called "contractors") with advanced Java and Web-development skills for short-term contracts. They are developing a Web-based system that will enable client companies to list job needs and search the database of independent contractors. The independent contractors also can post résumés and availabilities, and search the database of available jobs. Both contractors and companies pay fees to join the service. Some contractors and companies prefer to remain anonymous until they meet face-to-face. Develop the nonfunctional requirements and architecture design for the system.

PLANNING

ANALYSIS

DESIGN

☑ **Select Design**
☑ **Develop Physical Data Flow Diagrams**
☑ **Develop Physical Entity Relationship Diagrams**
☑ **Design Architecture**
☑ **Select Hardware and Software**
☐ Develop Use Scenarios
☐ Design Interface Structure
☐ Design Interface Standards
☐ Design Interface Prototype
☐ Evaluate User Interface
☐ Design User Interface
☐ Select Data Storage Format
☐ Denormalize Entity Relationship Diagram
☐ Performance Tune Data Storage
☐ Size Data Storage
☐ Develop Program Structure Chart
☐ Develop Program Specification

TASK CHECKLIST

PLANNING   ANALYSIS   DESIGN

CHAPTER 10

USER INTERFACE DESIGN

A user interface is the part of the system with which the users interact. It includes the screen displays that provide navigation through the system, the screens and forms that capture data, and the reports that the system produces (whether on paper, on the Web, or via some other media). This chapter introduces the basic principles and processes of interface design and discusses how to design the interface structure and standards.

OBJECTIVES

- Understand several fundamental user interface design principles.
- Understand the process of user interface design.
- Understand how to design the user interface structure.
- Understand how to design the user interface standards.
- Be able to design a user interface.

CHAPTER OUTLINE

IMPLEMENTATION

INTRODUCTION

Interface design is the process of defining how the system will interact with external entities (e.g., customers, suppliers, other systems). In this chapter, we focus on the design of *user interfaces* but it is also important to remember that sometimes there are *system interfaces* that exchange information with other systems (e.g., in the case of CD Selections, the Internet system must exchange data with the special order system and the in-store system). System interfaces are typically designed as part of a systems integration effort. They are defined in general terms as part of the physical entity relationship diagram (ERD) and in the nonfunctional requirements (operational requirements), and are designed in detail during data storage design (see Chapter 11) and program design (see Chapter 12).

The user interface design defines the way in which the users will interact with the system and the nature of the inputs and outputs that the system accepts and produces. The user interface includes three fundamental parts. The first is the *navigation mechanism*, the way in which the user gives instructions to the system and tells it what to do (e.g., *buttons, menus*). The second is the *input mechanism* the way in which the system captures information (e.g., *forms* for adding new customers). The third is the *output mechanism*, the way in which the system provides information to the user or to other systems (e.g., *reports, Web pages*). Each of these is conceptually different, but all are closely intertwined: all computer displays contain navigation mechanisms, and most contain input and output mechanisms.

This chapter first introduces several fundamental design principles and provides an overview of the user interface design process. It then provides an overview of the navigation, input, and output components that are used in interface design. This chapter focuses on the design of Web-based interfaces and *graphical user interfaces (GUI)* that use windows, menus, icons, and a mouse (e.g., Windows, Macintosh).[1] Although text-based interfaces are still commonly used on mainframes and UNIX systems, GUIs are probably the most common type of interfaces that you will use, with the possible exception of printed reports.[2]

PRINCIPLES FOR USER INTERFACE DESIGN

In many ways, user interface design is an art. The goal is to make the interface pleasing to the eye and simple to use, while minimizing the effort users need to accomplish their work. The system is never an end in itself; it is merely a means to accomplish the business of the organization.

We have found that the greatest problem facing experienced designers is using space effectively. Simply put, there often is much more information that needs to be presented on a screen or report or form than will fit comfortably. Analysts must bal-

---

[1] Many people attribute the origin of GUI interfaces to Apple or Microsoft. Some people know that Microsoft copied from Apple, which in turn "borrowed" the whole idea from a system developed at the Xerox Palo Alto Research Center (PARC) in the 1970s. Very few know that the Xerox system was based on a system developed by Doug Englebart of Standford, that was first demonstrated at the Western Computer Conference in 1968. Around the same time, Doug also invented the mouse, desktop videoconferencing, groupware, and host of other things we now take for granted. Doug is a legend in the computer science community and has won too many awards to count, but is relatively unknown by the general public.

[2] A good book on GUI design is Susan Fowler, *GUI Design Handbook*, New York: McGraw-Hill, 1998.

ance the need for simplicity and pleasant appearance against the need to present the information across multiple pages or screens, which decreases simplicity. In this section, we discuss some fundamental interface design principles, which are common for navigation design, input design, and output design[3] (Figure 10-1).

Layout

The first element of design is the basic *layout* of the screen, form, or report. Most software designed for personal computers follows the standard Windows or Macintosh approach for screen design. The screen is divided into three boxes. The top box is the navigation area through which the user issues commands to navigate through the system. The bottom box is the status area, which displays information about what the user is doing. The middle—and largest—box is used to display reports and present forms for data entry.

In many cases (particularly on the Web), multiple layout areas are used. Figure 10-2 shows a screen with five navigation areas, each of which is organized to provide different functions and navigation within different parts of the system. The top area provides the standard Internet Explorer navigation and command controls that change the contents of the entire system. The navigation area on the left edge navigates between sections and changes all content to its right. The other two section navigation areas at the top and bottom of the page provide other ways to navigate between sections. The content in the middle of the page displays the results (i.e. a report on a book) plus provides additional navigation within the page about this book.

Principle	Description
Layout	The interface should be a series of areas on the screen that are used consistently for different purposes—for example, a top area for commands and navigation, a middle area for information to be input or output, and a bottom area for status information.
Content awareness	Users should always be aware of where they are in the system and what information is being displayed.
Aesthetics	Interfaces should be functional and inviting to users through careful use of white space, colors, and fonts. There is often a trade-off between including enough white space to make the interface look pleasing without losing so much space that important information does not fit on the screen.
User experience	Although ease of use and ease of learning often lead to similar design decisions, there is sometimes a trade-off between the two. Novice users or infrequent users of software will prefer ease of learning, whereas frequent users will prefer ease of use.
Consistency	Consistency in interface design enables users to predict what will happen before they perform a function. It is one of the most important elements in ease of learning, ease of use, and aesthetics.
Minimal user effort	The interface should be simple to use. Most designers plan on having no more than three mouse clicks from the starting menu until users perform work.

FIGURE 10-1
Principles of User Interface Design

---

[3] A good book on the design of interfaces is Susan Weinschenk, Pamela Jamar, and Sarah Yeo, *GUI Design Essentials*, New York: John Wiley & Sons, 1997.

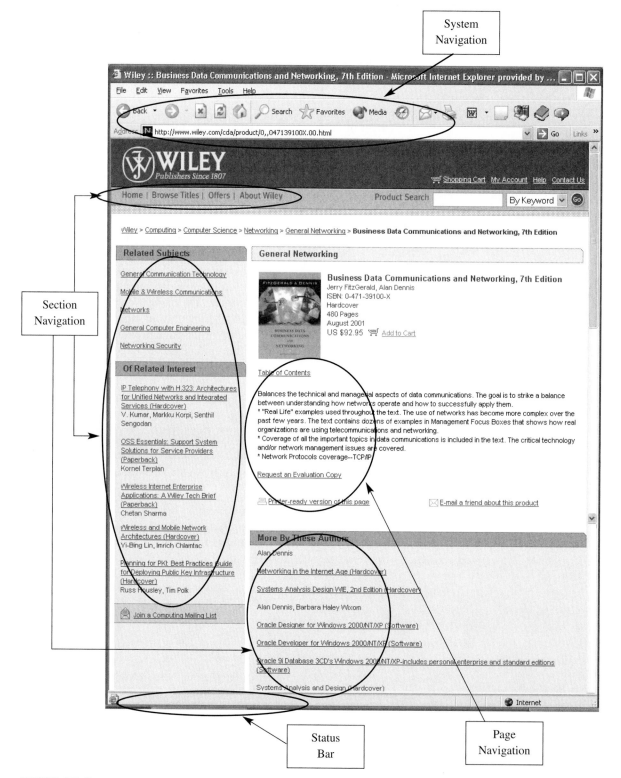

FIGURE 10-2

Layout with Multiple Navigation Areas

This use of multiple layout areas for navigation also applies to inputs and outputs. Data areas on reports and forms are often subdivided into subareas, each of which is used for different types of information. These areas are almost always rectangular in shape, although sometimes space constraints will require odd shapes. Nonetheless, the margins on the edges of the screen should be consistent. Each of the areas within the report or form is designed to hold different information. For example, on an order form (or order report), one part may be used for customer information (e.g., name, address), one part for information about the order in general (e.g., date, payment information), and one part for the order details (e.g., how many units of which items at what price each). Each area is self-contained so that information in one area does not run into another.

The areas and information within areas should have a natural intuitive flow to minimize users' movement from one area to the next. People in western nations (e.g., United States, Canada, Mexico) tend to read top to bottom, left to right, so related information should be placed so it is used in this order (e.g., address lines, followed by city, state/province, and then ZIP code/postal code.) Sometimes the sequence is in chronological order, or from the general to the specific, or from most frequently to least frequently used. In any event, before the areas are placed on a form or report, the analyst should have a clear understanding of what arrangement makes the most sense for how the form or report will be used. The flow between sections should also be consistent, whether horizontal or vertical (Figure 10-3). Ideally, the areas will remain consistent in size, shape, and placement for the forms used to enter information (whether on paper or on a screen) and the reports used to present it.

Content Awareness

Content awareness refers to the ability of an interface to make the user aware of the information it contains with the least amount of effort by the user. All parts of the interface, whether navigation, input, or output, should provide as much content awareness as possible, but it is particularly important for forms or reports that are used quickly or irregularly (e.g., a Web site).

Content awareness applies to the interface in general. All interfaces should have titles (on the screen frame, for example). Menus should show where the user is and, if possible, where the user came from to get there. For example, in Figure 10-2, the line below the top section navigation area shows that the user has browsed from the Wiley Home page, to the Computer Science section, to the Networking section, to General Networking section, and then to this specific book.

Content awareness also applies to the area within forms and reports. All areas should be clear and well defined (with titles if space permits) so that it is difficult for users to become confused about the information in any area. Then users can quickly locate the part of the form or report that is likely to contain the information they need. Sometimes the areas are marked using lines, colors, or headings (e.g., the section area in Figure 10-2); in other cases, the areas are only implied (e.g., the page controls at the bottom of Figure 10-2).

Content awareness also applies to the fields within each area. *Fields* are the individual elements of data that are input or output. The *field labels* that identify the fields on the interface should be short and specific—objectives that often conflict. There should be no uncertainty about the format of information within fields,

FIGURE 10-3
Flow between Interface Sections

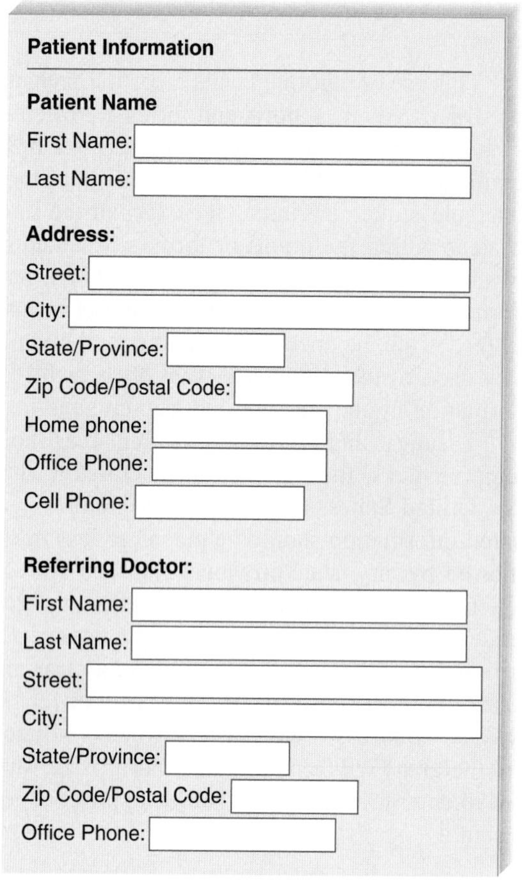

Patient Information

Patient Name

First Name:

Last Name:

Address:

Street:

City:

State/Province:

Zip Code/Postal Code:

Home phone:

Office Phone:

Cell Phone:

Referring Doctor:

First Name:

Last Name:

Street:

City:

State/Province:

Zip Code/Postal Code:

Office Phone:

(*a*) Vertical flow

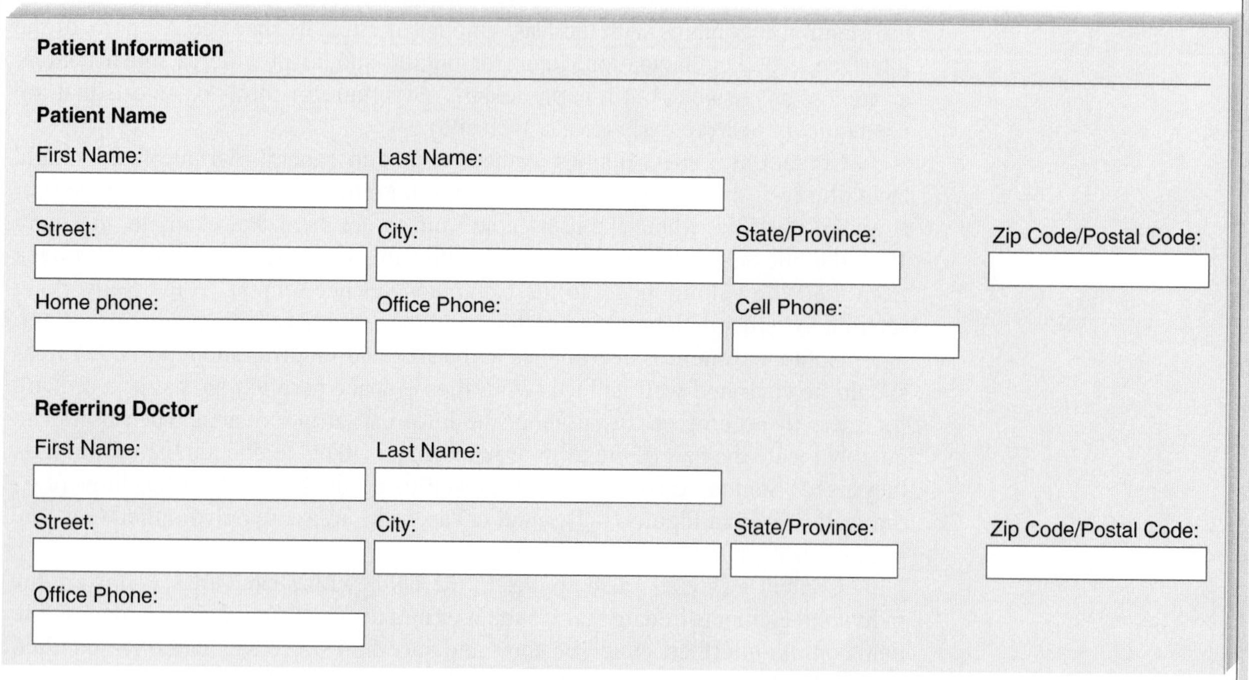

Patient Information

Patient Name

First Name: Last Name:

Street: City: State/Province: Zip Code/Postal Code:

Home phone: Office Phone: Cell Phone:

Referring Doctor

First Name: Last Name:

Street: City: State/Province: Zip Code/Postal Code:

Office Phone:

(*b*) Horizontal Flow

whether for entry or display. For example, a date of 10/5/02 means different things depending on whether you are in the United States (October 5, 2002) or in Canada (May 10, 2002). Any fields for which there is the possibility of uncertainty or multiple interpretations should provide explicit explanations.

Content awareness also applies to the information that a form or report contains. In general, all forms and reports should contain a preparation date (i.e., the date printed or the date data were completed) so that the age of information is obvious. Likewise, all printed forms and software should provide version numbers so that users, analysts, and programmers can identify outdated materials.

Figure 10-4, a form from the University of Georgia, illustrates the logical grouping of fields into areas with an explicit box (top left), as well as an implied area with no box (lower left). The address fields within the address area follow a clear, natural order. Field labels are short where possible (see the top left) but long where more information is needed to prevent misinterpretation (see the bottom left).

Aesthetics

Aesthetics refers to designing interfaces that are pleasing to the eye. Interfaces do not have to be works of art, but they do need to be functional and inviting to use. In most cases, "less is more," meaning that a simple, minimalist design is the best.

Space is usually at a premium on forms and reports, and often there is the temptation to squeeze as much information as possible onto a page or a screen. Unfortunately, this can make a form or report so unpleasant that users do not want to use it. In general, all forms and reports need a minimum amount of white space that is intentionally left blank.

What was your first reaction when you looked at Figure 10-4? This is the most unpleasant form at the University of Georgia, according to staff members. Its *density* is too high; it has too much information packed into too small a space with too little white space. Although it may be efficient in saving paper by using one page instead of two, it is not effective for many users.

In general, novice or infrequent users of an interface, whether on a screen or on paper, prefer interfaces with low density, often one with a density of less than 50% (i.e., less than 50% of the interface occupied by information). More experienced users prefer higher densities, sometimes approaching 90% occupied, because they know where information is located, and high densities reduce the amount of physical movement through the interface. We suspect the form in Figure 10-4 was designed for the experienced staff in the personnel office who use it daily rather than for the clerical staff in academic departments with less personnel experience who use the form only a few times a year.

The design of text is equally important. As a general rule, all text should be in the same font and about the same size. Fonts should be no less than 8 points in size, but a minimum of 10 points is preferred, particularly if the interface will be used by older people. Changes in font and size are used to indicate changes in the type of information that is presented (e.g., headings, status indicators). In general, italics and underlining should be avoided because they make text harder to read.

Serif fonts (i.e., those having letters with serifs or tails, such as Times Roman or the font you are reading right now) are the most readable for printed reports, particularly for small letters. Sans serif fonts (i.e., those without serifs, such as Helvetica or Arial or the ones used for the chapter titles in this book) are the most readable for computer screens and are often used for headings in printed reports. Never

EMPLOYEE PERSONNEL REPORT

UNIVERSITY OF GEORGIA

PAY TYPE

DOCUMENT NO. | PAGE | DATE | FY | DEPARTMENT PHONE | COLLEGE OR DIVISION

DEPARTMENT/PROJECT | PRI DEPT | HIGH DEGREE | INSTITUTION | YEAR

UGA EMPLOYMENT HISTORY
☐ (C) CURRENT ☐ (P) PREVIOUS
DATE

ACTION
MO DA YR

SOC.SEC.NUM. | LAST NAME | FIRST NAME/INITIAL | MIDDLE INITIAL/NAME | SUF

☐ (1) REGULAR   ☐ (3) TEMPORARY
☐ (2) UGA STUDENT   ☐ (4) NR-ALIEN
UGA % TIME

STREET OR ROUTE NO. (LINE 1) | NON-WORK PHONE | BIRTH DATE | SPOUSE'S NAME | CHAIR

☐ (E) EXEMPT   ☐ (N) NON-EXEMPT   ☐ (T) TIPPED
☐ (M) MALE   ☐ (S) SINGLE   ☐ (Y) FACULTY-RANK
☐ (F) FEMALE   ☐ (M) MARRIED   ☐ (N) NON-FACULTY

STREET OR ROUTE NO. (LINE 2) | UNIVERSITY PHONE | CITIZEN OF | I-9 VISA COUNTY

☐ (1) WHITE   ☐ (3) ORIENTAL/ASIAN   ☐ (5) HISPANIC
☐ (2) BLACK   ☐ (4) AMERICAN INDIAN   ☐ (6) MULTIRACIAL
                                       ☐ (9)

CITY | STATE | ZIP + 4 | UNIVERSITY BUILDING NAME | BLDG.NO/FLOOR/ROOM

COUNTY MONEY
(PER PAY PERIOD)

PAYROLL PAYMENT DISTRIBUTION
☐ (1) SEND TO DEPT (DIST CODE)
☐ (2) DIRECT DEPOSIT(SEND PR105 TO PAYROLL
☐ (3) PICK UP AT PAYROLL WINDOW

COOP. EXT. EMPLOYEES ONLY
UGA SALARY ____
COUNTY MONEY ____
TOTAL ____

FOR PAYROLL DEPT USE ONLY
FED EXM | STATE EXM

OASDI ____   RETIRE ____
HI ____   EIC ____

GDCP ____

TRX	HOME DEPT	SHORT TITLE	POSN NO.	APPT. BEGIN MO DA YR HR	APPT. END MO DA YR HR	JOBCLASS CODE	POSITION TITLE	POS % TIME	C N	FULL TIME ANNUAL SALARY	S C	SUPPLEMENT AMOUNT

PAYROLL AUTHORIZATION

FISCAL YEAR BUDGET

TRX	HOME DEPT	SHORT TITLE	POSN NO	ACCOUNT	EFT	FROM	MO DA YR	MO DA YR	MO DA YR	MO DA YR	MO DA YR	MO DA YR	MO DA YR
						THRU							
						AMOUNT PER PAY PERIOD OR HOURLY RATE							

TOTALS

☐ (A) NEW UGA EMPLOYEE   ☐ (B) LATERAL TRANSFER   ☐ (C) PROMOTION
☐ (D) REPLACEMENT POSN-NAME OF LAST INCUMBENT ____
☐ (E) APPOINTMENT TO NEW POSITION
☐ (F) CHANGE % TIME EMPLOYED FROM ____ TO ____
☐ (G) CONTINUATION WITHIN EXISTING BUDGET POSITION
☐ (H) REVISE DISTRIBUTION OF SALARY
☐ (I) TRANSFER FROM DEPT ____ TO ____
☐ (J) CHANGE PAY TYPE FROM ____ TO ____

☐ (K) CHANGE TITLE FROM ____
☐ (L) CHANGE NAME FROM ____
☐ (M) CHANGE SSN FROM ____
☐ (N) LEAVE W/O PAY FROM ____
☐ (O) CHG COUNTY $ FROM ____
☐ (P) TERMINATION-REASON ____
☐ (Q) OTHER (SPECIFY) ____

REMARKS

DEPARTMENT HEAD | DATE | VICE PRESIDENT | DATE | BUDGET REVIEW | DATE | BUDGET OFFICE | DATE

FIGURE 10-4
Form Example

use all capital letters, except possibly for titles—all-capitals text "shouts" and is harder to read.

Color and patterns should be used carefully and sparingly and only when they serve a purpose. About 10% of men are color blind, so the improper use of color can impair their ability to read information. A quick trip around the Web will demonstrate the problems caused by indiscriminate use of colors and patterns. Remember, the goal is pleasant readability, not art; color and patterns should be used to strengthen the message, not overwhelm it. Color is best used to separate and categorize items, such as showing the difference between headings and regular text, or to highlight important information. Therefore, colors with high contrast should be used (e.g., black and white). In general, black text on a white background is the most readable, with blue on red the least readable. (Most experts agree that background patterns on Web pages should be avoided.) Color has been shown to affect emotion, with red provoking intense emotion (e.g., anger) and blue provoking lowered emotions (e.g., drowsiness).

User Experience

There are two types of users for most computer systems: those with experience and those without. Interfaces should be designed for both types of users. Novice users are usually most concerned with *ease of learning*, how quickly they can learn new systems. Expert users are usually most concerned with *ease of use*, how quickly they can use the system, once they have learned how to use it. Often these two are complementary and lead to similar design decisions, but sometimes there are trade-offs. Novices, for example, often prefer menus that show all available system functions, because these promote ease of learning. Experts, on the other hand, sometimes prefer fewer menus that are organized around the most commonly used functions.

Systems that will end up being used by many people on a daily basis are more likely to have a majority of expert users (e.g., order entry systems). Although interfaces should try to balance ease of use and ease of learning, these types of systems should put more emphasis on ease of use rather than on ease of learning. Users should be able to access the commonly employed functions quickly, with few keystrokes or a small number of menu selections.

In many other systems (e.g., decision support systems), most people will remain occasional users for the lifetime of the system. In this case, greater emphasis may be placed on ease of learning rather than on ease of use.

Although ease of use and ease of learning often go hand in hand, sometimes they don't. Research shows that expert and novice users have different requirements and behavior patterns in some cases. For example, novices virtually never look at the bottom area of a screen that presents status information, but experts refer to the status bar when they need information. Most systems should be designed to support frequent users, except for systems that are to be used infrequently or those for which many new users or occasional users are expected (e.g., the Web). Likewise, systems that contain functionality that is used only occasionally must contain a highly intuitive interface, or an interface that contains explicit guidance regarding its use.

The balance of quick access to commonly used and well-known functions, and guidance through new and less well-known functions is challenging to the interface designer, and this balance often requires elegant solutions. Microsoft Office, for example, addresses this issue through the use of the "show me" functions that demonstrate the menus and buttons for specific functions. These features remain in the background until they are needed by novice users (or even experienced users when they use an unfamiliar part of the system).

Consistency

Consistency in design is probably the single most important factor in making a system simple to use because it enables users to predict what will happen. When interfaces are consistent, users can interact with one part of the system and then know how to interact with the rest—aside, of course, from elements unique to those parts. Consistency usually refers to the interface within one computer system, so that all parts of the same system work in the same way. Ideally, however, the system also should be consistent with other computer systems in the organization and with whatever commercial software is used (e.g., Windows). For example, many users are familiar with the Web, so the use of Web-like interfaces can reduce the amount of learning required by the user. In this way, the user can reuse Web knowledge, thus significantly reducing the learning curve for a new system.

Consistency occurs at many different levels. Consistency in the navigation controls conveys how actions in the system should be performed. For example, using the same icon or command to change an item clearly communicates how changes are made throughout the system. Consistency in terminology is also important. This refers to using the same words for elements on forms and reports (e.g., not customer in one place and client in another). We also believe that consistency in report and form design is important, although one study suggests that being too consistent can cause problems.[4] When reports and forms are very similar except for very minor changes in titles, users sometimes mistakenly use the wrong form and either enter incorrect data or misinterpret its information. The implication for design is to make the reports and forms similar but give them some distinctive elements (e.g., color, size of titles) that enable users to immediately detect differences.

Minimize User Effort

Finally, interfaces should be designed to minimize the amount of effort needed to accomplish tasks. This means using the fewest possible mouse clicks or keystrokes to move from one part of the system to another. Most interface designers follow the *three-clicks rule:* users should be able to go from the start or main menu of a sys-

---

YOUR
TURN

10-1 WEB PAGE CRITIQUE

Visit the Web home page for your university and navigate through several of its Web pages. Evaluate the extent to which they meet the six design principles.

---

[4] John Satzinger and Lorne Olfman, "User Interface Consistency Across End-User Application: The Effects of Mental Models," *Journal of Management Information-Systems*, Spring 1998, 14 (4) 167–193.

tem to the information or action they want in no more than three mouse clicks or three keystrokes.

USER INTERFACE DESIGN PROCESS

User interface design[5] is a five-step process that is iterative—analysts often move back and forth between steps rather than proceed sequentially from step 1 to step 5 (Figure 10-5). First, the analysts examine the DFDs and use cases developed in the analysis phase (see Chapters 5 and 6) and interview users to develop *use scenarios* that describe users' commonly employed patterns of actions so the interface enables users to quickly and smoothly perform these scenarios. Second, the analysts develop the *interface structure diagram (ISD)* that defines the basic structure of the interface. This diagram (or set of diagrams) shows all the interfaces (e.g., screens, forms, and reports) in the system and how they are connected. Third, the analysts design *interface standards* which are the basic design elements on which interfaces in the system are based. Fourth, the analysts create an *interface design prototype* for each of the individual interfaces in the system, such as navigations controls, input screens, output screens, forms (including preprinted paper forms), and reports. Finally, the individual interfaces are subjected to *interface evaluation* to determine if they are satisfactory and how they can be improved.

Interface evaluations almost always identify improvements, so the interface design process is repeated in a cyclical process until no new improvements are

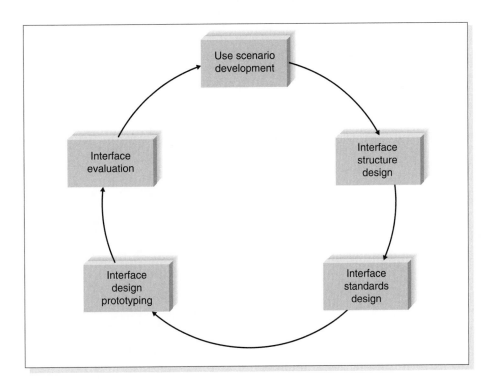

FIGURE 10-5
User Interface Design Process

---

[5] One of the best books on user interface design is Ben Schneiderman, *Designing the User Interface: Strategies for Effective Human-Computer Interaction*, 3rd ed., Reading MA: Addison-Wesley, 1998.

identified. In practice, most analysts interact closely with the users during the interface design process, so that users have many chances to see the interface as it evolves rather than waiting for one overall interface evaluation at the end of the interface design process. It is better for all concerned (both analysts and users) if changes are identified sooner rather than later. For example, if the interface structure or standards need improvements, it is better to identify changes before most of the screens that use the standards have been designed.

Use Scenario Development

A use scenario is an outline of the steps that the users perform to accomplish one aspect of their work. A use scenario is one commonly used path through a use case. For example, Figure 6-20 in Chapter 6 shows the level 1 DFD for the major Web section of the Internet system. This figure shows process 1.2 (Provide CD Information) as being distinct from process 1.5 (Check Out). We model the two processes separately and write the programs separately because they are separate processes within process 1 (Take Requests).

The DFD was designed to model all possible uses of the system—that is, its complete functionality or all possible paths through the use case. But use scenarios are just one path through the use case. In one use scenario, for example, a user will browse for many CDs, much like someone browsing through a real music store looking for interesting CDs. He or she will search for a CD, read the marketing materials for it, perhaps add it to the shopping cart, browse for more, and so on. Eventually, the user will want to place the request, perhaps removing some things from the shopping cart beforehand.

In another use scenario, a user will want to buy one specific CD. He or she will go directly to the CD, price it, and buy it immediately, much like someone running

---

Use Scenario: The Browsing Shopper
User is not sure what they want to buy and will browse for several CDs

1. User may search for a specific artist or browse through a music category (1.1).
2. User will likely read the basic information for several CDs, as well as the marketing material for some. He or she will likely listen to music samples and browse related CDs (1.2).
3. User will locate store(s) for the selected CD (1.3)
4. User will put the CD in the shopping cart (1.4) and will continue browsing (1.1).
5. Eventually the user will want to place the order, but will probably want to look through the shopping cart, possibly discarding some CDs first (1.5).

Use Scenario: The Hurry-up Shopper
User knows exactly what he or she wants and wants it quickly

1. User will search for a specific artist or CD (1.1).
2. User will look at the price and enough other information to confirm the CD is the desired CD (1.2).
3. The user will locate the closest store with availability (1.3)
4. User will want to place the order (1.4 & 1.5) or move on on to other Web sites.

FIGURE 10-6

Two Use Scenarios For the Take Requests Use Case

The numbers in parentheses refer to process numbers in the DFD

YOUR

TURN

10-2 USE SCENARIO DEVELOPMENT FOR THE WEB

Visit the Web site for your university and navigate through several of its Web pages. Develop two use scenarios for it.

into a store, making a beeline for the one CD he or she wants, and immediately paying and leaving the store. This user will enter the CD information in the search portion of the system, look at the resulting cost information, and immediately place an request or leave. Anything that slows him or her down will risk loss of the sale. For this use scenario, we need to ensure that the path through the DFD as presented by the interface is short and simple, with very few menus and mouse clicks.

Use scenarios are presented in a simple narrative description that is tied to the DFD. Figure 10-6 shows the two use scenarios just described. The key point in using use scenarios for interface design is not to document all possible use scenarios within a use case, because then you end up just repeating the DFD in a different form. The goal is to describe the handful of most commonly occurring use scenarios so the interface can be designed to enable the most common uses to be performed simply and easily.

Interface Structure Design

The interface structure defines the basic components of the interface and how they work together to provide functionality to users. An ISD is used to show how all the screens, forms, and reports used by the system are related and how the user moves from one to another. Most systems have several ISDs, one for each major part of the system.

An ISD is somewhat similar to a DFD in that it uses boxes and lines to show structure. However, unlike DFDs, there are no commonly used rules or standards for their development. With one approach, each interface (e.g., screen, form, report) on an ISD is drawn as a box and is given a unique number (at the top) and a unique name (in the middle). The numbers usually follow a tree-type structure, although this is not always done. Unlike the DFDs, however, the numbers do not mean that all the screens belong to "parents" higher in the tree; instead, they usually imply relationships between a menu and a submenu. The lines denote the ability to navigate from one menu to another.

YOUR

TURN

10-3 USE SCENARIO DEVELOPMENT FOR AN AUTOMATED TELLER MACHINE

Pretend you have been charged with the task of redesigning the interface for the ATM at your local bank. Develop two use scenarios for it.

Each box on the ISD also shows (at the bottom) the DFD process that is supported by the interface (Figure 10-7). Sometimes there is more than one interface for a given process (e.g., in Figure 10-7, interfaces 1.1 through 1.3 support process 1.1.1); in other cases, there is only one interface for each process (e.g., interfaces 3.1 through 3.3 support processes 1.1.3.1 through 1.1.3.3).

Each interface is linked to other interfaces by lines that show how users can transition from one interface to the next. In most cases, the interfaces form a hierarchy or a tree, but sometimes an interface is linked to one outside of the hierarchy,

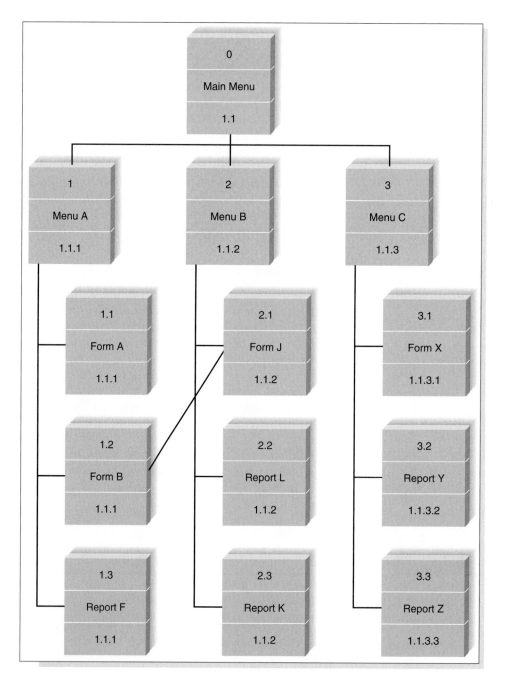

FIGURE 10-7
Example Interface Structure Diagram

as shown by the link from Form J to Form B (e.g., the ability to update a customer address while entering a new order).

The basic structure of the interface follows the basic structure of the business process itself as defined in the process model or object model. The analyst starts with the DFD and develops the fundamental flow of control of the system as it moves from process to process. There are usually several major parts to an information system, each of them distinct, in the same way there are several high-level processes in a DFD. In general—but not always—there is one ISD for each process on the level 1 DFD.

The analyst then examines the use scenarios to see how well the ISD supports them. Quite often, the use scenarios identify paths through the ISD that are more complicated than they should be. The analyst then reworks the ISD to simplify the ability of the interface to support the use scenarios, sometimes by making major changes to the menu structure, sometimes by adding shortcuts.

Interface Standards Design

The interface standards are the basic design elements that are common across the individual screens, forms, and reports within the system. Depending on the application, there may be several sets of interface standards for different parts of the system (e.g., one for Web screens, one for paper reports, one for input forms). For example, the part of the system used by data-entry operators may mirror other data-entry applications in the company, whereas a Web interface for displaying information from the same system may adhere to some standardized Web format. Likewise, each individual interface may not contain all the elements in the standards (e.g., a report screen may not have an "edit" capability), and they may contain additional characteristics beyond the standard ones, but the standards serve as the touchstone that ensures the interfaces are consistent across the system.

Interface Metaphor   First of all, the analysts must develop the fundamental *interface metaphor*(s) that defines how the interface will work. An interface metaphor is a concept from the real world that is used as a model for the computer system. The metaphor helps the user understand the system and enables the user to predict what features the interface might provide, even without actually using the system. Sometimes systems have one metaphor, whereas in other cases, there are several metaphors in different parts of the system.

In many cases, the metaphor is explicit. Quicken, for example, uses a checkbook metaphor for its interface, even to the point of having the users type information into an on-screen form that looks like a real check. In other cases, the metaphor is implicit or unstated, but it is there nonetheless. Many Windows systems use the paper form or table as a metaphor.

YOUR **TURN** | **10-4 INTERFACE STRUCTURE DESIGN**

Pretend you have been charged with the task of redesigning the interface for the ATM at your local bank. Design an ISD that shows how a user would navigate among the screens.

In some cases, the metaphor is so obvious that it requires no thought. The CD Selections Internet system, for example, will use the music store as the metaphor (e.g., shopping cart). In other cases, a metaphor is hard to identify. In general, it is better not to force a metaphor that really doesn't fit a system, because an ill-fitting metaphor will confuse users by promoting incorrect assumptions.

Interface Objects The template specifies the names that the interface will use for the major *interface objects*, the fundamental building blocks of the system such as the entities and data stores. In many cases, the object names are straightforward, such as calling the shopping cart the "shopping cart." In other cases, it is not simple. For example, CD Selections sells both CDs and tapes. When users search for items to buy, should they search for CDs or CDs & Tapes or CDs/Tapes or Music or Albums or something else? Obviously, the object names should be easily understood and help promote the interface metaphor.

In general, in cases of disagreements between the users and the analysts over names, whether for objects or actions (discussed later in this chapter), the users should win. A more understandable name always beats a more precise or more accurate name.

Interface Actions The template also specifies the navigation and command language style (e.g., menus) and grammar (e.g., object–action order; see "Navigation Design" later in this chapter). It gives names to the most commonly used *interface actions* in the navigation design (e.g., buy versus purchase, or modify versus change).

Interface Icons The interface objects and actions and also their status (e.g., deleted, overdrawn) may be represented by *interface icons*. Icons are pictures that will appear on command buttons as well as in reports and forms to highlight important information. Icon design is very challenging because it means developing a simple picture less than half the size of a postage stamp that needs to convey an often complex meaning. The simplest and best approach is to simply adopt icons developed by others (e.g., a blank page to indicate "create a new file," a diskette to indicate "save"). This has the advantage of quick icon development, and the icons may already be well understood by users because users have seen them in other software.

Commands are actions that are especially difficult to represent with icons because they are in motion, not static. Many icons have become well known from widespread use, but icons are not as well understood as first believed. Use of icons can sometimes cause more confusion than insight. (For example, did you know that a picture of a sweeping paintbrush in Microsoft Word means "format painter"?) Icon meanings become clearer with use, but sometimes a picture is not worth even one word; when in doubt, use a word, not a picture.

Interface Templates The *interface template* defines the general appearance of all screens in the information system and the paper-based forms and reports that are used. The template design, for example, specifies the basic layout of the screens (e.g., where the navigation area[s], status area, and form/report area[s] will be placed) and the color scheme(s) that will be applied. It defines whether windows will replace one another on the screen or will cascade on top of each other. The template defines a standard placement and order for common interface actions (e.g., "File, Edit, View" rather than "File, View, Edit"). In short, the template draws together the other major interface design elements: metaphors, objects, actions, and icons.

Interface Design Prototyping

An interface design prototype is a mock-up or a simulation of a computer screen, form, or report. A prototype is prepared for each interface in the system to show the users and the programmers how the system will perform. In the "old days," an interface design prototype was usually specified on a paper form that showed what would be displayed on each part of the screen. Paper forms are still used today, but more and more interface design prototypes are being built using computer tools instead of paper. The three most common approaches to interface design prototyping are storyboarding, HTML prototyping, and language prototyping.

Storyboard At its simplest, an interface design prototype is a paper-based *storyboard*. The storyboard shows hand-drawn pictures of what the screens will look like and how they flow from one screen to another, in the same way a storyboard for a cartoon shows how the action will flow from one scene to the next (Fig. 10-8). Storyboards are the simplest technique because all they require is paper (often a flip chant) and a pen—and someone with some artistic ability.

HTML Prototype One of the most common types of interface design prototypes used today is the *HTML prototype*. As the name suggests, an HTML prototype is built using Web pages created in HTML (hypertext mark-up language). The designer uses HTML to create a series of Web pages that show the fundamental parts of the system. The users can interact with the pages by clicking on buttons and entering pretend data into forms (but because there is no system behind the pages, the data are never processed). The pages are linked together so that as the user clicks on buttons, the requested part of the system appears. HTML prototypes are superior to storyboards in that they enable users to interact with the system and gain a better sense of how to navigate among the different screens. However, HTML has limitations—the screens shown in HTML will never appear exactly like the real screens in the system (unless, of course, the real system will be a Web system in HTML).

Language Prototype A *language prototype* is an interface design prototype built using the actual language or tool that will be used to build the system. Language prototypes are designed in the same ways as HTML prototypes (they enable the

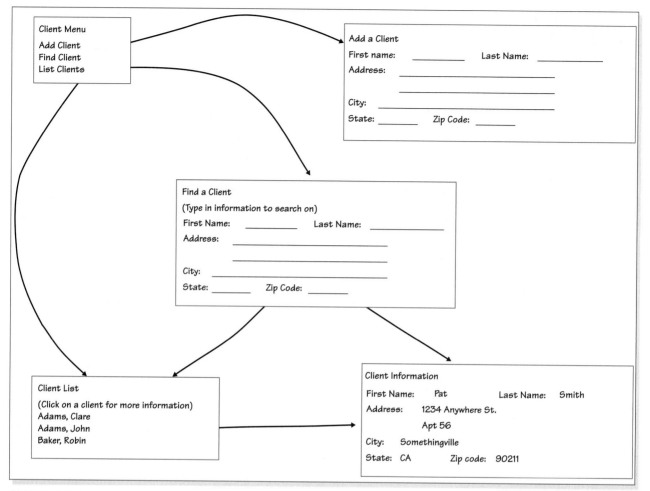

FIGURE 10-8
An Example Storyboard

user to move from screen to screen, but they perform no real processing). For example, in Visual Basic, it is possible to create and view screens without actually attaching program code to the screens. Language prototypes take longer to develop than do storyboards or HTML prototypes, but they have the distinct advantage of showing exactly what the screens will look like. The user does not have to guess about the shape or position of the elements on the screen.

Selecting the Appropriate Techniques Projects often use a combination of different interface design prototyping techniques for different parts of the system. Storyboarding is the fastest and least expensive but provides the least amount of detail. Language prototyping is the slowest, most expensive, and most detailed approach. HTML prototyping falls between the two extremes. Therefore, storyboarding is used for parts of the system in which the interface is well understood and when more expensive prototypes are thought to be unnecessary. HTML prototypes and language prototypes are used for parts of the system that are critical yet not well understood.

CONCEPTS

IN ACTION

10-A INTERFACE DESIGN PROTOTYPES FOR A DSS APPLICATION

I was involved in the development of several decision support systems (DSS) while working as a consultant. On one project, a future user was frustrated because he could not imagine what a DSS looked like and how one would be used. He was a key user, but the project team had a difficult time involving him in the project because of his frustration. The team used SQL Windows (one of the most popular development tools at the time) to create a language prototype that demonstrated the future system's appearance, proposed menu system, and screens (with fields, but no processing).

The team was amazed at the user's response to the prototype. He appreciated being given a context with which to visualize the DSS, and he soon began to recommend improvements to the design and flow of the system, and to identify some important information that was overlooked during the Analysis Phase. Ultimately, the user became one of the strongest supporters of the system, and the project team felt sure that the prototype lead to a much better product in the end. *Barbara Wixom*

QUESTIONS
1. Why do you think the team chose to use a language prototype rather than a storyboard or HTML prototype?
2. What trade-offs were involved in the decision?

Interface Evaluation

The objective of interface evaluation is to understand how to improve the interface design. Interface design is subjective; there are no formulas that guarantee a great user interface. Most interface designers intentionally or unintentionally design an interface that meets their personal preferences, which may or may not match the preferences of the users. The key message, therefore is to have as many people as possible evaluate the interface, and the more users the better. Most experts recommend involving at least 10 potential users in the evaluation process.

Many organizations save interface evaluation for the very last step in the SDLC before the system is installed. Ideally, however, interface evaluation should be performed while the system is being designed—before it is built—so that any major design problems can be identified and corrected before the time and cost of programming has been spent on a weak design. It is not uncommon for the system to undergo one or two major changes after the users see the first interface design prototype, because they identify problems that are overlooked by the project team.

As with interface design prototyping, interface evaluation can take many different forms, each with different costs and different amounts of detail. Four common approaches are heuristic evaluation, walk-through evaluation, interactive evaluation, and formal usability testing. As with interface design prototyping, the different parts of a system can be evaluated using different techniques.

Heuristic Evaluation A *heuristic evaluation* examines the interface by comparing it to a set of heuristics or principles for interface design. The project team develops a checklist of interface design principles—from the list at the start of this chapter, for example, as well as the lists of principles in the navigation, input, and output design sections. At least three members of the project team then individually work through the interface design prototype, examining each interface to ensure it satisfies each design principle on the formal checklist. After each member has gone through the prototype separately, they all meet as a team to discuss their evaluation and identify specific improvements that are required. This is the weakest type of evaluation.

Walk-through Evaluation   An interface design *walk-through evaluation* is a meeting conducted with the users who will ultimately have to operate the system. The project team presents the prototype to the users and walks them through the various parts of the interface. The project team shows the storyboard or actually demonstrates the HTML or language prototype and explains how the interface will be used. The users identify improvements to each of the interfaces that are presented.

Interactive Evaluation   With an *interactive evaluation* the users themselves actually work with the HTML or language prototype in one-on-one sessions with members of the project team (an interactive evaluation cannot be used with a storyboard). As the user works with the prototype (often by going through the use scenarios or just navigating at will through the system), he or she tells the project team members what he or she likes and doesn't like and what additional information or functionality is needed. As the user interacts with the prototype, team members record the situation when the user appears to be unsure what to do, makes mistakes, or misinterprets the meaning of an interface component. If the pattern of uncertainty, mistakes, or misinterpretations recurs across several evaluation sessions with several users, it is a clear indication that those parts of the interface need improvement.

Formal Usability Testing   Formal *usability testing* is commonly done with commercial software products and products developed by large organizations that will be widely used through the organization. As the name suggests, it a very formal—almost scientific—process that can be used only with language prototypes (and systems that have been completely built awaiting installation or shipping).[6] As with interactive evaluation, usability testing is done in one-on-one sessions in which a user works directly with the software. However, it is typically done in a special lab equipped with video cameras and special software that records each and every keystroke and mouse operation so they can be replayed to understand exactly what the user did.

   The user is given a specific set of tasks to accomplish (usually the use scenarios) and after some initial instructions, the project teams member(s) are not permitted to interact with the user to provide assistance. The user must work with the software without help, which can be hard on the user if he or she becomes confused with the system. It is critical that users understand that the goal is to test the interface, not their abilities, and if they are unable to complete the task, the interface—not the user—has failed the test. Several performance measures are used, such as time to complete the task, error rate, as well as user satisfaction.

   Formal usability testing is very expensive, because each one-user session (which typically lasts 1 to 2 hours) can take 1 to 2 days to analyze due to the volume of detail collected in the computer logs and videotapes. Most usability testing involves 5 to 10 users because fewer than 5 users makes the results depend

---

[6] A good source for usability testing is Jakob Nielsen and Robert Mack (eds.), *Usability Inspection Methods,* New York: John Wiley & Sons, 1994. See also www.useit.com/papers.

| YOUR | **10-6 Prototyping and Evaluation** |
| TURN | |

Pretend you have been charged with the task of redesigning the interface for the ATM at your local bank. What type of prototyping and interface evaluation approach would you recommend?

too much on the specific individual users who participated, and more than 10 users is often too expensive to justify (unless you work for a large commercial software developer).

NAVIGATION DESIGN

The navigation component of the interface enables the user to enter commands to navigate through the system and perform actions to enter and review information it contains. The navigation component also presents messages to the user about the success or failure of his or her actions. The goal of the navigation system is to make the system as simple as possible to use. A good navigation component is one the user never really notices. It simply functions the way the user expects, and thus the user gives it little thought.

Basic Principles

One of the hardest things about using a computer system is learning how to manipulate the navigation controls to make the system do what you want. Analysts usually must assume that users have not read the manual, have not attended training, and do not have external help readily at hand. All controls should be clear and understandable and placed in an intuitive location on the screen. Ideally, the controls should anticipate what the user will do and simplify his or her efforts. For example, many set-up programs are designed so that for a "typical" installation, the user can simply keep pressing the Next button.

Prevent Mistakes   The first principle of designing navigation controls is to prevent the user from making mistakes. A mistake costs time and creates frustration. Worse still, a series of mistakes can cause the user to discard the system. Mistakes can be reduced by labeling commands and actions appropriately and by limiting choices. Too many choices can confuse the user, particularly when they are similar and hard to describe in the short space available on the screen. When there are many similar choices on a menu, consider creating a second level of menu or a series of options for basic commands.

Never display a command that cannot be used. For example, many Windows applications gray-out commands that cannot be used; they are displayed on pulldown menus in a very light-colored font, but they cannot be selected. This shows that they are available (and keeps all menu items in the same place) but that they cannot be used in the current context.

When the user is about to perform a critical function that is difficult or impossible to undo (e.g., deleting a file), it is important to confirm the action with the user (and make sure the selection was not made by mistake). This is usually done by having the user respond to a confirmation message that explains what the user has requested and asks the user to confirm that this action is correct.

Simplify Recovery from Mistakes   No matter what the system designer does, users will make mistakes. The system should make it as easy as possible to correct these errors. Ideally, the system will have an "Undo" button that makes mistakes easy to override; however, writing the software for such buttons can be very complicated.

Use Consistent Grammar Order   One of the most fundamental decisions is the *grammar order*. Most commands require the user to specify an object (e.g., file, record, word), and the action to be performed on that object (e.g., copy, delete). The interface can require the user to first choose the object and then the action (an *object–action order*), or first the action and then the object (an *action–object order*). Most Windows applications use an object–action grammar order (e.g., think about copying a block of text in your word processor).

The grammar order should be consistent throughout the system, both at the data element level and at the overall menu level. Experts debate about the advantages of one approach over the other, but because most users are familiar with the object–action order, most systems today are designed using that approach.

Types of Navigation Controls

There are two traditional hardware devices that can be used to control the user interface: the keyboard and a pointing device, such as a mouse, trackball, or touch screen. In recent years, voice recognition systems have made an appearance, but they are not yet common. There are three basic software approaches for defining user commands: languages, menus, and direct manipulation.

Languages   With a *command language*, the user enters commands using a special language developed for the computer system (e.g., Disk Operating System [DOS] and Structured Query Language [SQL] both use command languages). Command languages sometimes provide greater flexibility than do other approaches because the user can combine language elements in ways not predetermined by developers. However, they put a greater burden on users because users must learn syntax and type commands rather than select from a well-defined, limited number of choices. Systems today use command languages sparingly, except in cases in which there are an extremely large number of command combinations, making it impractical to try to build all combinations into a menu (e.g., SQL queries for databases).

Natural language interfaces are designed to understand the user's own language (e.g., English, French, Spanish). These interfaces attempt to interpret what the user means, and often they present back to the user a list of interpretations from which to choose. An example of the use of natural language is Microsoft's Office Assistant, which enables users to ask free-form questions to get help.

Menus   The most common type of navigation system today is the menu. A *menu* presents the user with a list of choices, each of which can be selected. Menus are

YOUR	
TURN	

10-7 Design a Navigation System

Design a navigation system for a system into which users must enter information about customers, products, and orders. For all three information categories, users will want to change, delete, find one specific record, and list all records.

easier to learn than languages because a limited number of available commands are presented to the user in an organized fashion. Clicking on an item with a pointing device or pressing a key that matches the menu choice (e.g., a function key) takes very little effort. Therefore, menus are usually preferred to languages.

Menu design should be done with care, because the submenus behind a main menu are hidden from users until they click on the menu item. It is better to make menus broad and shallow (i.e., each menu containing many items with each item containing only one or two layers of menus) rather than narrow and deep (i.e., each menu containing only a few items, but each item leading to three or more layers of menus). A broad and shallow menu presents the user with the most information initially so that he or she can see many options and requires only a few mouse clicks or keystrokes to perform an action. A narrow and deep menu makes users hunt and seek for items hidden behind menu items and requires many more clicks or keystrokes to perform an action.

Research suggests that in an ideal world, any one menu should contain no more than eight items and it should take no more than two mouse clicks or keystrokes from any menu to perform an action (or three from the main menu that starts a system).[7] However, analysts sometimes must break this guideline in the design of complex systems. In this case, menu items are often grouped together and separated by a horizontal line (Fig. 10-9). Often menu items have *hot keys* that enable experienced users to quickly invoke a command with keystrokes in lieu of a menu choice (e.g., Control-F in Word invokes the Find command, whereas Alt-F opens the File menu).

Menus should put together similar categories of items so that the user can intuitively guess what each menu contains. Most designers recommend grouping menu items by interface objects (e.g., Customers, Purchase Orders, Inventory) rather than by interaction actions (e.g., New, Update, Format), so that all actions pertaining to one object are in one menu, all actions for another object are in a different menu, and so on. However, this is highly dependent on the specific interface. As Figure 10-9 shows, Microsoft Word groups menu items by interface objects (e.g., File, Table, Window) and by interface actions (e.g., Edit, Insert, Format) on the same menu.

Figure 10-9 illustrates several other types of menus. Fig. 10-10 summarizes several commonly used types of menus and explains when to use them.

---

[7] Kent L. Norman, *The Psychology of Menu Selection,* Norwood, NJ: Ablex Publishing, 1991.

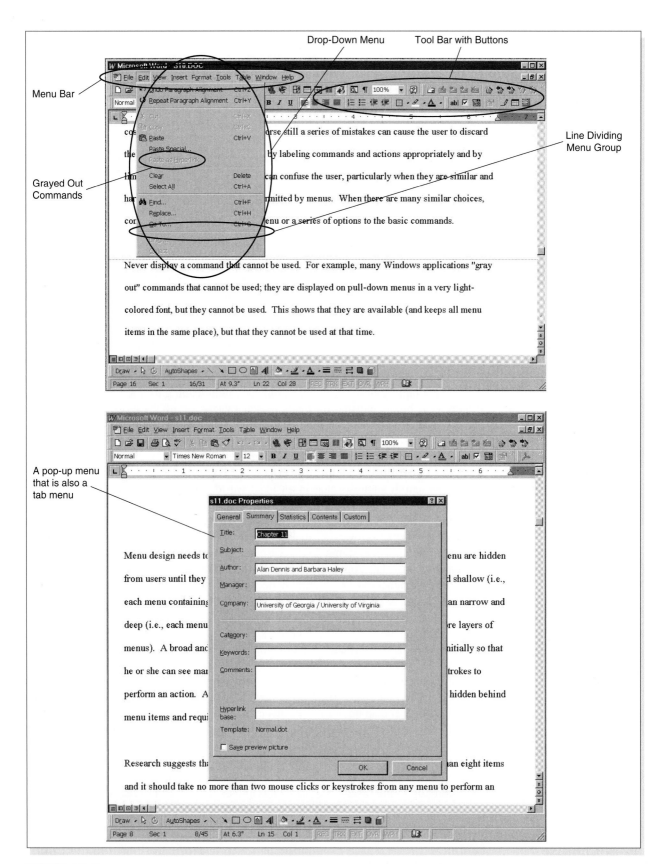

FIGURE 10-9
Common Types of Menus

Type of Menu	When to Use	Notes
Menu Bar List of commands at the top of the screen. Always on screen.	Main menu for system	• Use the same organization as the operating system and other packages (e.g., File, Edit, View) • Menu items are always one word, never two • Menu items lead to other menus, rather than performing action • Never allow users to select actions they can't perform (instead use grayed-out items)
Drop-Down Menu Menu that drops-down immediately below another menu. Disappears after one use.	Second level menu, often from menu bar	• Menu items are often multiple words • Avoid abbreviations • Menu items perform action or lead to another cascading drop-down menu, pop-up menu, or tab menu
Hyperlink Menu A set of items arranged as a menu, usually along one edge of the screen.	Main menu for Web-based system	• Most users are familiar with hyperlink menus on the left edge of the screen, although they can be placed along any edge • Menu items are usually only one or two words
Embedded Hyperlinks A set of items embedded and underlined in text.	As a link to ancillary, optional information	• Used sparingly to provide additional information because they can complicate navigation • Usually open a new window that is closed once the action is complete so the user can return to the original use scenario
Pop-up Menu Menu that pops-up and floats over the screen. Disappears after one use.	As a shortcut to commands for experienced users	• Often (not always) invoked by a right click in Windows-based systems • Often overlooked by novice users, so usually should duplicate functionality provided in other menus
Tab Menu Multi-page menu with one tab for each page that pops-up and floats over the screen. Remains on screen until closed.	When user needs to change several settings or perform several related commands	• Menu items should be short to fit on the tab label • Avoid more than one row of tabs because clicking on a tab to open it can change the order of the tabs and in virtually no other case does selecting from a menu rearrange the menu itself.
Tool Bar Menu of buttons (often with icons) that remains on the screen until closed	As a shortcut to commands for experienced users	• All buttons on the same tool bar should be the same size • If the labels very dramatically in size, then use two different sizes (small and large) • Buttons with icons should have a *tool tip*—an area that displays a text phase explaining the button when the user pauses the mouse over it
Image Map Graphical image in which certain areas are linked to actions or other menus.	Only when the graphical image adds meaning to the menu	• Image should convey meaning to show which parts perform action when clicked • Tool tips can be helpful

FIGURE 10-10
Types of Menus

Direct Manipulation With *direct manipulation*, the user enters commands by working directly with interface objects. For example, users can change the size of objects in Microsoft PowerPoint by clicking on objects and moving the sides, or they can move files in Windows Explorer by dragging the file names from one folder to another. Direct manipulation can be simple, but it suffers from two problems. First, users familiar with language- or menu-based interfaces don't always expect it. Second, not all commands are intuitive. (For example, how do you copy [not move] files in Windows Explorer? On the Macintosh, why does moving a folder to the trash delete the file if it is on the hard drive but eject the diskette if the file is on a diskette?)

Messages

Messages are the way in which the system responds to a user and informs him or her of the status of the interaction. There are many different types of messages, such as *error messages, confirmation messages, acknowledgment messages, delay messages,* and *help messages* (Figure 10-11). In general, messages should be clear, concise, and complete, which are sometimes conflicting objectives. All messages should be grammatically correct and free of jargon and abbreviations (unless they are users' jargon and abbreviations). Avoid negatives, because they can be confusing (e.g., replace "Are you sure you do not want to continue?" with "Do you want to quit?"). Likewise, avoid humor, because it wears off quickly after the same message appears dozens of times.

Messages should require the user to acknowledge them (by clicking, for example), rather than being displayed for a few seconds and then disappearing. The exceptions are messages that inform the user of delays in processing, which should disappear once the delay has passed. In general, messages are text, but sometimes standard icons are used. For example, Windows displays an hourglass when the system is busy.

All messages should be carefully crafted, but error messages and help messages require particular care. Messages (and especially error messages) should always explain the problem in polite, succinct terms (e.g., what the user did incor-

Type of Messages	When to Use	Notes
Error message Informs the user that he or she has attempted to do something to which the system cannot respond	When user does something that is not permitted or not possible	Always explain the reason and suggest corrective action. Traditionally, error messages have been accompanied by a beep, but many applications now omit it or permit users to remove it.
Confirmation message Asks the user to confirm that he or she really wants to perform the action selected	When user selects a potentially dangerous choice, such as deleting a file	Always explain the cause and suggest possible action. Often include several choices other than "OK" and "cancel."
Acknowledgment message Informs the user that the system has accomplished what it was asked to do	Seldom or never; users quickly become annoyed with all the unnecessary mouse clicks	Acknowledgment messages are typically included because novice users often like to be reassured that an action has taken place. The best approach is to provide acknowledgment information without a separate message on which the user must click. For example, if the user is viewing items in a list and adds one, then the updated list on the screen showing the added item is sufficient acknowledgment.
Delay message Informs the user that the computer system is working properly	When an activity takes more than seven seconds	This message should permit the user to cancel the operation in case he or she does not want to wait for its completion. The message should provide some indication of how long the delay may last.
Help message Provides additional information about the system and its components	In all systems	Help information is organized by table of contents and/or keyword search. *Context-sensitive help* provides information that is dependent on what the user was doing when help was requested. Help messages and on line documentation are discussed in Chapter 13.

FIGURE 10-11

Types of Messages

rectly) and explain corrective action as clearly and as explicitly as possible so the user knows exactly what needs to be done. In the case of complicated errors, the error message should display what the user entered, suggest probable causes for the error, and propose possible user responses. When in doubt, provide either more information than the user needs or the ability to get additional information. Error messages should provide a message number. Message numbers are not intended for users, but their presence makes it simpler for help desks and customer support lines to identify problems and help users because many messages use similar wording.

INPUT DESIGN

Inputs facilitate the entry of data into the computer system, whether highly structured data, such as order information (e.g., item numbers, quantities, costs), or unstructured information (e.g., comments). Input design means designing the screens used to enter the information, as well as any forms on which users write or type information (e.g., time cards, expense claims).

Basic Principles

The goal of input design is to simply and easily capture accurate information for the system. The fundamental principles for input design reflect the nature of the inputs (whether batch or online) and ways to simplify their collection.

Use Online and Batch Processing Appropriately   There are two general formats for entering inputs into a computer system: online processing and batch processing. With *online processing* (sometimes called *transaction processing*), each input item (e.g., a customer order, a purchase order) is entered into the system individually, usually at the same time as the event or transaction prompting the input. For example, when you take a book out from the library, buy an item at the store, or make an airline reservation, the computer system that supports that process uses online processing to immediately record the transaction in the appropriate database(s). Online processing is most commonly used when it is important to have *real-time information* about the business process. For example, when you reserve an airline seat, the seat is no longer available for someone else to use, so that piece of information must be recorded immediately.

With batch processing, all the inputs collected over some time period are gathered together and entered into the system at one time in a batch. Some business processes naturally generate information in batches. For example, most hourly payrolls are done using batch processing because time cards are gathered together in batches and processed at once. Batch processing also is used for transaction processing systems that do not require real-time information. For example, most stores send sales information to district offices so that new replacement inventory can be ordered. This information could be sent in real time as it is captured in the store, so that the district offices are aware within a second or two that a product is sold. If stores do not need up-to-the-second real-time data, they will collect sales data throughout the day and transmit it every evening in a batch to the district office. This batching simplifies the data communications process and often cuts communications costs; however, it does mean that inventories are not accurate in real time but rather are accurate only at the end of the day after the batch has been processed.

Capture Data at the Source   Perhaps the most important principle of input design is to capture the data in an electronic format at the original source or as close to the original source as possible. In the early days of computing, computer systems replaced traditional manual systems that operated on paper forms. As these business processes were automated, many of the original paper forms remained, either because no one thought to replace them or because it was too expensive to do so. Instead, the business process continued to contain manual forms that were taken to the computer center in batches to be typed into the computer system by a *data-entry operator*.

Many business processes still operate this way today. For example, many organizations have expense claim forms that are completed by hand and submitted to an Accounting Department, which approves them and enters them into the system in batches. There are three problems with this approach. First, it is expensive because it duplicates work (the form is filled out twice, once by hand and once by keyboard). Second, it increases processing time because the paper forms must be physically moved through the process. Third, it increases the cost and probability of error, because it separates the entry from the processing of information; someone may misread the handwriting on the input form, data could be entered incorrectly, or the original input may contain an error that invalidates the information.

Most transaction processing systems today are designed to capture data at its source. *Source data automation* refers to using special hardware devices to automatically capture data without requiring anyone to type it. Stores commonly use bar code readers that automatically scan products and that enter data directly into the computer system. No intermediate formats, such as paper forms, are used. Similar technologies include *optical character recognition*, which can read printed numbers and text (e.g., on checks); *magnetic stripe readers*, which can read information encoded on a stripe of magnetic material similar to a diskette (e.g., credit cards); and *smart cards* that contain microprocessors, memory chips, and batteries (much like credit card–size calculators). As well as reducing the time and cost of data entry, these systems reduce errors because they are far less likely to capture data incorrectly. Today, portable computers and scanners allow data to be captured at the source even in mobile settings (e.g., air courier deliveries, use of rental cars).

A lot of information, however, cannot be collected by these automatic systems, so the next best option is to capture data immediately from the source using a trained entry operator. Many airline and hotel reservations, loan applications, and catalog orders are recorded directly into a computer system while the customer provides the operator with answers to questions. Some systems eliminate the operator altogether and allow users to enter their own data. For example, several universities (e.g., the Massachusetts Institute of Technology [MIT]) no longer accept paper-based applications for admissions; all applications are typed by students into electronic forms.

The forms for capturing information (on a screen, on paper, etc.) should well support the data source. That is, the order of the information on the form should match the natural flow of information from the data source, and data-entry forms should match paper forms used to initially capture the data.

Minimize Keystrokes   Another important principle is to minimize keystrokes. Keystrokes cost time and money, whether they are performed by a customer, user, or trained data-entry operator. The system should never ask for information that can be obtained in another way (e.g., by retrieving it from a database or by per-

10-8 Career Services

Pretend that you are designing the new interface for a career services system at your university that accepts student résumés and presents them in a standard format to recruiters. Describe how you could incorporate the basic principles of input design into your interface design. Remember to include the use of online versus batch data input, the capture of information, and plans to minimize keystrokes.

forming a calculation). Likewise, a system should not require a user to type information that can be selected from a list; selecting reduces errors and speeds entry.

In many cases, data have values that often recur. These frequent values should be used as the *default value* for the data so that the user can simply accept the value and not have to retype it time and time again. Examples of default values are the current date, the area code held by the majority of a company's customers, and a billing address that is based on the customer's residence. Most systems permit changes to default values to handle data entry exceptions as they occur.

10-B ENTERPRISE RESOURCE PLANNING USER INTERFACES DRIVE USERS NUTS

It used to take workers at Hydro Agri's Canadian Fertilizer stores about 20 seconds to process an order. After installing SAP, it now takes 90 seconds. Entering an order requires users to navigate through 6 screens to find the data fields that were on one screen in the old system. The problem became so critical during the spring planting rush that the project team installing the SAP system was pressed into service to take telephone orders.

Many other customers have complained about similar problems in SAP and the other leading ERP [enterprise resource planning] systems. Ontario-based Algoma Steel uses PeopleSoft and now has to use a dozen screens to enter employee data that were contained in two screens in their old custom-built personnel system. A-dec, a dental equipment maker based in Oregon, discovered the hard way that its Baan inventory system was still counting products that had been shipped in its on-hand inventories; the system required users to confirm the order shipments before the inventories were recorded as shipped—but didn't automatically take them to the confirmation screen.

So why have companies implemented ERP systems? The driving force behind most implementations was not to simplify the users' jobs but instead to improve the quality of the data, simplify system maintenance, and/or beat the Year 2000 problem. Ease-of-use wasn't a consideration, and what makes ERP systems so hard to use is that in the attempt to make them one-size-fits-all, developers had to include many little-used data items and processes. Instead of having a small custom system collecting only the data needed by the company itself (which could be condensed to fit one or two screens), companies now find themselves using a system designed to collect all the data items that any company could possibly use—data items that now require 6 to 12 screens.

Source: "ERP User Interfaces Drive Workers Nuts," *Computerworld,* November 2, 1998, 32(44): 1, 24, by Craig Stedman.

QUESTION

Suppose you were a systems analyst at one of the leading ERP vendors (e.g., SAP, PeopleSoft, Baan). How could you apply the interface design principles and techniques in this chapter to improve the ease of use of your system?

Types of Inputs

Each data item that has to be input is linked to a field on the form into which its value is typed. Each field also has a field label, which is the text beside, above, or below the field that tells the user what type of information belongs in the field. Often the field label is similar to the name of the data element, but they do not have to have identical wording. In some cases, a field will display a template over the entry box to show the user exactly how data should be typed. There are many different types of inputs, in the same way that there are many different types of fields (see Fig. 10-12).

Text   As the name suggests, a *text box* is used to enter text. Text boxes can be defined to have a fixed length or can be scrollable and accept a virtually unlimited amount of text. In either case, boxes can contain single or multiple lines of textual information. Never use a text box if you can use a selection box (see later).

Text boxes should have field labels placed to the left of the entry area, with their size clearly delimited by a box (or a set of underlines in a non-GUI interface). If there are multiple text boxes, their field labels and the left edges of their entry boxes should be aligned. Text boxes should permit standard GUI functions such as cut, copy, and paste.

Numbers   A *number box* is used to enter numbers. Some software can automatically format numbers as they are entered, so that 3452478 becomes $34,524.78. Dates are a special form of numbers that sometimes have their own type of number box. Never use a number box if you can use a selection box (see next).

Selection Box   A *selection box* enables the user to select a value from a predefined list. The items in the list should be arranged in some meaningful order, such as alphabetic for long lists, or in order of most frequently used. The default selection value should be chosen with care. A selection box can be initialized as "unselected" or, better still, start with the most commonly used item already selected.

There are six commonly used types of selection boxes: *check boxes, radio buttons, on-screen list boxes, drop-down list boxes, combo boxes*, and *sliders* (Fig. 10-13). The choice among the types of text selection boxes generally comes down to one of screen space and the number of choices the user can select. If screen space is limited and only one item can be selected, then a drop-down list box is the best choice, because not all list items need to be displayed on the screen. If screen space is limited but the user can select multiple items, an on-screen list box that displays only a few items can be used. Check boxes (for multiple selections) and radio buttons (for single selections) both require all list items to be displayed at all times, thus requiring more screen space, but since they display all choices, they are often simpler for novice users.

Input Validation

All data entered into the system must be validated to ensure accuracy. Input *validation* (also called *edit checks*) can take many forms. Ideally, computer systems should not accept data that fail any important validation check to prevent invalid information from entering the system. However, this can be very difficult, and invalid data often slip by data-entry operators and the users providing the information. It is up the system to identify invalid data and either make changes or notify someone who can resolve the information problem.

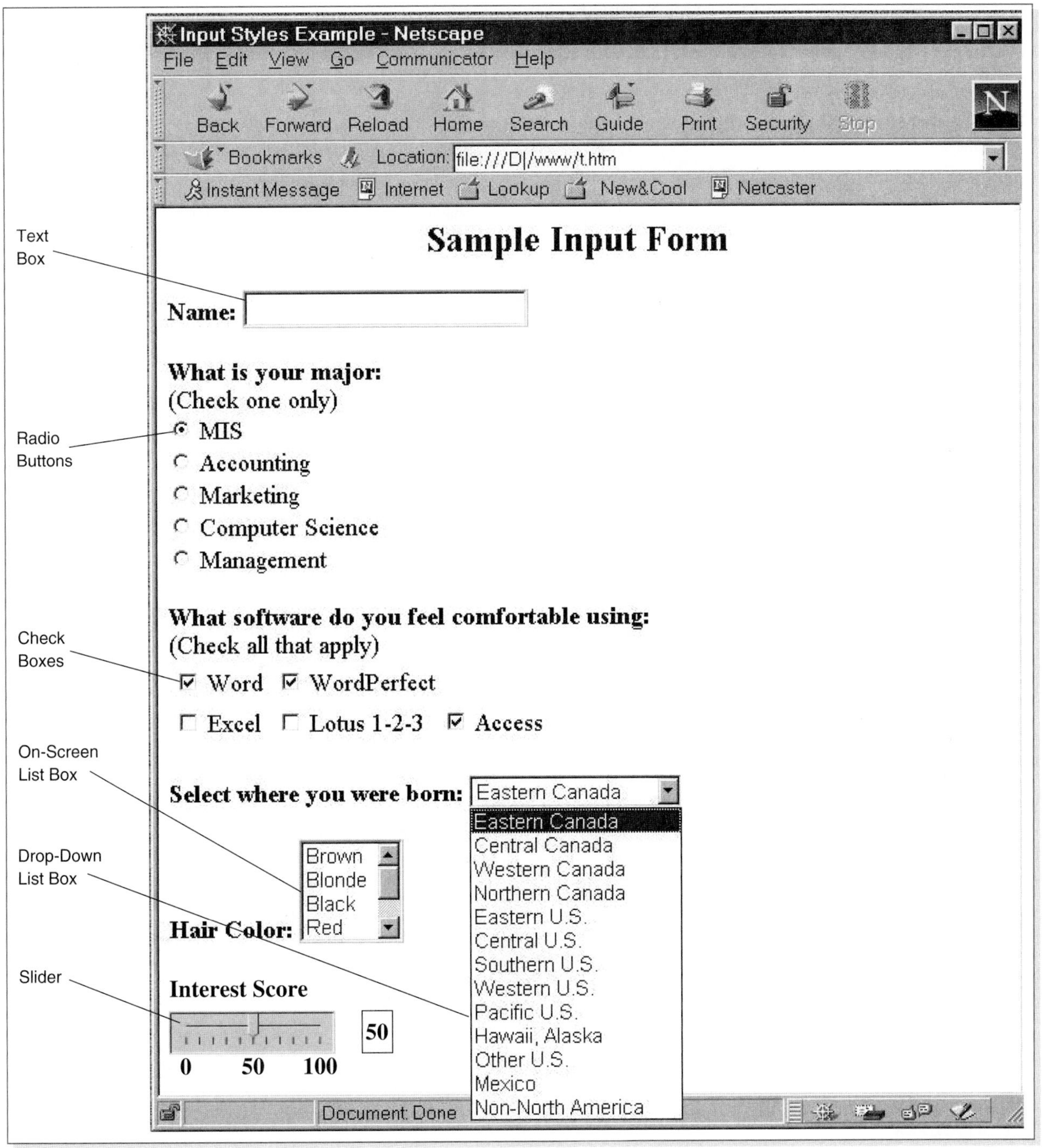

Text
Box

Radio
Buttons

Check
Boxes

On-Screen
List Box

Drop-Down
List Box

Slider

FIGURE 10-12
Types of Input Boxes

There are six different types of validation checks as explained in Fig. 10-14: *completeness check, format check, range check, check digit check, consistency check*, and *database check*. Every system should use at least one validation check on all entered data and ideally will perform all appropriate checks where possible.

Type of Box	When to Use	Notes
Check box Presents a complete list of choices, each with a square box in front	When several items can be selected from a list of items	Check boxes are not mutually exclusive. Do not use negatives for box labels. Check box labels should be placed in some logical order, such as that defined by the business process, or failing that, alphabetically or most commonly used first. Use no more than ten check boxes for any particular set of options. If you need more boxes, group them into subcategories.
Radio button Presents a complete list of mutually exclusive choices, each with a circle in front	When only one item can be selected from a set of mutually exclusive items	Use no more than six radio buttons in any one list; if you need more, use a drop-down list box. If there are only two options, one check box is usually preferred to two radio buttons, unless the options are not clear. Avoid placing radio buttons close to check boxes to prevent confusion between different selection lists.
On-screen list box Presents a list of choices in a box	Seldom or never—only if there is insufficient room for check boxes or radio buttons	This type of box can permit only one item to be selected (in which case it is an ugly version of radio buttons). This type of box can also permit many items to be selected (in which case it is an ugly version of check boxes), but users often fail to realize they can choose multiple items. This type of box permits the list of items to be scrolled, thus reducing the amount of screen space needed.
Drop-down list box Displays selected item in one-line box that opens to reveal list of choices	When there is insufficient room to display all choices	This type of box acts like radio buttons but is more compact. This type of box hides choices from users until it is opened, which can decrease ease of use; conversely, because it shelters novice users from seldom-used choices, it can improve ease of use. This type of box simplifies design if the number of choices is unclear, because it takes only one line when closed.
Combo box A special type of drop-down list box that permits user to type as well as scroll the list	Shortcut for experienced users	This type of box acts like drop-down list but is faster for experienced users when the list of items is long.
Slider Graphic scale with a sliding pointer to select a number	Entering an approximate numeric value from a large continuous scale	The slider makes it difficult for the user to select a precise number. Some sliders also include a number box to enable the user to enter a specific number.

FIGURE 10-13
Types of Selection Boxes

YOUR TURN

10-9 Career Services

Consider a Web form that a student would use to input student information and rèsumè information into a Career Services application at your university. First, sketch out how this form would look and iden-tify the fields that the form would include. What types of validity checks would you use to make sure that the cor-rect information is entered into the system?

Type of Validation	When to Use	Notes
Completeness check Ensures all required data have been entered	When several fields must be entered before the form can be processed	If required information is missing, the form is returned to the user unprocessed.
Format check Ensures data are of the right type (e.g., numeric) and in the right format (e.g., month, day, year)	When fields are numeric or contain coded data	Ideally, numeric fields should not permit users to type text data, but if this is not possible, the entered data must be checked to ensure it is numeric. Some fields use special codes or formats (e.g., license plates with three letters and three numbers) that must be checked.
Range check Ensures numeric data are within correct minimum and maximum values	With all numeric data, if possible	A range check permits only numbers between correct values. Such a system can also be used to screen data for "reasonableness"—e.g., rejecting birthdates prior to 1880 because people do not live to be a great deal over 100 years old (most likely, *1980* was intended).
Check digit check Check digits are added to numeric codes	When numeric codes are used	Check digits are numbers added to a code as a way of enabling the system to quickly validate correctness. For example, U.S. Social Security Numbers and Canadian Social Insurance Numbers assign only eight of the nine digits in the number. The ninth number—the check digit—is calculated using a mathematical formula from the first eight numbers. When the identification number is typed into a computer system, the system uses the formula and compares the result with the check digit. If the numbers don't match, then an error has occurred.
Consistency checks Ensure combinations of data are valid	When data are related	Data fields are often related. For example, someone's birth year should precede the year in which he or she was married. Although it is impossible for the system to know which data are incorrect, it can report the error to the user for correction.
Database checks Compare data against a database (or file) to ensure they are correct	When data are available to be checked	Data are compared against information in a database (or file) to ensure they are correct. For example, before an identification number is accepted, the database is queried to ensure that the number is valid. Because database checks are more "expensive" than the other types of checks (they require the system to do more work), most systems perform the other checks first and perform database checks only after the data have passed the previous checks.

FIGURE 10-14
Types of Input Validation

OUTPUT DESIGN

Outputs are the reports that the system produces, whether on the screen, on paper, or in other media, such as the Web. Outputs are perhaps the most visible part of any system, because a primary reason for using an IS is to access the information that it produces.

Basic Principles

The goal of the output mechanism is to present information to users so they can accurately understand it with the least effort. The fundamental principles for output design reflect how the outputs are used and ways to make it simpler for users to understand them.

Understand Report Usage   The first principle in designing reports is to understand how they are used. Reports can be used for many different purposes. In some cases—but not very often—reports are read cover to cover because all information is needed. In most cases, reports are used to identify specific items or are used as references to find information, so the order in which items are sorted on the report or grouped within categories is critical. This is particularly important for the design of electronic or Web-based reports. Web reports that are intended to be read end to end should be presented in one long scrollable page, whereas reports that are primarily used to find specific information should be broken into multiple pages, each with a separate link. Page numbers and the date on which the report was prepared also are important for reference reports.

The frequency of the report may also play an important role in its design and distribution. *Real-time reports* provide data that are accurate to the second or minute at which they were produced (e.g., stock market quotes). *Batch reports* are those that report historical information that may be months, days, or hours old, and they often provide additional information beyond the reported information (e.g., totals, summaries, historical averages).

There are no inherent advantages to real-time reports over batch reports. The only advantages lie in the time value of the information. If the information in a report is time critical (e.g., stock prices, air traffic control information), then real-time reports have value. This is particularly important because real-time reports often are expensive to produce; unless they offer some clear business value, they may not be worth the extra cost.

Manage Information Load   Most managers get too much information, not too little. The goal of a well-designed report is to provide all the information needed to support the task for which it was designed. This does not mean that the report should provide all the information available on the subject—just what the users decide they need to perform their jobs. In some cases, this may result in the production of several different reports on the same topics for the same users, because they are used in different ways. This is not bad design.

For users in westernized countries, the most important information generally should be presented first in the top left corner of the screen or paper report. Information should be provided in a format that is usable without modification. The user should not need to re-sort the report's information, highlight critical information to find it more easily amid a mass of data, or perform additional mathematical calculations.

Minimize Bias   No analyst sets out to design a biased report. The problem with *bias* is that it can be very subtle; analysts can introduce it unintentionally. Bias can be introduced by the way in which lists of data are sorted, because entries that appear first in a list may receive more attention than those appearing later in the list. Data often are sorted in alphabetic order, making those entries starting with the letter A more prominent. Data can be sorted in chronological order (or reverse chronological order), placing more emphasis on older (or most recent) entries. Data may be sorted by numeric value, placing more emphasis on higher or lower values. For example, consider a monthly sales report by state. Should the report be listed in alphabetic order by state name, in descending order by the amount sold, or in some other order (e.g., geographic region)? There are no easy answers to this, except to say that the order of presentation should match the way in which the information is used.

Graphic displays and reports can present particularly challenging design issues.[8] The scale on the axes in graphs is particularly subject to bias. For most types of graphs, the scale should always begin at zero; otherwise, comparisons among values can be misleading. For example, in Fig. 10-15, have sales increased by very much since 1993? The numbers in both charts are the same, but the visual images the two present are quite different. A glance at Fig. 10-15a would suggest

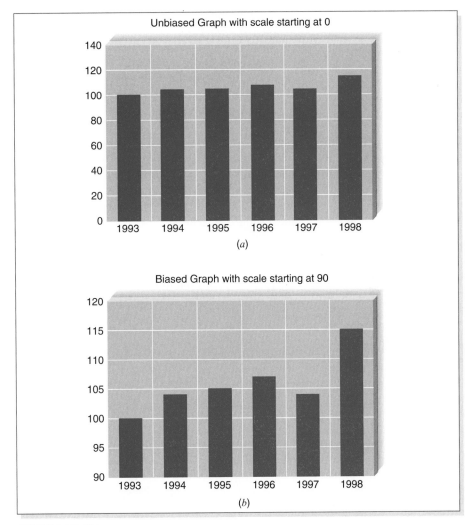

FIGURE 10-15
Bias in Groups: (a) Unbiased Graph with Scale Starting at 0; (b) Biased Graph with Scale Starting at 90.

[8] Some of the best books on the design of charts and graphical displays are by Edward R. Tufte, published by Graphics Press, Cheshire Connecticut: *The Visual Display of Quantitative Information, Envisioning Information,* and *Visual Explanations.* Another good book is by William Cleveland, *Visualizing Data,* Hobart Press, 1993.

10-10 Finding Bias

Read through recent copies of a newspaper or popular press magazine such as *Time, Newsweek,* or *BusinessWeek* and find four graphs. How many are biased and how many are unbiased?

only minor changes, whereas a glance at Fig. 10-15b might suggest that there have been some significant increases. In fact, sales have increased by a total of 15% over 5 years, or 3% per year. Fig. 10-15a presents the most accurate picture; Fig. 10-15b is biased because the scale starts very close to the lowest value in the graph and misleads the eye into inferring that there have been major changes (i.e., more than doubling from "two lines" in 1993 to "five lines" in 1998). Fig. 10-15b is the default graph produced by Microsoft Excel.

Types of Outputs

There are many different types of reports, such as *detail reports, summary reports, exception reports, turnaround documents*, and *graphs* (Fig. 10-16). Classifying reports is challenging because many reports have characteristics of several different types. For example, some detail reports also produce summary totals, making them summary reports.

Media

There are many different types of media used to produce reports. The two dominant medium in use today are paper and electronic. Paper is the more traditional media.

CONCEPTS

IN ACTION

10-C Selecting the Wrong Students

I helped a university department develop a small decision support system to analyze and rank students who applied to a specialized program. Some of the information was numeric and could easily be processed directly by the system (e.g., grade point average, standardized test scores). Other information required the faculty to make subjective judgements among the students (e.g., extracurricular activities, work experience). The users entered their evaluations of the subjective information via several data analysis screens in which the students were listed in alphabetical order.

In order to make the system "easier to use," the reports listing the results of the analysis were also presented in alphabetical order by student name, rather than in order from the highest ranked student to the lowest ranked student. In a series of tests prior to installation, the users selected the wrong students to admit in 20 percent of the cases. They assumed, wrongly, that the students listed first were the highest ranked students and simply selected the first students on the list for admission. Neither the title on the report, nor the fact that all the students, names were in alphabetical order made them realize that they had read the report incorrectly. *Alan Dennis*

Question

What concerns could this problem raise about the rest of the system?

Type of Reports	When to Use	Notes
Detail report Lists detailed information about all the items requested	When user needs full information about the items	This report is usually produced only in response to a query about items matching some criteria. This report is usually read cover to cover to aid understanding of one or more items in depth.
Summary report Lists summary information about all items	When user needs brief information on many items	This report is usually produced only in response to a query about items matching some criteria, but it can be a complete database. This report is usually read for the purpose of comparing several items to each other. The order in which items are sorted is important.
Turnaround document Outputs that "turn around" and become inputs	When a user (often a customer) needs to return an output to be processed	Turnaround documents are a special type of report that are both outputs and inputs. For example, most bills sent to consumers (e.g., credit card bills) provide information about the total amount owed and also contain a form that consumers fill in and return with payment.
Graphs Charts used in addition to and instead of tables of numbers	When users need to compare data among several items	Well-done graphs help users compare two or more items or understand how one has changed over time. Graphs are poor at helping users recognize precise numeric values and should be replaced by or combined with tables when precision is important. Bar charts tend to be better than tables of numbers or other types of charts when it comes to comparing values between items (but avoid three-dimensional charts that make comparisons difficult). Line charts make it easier to compare values over time, whereas scatter charts make it easier to find clusters or unusual data. Pie charts show proportions or the relative shares of a whole.

FIGURE 10-16
Types of Reports

For almost as long as there have been human organizations, there have been reports on paper or similar media (e.g., papyrus, stone). Paper is relatively permanent, easy to use, and accessible in most situations. It also is highly portable, at least for short reports.

Paper also has several rather significant drawbacks. It is inflexible. Once the report is printed, it cannot be sorted or reformatted to present a different view of the information. Likewise, if the information on the report changes, the entire report must be reprinted. Paper reports are expensive, hard to duplicate, and require considerable supplies (paper, ink) and storage space. Paper reports are also hard to quickly move long distances (e.g., from a head office in Toronto to a regional office in Bermuda).

Many organizations are therefore moving to electronic production of reports, whereby reports are "printed" but stored in electronic format on the Web so that users can easily access them. Often, the reports are available in more predesigned formats than are their paper-based counterparts because the cost of producing and storing different formats is minimal. Electronic reports also can be produced on demand as needed, and they enable the user to search more easily for certain words. Some users still print the electronic report on their own printers, but the reduced cost of electronic delivery over distance and the ease of enabling more users to access the reports than when they were only in paper form usually offsets the cost of local printing.

CONCEPTS

IN ACTION

10-D Cutting Paper to Save Money

One of the Fortune 500 firms with which I have worked had an 18-story office building for its world headquarters. It deveoted two full floors of this building to nothing more than storing "current" paper, reports (a separate warehouse was maintained outside the city for "archived" reports such as tax documents). Imagine the annual cost of office space in the headquarters building tied up in these paper reports. Now imagine how a staff member would gain access to the reports, and you can quickly understand the driving force behind

electronic reports, even if most users end up printing them. Within one year of switching to electronic reports (for as many reports as practical) the paper report storage area was reduced to one small storage room.

Alan Dennis

Question

What types of reports are most suited to electronic format? What types of reports are less suited to electronic reports?

APPLYING THE CONCEPTS AT CD SELECTIONS

In the CD Selections case, there are three different user interfaces to be designed: Process 1: Take Requests; Process 2: Maintain Marketing Materials; and Process 3: Process In-Store Holds (see Fig. 6-19 in Chapter 6). In this section, we focus only on Process 1, the Web portion used by customers.

Use Scenario Development

The first step in the interface design process was to develop the key use scenarios for the Internet system. Alec Adams, Senior Systems Analyst at CD Selections and Project Manager for the Internet system, began by examining the DFD and thinking about the types of users and how they would interact with the system. As discussed previously, Alec identified two use scenarios: the browsing shopper and the hurry-up shopper (see Fig. 10-6). Alec also thought of several other use scenarios for the Web site in general, but he omitted them because they were not common. Likewise, he thought of several use scenarios that did not lead to sales (e.g., fans looking for information about their favorite artists and albums), and he omitted them as well as they were not important in the design of the Web site.

Interface Structure Design

Next, Alec created an ISD for the Web system. He began with the DFDs to ensure that all functionality defined for the system was included in the ISD. Fig. 10-17 shows the ISD for the Web portion of process 1. In practice, some of the processes on the level 1 DFD for this part of the system (Fig. 6-20) might be decomposed into several level 2 DFDs. However, to keep things simple, in this chapter we show an ISD that links to the level 1 DFD in Fig. 6-20, rather than attempting create more DFDs and link the ISD to them.

The Initial Interface Structure Design   The Internet system will have a main menu or home page (interface number 0) that will enable users to navigate to a search

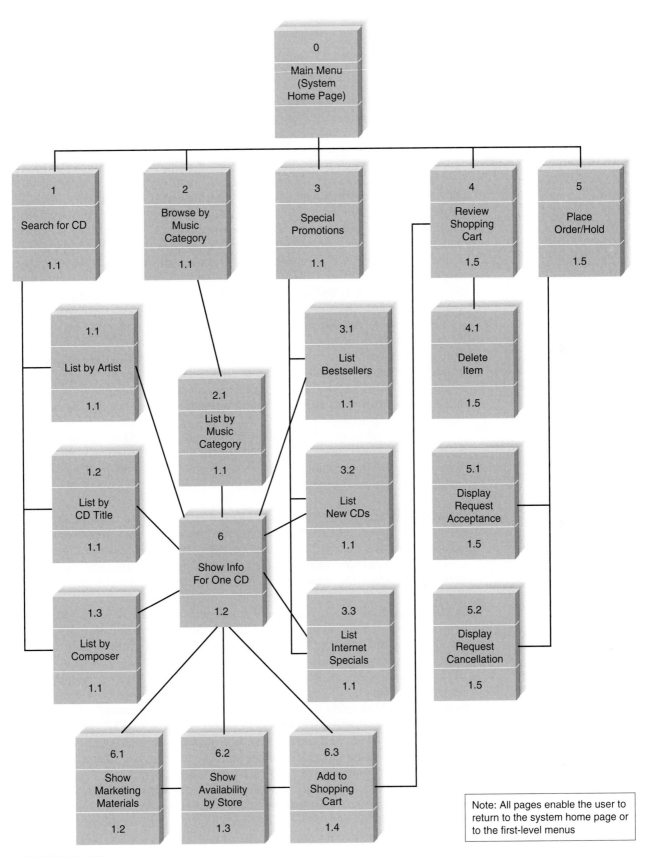

FIGURE 10-17
CD Selections Initial ISD for the Web Portion of the Internet System

page (1), which would enable the user to enter search criteria to produce a list of CDs based on artist (1.1), title (1.2), or composer (1.3). The home page would also have a link to a browse page (2) that would enable the user to enter a music category and sort criteria (e.g., alphabetically by artist or title, then by date published) to produce a list of CDs within that category (2.1). The home page would also have a link to a special promotions page (3) that would lead to lists of best-sellers (3.1), newly released CDs (3.2), or Internet special promotions (3.3).

Each of these lists of CDs would enable the user to click on a specific CD title and view detailed information about it (6), such as artists, tracks, and eventually music samples (in version 2 of the system). Alec decided to provide the additional marketing materials (e.g., reviews) on a separate page (6.1), rather than including them on the main entry for each CD, to prevent overcrowding and long time delays on the Web. Both the CD page (6) and marketing material page (6.1) would let the user locate stores with the CD (6.2) and then put the CD in the selected store into the shopping cart (6.3) and then show the shopping cart (4). The user could then return to the home page (0) to enter a new search.

The system home page (0) lets users manage the shopping cart by showing the current items it contains (4) and enabling users to delete them (4.1). The home page also has the link to placing the request, which would display the current request information, and enable users to enter their name and e-mail (5), which would be accepted (5.1) or not, in the event that someone had purchased the last CD in the store between the time when the CD was placed in the shopping cart and the request actually placed (5.2).

Alec also envisioned that by using frames, the user would be able to return to the home page (interface 0) or any of the first-level menu (interfaces 1 through 5) from any screen. Documenting these would give the ISD too many lines, so Alec simply put a note describing it on the ISD.

The Revised Interface Structure Design   Alec then examined the use scenarios to see how well the initial ISD enabled different types of users to work through the system. He started with the "Browsing Shopper" scenario and followed it through the ISD, imagining what would appear on each screen and pretending to navigate through the system. He found that the ISD worked well but noticed one anomaly. The information presented on the "Review Shopping Cart" screen (a list of CDs in the shopping cart, their prices, and store locations, plus the ability to delete items) was very similar to that on the "Place Order" screen (a list of CDs in the shopping cart, their prices, and store locations, plus a form on which to enter order information (e.g., customer name, e-mail address).

Therefore, Alec decided to combine the two screens into one that presented the CD information, enabled the user to delete CDs, and contained the request form. This would simplify the interface and also save programming time later on. However, he was reluctant to combine the menu items (i.e., HTML links) leading to the screen. He believed that users who were uncertain about ordering and wanted to review the shopping cart first would be reluctant to click on a "Place Order" button. Likewise, to someone ready to check out, clicking on a "Review Shopping Cart" button would not be intuitive. Therefore, he decided to keep both links with names to the same page. This was rather unusual, but sometimes the unusual is the best solution. The revised ISD, presented in Fig. 10-18, shows the two interfaces coupled with the other three interfaces linked off of them (4.1, 4.2, 4.3).

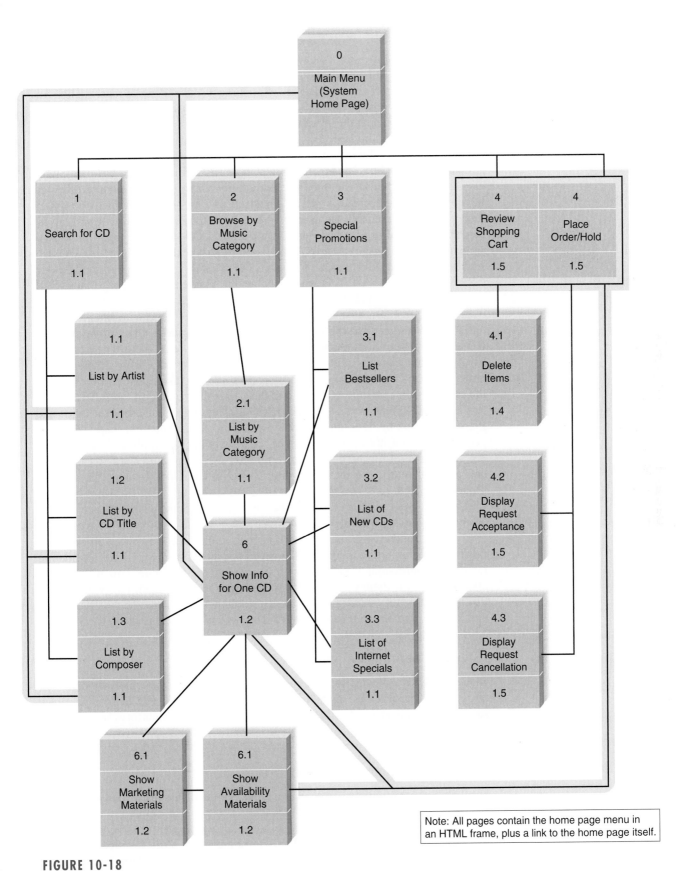

FIGURE 10-18
CD Selections Revised ISD for the Web Portion of the Internet Sales Systems

As an aside, it is important to note that we have not combined the functionality of these two processes ("Review Shopping Cart," "Place Order"). The ISD describes the screens, not the business processes—that's the role of the DFD. The ISD merely shows how the screens flow.

Alec next explored the "Hurry-up Shopper" scenario. In this case, the ISD did not work as well. Moving from the home page to the search page to the list of matching CDs to the CD page with price and other information takes three mouse clicks. This falls within the three-clicks rule, but for someone in a hurry, this may be too many. Alec decided to add a "quick search" option to the home page (interface 0) that would enable the user to enter one search criterion (e.g., searching by just artist name or title, rather than doing a more detailed search, as would be possible on the search page) that would with one click take the user to the one CD that matched the criterion (interface 6) or to a list of CDs if there were more than one (interfaces 1.1, 1.2, 1.3). This would enable an impatient user to get to the CD of interest in one or two clicks.

Once the CD is displayed on the screen (6), the "Hurry-up Shopper" scenario would suggest that the user would immediately find a store and request the CD, do a new search, or abandon the Web site and surf elsewhere. This suggested two important changes. First, there had to be an easy way to go to the "Place Order" screen (4). As the ISD stands (Fig 10-18), the user must add the item to the shopping cart (resulting in 6.3) and then click on the link on the HTML frame to get to the "Place Order" screen. Although the ability of users to notice the Place Order link in the frame would await the interface evaluation stage, Alec suspected, on the basis of past experience, that a significant number of users would not see it. Therefore, he decided to eliminate the "Shopping Cart" screen (6.3) and add two buttons to the "Show Info for One CD Order" screen (6) and the "Marketing Materials" screen (6.2) called "Place Order" and "Add to Shopping Cart" that would take the user to the "Review Shopping Cart" "Place Order" screen (4) (see Fig. 10-18).

The second change was to the HTML frame. Since the "hurry-up shopper" might want to search for another CD instead of requesting the CD, Alec decided to include the quick-search item from the home page on the frame. This would make all searches immediately available from anywhere in the system. This would mean that all functionality on the home page (0) would now be carried on the frame. Alec updated the note on the bottom of the ISD to reflect the change (see Fig. 10-18).

Interface Standards Design

Once the ISD was complete, Alec moved on to develop the interface standards for the system. The interface metaphor was straightforward: a CD Selections music store. The key interface objects and actions were equally straightforward, as was the use of the CD Selections logo icon (Fig. 10-19).

Interface Template Design

For the interface template, Alec decided on a simple, clean design that had a modern background pattern, with the CD Selections logo in the upper left corner. The template had two navigation areas: one menu across the top for navigation within the entire Web site (e.g., overall Web site home page, store locations) and one menu down the left edge for navigation within the Internet system. The left edge

Interface Metaphor: a CD Selections Music Store

Interface Objects

- **CD:** all items, whether CD, tape, or DVD, unless it is important to distinguish among them
- **Artist:** person or group who records the CD
- **Title:** title or name of CD
- **Composer:** person or group who wrote the music for the CD (primarily used for classical music)
- **Music Category:** type of music. Current categories include: Rock, Jazz, Classical, Country, Alternative, Soundtracks, Rap, Folk, Gospel
- **CD List:** list of CD(s) that match the specified criteria
- **Shopping Cart:** place to store CDs until they are requested

Interface Actions

- **Search for:** display a CD list that matches specified criteria
- **Browse:** display a CD list sorted in order by some criteria
- **Order:** authorize special order or place hold

Interface Icons

- **CD Selections Logo** will be used on all screens

FIGURE 10-19
CD-Selection Interface Standards

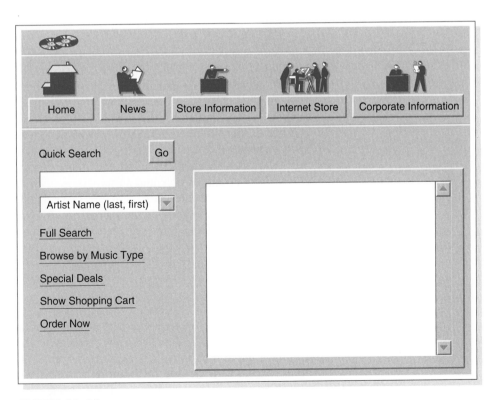

FIGURE 10-20
CD Selections Interface Template for the Web Portion of the Internet Order System

menu contained the links to the five top-level screens (interfaces 1, 2, 3, 4 in Fig. 10-18), as well as the "quick search" option. The center area of the screen is used for displaying forms and reports when the appropriate link is clicked. See Fig. 10-20. At this point, Alec decided to seek some quick feedback on the interface structure and standards before investing time in prototyping the interface designs. Therefore, he met with Margaret Mooney, the project sponsor, and Chris Campbell, the consultant, to discuss the emerging design. Making changes at this point would be much simpler than after doing the prototype. Margaret and Chris had a few suggestions, so after the meeting Alec made the changes and moved into the design prototyping step.

Design Prototyping

Alec decided to develop an HTML prototype of the system. The Internet sales system was new territory for CD Selections and a strategic investment in a new business model, so it was important to make sure that no key issues were overlooked. The HTML prototype would provide the most detailed information and enable interactive evaluation of the interface.

Alec began designing the prototype starting with the home screen and gradually worked his way through all the screens. The process was very iterative and he made many changes to the screens as he worked. Once he had an initial prototype designed, he posted it on CD Selections's internal intranet and solicited comments from several friends with lots of Web experience. He revised it based on the comments he received. Fig. 10-21 presents some screens from the prototype.

Interface Evaluation

The step was interface evaluation. Alec decided on a two-phase evaluation. The first evaluation was to be an interactive evaluation conducted by Margaret Mooney, her marketing managers, selected staff members, selected store managers, and Chris Campbell. They worked hands-on with the prototype and identified several ways to improve it. Alec modified the HTML prototype to reflect the changes suggested by the group and asked Margaret and Chris to review it again.

The second evaluation was another interactive evaluation, this time by a series of two focus groups of potential customers, one with little Internet experience, the other with extensive Internet experience. Once again several minor changes were identified. Alec again modified the HTML prototype and asked Margaret and Chris to review it again. Once they were satisfied, the interface design was complete.

SUMMARY

User Interface Design Principles

The first element of the user interface design is the layout of the screen, form, or report, which is usually depicted using rectangular shapes with a top area for navigation, a central area for inputs and outputs, and a status line at the bottom. The design should help the user be aware of content and context, both between different parts of the system as they navigate through it and within any one form or report. All interfaces should be aesthetically pleasing (not works of art) and need to include significant white space, use colors carefully, and be consistent with fonts.

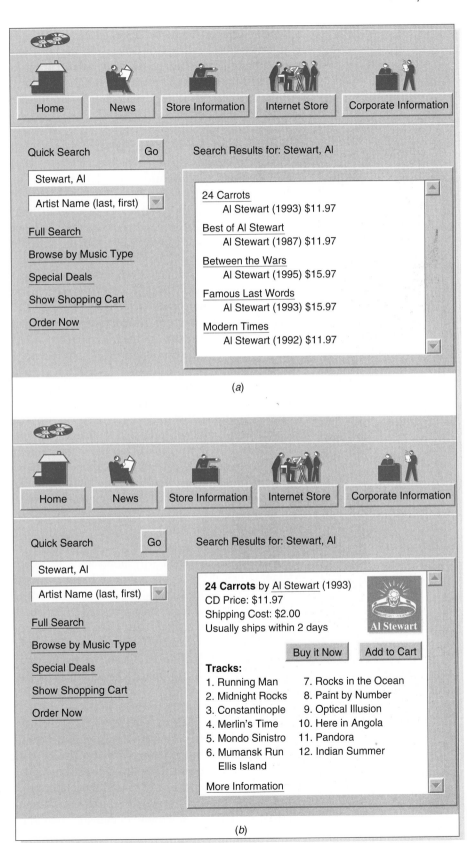

FIGURE 10-21

Sample interfaces from the CD Selections Design Prototype: (a) List by Artist (Interface 1.1); (b) Information for One CD (Interface 6)

Most interfaces should be designed to support both novice/first-time users, as well as experienced users. Consistency in design (both within the system and across other systems used by the users) is important for the navigation controls, terminology, and the layout of forms and reports. Finally, all interfaces should attempt to minimize user effort, for example, by requiring no more than three clicks from the main menu to perform an action.

The User Interface Design Process

First, analysts develop use scenarios that describe commonly used patterns of actions that the users will perform. Second, they design the interface structure via an ISD based on the DFD. The ISD is then tested with the use scenarios to ensure that it enables users to quickly and smoothly perform these scenarios. Third, analysts define the interface standards in terms of interface metaphor(s), objects, actions, and icons. These elements are drawn together by the design of a basic interface template for each major section of the system. Fourth, the designs of the individual interfaces are prototyped, either through a simple storyboard, an HTML prototype, or a prototype using the development language of the system itself (e.g., Visual Basic). Finally, interface evaluation is conducted using heuristic evaluation, walkthrough evaluation, interactive evaluation, or formal usability testing. This evaluation almost always identifies improvements, so the interfaces are redesigned and evaluated further.

Navigation Design

The fundamental goal of the navigation design is to make the system as simple to use as possible, by preventing the user from making mistakes, simplifying the recovery from mistakes, and using a consistent grammer order (usually object–action order). Command languages, natural languages, and direct manipulation are used in navigation, but the most common approach is menus (menu bar, drop-down menu, hyperlink menu, embedded hyperlinks, pop-up menu, tab menu, buttons and toolbars, and image maps). Error messages, confirmation messages, acknowledgment messages, delay messages, and help messages are common types of messages.

Input Design

The goal of input design is to simply and easily capture accurate information for the system, typically by using online or batch processing, capturing data at the source, and minimizing keystrokes. Input design includes both the design of input screens and all preprinted forms that are used to collect data before it is entered into the information system. There are many types of inputs, such as text fields, number fields, check boxes, radio buttons, on-screen list boxes, drop-down list boxes, and sliders. Most inputs are validated by using some combination of completeness checks, format checks, range checks, check digits, consistency checks, and database checks.

Output Design

The goal of output design is to present information to users so they can accurately understand it with the least effort, usually by understanding how reports will be used and designing them to minimize information overload and bias. Output design means designing both screens and reports in other media, such as paper and the Web. There are many types of reports, including detail reports, summary reports, exception reports, turnaround documents, and graphs.

KEY TERMS

Acknowledgment message
Action–object order
Aesthetics
Bar code reader
Batch processing
Batch report
Bias
Button
Check box
Check digit check
Combo box
Command language
Completeness check
Confirmation message
Consistency
Consistency check
Content awareness
Database check
Data-entry operator
Default value
Delay message
Density
Detail report
Direct manipulation
Drop-down list box
Drop-down menu
Ease of learning
Ease of use
Edit check
Error message
Exception report
Field

Field label
Form
Format check
Grammar order
Graph
Graphical user interface (GUI)
Help message
Heuristic evaluation
Hot key
HTML prototype
Image map
Information load
Input mechanism
Interactive evaluation
Interface action
Interface design prototype
Interface evaluation
Interface icon
Interface metaphor
Interface object
Interface standards
Interface structure diagram (ISD)
Interface template
Language prototype
Layout
Magnetic stripe readers
Menu
Natural language
Navigation mechanism
Number box
Object-action order
Online processing

On-screen list box
Optical character recognition
Output mechanism
Pop-up menu
Radio button
Range check
Real-time information
Real-time report
Report
Selection box
Slider
Smart card
Source data automation
Storyboard
Summary report
System interface
Tab menu
Text box
Three-clicks rule
Toolbar
Tool tip
Transaction processing
Turnaround document
Usability testing
Use scenario
User experience
User interface
Validation
Walk-through evaluation
Web page
White space

QUESTIONS

1. Explain three important user interface design principles.
2. What are three fundamental parts of most user interfaces?
3. Why is content awareness important?
4. What is white space, and why is it important?
5. Under what circumstances should densities be low? High?
6. How can a system be designed to be used by both experienced and first-time users?
7. Why is consistency in design important? Why can too much consistency cause problems?

8. How can different parts of the interface be consistent?
9. Describe the basic process of user interface design.
10. What are use scenarios, and why are they important?
11. What is an interface structure chart (ISD), and why is it used?
12. Why are interface standards important?
13. Explain the purpose and contents of interface metaphors, interface objects, interface actions, interface icons, and interface templates.
14. Why do we prototype the user interface design?
15. Compare and contrast the three types of interface design prototypes.

16. Why is it important to perform an interface evaluation before the system is built?
17. Compare and contrast the four types of interface evaluation.
18. Under what conditions is heuristic evaluation justified?
19. What type of interface evaluation did you perform in the "Your Turn 10.1"?
20. Describe three basic principles of navigation design.
21. How can you prevent mistakes?
22. Explain the differences between object–action order and action–object order.
23. Describe four types of navigation controls.
24. Why are menus the most commonly used navigation control?
25. Compare and contrast four types of menus.
26. Under what circumstances would you use a drop-down menu versus a tab menu?
27. Describe five types of messages.
28. What are the key factors in designing an error message?
29. What is context-sensitive help? Does your word processor have context-sensitive help?

30. Explain three principles in the design of inputs.
31. Compare and contrast batch processing and online processing. Describe one application that would use batch processing and one that would use online processing.
32. Why is capturing data at the source important?
33. Describe four devices that can be used for source data automation.
34. Describe five types of inputs.
35. Compare and contrast check boxes and radio buttons. When would you use one versus the other?
36. Compare and contrast on-screen list boxes and drop-down list boxes. When would you use one versus the other?
37. Why is input validation important?
38. Describe five types of input validation methods.
39. Explain three principles in the design of outputs.
40. Describe five types of outputs.
41. When would you use electronic reports rather than paper reports, and vice versa?
42. What do you think are three common mistakes that novice analysts make in interface design?
43. How would you improve the form in Fig. 10-4?

EXERCISES

A. Develop two use scenarios for a Web site that sells some retail products (e.g., books, music, clothes).
B. Draw an ISD for a Web site that sells some retail products (e.g., books, music, clothes).
C. Describe the primary components of the interface standards for a Web site that sells some retail products (metaphors, objects, actions, icons, and template).
D. Develop two use scenarios for the DFD in exercise C in chapter 6.
E. Draw an ISD for the DFD in exercise C in chapter 6.
F. Develop the interface standards (omitting the interface template) for the DFD in exercise C in Chapter 6.
G. Develop two use scenarios for the DFD in exercise E in Chapter 6.
H. Develop the interface standards (omitting the interface template) for the DFD in exercise E in Chapter 6.
I. Design an interface template for Exercise E.
J. Draw an ISD for the DFD in exercise E in Chapter 6.
K. Design a storyboard for exercise E in Chapter 6.

L. Develop an HTML prototype for Exercise E in this chapter.
M. Develop an HTML prototype for Exercise F in this chapter.
N. Develop the interface standards (omitting the interface template) for the DFD in Exercise I in Chapter 6.
O. Draw an ISD for the DFD in Exercise I in Chapter 6.
P. Design a storyboard for Exercise I in Chapter 6.
Q. Develop two use scenarios for the DFD in Exercise I in Chapter 6.
R. Ask Jeeves (http://www.askjeeves.com) is an Internet search engine that uses natural language. Experiment with it and compare it with search engines that use key words.
S. Draw an ISD for "Your Turn 10.7" using the opposite grammar order from your original design (if you didn't do it, draw two ISDs, one in each grammar order). Which is "best"? Why?
T. In "Your Turn 10.7" you probably used menus. Design the navigation system again using a command language.

MINICASES

1. Tots to Teens is a catalog retailer specializing in children's clothing. A project has been underway to develop a new order-entry system for the company's catalog clerks. The old system had a character-based user interface that corresponded to the system's COBOL underpinnings. The new system will feature a graphical user interface more in keeping with up-to-date PC products in use today. The company hopes that this new user interface will help reduce the turnover they have experienced with their order-entry clerks. Many newly hired order entry staff found the old system very difficult to learn and were overwhelmed by the numerous mysterious codes that had to be used to communicate with the system.

   A user interface walk-through evaluation was scheduled for today to give the users a first look at the new system's interface. The project team was careful to invite several key users from the order-entry department. In particular, Norma was included because of her years of experience with the order-entry system. Norma was known to be an informal leader in the department; her opinion influenced many of her associates. Norma had let it be known that she was less than thrilled with the ideas she had heard for the new system. Due to her experience and good memory, Norma worked very effectively with the character-based system and was able to breeze through even the most convoluted transactions with ease. Norma had trouble suppressing a sneer when she heard talk of such things as "icons" and "buttons" in the new user interface.

   Cindy was also invited to the walk-through because of her influence in the order-entry department. Cindy has been with the department for just one year, but she quickly became known because of her successful organization of a sick child day-care service for the children of the department workers. Sick children are the number-one cause of absenteeism in the department, and many of the workers could not afford to miss workdays. Never one to keep quiet when a situation needed improvement, Cindy has been a vocal supporter of the new system.

   a. Drawing upon the design principles presented in the text, describe the features of the user interface that will be most important to experienced users like Norma.

   b. Drawing upon the design principles presented in the text, describe the features of the user interface that will be most important to novice users like Cindy.

2. The members of a systems development project team have gone out for lunch together, and as often happens, the conversation turned to work. The team has been working on the development of the user interface design, and so far, work has been progressing smoothly. The team should be completing work on the interface prototypes early next week. A combination of storyboards and language prototypes have been used in this project. The storyboards depict the overall structure and flow of the system, but the team developed language prototypes of the actual screens because they felt that seeing the actual screens will be valuable for the users.

   Chris (the youngest member of the project team): I read an article last night about a really cool way to evaluate a user interface design. It's called usability testing, and it's done by all the major software vendors. I think we should use it to evaluate our interface design.

   Heather (system analyst): I've heard of that, too, but isn't it really expensive?

   Mark (project manager): I'm afraid it is expensive, and I'm not sure we can justify the expense for this project.

   Chris: But we really need to know that the interface works. I thought this usability testing technique would help us prove we have a good design.

   Amy (systems analyst): It would, Chris, but there are other ways too. I assumed we'd do a thorough walk-through with our users and present the interface to them at a meeting. We can project each interface screen so that the users can see it and give us their reaction. This is probably the most efficient way to get the users' response to our work.

   Heather: That's true, but I'd sure like to see the users sit down and work with the system. I've always learned a lot by watching what they do, seeing where they get confused, and hearing their comments and feedback.

   Ryan (systems analyst): It seems to me that we've put so much work into this interface design that all we really need to do is review it ourselves. Let's just make a list of the design principles we're most concerned about and check it ourselves to make sure we've followed them consistently. If we have, we should be fine. We want to get moving on the implementation, you know.

   Mark: These are all good ideas. It seems like we've all got a different view of how to evaluate the interface design. Let's try and sort out the technique that is best for our project.

Develop a set of guidelines that can help a project team like the one above select the most appropriate interface evaluation technique for their project.

3. The menu structure for Holiday Travel Vehicle's existing character-based system is shown below. Develop and prototype a new interface design for the system's functions using a graphical user interface. Assume the new system will need to include the same functions as those shown in the menus provided. Include any messages that will be produced as a user interacts with your interface (error, confirmation, status, etc.). Also, prepare a written summary that describes how your interface implements the principles of good interface design as presented in the textbook.

Holiday Travel Vehicles

Main Menu

1 Sales Invoice
2 Vehicle Inventory
3 Reports
4 Sales Staff

Type number of menu selection here:____

Holiday Travel Vehicles

Sales Invoice Menu

1 Create Sales Invoice
2 Change Sales Invoice
3 Cancel Sales Invoice

Type number of menu selection here:____

Holiday Travel Vehicles

Vehicle Inventory Menu

1 Create Vehicle Inventory Record
2 Change Vehicle Inventory Record
3 Delete Vehicle Inventory Record

Type number of menu selection here:____

Holiday Travel Vehicles

Reports Menu

1 Commission Report
2 RV Sales by Make Report
3 Trailer Sales by Make Report
4 Dealer Options Report

Type number of menu selection here:____

Holiday Travel Vehicles

Sales Staff Maintenance Menu

1 Add Salesperson Record
2 Change Salesperson Record
3 Delete Salesperson Record

Type number of menu selection here:____

---

4. One aspect of the new system under development at Holiday Travel Vehicles will be the direct entry of the sales invoice into the computer system by the salesperson as the purchase transaction is being completed. In the current system, the salesperson fills out a paper form (shown on p. 353)

Design and prototype an input screen that will permit the salesperson to enter all the necessary information for the sales invoice. The following information may be helpful in your design process. Assume that Holiday Travel Vehicles sells recreational vehicles and trailers from four different manufacturers. Each manufacturer has a fixed number of names and models of RVs and trailers. For the purposes of your prototype, use this format:

Mfg-A	Name-1 Model-X		Mfg-C	Name-1 Model-X
Mfg-A	Name-1 Model-Y		Mfg-C	Name-1 Model-Y
Mfg-A	Name-1 Model-Z		Mfg-C	Name-1 Model-Z
Mfg-B	Name-1 Model-X		Mfg-C	Name-2 Model-X
Mfg-B	Name-1 Model-Y		Mfg-C	Name-3 Model-X
Mfg-B	Name-2 Model-X		Mfg-D	Name-1 Model-X
Mfg-B	Name-2 Model-Y		Mfg-D	Name-2 Model-X
Mfg-B	Name-2 Model-Z		Mfg-D	Name-2 Model-Y

Also, assume there are 10 different dealer options that could be installed on a vehicle at the customer's request. The company currently has 10 salespeople on staff.

Holiday Travel Vehicles
Sales Invoice

Invoice #: _____
Invoice Date: _____

Customer Name: _____
Address: _____
City: _____
State: _____
Zip: _____
Phone: _____

New RV/TRAILER
(circle one)   Name: _____
Model: _____
Serial #: _____ Year: _____
Manufacturer: _____

Trade-in RV/TRAILER
(circle one)   Name: _____
Model: _____
Year: _____
Manufacturer: _____

Options:	Code	Description	Price
	_____	_____	_____
	_____	_____	_____
	_____	_____	_____
	_____	_____	_____

Vehicle Base Cost: _____
Trade-in Allowance: _____   _____
Total Options: _____   (Salesperson Name)
Tax: _____
License Fee: _____   _____
Final Cost: _____   (Customer Signature)

PLANNING

ANALYSIS

DESIGN

- ☑ **Select Design**
- ☑ **Develop Physical Data Flow Diagrams**
- ☑ **Develop Physical Entity Relationship Diagrams**
- ☑ **Design Architecture**
- ☑ **Select Hardware and Software**
- ☑ **Develop Use Scenarios**
- ☑ **Design Interface Structure**
- ☑ **Design Interface Standards**
- ☑ **Design Interface Prototype**
- ☑ **Evaluate User Interface**
- ☑ **Design User Interface**
- ☐ **Select Data Storage Format**
- ☐ **Denormalize Entity Relationship Diagram**
- ☐ **Performance Tune Data Storage**
- ☐ **Size Data Storage**
- ☐ Develop Program Structure Chart
- ☐ Develop Program Specification

TASK CHECKLIST

PLANNING   ANALYSIS   DESIGN

CHAPTER 11

DATA STORAGE

DESIGN

A project team designs the data storage component of a system using a two-step approach: slecting the format of the data storage and then optimizing it to perform efficiently. This chapter first describes the different ways in which data can be stored and several important characteristics that should be considered when choosing among data storage formats. Because one of the most popular data storage formats today is the relational database, the rest of this chapter focuses on the optimization of relational databases from both storage and data access perspectives.

OBJECTIVES

- Become familiar with several file and database formats.
- Understand several goals of data storage.
- Be able to optimize a relational database for data storage and data access.
- Become familiar with indexes.
- Be able to estimate the size of a database.

CHAPTER OUTLINE

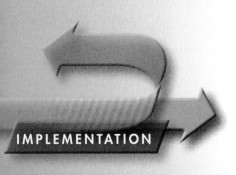

IMPLEMENTATION

INTRODUCTION

As explained in Chapter 9, the work done by any application program can be divided into four general functions: data storage, data access logic, application logic, and presentation logic. The data storage function manages how data is stored and handled by the programs that run the system. This chapter describes how a project team designs data storage using a two-step approach: selecting the format of the data storage and then optimizing it to perform efficiently.

Applications are of little use without the data that they support. How useful is a multimedia application that can't support images or sound? Why would someone log into a system to find information if it took him or her less time to locate the information manually? The design phase of the SDLC includes two steps to data storage design that decrease the chances of ending up with inefficient systems, long system response times, and users who can't get to the information that they need in the way that they need it—all of which can affect the success of the project.

The first part of this chapter describes a variety of data storage formats and explains how to select the appropriate one for your application. There are two basic types of formats used to store data for application systems: files and databases. Each has a variety of flavors; for instance, databases can be object oriented, relational, multidimensional, and so on. Each type has certain characteristics that make it more appropriate for some types of systems over others.

Once the data storage format is selected to support the system, it must be designed to optimize its processing efficiency, which is the focus of the second part of this chapter. One of the leading complaints by end users is that the final system is too slow, so to avoid such complaints project team members must allow time during the design phase to carefully make sure that the file or database performs as fast as possible. At the same time, the team must keep hardware costs down by minimizing the storage space that the application will require. The goals of maximizing access to data and minimizing the amount of space taken to store data can conflict, so designing data storage efficiency usually requires trade-offs.

DATA STORAGE FORMATS

There are two main types of data storage formats: files and databases. *Files* are electronic lists of data that have been optimized to perform a particular transaction. For example, Figure 11-1 shows a patient appointment file with information about patient's appointments, in the form in which it is used, so that the information can be accessed and processed quickly by the system.

A *database* is a collection of groupings of information that are related to each other in some way (e.g., through common fields). Logical groupings of information could include such categories as customer data, information about an order, and product information. A *database management system (DBMS)* is software that creates and manipulates these databases (Figure 11-2). Such *end-user DBMS*s as Microsoft Access support small-scale databases that are used to enhance personal productivity, and *enterprise DBMS*s, such as DB2, Jasmine, and Oracle, can manage huge volumes of data and support applications that run an entire company. An end-user DBMS is significantly less expensive and easier for novice users to use than its enterprise counterpart, but it does not have the features or capabilities that are necessary to support mission-critical or large-scale systems.

Appointment Date	Appointment Time	Duration	Reason	Patient ID	First Name	Last Name	Phone Number	Doctor ID	Doctor Last Name
11/23/2003	2:30	.25 hour	Flu	758843	Patrick	Dennis	548-9456	V524625587	Vroman
11/23/2003	2:30	1 hour	Physical	136136	Adelaide	Kin	548-7887	T445756225	Tantalo
11/23/2003	2:45	.25 hour	Shot	544822	Chris	Pullig	525-5464	V524625587	Vroman
11/23/2003	3:00	1 hour	Physical	345344	Felicia	Marston	548-9333	B544742245	Brousseau
11/23/2003	3:00	.5 hour	Migraine	236454	Thomas	Bateman	667-8955	V524625587	Vroman
11/23/2003	3:30	.5 hour	Muscular	887777	Ryan	Nelson	525-4772	V524625587	Vroman
11/23/2003	3:30	.25 hour	Shot	966233	Peter	Todd	667-2325	T445756225	Tantalo
11/23/2003	3:45	.75 hour	Muscular	951657	Mike	Morris	663-8944	T445756225	Tantalo
11/23/2003	4:00	1 hour	Physical	223238	Ellen	Whitener	525-8874	B544742245	Brousseau
11/23/2003	4:00	.5 hour	Flu	365548	Jerry	Starsia	548-9887	V524625587	Vroman
11/23/2003	4:30	1 hour	Minor surg	398633	Susan	Perry	525-6632	V524625587	Vroman
11/23/2003	4:30	.5 hour	Migraine	222577	Elizabeth	Gray	667-8400	T445756225	Tantalo
11/24/2003	8:30	.25 hour	Shot	858756	Elias	Awad	663-6364	T445756225	Tantalo
11/24/2003	8:30	1 hour	Minor surg	232158	Andy	Ruppel	525-9888	V524625587	Vroman
11/24/2003	8:30	.25 hour	Flu	244875	Rick	Grenci	548-2114	B544742245	Brousseau
11/24/2003	8:45	.5 hour	Muscular	655683	Eric	Meier	667-0254	T445756225	Tantalo
11/24/2003	8:45	1 hour	Physical	447521	Jane	Pace	548-0025	B544742245	Brousseau
11/24/2003	9:30	.5 hour	Flu	554263	Trey	Maxham	663-8547	V524625587	Vroman

FIGURE 11-1
Appointment File

The next section describes several different kinds of files and databases that can be used to handle a system's data storage requirements.

Files

A file contains an electronic list of information that is formatted for a particular transaction, and the information is changed and manipulated by programs that are written for those purposes. Typically, files are organized sequentially, and new records are added to the file's end. These records can be associated with other records using a pointer, which is information about the location of the related record. A pointer is placed at the end of each record, and it "points"to the next record in a series or set. Sometimes files are called *linked lists* because of the way the records are linked together using the pointers. There are several types of files that differ in the way they are used to support an application: master files, look-up files, transaction files, audit files, and history files.

Master files store core information that is important to the business and, more specifically, to the application, such as order information or customer mailing information. They usually are kept for long periods of time, and new records are appended to the end of the file as new orders or new customers are captured by the system. If changes need to be made to existing records, programs must be written to update the old information.

Look-up files contain static values, such as a list of valid codes or the names of the U.S. states. Typically, the list is used for validation. For example, if a cus-

Appointment Date	Appointment Time	Duration	Reason	Patient ID	Doctor ID
11/23/2003	2:30	.5 hour	Flu	758843	V524625587
11/23/2003	2:30	1 hour	Physical	136136	T445756225
11/23/2003	2:45	.25 hour	Shot	544822	V524625587
11/23/2003	3:00	1 hour	Physical	345344	B544742245
11/23/2003	3:00	.5 hour	Migraine	236454	V524625587
11/23/2003	3:30	.5 hour	Muscular	887777	V524625587
11/23/2003	3:30	.25 hour	Shot	966233	T445756225
11/23/2003	3:45	.75 hour	Muscular	951657	T445756225
11/23/2003	4:00	1 hour	Physical	223238	B544742245
11/23/2003	4:00	.5 hour	Flu	365548	V524625587
11/23/2003	4:30	1 hour	Minor surg	398633	V524625587
11/23/2003	4:30	.5 hour	Migraine	222577	T445756225
11/24/2003	8:30	.25 hour	Shot	858756	T445756225
11/24/2003	8:30	1 hour	Minor surg	232158	V524625587
11/24/2003	8:30	.25 hour	Flu	244875	B544742245
11/24/2003	8:45	.5 hour	Muscular	655683	T445756225
11/24/2003	8:45	1 hour	Physical	447521	B544742245
11/24/2003	9:30	.5 hour	Flu	554263	V524625587

Tables related using patient id

Tables related using doctor id

Patient ID	First Name	Last Name	Phone Number
136136	Adelaide	Kin	548-7887
222577	Elizabeth	Gray	667-8400
223238	Ellen	Whitener	525-8874
232158	Andy	Ruppel	525-9888
236454	Thomas	Bateman	667-8955
244875	Rick	Grenci	548-2114
345344	Felicia	Marston	548-9333
365548	Jerry	Starsia	548-9887
398633	Susan	Perry	525-6632
447521	Jane	Pace	548-0025
544822	Chris	Pullig	525-5464
554263	Trey	Maxham	663-8547
655683	Eric	Meier	667-0254
758843	Patrick	Dennis	548-9456
858756	Elias	Awad	663-6364
887777	Ryan	Nelson	525-4772
951657	Mike	Morris	663-8944
966233	Peter	Todd	667-2325

Doctor ID	Last Name
B544742245	Brousseau
T445756225	Tantalo
V524625587	Vroman

FIGURE 11-2

Appointment Database

tomer's mailing address is entered into a master file, the state name is validated against a look-up file that contains U.S. states to make sure that the operator entered the value correctly.

A *transaction file* holds information that can be used to update a master file. The transaction file can be destroyed after changes are added, or the file may be saved in case the transactions need to be accessed again in the future. Customer address changes, for one, would be stored in a transaction file until a program is run that updates the customer address master file with the new information.

For control purposes, a company might need to store information about how data changes over time. For example, as human resources clerks change employee salaries in a human resources system, the system should record the person who made the changes to the salary amount, the date, and the actual change that was made. An *audit file* records "before" and "after" images of data as it is altered so that an audit can be performed if the integrity of the data is questioned.

Sometimes files become so large that they are unwieldy, and much of the information in the file is no longer used. The *history file* (or archive file) stores past transactions (e.g., old customers, past orders) that are no longer needed by system users. Typically, the file is stored off-line, yet it can be accessed on an as-needed basis. Other files, such as master files, can then be streamlined to include only active or very recent information.

Databases

There are many different types of databases that exists on the market today. In this section, we provide a brief description of four databases with which you may come into contact: legacy, relational, object, and multidimensional. You will likely encounter a variety of ways to classify databases in your studies, but in this book we classify databases in terms of how they store and manipulate data.

Legacy Databases   The name *legacy database* is given to those databases that are based on older, sometimes outdated technology that is seldom used to develop new applications; however, you may come across them when maintaining or migrating from systems that already exist within your organization. Two examples of legacy databases include hierarchical databases and network databases. *Hierarchical databases* (e.g., IDMS) use hierarchies, or inverted trees, to represent relationships (similar to the one-to-many [1 : M] relationships that were described in Chapter 7). The record at the top of the tree has zero or more child records, which in turn can serve as parents for other records (Figure 11-3).

YOUR TURN

11-1 STUDENT ADMISSIONS SYSTEM

Pretend that you are building a Web-based system for the admissions office at your university that will be used to accept electronic applications from students. All the data for the system will be stored in a variety of files.

QUESTION:

Give an example using the above system for each of the following file types: master, look-up, transaction, audit, and history. What kind of information would each file contain and how would the file be used?

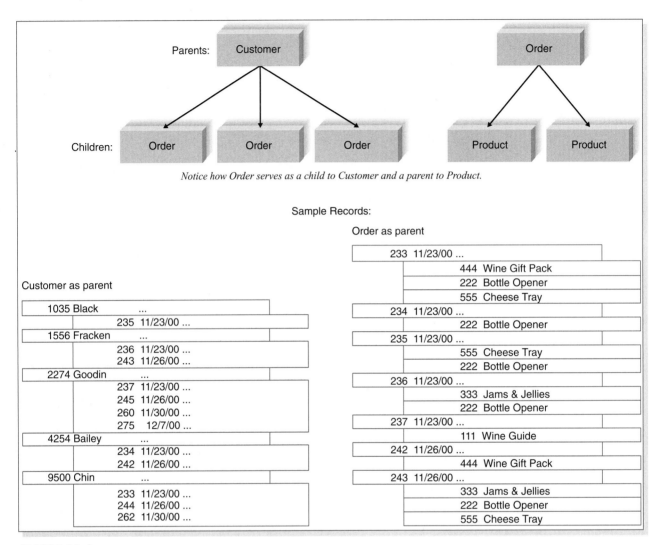

FIGURE 11-3
Hierarchical Database

Hierarchical databases cannot efficiently represent many-to-many (M : M) relationships or nonhierarchical associations—a major drawback—so network databases were developed to address this limitation (and others) of hierarchical technology. *Network databases* (e.g., IDMS/R, DBMS 10) are collections of records that are related to each other through *pointers*. Basically, a record is a *member* in one or more *sets*, and the pointers link the members in a set together (Figure 11-4).

Both kinds of legacy systems can handle data quite efficiently, but they require a great deal of programming effort. The application system software needs to contain code that manipulates the database pointers; in other words, the application program must understand how the database is built and be written to follow the structure of the database. When the database structure is changed, the application program must be rewritten to change the way it works, which makes applications using the databases difficult to build and maintain. Years ago when hardware was expensive and programmer time was cheap, hierarchical and network databases

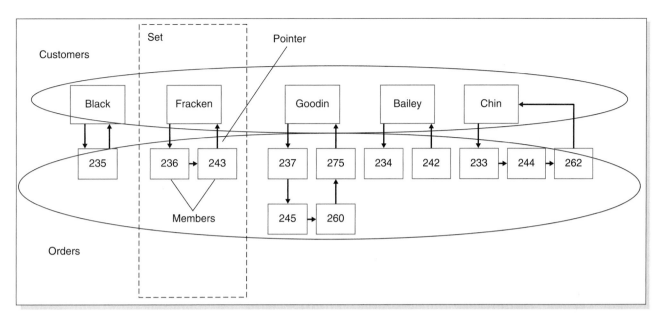

FIGURE 11-4
Network Database

were good solutions for large systems; however, as hardware costs dropped and people costs skyrocketed, the solutions became much less cost effective.

Relational Databases *The relational database* is the most popular kind of database for application development today. Although it is less "machine efficient" than its legacy counterparts, it is much easier to work with from a development perspective. A relational database is based on collections of tables, each of which has a *primary key*—a field(s) whose value is different for every row of the table. The tables are related to each other by placing the primary key from one table into the related table as a *foreign key* (Figure 11-5).

Most relational database management systems (RDBMSs) support *referential integrity*, or the idea of ensuring that values linking the tables together through the primary and foreign keys are valid and correctly synchronized. For example, if an order-entry clerk using the tables in Figure 11-5 attempted to add order 254 for customer number 1111, he or she would have made a mistake because no customer exists in the Customer table with that number. If the RDBMS supported referential integrity, it would check the customer numbers in the Customer table, discover that the number 1111 is invalid, and return an error to the entry clerk. The clerk would then go back to the original order form an recheck the customer information. Can you imagine the problems that would occur if the RDBMS let the entry clerk add the order with the wrong information? There would be no way to track down the name of the customer for order 254.

Tables have a set number of columns and a variable number of rows that contain occurrences of data. *Structured Query Language (SQL)* is the standard language for accessing the data in the tables, and it operates on complete tables, as opposed to the individual records in the tables. Thus, a query written in SQL is applied to all the records in a table all at once, which is different from a lot of programming languages that manipulate data record by record. When queries must

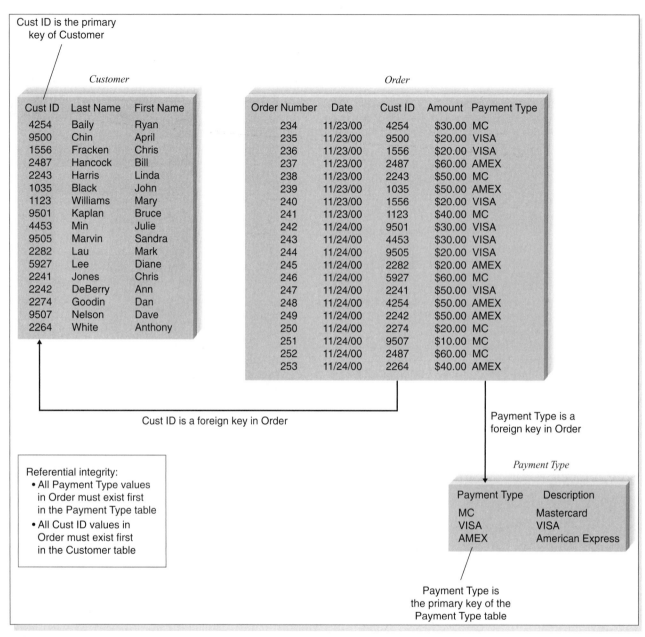

Cust ID is the primary key of Customer

Customer

Cust ID	Last Name	First Name
4254	Baily	Ryan
9500	Chin	April
1556	Fracken	Chris
2487	Hancock	Bill
2243	Harris	Linda
1035	Black	John
1123	Williams	Mary
9501	Kaplan	Bruce
4453	Min	Julie
9505	Marvin	Sandra
2282	Lau	Mark
5927	Lee	Diane
2241	Jones	Chris
2242	DeBerry	Ann
2274	Goodin	Dan
9507	Nelson	Dave
2264	White	Anthony

Order

Order Number	Date	Cust ID	Amount	Payment Type
234	11/23/00	4254	$30.00	MC
235	11/23/00	9500	$20.00	VISA
236	11/23/00	1556	$20.00	VISA
237	11/23/00	2487	$60.00	AMEX
238	11/23/00	2243	$50.00	MC
239	11/23/00	1035	$50.00	AMEX
240	11/23/00	1556	$20.00	VISA
241	11/23/00	1123	$40.00	MC
242	11/24/00	9501	$30.00	VISA
243	11/24/00	4453	$30.00	VISA
244	11/24/00	9505	$20.00	VISA
245	11/24/00	2282	$20.00	AMEX
246	11/24/00	5927	$60.00	MC
247	11/24/00	2241	$50.00	VISA
248	11/24/00	4254	$50.00	AMEX
249	11/24/00	2242	$50.00	AMEX
250	11/24/00	2274	$20.00	MC
251	11/24/00	9507	$10.00	MC
252	11/24/00	2487	$60.00	MC
253	11/24/00	2264	$40.00	AMEX

Cust ID is a foreign key in Order

Payment Type is a foreign key in Order

Referential integrity:
- All Payment Type values in Order must exist first in the Payment Type table
- All Cust ID values in Order must exist first in the Customer table

Payment Type

Payment Type	Description
MC	Mastercard
VISA	VISA
AMEX	American Express

Payment Type is the primary key of the Payment Type table

FIGURE 11-5
Relational Database

include information from more than one table, the tables first are joined together on the basis of their primary key and foreign key relationships and treated as if they were one large table. Examples of RDBMS software are Microsoft Access, Oracle, DB2, Sybase, Informix, and Microsoft SQL Server.

Object Databases The next type of database is the *object database*, or object-oriented database. (See Chapter 15 for more information on object-oriented approaches.) The basic premise of object orientiation is that all things should be treated as objects that have both data and processes. Changes to one object have no effect on other objects because the data and processes are self-contained, or encapsulated, within each one. This *encapsulation* allows objects to be reused to build many different systems because they can be inserted and removed from applications with few ripple effects. For examples, a customer object could be defined one time as having data (e.g., customer number, customer name) and processes (e.g., inserting a customer, deleting a customer), and then this customer object could be used to build any system that involves a customer.

In object databases, the combination of data and processes is represented by *object classes*, which are the major categories of objects in the system, and a class can contain a variety of *subclasses*, or special cases of that class. For example, a person class can have subclasses of employee and customer because employee and customer are special cases of person. An instance of data in object databases is referred to as an *instantiation* (e.g., *John Smith* is an instantiation of the *customer* object), and the relationships among classes are maintained using pointers (Figure 11-6).

Object-oriented database management systems (OODBMSs) are mainly used to support multimedia applications or systems that involve complex data (e.g., graphics, video, and sound). Telecommunications, financial services, health care, and transporation have been the most receptive to object databases, and they are becoming popular technology for supporting electronic commerce, online catalogs, and large Web multimedia applications.

Although pure OODBMSs like Jasmine exist, most organizations invest in *hybrid OODBMS* technology, which includes databases with both object and relational features. For instance, Oracle, a leader in the relational database market, incorporates object functionality and capabilities into its relational product.

Although the market for OOBDMSs is expected to grow, the $164 million market for the technology is dwarfed by that for its relational database counterpart

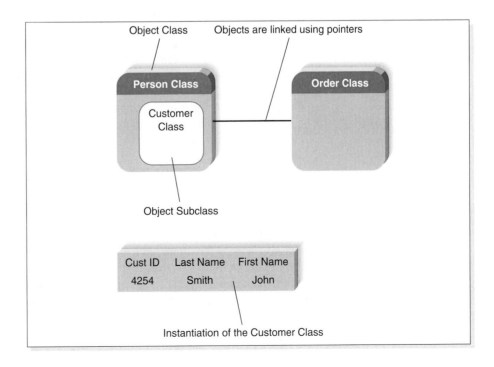

FIGURE 11-6
Object Database

($6.2 billion).[1] For one, there are many more experienced developers and tools in the relational database arena. Also, relational users find that OODBMS technology comes with a very steep learning curve.

Multidimensional Databases   One of the newest members in the database arena is the *multidimensional database*, which has been driven in large part by the rise of data warehousing. Data warehousing is the practice of taking data from a company's transaction processing systems, transforming the data (e.g., cleaning them up, totaling them, aggregating them), and then storing the data for use in a data warehouse (i.e., a large database) that supports *decision support systems (DSS)*. A data warehouse itself usually relies on relational technology as its storage format; however, companies can create *data marts*, which are smaller databases based on data warehouse data. Typically, a data mart receives downloads of data from the data warehouse regularly, and it supports DSS for a specific department or functional area of the company. For example, the Marketing Department may have a data mart that supports their campaign management DSS. Data marts are usually created using multidimensional databases.

In most cases, DSS is designed not to search for a particular record (e.g., "What did John Smith order on July 5, 2001?") but rather to display information that is *aggregated* (e.g., totaled or averaged) across many records (e.g., "What was the average sales by quarter for product A?") Thus, data marts that support a DSS require that data be stored in a format in which they can be easily aggregated and manipulated across a variety of dimensions (e.g., time, product, region, sales rep). Unfortunately, legacy, object, and relational databases are designed and optimized to provide access to individual records, not to store data to support aggregations of data on multiple dimensions.

When data is first loaded into a multidimensional database, the database precalculates the data across multiple dimensions and stores the answers using arrays or some other techniques. Although the initial loading of the data can be quite slow because of all the calculations that must take place, data access is extremely fast because the "answers" already exist in the arrays. For example, the cube in Figure 11-7 represents a multidimensional database that contains data that has been organized by customer, payment type, and order date. Precalculated quantitative information (e.g., totals, averages) is stored at the intersection of the dimensions (in each block), and the DSS directly accesses those blocks. Because blocks contain precalculated information, there is much less processing that needs to occur to provide the DSS with aggregated results.

Selecting a Storage Format

Each of the file and database data storage formats has its strengths and weaknesses, and no one format is inherently better than the others are. In fact, sometimes a project team will choose multiple data storage formats (e.g., a relational database for one data store, a file for another, and a multidimensional database for a third). Thus, it is important to understand the strengths and weaknesses of each format and when to use each one. Figure 11-8 summarizes the characteristics of each and the characteristics that can help identify when each types of database is more appropriate.

---

[1] Criag Stedman, "Objects Still Face a Tough Sell," *Computerworld,* February 9, 1998, 32(6):33.

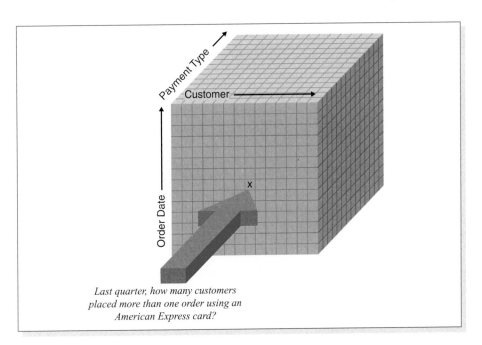

Last quarter, how many customers
placed more than one order using an
American Express card?

FIGURE 11-7
Multidimensional Database

	Files	Legacy DBMS	Relational DBMS	Object-Oriented DBMS	Multi-dimensional DBMS
Major strengths	Files can be designed for fast performance; good for short-term data storage	Very mature products	Leader in the data-base market; can handle diverse data needs	Able to handle complex data	Configured to answer decision support questions quickly
Major weaknesses	Redundant data; data must be updated using programs	Not able to store data as efficiently; limited future	Cannot handle complex data	Technology is still maturing; skills are hard to find	Highly specialized use; skills are hard to find
Data types supported	Simple	*Not recommended for new systems*	Simple	Complex (e.g., video, audio, images)	Aggregated
Types of application systems supported	Transaction processing	*Not recommended for new systems*	Transaction processing and decision making	Transaction processing	Decision making
Existing data formats	Organization dependent	Organization dependent	Organization dependent	Organization dependent	Organization dependent
Future needs	Poor future prospects	Poor future prospects	Good future prospects	Uncertain future prospects	Uncertain future prospects

DBMS = database management system.

FIGURE 11-8
Comparison of Data Storage Formats

Data Types The first issue is the type of data that will need to be stored in the system. Most applications need to store simple data types, such as text, dates, and numbers, and all DBMSs are equipped to handle this kind of data. The best choice for simple data storage, however, usually is the relational database because the technology has matured over time and has continuously improved to handle simple data very effectively.

Increasingly, applications are incorporating complex data, such as video, images, or audio, and object databases are best able to handle data of this type. Complex data are stored as objects that can be manipulated much faster than with other storage formats. Other applications require aggregated data (i.e., information that has been summed, averaged, or combined in some way). Multidimensional databases are specially designed to store data so that they can be "sliced and diced" and examined across important business dimensions. If the system is being built for analytical decision support, then this option likely will be most appropriate.

Type of Application System There are many different kinds of application systems that can be developed. *Transaction processing systems* are designed to accept and process many simultaneous requests (e.g., order entry, distribution, payroll). In transaction processing systems, the data are continuously updated by a large number of users, and the queries that are asked of the systems typically are predefined or targeted at a small subset of records (e.g., "List the orders that were backordered today"; "What products did customer #1234 order on May 12, 2001?").

Another set of application systems are those designed to support decision making, such as business intelligence management information systems (MISs), executive information systems (EISs), and expert systems (ESs). These decision support systems (DSS) are built to support decision makers who need to examine large amounts of read-only historical data. The questions that they ask are often ad hoc, and they include hundreds or thousands of records at a time (e.g., "List all customers in the west region who purchased a product costing more than $500 at least three times"; "What products had increased sales in the summer months but have not been classified as summer merchandise?").

Transaction processing and DSSs thus have very different data storage needs. Transaction processing systems need data storage formats that are tuned for a lot of data updates and fast retrieval of predefined, specific questions. Files, relational databases, and object databases can all support these kinds of requirements. By contrast, systems to support decision making are usually only reading data (not updating it), often in ad hoc ways. The best choices for these systems usually are relational databases and multidimensional databases because these formats can be configured specially for needs that may be unclear and less apt to change the data.

Existing Storage Formats The data storage format primarily should be selected on the basis of the kind of data and application system being developed. However, project teams should consider the existing data storage formats in the organization when making design decisions. In this way, they can better understand the technical skills that already exist and how steep the learning curve will be when the data storage format is adopted. For example, a company that is familiar with relational databases will have little problem adopting a relational database for the project, whereas an object database may require substantial developer training. In the latter situation, the project team may have to plan for more time to integrate the object database with the company's relational systems.

Future Needs   Not only should a project team consider the data storage technology within the company but they also should be aware of current trends and technologies that are being used by other organizations. A large number of installations of a type of data storage format suggests that the selection of that format is "safe," in that skills and products are available. For example, it would probably be easier and less expensive to find relational database expertise when implementing a system than to find help with a multidimensional data storage format. Legacy database skills, too, likely would be difficult to come by.

OPTIMIZING DATA STORAGE

Once the data storage format is selected, the second step of data storage design is to optimize it for processing efficiency. The methods of optimization will vary on the basis of the format that you select; however, the basic concepts will remain the same. Once you understand how to optimize a particular type of data storage, you will have some idea as to how to approach the optimization of other formats. This section focuses on the optimization of the most popular data storage format: relational databases.

There are two primary dimensions in which to optimize a relational database: for storage efficiency and for speed of access. Unfortunately, these two goals often conflict because the best design for access speed may take up a great deal of storage space as compared with other less speedy designs. This section describes how to optimize data storage for storage efficiency using normalization (Chapter 7). The next section presents design techniques, such as denormalization and indexing that will quicken the performance of the system. Ultimately, the project team will go through a series of trade-offs until the ideal balance between both optimization dimensions is reached. Finally, the project team must estimate the size of the data storage needed to ensure there is enough capacity on the server(s).

Optimizing Storage Efficiency

The most efficient tables in a relational database in terms of storage space have no redundant data and very few null values because the presence of these suggest that space is being wasted (and more data to store means higher data storage hardware costs). For example, notice that the sample records table in Figure 11-9 repeats customer information, such as name and state, each time a customer places an order, and it contains many null values in the last four columns. These nulls occur whenever a customer places an order for less than three items (the maximum number on an order).

In addition to wasting space, redundancy and null values also allow more room for error and increase the likelihood that problems will arise with the integrity of the data. What if customer 1135 moves from Maryland to Georgia? In the case of Figure 11-9, a program must be written to ensure that all instances of that customer are updated to show "GA" as the new state of residence. If some of the instances are overlooked, then the table will contain an update anomaly where by some of the records contain the correctly updated value for state and other records contain the old information.

Nulls threaten data integrity because they are difficult to interpret. A blank value in the customer order table's product fields could mean (1) the customer did

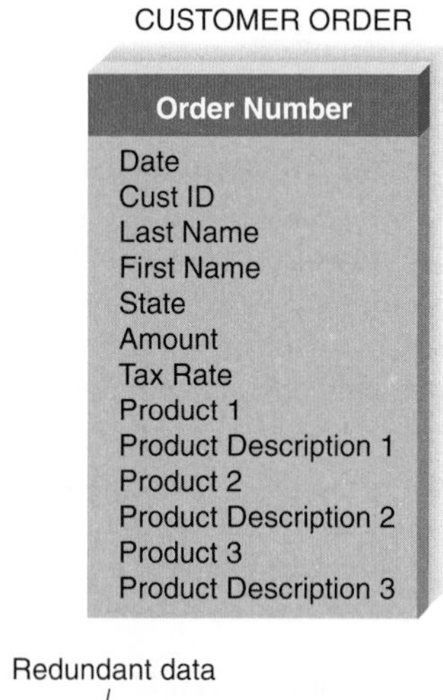

CUSTOMER ORDER

| **Order Number** |
| Date |
| Cust ID |
| Last Name |
| First Name |
| State |
| Amount |
| Tax Rate |
| Product 1 |
| Product Description 1 |
| Product 2 |
| Product Description 2 |
| Product 3 |
| Product Description 3 |

Redundant data   Null cells

Order Number	Date	Cust ID	Last Name	First Name	State	Amount	Tax Rate	Product	Product Desc	Product	Product Desc	Product	Product Desc
239	11/23/00	1135	Black	John	MD	$50.00	0.05	555	Cheese Tray				
260	11/24/00	1135	Black	John	MD	$40.00	0.05	444	Wine Gift Pack				
273	11/27/00	1135	Black	John	MD	$20.00	0.05	222	Bottle Opener				
241	11/23/00	1123	Williams	Mary	CA	$40.00	0.08	444	Wine Gift Pack				
262	11/24/00	1123	Williams	Mary	CA	$20.00	0.08	222	Bottle Opener				
287	11/27/00	1123	Williams	Mary	CA	$20.00	0.08	222	Bottle Opener				
290	11/30/00	1123	Williams	Mary	CA	$50.00	0.08	555	Cheese Tray				
234	11/23/00	2242	DeBerry	Ann	DC	$50.00	0.065	555	Cheese Tray				
237	11/7/00	2242	DeBerry	Ann	DC	$50.00	0.065	111	Wine Guide	444	Wine Gift Pack		
238	11/10/00	2242	DeBerry	Ann	DC	$40.00	0.065	444	Wine Gift Pack				
245	11/11/00	2242	DeBerry	Ann	DC	$20.00	0.065	222	Bottle Opener				
250	11/18/00	2242	DeBerry	Ann	DC	$20.00	0.065	222	Bottle Opener				
252	11/22/00	2242	DeBerry	Ann	DC	$60.00	0.065	222	Bottle Opener	444	Wine Gift Pack		
253	11/23/00	2242	DeBerry	Ann	DC	$60.00	0.065	222	Bottle Opener	444	Wine Gift Pack		
297	11/24/00	2242	DeBerry	Ann	DC	$30.00	0.065	333	Jams & Jellies				
243	11/11/00	4254	Bailey	Ryan	MD	$50.00	0.05	555	Cheese Tray				
246	11/18/00	4254	Bailey	Ryan	MD	$30.00	0.05	333	Jams & Jellies				
248	11/22/00	4254	Bailey	Ryan	MD	$60.00	0.05	222	Bottle Opener	333	Jams & Jellies	111	Wine Guide
235	11/17/00	9500	Chin	April	KS	$20.00	0.05	222	Bottle Opener				
242	11/23/00	9500	Chin	April	KS	$30.00	0.05	333	Jams & Jellies				
244	11/24/00	9500	Chin	April	KS	$20.00	0.05	222	Bottle Opener				
251	11/27/00	9500	Chin	April	KS	$10.00	0.05	111	Wine Guide				

FIGURE 11-9
Optimizing Data Storage

not want more than one or two products on his or her order, (2) the operator forgot to enter in all three products on the order, or (3) the customer canceled part of the order and the products were deleted by the operator. It is impossible to be sure of the actual meaning of the nulls.

For both of these reasons—wasted storage space and data integrity threats—project teams should remove redundancy and nulls from data storage design. During the design phase, the logical data model is used to examine the data storage design and optimize it for storage efficiency. If you follow the modeling instructions and guidelines that were presented in Chapter 7, you will have little trouble creating a design that is highly optimized in this way because a well-formed logical data model does not contain redundancy or many null values.

Sometimes, however, a project team needs to start with a logical model that was poorly constructed or with a model that was created for files or a nonrelational type of data storage format. In these cases, the project team should follow a series of steps that serve to check the logical data model for storage efficiency. These steps

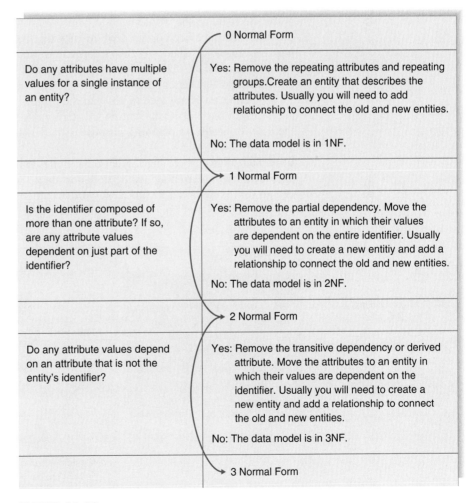

FIGURE 11-10
Normalization Steps

make up a process called *normalization* and it is described in detail in Chapter 7. As a reminder, the steps of normalization are listed in Figure 11-10. Normalization is the best way to optimize data storage for efficiency.

Optimizing Access Speed

After you have optimized your data model design for data storage efficiency, the end result is data that is spread out across a number of tables. When data from multiple tables must be accessed or queried, the tables must be first joined together. For example, in Figure 11-2 before the office manager can print out a list of appointments with patient and doctor names on it, first the patient and doctor tables need to be joined with the appointment table on the basis of the patient ID and doctor ID fields. Only then can appointment, patient, and doctor information be included in the query's output. Joins can take a lot of time, especially if the tables are large or if many tables are involved.

Consider an order system that stores information about 10,000 different products, 25,000 customers, and 100,000 orders, each order averaging three products. If an analyst wanted to investigate whether there were regional differences in buying preferences, he or she would need to combine all of the tables to be able to look at products that have been ordered while knowing the location of the customers placing the orders. A query of this information would result in a huge table with 300,000 rows (i.e., the number of products that have been ordered) and many columns representing columns from all three tables combined.

There are several techniques that the project team can use to try to speed up access to the data: denormalization, clustering, indexing, and estimating the size of the data for hardware planning purposes.

Denormalization   After the logical data model is optimized in terms of data storage, the project team may decide to denormalize, or add redundancy back into the design that is depicted in the physical data model. *Denormalization* reduces the number of joins that must be performed in a query, thus speeding up data access.

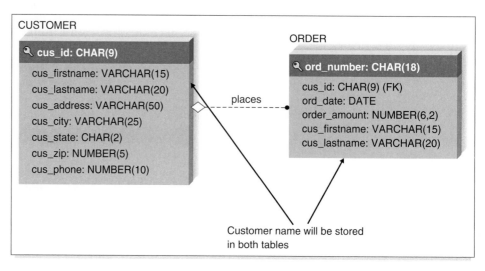

FIGURE 11-11
Denormalized Data Model

Figure 11-11 shows a denormalized physical data model for customer orders. The customer last name was added back into the Order table because the project team learned during the analysis phase that queries about orders usually require the customer last name field. Instead of joining the Order table repeatedly to the Customer table, the system now needs to access only the Order table because it contains all the relevant information.

Of course, denormalization should be applied sparingly for the reasons described in the previous section, but it is ideal in situations in which information is queried frequently yet updated rarely. There are four cases in which you may rely on denormalization to reduce joins and improve performance. First, denormalization can be applied in the case of *look-up tables*, which are tables that contain descriptions of values (e.g., a table of product descriptions, a table of payment types). Because descriptions of codes rarely change, it may be more efficient to include the description along with its respective code in the main table to eliminate the need to join the look-up table each time a query is performed (Figure 11-12).

Second, 1 : 1 relationships are good candidates for denormalization. Although two entities logically should be separated, from a practical standpoint the information from both entities may be regularly accessed together. Think about an order and its shipping information. Logically, it may make sense to separate the attributes related to shipping into a separate entity, but as a result the queries regarding shipping likely will always need a join to the Order table. If the project team finds that certain shipping information, such as state and shipping method, are needed when orders are accessed, they may decide to combine the entities or include some shipping attributes in the order entity (Figure 11-12).

Third, at times it will be more efficient to include a parent entity's attributes in its child entity on the physical data model. For example, consider a customer table and an order table that share a 1 : N relationship, with customer as the parent and order as the child. If queries regarding orders continuously require customer information, the most popular customer fields can be placed in order to reduce the required joins to the customer table, as was done with customer last name in Figure 11-12.

Finally, denormalization is used when using a popular data modeling technique called *star schema design*[2]. Learning how to model using star schema is beyond the scope of this book, but there are a number of web resources and books available that are listed on the textbook Web site. Basically, star schema is a way to model data whereby data is denormalized to speed up data access for DSS. It uses two kinds of tables—fact tables and dimension tables—to store numerical, additive

YOUR	**12-3 Denormalizing a Student Activity File**
TURN	

Consider the logical data model that you created in Chapter 7 for "Your Turn 7-7." Examine the model and describe possible opportunities for denormalization.

Question:

How would you denormalize the physical data model, and what are the benefits of your changes?

---

[2] A good book on star schema design is that by Ralph Kimball, *The Data Warehouse Toolkit,* New York: John Wiley & Sons, 1996.

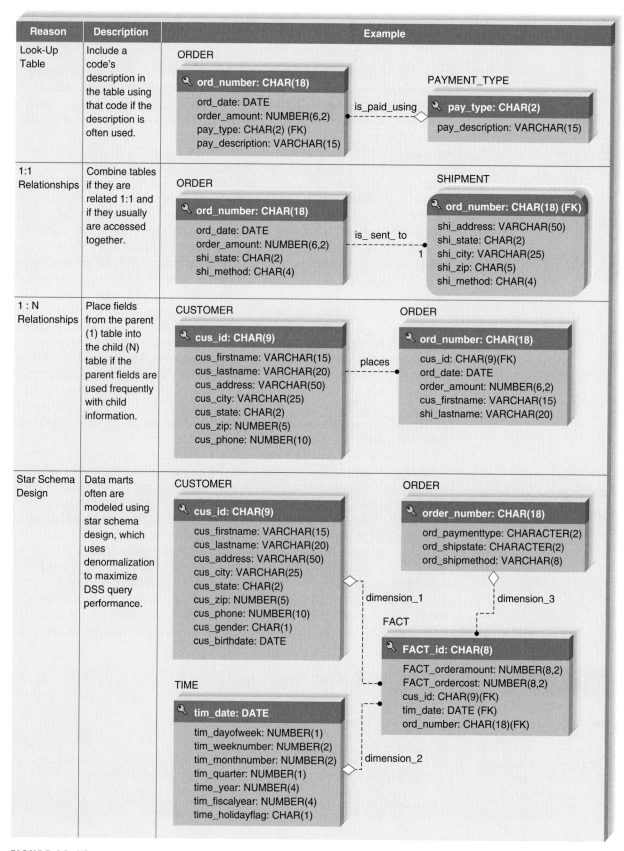

FIGURE 11-12
Reasons to Denormalize

data and descriptive data, respectively. Star schema modeling is the way in which relational databases can be designed to emulate a multidimensional database. See Figure 11-12 for an example of a star schema design of a customer order database. The fact table contains order amount and cost (i.e., additive data), and the dimension tables contain information describing different dimensions of an order: the customer, the order itself, and time.

Clustering   Speed of access also is influenced by the way in which the data is retrieved. Think about going shopping in a grocery store. If you have a list of items to buy but you are unfamiliar with the store's layout, then you need to walk down every aisle to make sure that you don't miss anything from your list. Likewise, if records are arranged on a hard disk in no particular order (or in an order that is irrelevant to your data needs), then any query of the records results in a *table scan* in which the DBMS has to access every row in the table before retrieving the result set. Table scans are the most inefficient of data retrieval methods.

One way to improve access speed is to reduce the number of times that the storage medium must be accessed during a transaction. One method is to *cluster* records together physically so that like records are stored close together. With *intrafile clustering*, similar records in the table are stored together in some way, such as in order by primary key or, in the case of a grocery store, by item type. Thus, whenever a query looks for records, it can go directly to the right spot on the hard disk (or other storage medium) because it knows in what order the records are stored, just as we can walk directly to the bread aisle to pick up a loaf of bread. *Interfile clustering* combines records from more than one table that typically are retrieved together. For example, if customer information is usually accessed with the related order information, then the records from the two tables may be physically stored in a way that preserves the customer order relationship. Returning to the grocery store scenario, an interfile cluster would be similar to storing peanut butter, jelly, and bread next to each other in the same aisle because they are usually purchased together, not because they are similar types of items. Of course, each table can have only one clustering strategy because the records can be arranged physically in only one way.

Indexing   A time saver that you are familiar with is an index located in the back of a textbook, which points you directly to the page or pages that contain your topic of interest. Think of how long it would take you to find all of the times that *relational database* appears in this textbook if you didn't have the index to rely on. An *index* in data storage is like an index in the back of a textbook; it is a minitable that contains values from one or more columns in a table and the location of the values within the table. Instead of paging through the entire textbook, you can move directly to the right pages and get the information you need. Indexes are one of the most important ways to improve database performance. Whenever you have performance problems, the first place to look is an index.

A query can use an index to find the locations of only those records that are included in the query answer, and a table can have an unlimited number of indexes. Figure 11-13 shows an index that orders records by payment type. A query that searches for all of the customers who used American Express can use this index to find the locations of the records that contain American Express as the payment type without having to scan the entire order table.

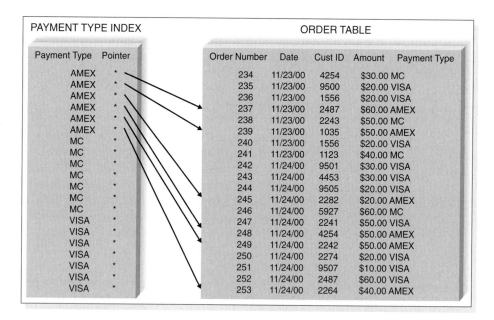

FIGURE 11-13
Payment Type Index

Project teams can make indexes perform even faster by placing them into the main memory of the data storage hardware. Retrieving information from memory is much faster than from another storage medium like a hard disk—think about how much faster it is to retrieve a phone number that you have memorized versus one that you need to look up in a phone book. Similarly, when a database has an index in memory, it can locate records very, very quickly.

Of course, indexes require overhead in that they take up space on the storage medium. Also, they need to be updated as records in tables are inserted, deleted, or changed. Thus, although queries lead to faster access to the data, they slow down the update process. In general, you should create indexes sparingly for transaction systems or systems that require a lot of updates, but apply indexes generously when designing systems for decision support (Figure 11-14).

Usually CASE tools allow you to define indexes and clustering strategies within the design of the physical data model. Figure 11-15 shows the index screen

---

CONCEPTS

IN ACTION

11-A MAIL ORDER INDEX

A Virginia-based mail-order company sends out approximately 25 million catalogs each year using a customer table with 10 million names. Although the primary key of the customer table is customer identification number, the table also contains an index of customer last names. Most people who call to place orders remember their last name but not their customer identification number, so this index is used frequently.

An employee of the company explained that indexes are critical to reasonable response times. A fairly complicated query was written to locate customers by the state in which they lived, and it took over 3 weeks to return an answer. A customer state index was created, and that same query provided a response in 20 minutes: that's 1,512 times faster!

QUESTION:

As an analyst, how can you make sure that the proper indexes have been put in place so that users are not waiting for weeks to receive the answers to their questions?

- Use indexes sparingly for transaction systems.
- Use many indexes to improve response times in decision support systems.
- For each table, create a unique index that is based on the primary key.
- For each table, create an index that is based on the foreign key to improve the performance of joins.
- Create an index for fields that are used frequently for grouping, sorting, or criteria.

FIGURE 11-14
Guidelines for Creating Indexes

in one CASE tool (ER*win*) for the order table. In this example, three indexes have been designed for the table, and during the implementation phase, the CASE tool will generate the code that is necessary to construct these indexes in the DBMS.

Estimating Storage Size

Even if you have denormalized your physical data model, clustered records, and created indexes appropriately, the system will perform poorly if the database server cannot handle its volume of data. Therefore, one last way to plan for good perfor-

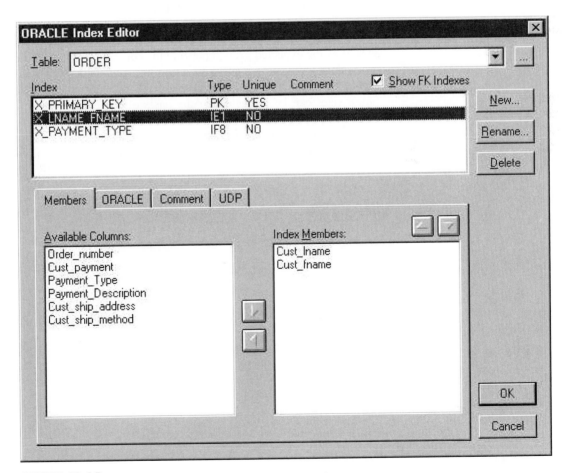

FIGURE 11-15
Indexes in ERwin

mance is to apply *volumetrics*, which means estimating the amount of data that the hardware will need to support. You can incorporate your estimates into the database server hardware specification to make sure that the database hardware is sufficient for the project's needs. The size of the database will be determined by the amount of raw data in the tables and the *overhead* requirements of the DBMS. To estimate size, you will need to have a good understanding of the initial size of your data as well as its expected growth rate over time.

Raw data refers to all the data that are stored within the tables of the database, and it is calculated based on a bottom-up approach. First, write down the estimated average width for each column (field) in the table and sum the values for a total record size (Figure 11-16). For example, if a variable-width last name column is assigned a width of 20 characters, you can enter 13 as the average character width of the column. In Figure 11-16, the estimated record size is 49.

Next, calculate the overhead for the table as a percentage of each record. Overhead includes the room needed by the DBMS to support such functions as administrative actions and indexes, and it should be assigned on the basis of past experience, recommendations from technology vendors, or parameters that are built into software that was written to calculate volumetrics. For example, your DBMS vendor may recommend that you allocate 30% of the records' raw data size for overhead storage space, creating a total record size of 63.7 characters in the Figure 11-16 example.

Finally, record the number of initial records that will be loaded into the table, as well as the expected growth per month. This information should have been collected during the analysis phase as a nonfunctional data requirement. According to Figure 11-16, the initial space required by the first table is 3,185,000 characters, and future sizes can be projected on the basis of the growth figure. These steps are repeated for every table to get a total size for the entire database.

Many CASE tools will provide you with database size information based how you set up the physical data model, and they will calculate volumetrics estimates automatically. Figure 11-17 shows a volumetrics screen for ER*win*.

Field	Average Size (Characters)
Order number	8
Date	7
Cust ID	4
Last name	13
First name	9
State	2
Amount	4
Tax rate	2
Record size	49
Overhead	30%
Total record size	63.7
Initial table size	50,000
Initial table volume	3,185,000
Growth rate/month	1,000
Table volume @ 3 years	5,478,200

FIGURE 11-16
Calculating Volumetrics

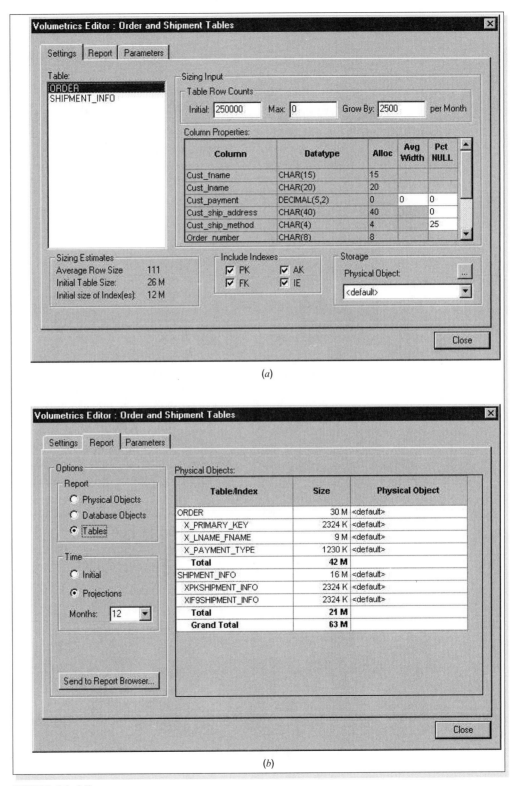

FIGURE 11-17

Volumetrics Screen in ERwin: (a) Information about Columns and Rows Is Entered into the ERwin; (b) Report Is Generated or the Basis of the Information.

Ultimately, the size of the database needs to be shared with the design team so that the proper technology can be put in place to support the system's data and potential performance problems can be addressed long before they affect the success of the system.

APPLYING THE CONCEPTS AT CD SELECTIONS

The CD Selections Internet system needs to effectively present CD information to users and capture order data. Alec Adams, Senior Systems Analyst and Project Manager for the sales system, recognized that these goals were dependent on a good design of the data storage component for the new application. He approached data storage design in two steps: by first selecting the data storage format and then later optimizing it for processing efficiency.

Data Format Selection

The project team met to discuss two issues that would drive the data storage format selection: what kind of data would be in the system and how that data would be used by the application system. Using a white board, they listed the ideas presented in Figure 11-18. The project team agreed that the bulk of the data in the system would be the text and numbers that are exchanged with Web users regarding customers and orders. A relational database would be able to handle the data effectively, and the technology would be well received at CD Selections because relational technology is already in place throughout the organization.

Data	Type	Use	Suggested Format
Customer information	Simple (mostly text)	Transactions	Relational
Order information	Simple (text and numbers)	Transactions	Relational
Marketing information	Both simple and complex (eventually the system will contain audio clips, video, etc.)	Transactions	Object add-on?
Information that will be exchanged with the distribution system	Simple text, formatted specifically for importing into the distribution system	Transactions	Transaction file
Temporary information	The Web component will likely need to hold information for temporary periods of time (e.g., the shopping cart will store order information before the order is actually placed)	Transactions	Transaction file

FIGURE 11-18

Types of Data in Internet System

However, they realized that relational technology was not optimized to handle complex data, such as the images, sound, and video that the marketing facet of the system ultimately will require. Alec asked Brian, one of the staff members on the Internet system project, to look into relational databases that offered object add-on products. It might be possible for the team to invest in a relational database foundation and then buy functionality to handle the complex data at a later date.

The team noted that it also must design two transaction files to handle the interface with the distribution system and the Web shopping cart program. The Internet system would regularly download order information to the special order system using a transaction file containing all the required information for that system. Also, the team must design the file that stored temporary order information on the Web server as customers navigated through the Web site. The file would contain the fields that ultimately would be transferred to an order record.

Of course, Alec realized that other data needs would arise over time, but he felt confident that the major data issues were identified (e.g., like the capability to handle complex data) and that the data storage design would be based on the proper storage technologies.

Data Storage Optimization

After the meeting, Alec asked Brian to stay behind to discuss the logical data model. Now that the team members had a good idea of the type of data storage formats that would be used, they were ready for the second step of data design: optimizing the data for performance efficiency. Brian was the analyst in charge of the logical data model, and Alec wanted to be sure that the data model was optimized for storage efficiency before the team discussed access speed issues.

Brian assured Alec that the current data model was in third normal form. He was confident of this because the project team followed the data modeling guidelines that led to a well-formed logical model. Of course, a week or so earlier, he did apply the three rules of normalization to the data model as a check to make sure that no design errors were overlooked.

Brian then asked about the file formats for the two transaction files identified in the earlier meeting. Alec suggested that he normalize the files to better understand the various tables that would be involved in the import procedure.

The last step of data storage design was to optimize the design for data access speed. Alec met with the analysts on the data storage design team and talked about the techniques that were available to speed up access to data in the system. Together the team listed all the data that would be supported by the Internet system and discussed how all the data would be used. They developed the strategy laid out in Figure 11-19 to identify the specific techniques to put in place.

Data Sizing

Ultimately, clustering strategies, indexes, and denormalization decisions were applied to the physical data model, and a volumetrics report was run from the CASE tool to estimate the initial and projected size of the database. The report suggested that an initial data storage capacity of about 450 megabytes would be needed for the expected 1-year life of the first version of the system. Additional storage capacity would be needed for the second version, which would include sound files for samples of the songs, but for the moment not much data storage would be needed.

Target	Comments	Suggestions to Improve Data Access Speed
All tables	Basic table manipulation	Investigate if records should be clustered physically by primary key Create indexes for primary keys Create indexes for foreign key fields
All tables	Sorts and grouping	Create indexes for fields that are frequently sorted or grouped
CD information	Users will need to search CD information by title, artist, and category	Create indexes for CD title, artist, and category
Entire physical model	Investigate denormalization opportunities for all fields that are not updated very often	Investigate 1:1 relationships Investigate look-up tables Investigate 1:M relationships

FIGURE 11-19
Internet System Performance

Alec gave the estimates to the analyst in charge of managing the server hardware acquisition so that the person could make sure that the technology could handle the expected volume of data for the Internet system. The estimates would also go to the DBMS vendor during the implementation of the software so that the DBMS could be configured properly.

SUMMARY

File Data Storage Formats
There are two basic types of data storage formats: files and databases. Files are electronic lists of data that have been optimized to perform a particular transaction, and there are five different types: master, look-up, transaction, audit, and history. Master files typically are kept for long periods of time because they store important business information, such as order information or customer mailing information. Look-up files contain static values that are used to validate fields in the master files, and transaction files temporarily hold information that will be used for a master file update. An audit file records "before" and "after" images of data as they are altered so that an audit can be performed if the integrity of the data is questioned. Finally, the history file stores past transactions (e.g., old customers, past orders) that are no longer needed by the system.

Database Data Storage Formats
A database is a collection of groupings of information that are related to each other in some way, and a DBMS (database management system) is software that creates and manipulates these databases. There are four types of databases that are likely to be encountered during a project: legacy, relational, object, and multidimensional. The legacy databases (e.g., hierarchical databases and network databases) use older, sometimes outdated technology and are rarely used to develop new applications. The relational database is the most popular kind of database for application development today, and it is based on collections of tables that are related to each other through common fields, known as foreign keys. Object databases contain data and processes that are represented by object classes, and relationships between object

classes are shown by encapsulating one object class within another and are mainly used in multimedia applications (e.g., graphics, video, and sound). One of the newest members in the database arena is the multidimensional database, which has become popular with the increase in data warehousing. It stores precalculated quantitative information (e.g., totals, averages) at the intersection of dimensions (e.g., time, salesperson, product) to support applications that require data to be sliced and diced.

Selecting a Data Storage Format

The application's data should drive the storage format decision. Relational databases support simple data types very effectively, whereas object databases are best for complex data. Multidimensional databases are tuned to store aggregated, quantitative information. The type of system also should be considered when choosing among data storage formats. Relational databases have matured to support transactional systems, whereas multidimensional databases have been designed to perform best in decision support environments. Although less critical to the format selection decision, the project team needs to consider the kind of technology that exists within the organization and the kind of technology likely to be used in the future.

Optimizing Data Storage

There are two primary dimensions in which to optimize a relational database: for storage efficiency and for speed of access. The most efficient relational database tables in terms of data storage are those that have no redundant data and very few null values. Normalization is the process whereby a series of rules are applied to the logical data model to determine how well optimized it is for storage efficiency.

Once you have optimized your logical data design for storage efficiency, the data may be spread out across a number of tables. To improve speed, the project team may decide to denormalize—or add redundancy back into—the design that is depicted in the physical data model. Denormalization reduces the number of joins that must be performed in a query, thus speeding up data access. Denormalization is best in situations in which data are accessed frequently and updated rarely. There are three modeling situations that are good candidates for denormalization: look-up tables, entities that share one-to-one (1 : 1) relationships, and entities that share one-to-many (1 : M) relationships. In all three cases, attributes from one entity are moved or repeated in another entity to reduce the joins that must occur during data access.

Clustering occurs when similar records are stored close together on the storage medium to speed up data retrieval. In intrafile clustering, similar records in the table are stored together in some way, such as in sequence. Interfile clustering combines records from more than one table that typically are retrieved together. Indexes also can be created to improve the access speed of a system. An index is a minitable that contains values from one or more columns in a table and information about where the values can be found. Instead of performing a table scan, which is the most inefficient way to retrieve data from a table, an index points directly to the records that match the requirements of a query.

Finally, the speed of the system can be improved if the right hardware is purchased to support its data. Analysts can use volumetrics to estimate the current and future size of data in the database and then share these numbers with the people who are responsible for buying and configuring the database hardware.

KEY TERMS

Aggregated	History file	Object database
Audit file	Hybrid object-oriented DBMS	Object-oriented DBMS
Cluster	Index	Overhead
Data mart	Instantiation	Primary key
Data warehousing	Interfile cluster	Pointer
Database	Intrafile cluster	Raw data
Database management system	Legacy database	Referential integrity
(DBMS)	Linked list	Relational database
Decision support system (DSS)	Look-up file	Set
Denormalization	Look-up table	Star schema design
Encapsulation	Master file	Structured Query Language (SQL)
End-user DBMS	Member	Subclass
Enterprise DBMS	Multidimensional database	Table scan
File	Network database	Transaction file
Foreign key	Normalization	Transaction processing system
Hierarchical database	Object class	Volumetrics

QUESTIONS

1. Describe the two steps to data storage design.
2. How are a file and a database different from each other?
3. What is the difference between an end-user database and an enterprise database? Provide an example of each one.
4. Name five types of files, and describe the primary purpose of each type.
5. Name two types of legacy databases and the main problems associated with each type.
6. What is the most popular kind of database today? Provide three examples of products that are based on this technology.
7. What is referential integrity, and how is it implemented in a relational database?
8. What is the biggest strength of the object database? Describe two of its weaknesses.
9. How does the multidimensional database store data?
10. What are the two most important factors in determining the type of data storage format that should be adopted for a system? Why are these factors so important?
11. Why should you consider the storage formats that already exist in an organization when deciding on a storage format for a new system?
12. Name three ways that null values in a database can be interpreted. Why is this problematic?
13. What are the two dimensions in which to optimize a relational database?
14. What is the purpose of normalization?
15. Describe three situations that can be good candidates for denormalization.
16. Describe several techniques that can improve performance of a database.
17. What is the difference between interfile and intrafile clustering? Why are they used?
18. What is an index, and how can it improve the performance of a system?
19. Describe what should be considered when estimating the size of a database.
20. Why is it important to understand the initial and projected size of a database during the design phase?
21. What are the key issues in deciding between using perfectly normalized databases and denormalized databases?

EXERCISES

A. Using the Web or other resources, identify a product that can be classified as an end-user database and a product that can be classified as an enterprise database. How are the products described and marketed? What kinds of applications and users do they support? In what kinds of situations would an organization choose to implement an end-user database over an enterprise database?

B. Visit a commercial Web site (e.g., CDnow, Amazon.com). If files were being used to store the data supporting the application, what types of files would be needed? What data would they contain?

C. Using the Web, review one of the products listed below. What are the main features and functions of the software? In what companies has the database management system (DBMS) been implemented, and for what purposes? According to the information that you found, what are three strengths and weaknesses of the product?

- Relational DBMS
- Object DBMS
- Multidimensional DBMS

D. You have been given a file that contains the following fields relating to CD information. Using the steps of normalization, create a logical data model that represents this file in third normal form. The fields include the following:

- Musical group name
- Musicians in group
- Date group was formed
- Group's agent
- CD title 1
- CD title 2
- CD title 3
- CD 1 length
- CD 2 length
- CD 3 length

The assumptions are as follows:

- Musicians in group contains a list of the members of the people in the musical group.
- Musical groups can have more than one CD, so both group name and CD title are needed to uniquely identify a particular CD.

E. Jim Smith's dealership sells Fords, Hondas, and Toyotas. The dealership keeps information about each car manufacturer with whom it deals so that employees can get in touch with manufacturers easily. The dealership staff also keeps information about the models of cars that the dealership carries from each manufacturer. They keep such information as list price, the price the dealership paid to obtain the model, and the model name and series (e.g., Honda Civic LX). They also keep information about all sales that they have made (for instance, they will record the buyer's name, the car he or she bought, and the amount he or she paid for the car). So that staff can contact the buyers in the future, contact information is also kept (e.g., address, phone number). Create a logical data model. (You may have done this already in Chapter 7.) Apply the rules of normalization to the model to check the model for processing efficiency.

F. Describe how you would denormalize the model that you created in question E. Draw the new physical model on the basis of your suggested changes. How would performance be affected by your suggestions?

G. Examine the physical data model that you created in question F. Develop a clustering and indexing strategy for this model. Describe how your strategy will improve the performance of the database.

H. Investigate the volumetric interface with the computer-aided software engineering (CASE) tool that you are using for class. What information do you as an analyst need to input into the tool? How are size estimates calculated? If your CASE tool does not accept volumetric information, how can you calculate the size of the database?

I. Calculate the size of the database that you created in question F. Provide size estimates for the initial size of the database as well as for the database in one year's time. Assume that the dealership sells 10 models of cars from each manufacturer to approximately 20,000 customers a year. The system will be set up initially with one year's worth of data.

MINICASES

1. In the new system under development for Holiday Travel Vehicles, seven tables will be implemented in the new relational database. These tables are: New Vehicle, Trade-in Vehicle, Sales Invoice, Customer, Salesperson, Installed Option, and Option. The expected average record size for these tables and the initial record count per table are given next.

Table Name	Average Record Size	Initial Table Size (records)
New Vehicle	65 characters	10,000
Trade-in Vehicle	48 characters	7,500
Sales Invoice	76 characters	16,000
Customer	61 characters	13,000
Salesperson	34 characters	100
Installed Option	16 characters	25,000
Option	28 characters	500

Perform a volumetrics analysis for the Holiday Travel Vehicles system. Assume that the DBMS that will be used to implement the system requires 35% overhead to be factored into the estimates. Also, assume a growth rate for the company of 10% per year. The systems development team wants to ensure that adequate hardware is obtained for the next 3 years.

DESIGN

- ☑ **Select Design**
- ☑ **Develop Physical Data Flow Diagrams**
- ☑ **Develop Physical Entity Relationship Diagrams**
- ☑ **Design Architecture**
- ☑ **Select Hardware and Software**
- ☑ **Develop Use Scenarios**
- ☑ **Design Interface Structure**
- ☑ **Design Interface Standards**
- ☑ **Design Interface Prototype**
- ☑ **Evaluate User Interface**
- ☑ **Design User Interface**
- ☑ **Select Data Storage Format**
- ☑ **Denormalize Entity Relationship Diagram**
- ☑ **Performance Tune Data Storage**
- ☑ **Size Data Storage**
- ☐ Develop Program Structure Chart
- ☐ Develop Program Specification

TASK CHECKLIST

CHAPTER 12

PROGRAM

DESIGN

A nother important step of the design phase is designing the programs that will perform the system's application logic. Programs can be quite complex, so analysts must create instructions and guidelines for programmers that clearly describe what the program must do. This chapter presents two techniques for describing programs that are typically used together. The first, the structure chart, depicts a program at a high level in graphic form. The second, the program specification, is a set of written instructions at a lower level of detail. Together these techniques communicate how the application logic for the system needs to be coded.

OBJECTIVES

- Be able to create a structure chart.
- Be able to write a program specification.
- Understand the use of pseudocode.
- Become familiar with event-driven programming.

CHAPTER OUTLINE

IMPLEMENTATION

INTRODUCTION

Program design is the part of the design phase of the SDLC during which analysts create instructions for the programmers about how code should be written and how pieces of code should fit together to form a program. Some people may think that program design is becoming less important as project teams rely increasingly on packaged software or libraries of preprogrammed code to build systems. However, program design techniques are still very important for two reasons. First, even pre-existing code needs to be understood, organized, and pieced together. Second, it is still common for the project team to have to write some code (if not all) and produce original programs that support the application logic of the system.

It can be tempting to jump right into the implementation phase by coding without much thought or planning, but this can lead to disastrous results, such as inefficient programs, code that does not work with other code, and a system that doesn't do what it's supposed to do. Instead analysts should first take time in the design phase to create a maintainable system. In other words, analysts should create a design that is modular and flexible. To do this, analysts can design programs in a *top-down, modular approach*, using a variety of program design techniques.

Think about giving someone directions to your house (Figure 12-1). Before getting to the details, such as naming streets and identifying landmarks, it is best to first orient the person to your general location (e.g., the state you live in, the part of town). As he or she becomes comfortable with where to go at a high level, you can become more detailed in your instructions. This top-down approach helps orient the other person and conveys the big picture of where you live, making the detailed directions much easier to understand.

Also, directions can be communicated in *modules*:

First drive from your house to the highway.
Then drive from the highway to the appropriate exit.
Next, locate my neighborhood.
Finally, drive to my house.

Each line, or module, can change without affecting the rest of the directions. For example, if one friend is traveling to your house from the north and another is traveling from the south, it is likely that the last two modules of directions (i.e., to the neighborhood and to the house) will not change even though the first two modules will differ for each friend. The modular approach makes the directions much easier to develop and change.

Good program design is similar to the top-down modular approach that we described. First, analysts create a high-level diagram that shows the various components of a program, how the components should be organized, and how the components interrelate. This diagram, known as the structure chart, illustrates the organization and interactions of the different pieces of code within the program to the analysts and programmers so that the program can be developed by many programmers working independently. The diagram can be used when the project team plans to write code from scratch or when existing pieces of code will be assembled to build the system. Process models provide a good starting point for understanding what this structure chart needs to include.

Once the overall program is defined at a high level using a structure chart, program specifications are written to describe exactly what needs to be included in

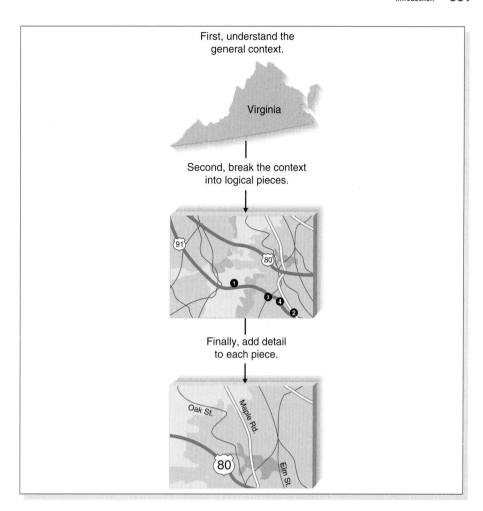

First, understand the
general context.

Virginia

Second, break the context
into logical pieces.

Finally, add detail
to each piece.

Oak St.

Maple Rd.

Elm St.

80

FIGURE 12-1
Using a Top-Down Modular Approach

each program module. The specifications include basic module information (e.g., a name, calculations that need to be performed, and the target programming language), special instructions for the programmer, and pseudocode. Pseudocode is a technique similar to structured English that is used to communicate what needs to be written using programming structures and a generic language that is not program language specific. Program specifications leave the implementation details to the programmers, but they communicate the basic logic and programming structures to help reduce logical and syntactical errors during the implementation phase. Some RAD approaches deemphasize program specifications.

You will notice as you read through this chapter that the design techniques that are described are based on information and techniques from earlier phases of the SDLC. For example, the components of the structure chart typically mirror the processes found on the data flow diagrams, and the process descriptions suggest the ways in which structure chart components should interrelate. Data models are used to understand the data that pass throughout the diagram. Also, analysts use the techniques and information to develop the program specifications, especially when writing pseudocode. Often during design the analysts detect problems or inconsistencies with the analysis deliverables, and they must fine-tune or clarify previous work as they move forward.

12-A WINNING BY DESIGN

The rapid development face-off was a competition between rapid application development (RAD) teams from the leading consulting firms in the United States. The goal was to see which team could develop a specific system in the least amount of time. Most teams used a very short program design step and quickly began programming.

The Ernst & Young (E&Y) team members used a different approach. They spent much more time in the pro-

gram design step to ensure the system was well designed before they moved into programming. At first, the E&Y team fell behind while its competitors jumped ahead. But E&Y ended up winning because the team spent much less time programming by following its well-designed blueprint.

QUESTION:

What are several reasons why planning ahead may have helped E&Y win?

In recent years, programmers have moved away increasingly from structured programming languages and have migrated to event-driven programming, which is described later in this chapter; however, this trend has not reduced the value of the structured design approaches that we just described. The point is that planning before doing almost always improves the ultimate process when it comes to programming, and analysts should never begin writing code without having a complete understanding of what code must do. Unfortunately, our experiences suggest that many project teams are much too quick at jumping into writing program code without first organizing and defining the basic program modules and how they interact.

At the end of program design, the project team compiles the *program design* document, which includes all of the structure charts and program specifications that will be used to implement the system. The program design is used by programmers to write code. This chapter first describes the structure chart, a helpful tool that illustrates the overall organization of a program. Then it presents the program specification, which contains detailed information about each module of code. Much of the information in this chapter is based on a book written by Meiler Page-Jones,[1] which we highly recommend you read if you are interested in additional information on structured program design.

STRUCTURE CHART

The *structure chart* is an important technique that helps the analyst design the program for the new system. The structure chart shows all the components of code that must be included in a program at a high level, arranged in a hierarchical format that implies *sequence* (in what order components are invoked), *selection* (under what condition a module is invoked), and *iteration* (how often a component is repeated). The components are usually read from top to bottom, left to right, and they are numbered using a hierarchical numbering scheme in which lower levels have an additional level of numbering (e.g., the third level of modules would be numbered 1.1.1, 1.1.2, 1.1.3…).

---

[1] Meiler Page-Jones, *The Practical Guide to Structured Systems Design,* New York: Yourdon Press, 1980.

Structure charts historically have been used to create transaction-based mainframe applications, which have many lines of code that must be carefully monitored. They help analysts create programs that are easy to understand and maintain because the use of self-contained modules keeps changes from rippling throughout the programs. We believe structure charts can be helpful in the building of many types of systems because they emphasize structure and reusability, characteristics of any good program.

Suppose that an academic system needs a program that will print a listing of students along with their grade point averages (GPAs), both for the current semester and overall. First, the program must retrieve the student grade records, then it must calculate the current and cumulative GPAs, and finally the grade list can be printed. The structure chart shown in Figure 12-2 communicates the basic components of this program and shows the interrelatedness of the modules. For example, by looking at this structure chart, a programmer can tell that there are four main code modules involved in creating a student grade listing: getting the student grade records, calculating current GPA, calculating cumulative GPA, and printing the listing. Also, there are various pieces of information that are either required by each module or created by it (e.g., the grade record, the cumulative GPA). The following sections describe each component of the structure chart using this example.

Syntax

Module   A structure chart is composed of *modules* (lines of program code that perform a single function) that work together to form a program (Figure 12-3). The modules are depicted by a rectangle and connected by lines, which represent the passing of control. A *control module* is a higher-level component that contains the logic for performing other modules, and the components that it calls and controls are considered *subordinate modules*. For example, in Figure 12-2, module 1.0 is the control module that directs modules 1.1 through 1.4 as its subordinates.

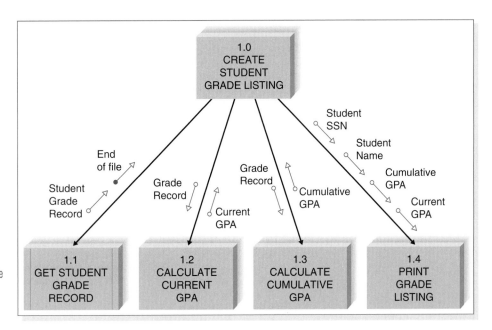

FIGURE 12-2
Structure Chart Example (GPA = grade point average; SSN = Social Security Number)

Structure Chart Element	Purpose	Symbol
Every *module*: • Has a number • Has a name • Is a control module if it calls other modules below it • Is a subordinate module if it is controlled by a module at a higher level	Denotes a logical piece of the program	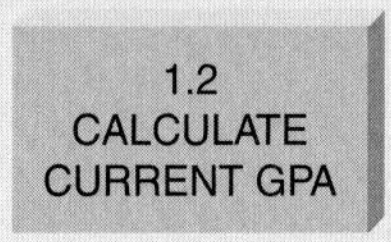 1.2 CALCULATE CURRENT GPA
Every *library module* has: • A number • A name • Multiple instances within a diagram	Denotes a logical piece of the program that is repeated within the structure chart	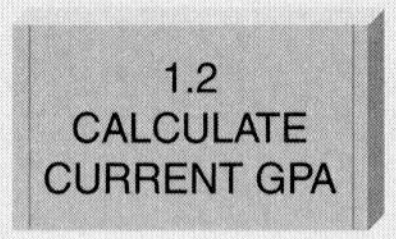 1.2 CALCULATE CURRENT GPA
A *loop*: • Is drawn using a curved arrow • Is placed around lines of one or more modules that are repeated	Communicates that a module(s) is repeated	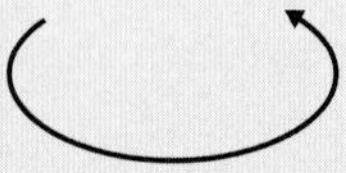
A *conditional line*: • Is drawn using a diamond • includes modules that are invoked based on some condition	Communicates that subordinate modules are invoked by the control module based on some condition	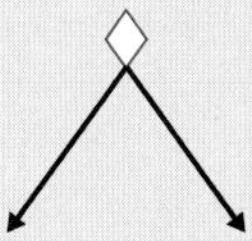
A *data couple*: • Contains an arrow • Contains an empty circle • Names the type of data that is being passed • Can be passed up or down • Has a direction that is denoted by the arrow	Communicates that data is being passed from one module to another	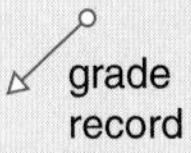 grade record
A *control couple*: • Contains an arrow • Contains a filled-in circle • Names the message or flag that is being passed • Should be passed up, not down • Has a direction that is denoted by the arrow	Communicates that a message or a system flag is being passed from one module to another	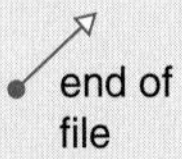 end of file
An *off-page connector*: • Is denoted by the hexagon • Has a title • Is used when the diagram is too large to fit everything on the same page	Identifies when parts of the diagram are continued on another page of the structure chart	 PRINT GRADE LISTING
An *on-page connector*: • Is denoted by the circle • Has a title • Is used when the diagram is too large to fit everything in the same spot on a page	Identifies when parts of the diagram are continued somewhere else on the same page of the structure chart	 PRINT GRADE LISTING

FIGURE 12-3
Structure Chart Elements

At times, modules are reused. These modules, called *library modules*, have vertical lines on both sides of the rectangle to communicate that they will appear several times on the structure chart (see Figure 12-3). The library module in Figure 12-2 is module 1.1, get student grade record, and this module is a generic module that will be depicted several times in other parts of the diagram. Library modules are highly encouraged because their reusability can save programmers from rewriting the same piece of code over and over again.

The lines that connect the modules communicate the passing of control. In Figure 12-2, the control is linear, whereby all of the modules are performed in order from top to bottom, left to right. There are two symbols that describe special types of control that can appear on the structure chart. The curved arrow, or *loop*, indicates that the execution of some or all subordinate modules is repeated, and a *conditional line* (depicted by a diamond) denotes that execution of one or more of the subordinate modules occurs in some cases but not in others (see Figure 12-3).

Look at the structure chart in Figure 12-4 and see how the loops and conditional lines affect the meaning of the diagram. First, the loop through the lines to modules 1.1, 1.2, and 1.3 means that before the next two modules are invoked, the first three modules will be repeated until their functionality is completed (i.e., all of the student grades will be read and the two GPAs will calculated before moving to the print modules). Second, the lines connected by the conditional line convey that both the dean's list report and grade listing are not printed each time this program is run but instead are performed based upon some condition. Therefore, there are times when one or both of the print modules may not be invoked.

Another new symbol found on the structure chart in Figure 12-4 is the *connector* (see Figure 12-3). Structure charts can become quite unwieldy, especially when they depict a large or complex program. A circle is used to connect parts of the structure chart when there are space constraints and a diagram needs to be continued on another part of the page (i.e., an *on-page connector*), and a hexagon is used to continue the diagram on another page entirely (i.e., an *off-page connector*). In Figure 12-4, notice that modules 1.4 and 1.5 are depicted on another page of the diagram.

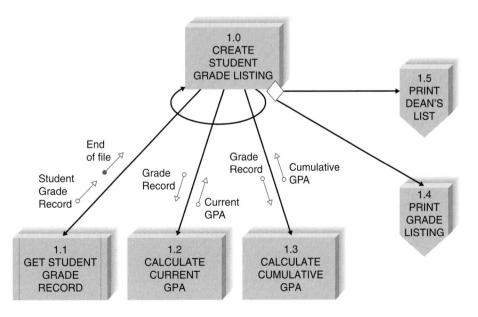

FIGURE 12-4
Revised Structure Chart Example

Couples   *Couples*, shown using arrows, are drawn on the structure chart to show that information is passed between modules, with the arrowhead indicating which way the information is being sent (see Figure 12-3). *Data couples*, shown using arrows with empty circles, are used to represent the passing of pieces of data or data structures to other modules. For example, in Figure 12-4, a student grade record must be sent to module 1.2 for the GPA to be calculated, so a data couple is used to show the grade data structure being passed along.

Control couples, drawn using arrows with filled-in circles, are used to pass parameters or system-related messages back and forth among modules. If some type of parameter needed to be passed (e.g., the customer is a new customer; the end of a file has been reached), a control couple (also called a *flag*) would be used. In Figure 12-4, module 1.1 sends an end-of-file parameter when the program reaches the end of the student grade file.

In general, control flags should be passed from subordinates to control modules but not the other way around. Control flags are passed so that the control modules can make decisions about how the program will operate (e.g., module 1.1 passes the end-of-file marker to indicate that all records have been processed). Passing a control flag from higher to lower modules suggests that a lower-level module has control over the higher-level module.

The presence of couples signals that modules on the structure chart depend on each other in some way. A general rule is to be very conservative when applying couples to your diagram. In a later section, we will discuss style guidelines for couples to help you determine "good" from "bad" coupling situations.

Building the Structure Chart

Now that you understand the individual components of the structure chart, the next step is to learn how to put them together to form an effective design for the new system. Many times, process models are used as the starting point for structure charts. There are three basic kinds of processes on a process model: afferent, central, and efferent. *Afferent processes* are processes that provide inputs into the system, *central processes* perform critical functions in the operation of the system, and *efferent processes* deal with system outputs. Identify these three kinds of processes in Figure 12-5.

Each process of a DFD tends to represent one module on the structure chart, and if leveled DFDs are used, then each DFD level tends to correspond to a different level of the structure chart hierarchy (e.g., the process on the context-level DFD would correspond to the top module on the structure chart).

The difficulty comes when determining how the components on the structure chart should be organized. As we mentioned earlier, the structure chart communicates sequence, selection, and iteration, but none of these concepts is depicted explicitly in the process models. It is up to the analyst to make assumptions from the DFDs and read the process model descriptions to really understand how the structure chart should be drawn.

Transaction Structure   Luckily, there are two basic arrangements, or structures, for combining structure chart modules. The first arrangement is used when modules each perform one of a group of individual transactions. This *transaction structure* contains a control module that calls subordinate modules, each of which handles a particular transaction. Pretend that Figure 12-6 illustrates the highest

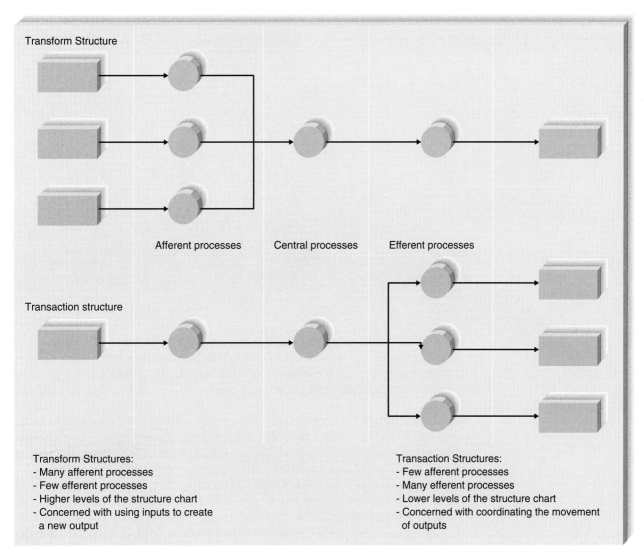

Transform Structure

Afferent processes   Central processes   Efferent processes

Transaction structure

Transform Structures:
- Many afferent processes
- Few efferent processes
- Higher levels of the structure chart
- Concerned with using inputs to create
  a new output

Transaction Structures:
- Few afferent processes
- Many efferent processes
- Lower levels of the structure chart
- Concerned with coordinating the movement
  of outputs

FIGURE 12-5
Transform and Transaction Structures

level of a student grade system. Module 1 is the control module that accepts a user's selection for what activity needs to be performed (e.g., maintain grade), and, depending on the choice, one of the subordinate modules (1.1 through 1.4) is invoked. Transaction structures often occur where the actual system contains menus or submenus, and they are usually found higher up in the levels of a structure chart.

If the project team has used leveled DFDs to illustrate the processes for the system, the high levels of the DFD usually represent activities that belong in a transaction structure. In the current example, student grade system could correspond to the single process on the context-level DFD, and the four modules (1.1 through 1.4) would be the four processes on the level 0 diagram. If a leveled DFD approach is not used, then it may be a bit more difficult to differentiate a control module from its subordinates using the process model. One hint is to look for points

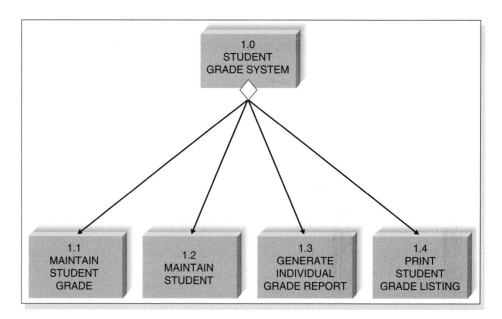

FIGURE 12-6
Transaction Structure

on the DFD in which a single data flow enters a process that produces multiple data flows as output—this usually indicates a transaction structure. See Figure 12-4 for an example of a process model that has transaction structure; notice how it contains many efferent processes and few afferent processes.

Transform Structure A second type of module structure, called a *transform structure*, has a control module that calls several subordinate modules in sequence, after which something "happens." These modules are related because together they form a process that transforms some input into an output. Often, each module accepts an input from the module proceeding it, works on the input, then passes it to the next module for more processing. For example, Figure 12-4 shows a control module that calls five subordinates. The control module describes what the subordinates will do (e.g., create student grade listing), and the subordinates are invoked from left to right and transform the student grade records into two types of listings for student grades.

In a leveled DFD, the lowest levels usually represent transform structures. If a leveled DFD approach is not used, then you should look for the processes on the DFD for which an input is changed into an output of a different form. In this situation, the process in which the change is made likely will become a control module. All the processes leading up to the control module are subordinates that are performed first by the control module, followed by the processes that come after the control module. See Figure 12-4 for an example of transform structure; notice how there are many afferent processes and few efferent processes.

Applying the Concepts at CD Selections

Now that you are familiar with the basic components of the structure chart, the best way to learn how to build the diagram is to walk through an example that shows how to create one. Creating a structure chart is usually a four-step process. First, the analyst identifies the top-level modules and then decomposes them into lower levels. (This process is similar in some ways to identifying high-level processes in a DFD and then decomposing them into lower-level processes.) Second, the analyst adds the control connections among modules, such as loops and conditional lines that show when modules call subordinates. Third, he or she adds couples, the information that modules pass among themselves. Finally, the analyst reviews the structure chart and revises it again and again until it is complete.

The goal for this example is to create a structure chart that contains the modules of code that need to be programmed and shows how they need to be organized. The process model from the analysis phase can be used as its starting point. Although it may neither map exactly into the future program nor contain enough levels of detail, the DFD will form a good rough draft structure chart that can then be changed and improved. The requirements definition and use cases will provide additional detail. Let's walk through a structure chart example for CD Selections.

Step 1: Identify Modules and Levels   First, identify the modules that belong on the diagram by converting the DFD processes into structure chart modules. Modules should perform only one function, so if for some reason a process contains more than one function, it should be broken into more than one module.

The various levels of the DFD generally translate into different levels of the structure chart. Look back at the DFDs that we created for CD Selections in Chapter 6 (Figures 6-16, 6-19, and 6-20). The context-level DFD (the overall system) is placed at the top of the structure chart in Figure 12-7 to represent the overall control module of the system that manages the highest level of system functions. Then, the level 0 DFD processes are placed below it as subordinates. You should recognize that this particular structure of modules is a transaction structure because the subordinates represent different functions that can be called by the control module.

This pattern continues through all the DFD levels. For example, the level 1 DFD that we created for the *take request* process is placed below the take request process control module. The subordinate modules are *find CD, provide CD information, find store,* and so on—the five processes from the *take request* process level 1 DFD. Note that this structure of modules is a transform structure because the subordinate modules are carried out in a sequence to perform the process that is represented by the control module, *take request* (Figure 12-7).

Likely, you will need to include additional levels of detail to the structure chart, until modules have enough detail so that they each perform only one function. Additional detail for the structure can be found within the use cases (Chapter 5) and requirements definition (Chapter 4) for the system. For example, if you read the use case for the *take request* process in Figure 5-5, notice that step 1 includes searching for CDs based on title, artist, and category. Step 5 of the use case explains that the check-out process includes confirming the CDs, calculating a total, and accepting customer information. Modules have been added to the last row of Figure 12-7 to reflect our detailed understanding of these processes.

Finally, you must determine if any modules on the diagram are reusable; if they are, they should be represented as library modules. In this particular portion of the structure chart, there are no library modules. If, however, any of the modules were reused in the diagram, vertical lines would be added to the sides of the modules to indicate their reuse.

Step 2: Identify Special Connections   The next step is to add loops and conditional lines to represent modules that are repeated or optional. For example, a customer

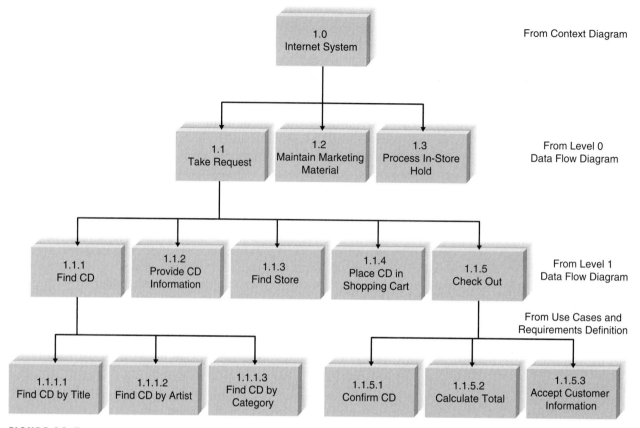

FIGURE 12-7

Step 1: Identify Modules and Levels for the Structure Chart

of the Internet system can make multiple CD requests. Thus, we place a curved arrow around the line under the *take request* process to show that modules 1.1.1 through 1.1.4 can be repeated several times. Only when these four modules are completed will the *check-out* process occur. Only then will the system begin to accept customer information. Can you think of other modules on the structure chart that will be iterated? According to Figure 12-8, one module under the *check-out* process also can happen several times before the system will accept information from the customer.

A diamond is placed below a control module that directs subordinates, which may or may not be performed. For example, customers may choose to search for a CD by title or artist or category—they do not necessarily use all three search alternatives, so a diamond is added below the *find CD* module to communicate this to the programmer. What other part of the structure chart contains subordinates that are invoked conditionally?

Step 3: Add Couples Next, we must identify the information that has to pass among the modules. This information can be data attributes (denoted by an arrow with an empty circle) or special control parameters (denoted by an arrow with a filled-in circle). The arrowheads on the arrows indicate which way information is

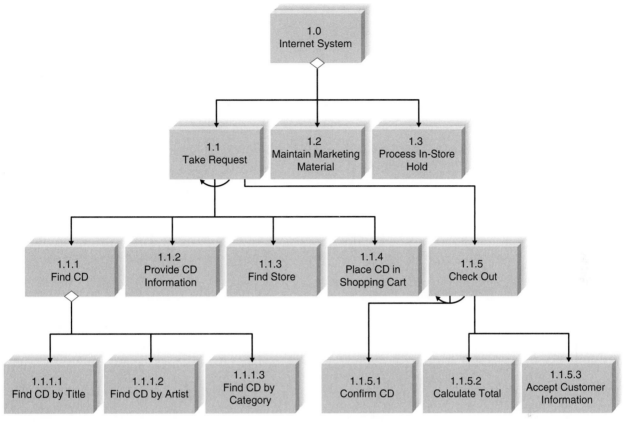

FIGURE 12-8
Step 2: Add Special Connections to the Structure Chart

passed along. The DFD data flows provide us with some guidance about the couples to add because the information that flows in the form of data flows in and out of the DFD processes likely will also flow in and out of the corresponding structure chart modules.

Let's begin with the *find CD* control module. The DFD in Figure 6-20 shows that the customer provides a search request (i.e., a title, artist, or category) and expects information to display on the basis of this input. Therefore, it should seem reasonable for the *find CD* module of the structure chart to pass a data couple with the request information to the appropriate subordinate (e.g., a title is sent to the *find CD by title* subordinate). The DFD contains a search response data flow that comes out of the *find CD* process, and a data couple with the CD information is added to the structure chart to represent the value that resulted from the search.

Stop here and see if you can identify the data and/or control couples that belong in the *check-out* portion of the structure chart.

First, data couples with order information are sent down to the *confirm CD* and *calculate total* modules, and a data couple with customer information is sent to the *accept customer information* module. A control couple is passed from the *confirm CDs* module to its control module to communicate when the customer has finalized the order. Figure 12-9 shows what the structure chart looks like with the additions that we have made.

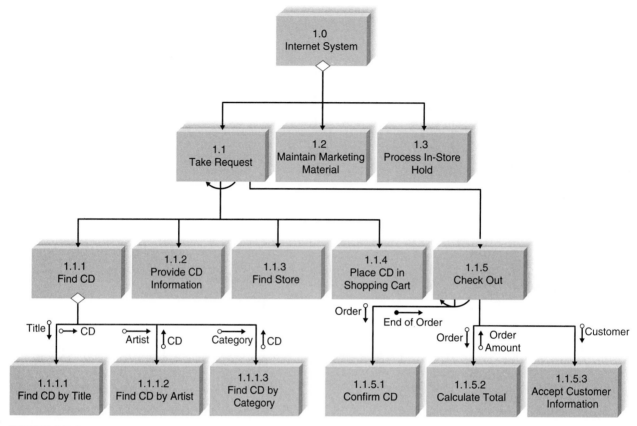

FIGURE 12-9
Step 3: Add Couples to the Structure Chart

Revise Structure Chart By now we have created the initial version of the structure chart based on the DFDs, use cases, and requirements definition, but rarely is a structure chart completed in one attempt. There are many gray areas and decisions that need to be confirmed by other information gleaned during analysis. There are several tools that can help when we are fine-tuning the structure chart. First, we can look at the process descriptions in the CASE repository to see if there are any details of the processes that haven't yet been captured on the diagram. The process descriptions may uncover couples that were overlooked or explain more about how modules should be broken down. Second, we can examine the data model to confirm that the right records and specific fields have been passed using the data couples. This exercise also will confirm that data being passed are actually being captured by the system.

As with most diagrams about which you have learned, the structure chart will evolve and contain more detail as new information is uncovered over the course of the project. Structure charts are not easy. The example that we have presented is much more straightforward than charts found in the real world. The following section explains some guidelines and good practices that you should apply to the chart as you work to improve it.

12-1 STRUCTURE CHART

Using the structure chart in Figure 12-9 as a starting point, add modules that correspond to other parts of the take request process. Use the data flow diagrams in Chapter 6 and the use case in Chapter 5 to help you.

QUESTION:

Do the modules that you added have a transform structure or a transaction structure—or both? Add any special connections to the chart as appropriate. Add necessary data and control couples.

Design Guidelines

As you construct a structure chart, there are several guidelines that you can use to improve its quality. High-quality structure charts result in programs that are modular, reusable, and easy to implement. Measures of good design include cohesion, coupling, and appropriate levels of fan-in and/or fan-out.

Build Modules with High Cohesion   *Cohesion* refers to how well the lines of code within each structure chart module relate to each other. Ideally, a module should perform only one task, making it highly cohesive. Cohesive modules are easy to understand and build because their code performs one function, and they are built to perform that function very efficiently. The more tasks that a module has to perform, the more complex the logic in the code must be to implement the tasks correctly. Typically, you can detect modules that are not cohesive from titles that have an *and* in them, signaling that the module performs multiple tasks.

Look back at the example in Figure 12-2. Currently, each module has good cohesion. Imagine, however, that module 1.3 actually said *calculate current and cumulative GPA* and that module 1.4 was *print grade listing and dean's list.* The *and* in both cases would signal a problem. These modules would not be considered cohesive because they each perform two different tasks, limiting the flexibility of the modules and making the modules much more difficult to build and understand. If the program had to calculate only the current GPA while the module performed both functions, it would require much more complex logic in the code to make that happen.

Another signal of poor cohesion is the presence of control flags that are passed down to subordinate modules; their presence suggests that the subordinate has multiple functions from which one is chosen. Placing this kind of power in a subordinate module is not advisable because it requires complex logic within the module to determine what functions to perform. In the previous example, if our subordinate module was *print grade listing and dean's list,* then a control flag would need to be sent to the subordinate module so it could determine which report or both to print; the subordinate would have to make decisions regarding how to perform its functions.

There are various types of cohesion, some of which are better than others. For example, *functional cohesion* occurs when a module performs one problem-related task, and this form of cohesion is highly desirable. By contrast, *temporal cohesion* takes place when functions within a module may not have much in common other

	Type	Definition	Example
Good	Functional	Module performs one problem-related task	Calculate Current GPA The module calculates current GPA only
	Sequential	Output from one task is used by the next	Format and Validate Current GPA Two tasks are performed, and the formatted GPA from the first task is the input for the second task
	Communicational	Elements contribute to activities that use the same inputs or outputs	Calculate Current and Cumulative GPA Two tasks are performed because they both use the student grade record as input
	Procedural	Elements are involved in different or unrelated activities	Print Grade Listing The module includes the following: calculate student GPA, print student record, calculate cumulative GPA, print cumulative GPA
	Temporal	Activities are related in time	Initialize Program Variables Although the tasks occur at the same time, each task is unrelated
	Logical	List of activities; which one to perform is chosen outside of module	Perform Customer Transaction This module will open a checking account, open a savings account, or calculate a loan, depending on the message that is sent by its control module
Bad	Coincidental	No apparent relationship	Perform Activities This module performs different functions that have nothing to do with each other: update customer record, calculate loan payment, print exception report, analyze competitor pricing structure

FIGURE 12-10

Types of Cohesion (GPA = grade-point average)

than being invoked at the same time, and *coincidental cohesion* occurs when there is no apparent relationship among a module's functions (definitely something to avoid). Figure 12-10 lists seven types of cohesion, along with examples of each type. If you have difficulty differentiating different types of cohesion, use the decision tree in Figure 12-11 for guidance.

Factoring is the process of separating out a function from one module into a module of its own. If you find that a module is not cohesive or that it displays characteristics of a "bad" form of cohesion, you can apply factoring to create a better structure. For example, a more cohesive design for the *print grade listing and dean's list* example would be to factor out print dean's list and print grade listing into two separate modules. A control flag is not needed using this approach because subordinate modules would not have to make any kind of decision; each would perform one task—to print a report.

Build Loosely Coupled Modules   *Coupling* involves how closely modules are interrelated, and the second guideline for good structure chart design states that modules should be loosely coupled. In this way, modules are independent from each other, which keeps code changes from rippling throughout the program. The numbers and kinds of couples on the structure chart reveal the presence of coupling between modules. Basically, the fewer the arrows on the diagrams, the easier it will be to make future alterations to the program.

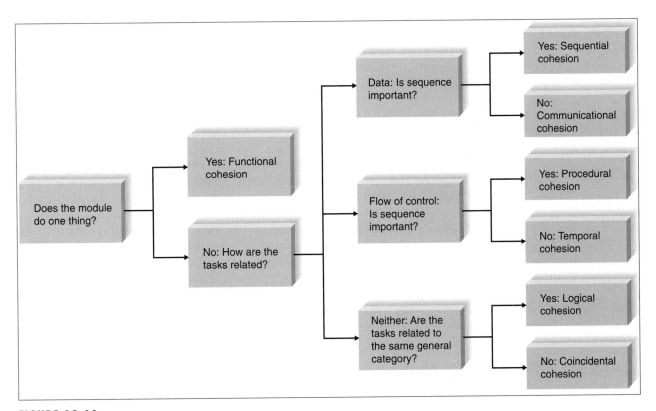

FIGURE 12-11
Cohesion Decision Tree (Adapted from Page-Jones, 1980)

Notice the coupling in the structure chart in Figure 12-4. The data couples (e.g., grade record) denote data that are passed among modules, and the control couple (e.g., end of file) shows that a message is being sent. Although the modules are communicating with one another, notice that the communication is quite limited (only one data couple passed in and out of the module), and there are no superfluous couples (data that are passed for no reason).

There are five types of coupling, each falling on different parts of a good to bad continuum. *Data coupling* occurs when modules pass parameters or specific pieces of data to each other, and this is a form of coupling that you want to see on your structure chart. A bad coupling type is *content coupling*, whereby one module actually refers to the inside of another module. Figure 12-12 presents the types of coupling and examples of each type.

Create High Fan-In *Fan-in* describes the number of control modules that communicate with a subordinate, so a module with high fan-in has many different control modules that call it. This is a very good situation because high fan-in indicates that a module is reused in many places on the structure chart, which suggests that the module contains well-written, generic code. (Fan-in also occurs when library modules are used.) Structures with high fan-in improve the reusability of modules and make it easier for programmers to recode when changes are made or mistakes are uncovered because a change can be made in one place. Figures 12-13a and 12-13b show two different approaches for representing the functionality of reading an employee record. Example *a* is better.

Avoid High Fan-Out Although we desire a subordinate to have multiple control modules, we want to avoid a large number of subordinates associated with a single control. Think of the management concept "span of control," which states that there is a limit to the number of employees that a boss can effectively manage. This concept applies to structure charts as well in that a control module will become much less effective when given large numbers of modules to control. The general rule of thumb is to limit a control module's subordinates to approximately seven. One exception to this is a control module within a transaction structure. If a control module coordinates the invocation of subordinates, each of which performs unique functions, then it usually can handle whatever number of transactions exist. Figures 12-13c and 12-13d show high and low *fan-out* situations, respectively.

Assess the Chart for Quality Finally, we have compiled a checklist (Figure 12-14) that may help you assess the quality of your structure chart. In addition, you should be aware that some CASE tools will critique the quality of your structure chart using predetermined heuristics. Visible Analyst Workbench, for example, checks to make sure that all modules are labeled and connected and that data couples are labeled. It then reviews the connections between modules for correctness of connection, complexity of interface, and completeness of design. The analyzer gives warnings for low fan-in and high fan-out situations.

PROGRAM SPECIFICATION

Once the analyst has communicated the big picture of how the program should be put together, he or she must describe the individual modules in enough detail so that

TYPE		DEFINITION	EXAMPLE
Good Data		Modules pass fields of data or messages	Update Student Record Student ID / Grade Record — Current GPA Calculate Current GPA All couples that are passed are used by the receiving module
	Stamp	Modules pass record structures	Update Student Record Student ID / Grade Record — Current GPA Calculate Current GPA The entire student record is not used by the receiving module, only the *student ID* field
	Control	Module passes a piece of information that intends to control logic	Update Student Record Student ID Grade Record — Current or Cummulative GPA Current or Cummulative Flag Calculate Current or Cumulative GPA The receiving module has to determine which GPA to calculate
	Common	Modules refer to the same global data area	Typically, common coupling cannot be shown on the structure chart; it occurs when modules access the same data areas; and errors made in those areas can ripple through all the modules that use the data
Bad	Content	Module refers to the inside of another module	Module A: Update Student If student = new Then go to Module B Module B: Create Student At all costs avoid modules referring to each other in this way

FIGURE 12-12

Types of Coupling (GPA = grade-point average)

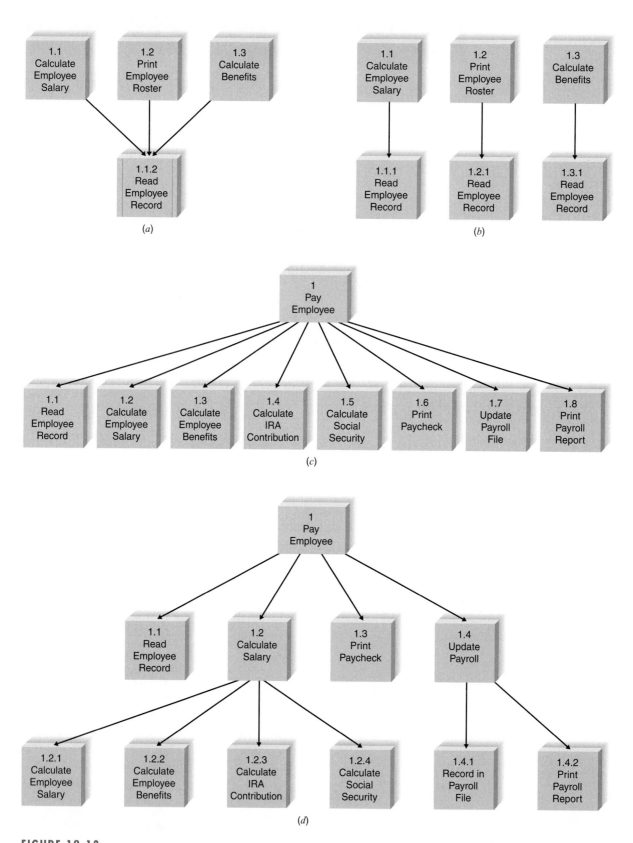

FIGURE 12-13

Examples of Fan-up and Fan-Out: (a) High Fan-In; (b) Low Fan-In; (c) High Fan-Out; (d) Low Fan-Out (IRA = individual retirement account)

> ✔ Library modules have been created whenever possible.
> ✔ The diagram has a high fan-in structure.
> ✔ Control modules have no more than seven subordinates.
> ✔ Each module performs only one function (high cohesion).
> ✔ Modules sparingly share information (loose coupling).
> ✔ Data couples that are passed are actually used by the accepting module.
> ✔ Control couples are passed from "low to high."
> ✔ Each module has a reasonable amount of code associated with it.

FIGURE 12-14
Checklist for Structure Chart Quality

programmers can take over and begin writing code. Modules on the structure chart are described using *program specifications*, written documents that include explicit instructions on how to program pieces of code. Typically, project team members write one program specification for each module on the structure chart and then pass them along to programmers, who write the code during the implementation phase of the project. Specifications must be very clear and easy to understand, or programmers will be slowed down trying to decipher vague or incomplete instructions.

Also, program specifications can pinpoint design problems that exist in the structure chart. At a high level, the structure chart may make sense, but when the analyst actually begins writing the detail behind the modules, he or she may find better ways for arranging the modules or uncover missing or unnecessary couples.

Syntax

There is no formal syntax for a program specification, so every organization uses its own format, often a form like the one in Figure 12-15. Most program specification forms contain four components that convey the information that programmers will need to write the appropriate code.

Program Information   The top of the form in Figure 12-15 contains basic program information, such as the name of the module, its purpose, the deadline, programmer, and the target programming language. This information is used to help manage the programming effort.

Events   The second section of the form is used to list the events that trigger the functionality in the program. An *event* is a thing that happens or takes place. Clicking the mouse generates a mouse event, pressing a key generates a keystroke event—in fact, almost everything the user does causes an event to occur.

In the past, programmers used procedural programming languages (e.g., COBOL, C) that contained instructions that were implemented in a predefined order, as determined by the computer system, and users were not allowed to deviate from the order. With structured programming, the event portion of the program specification is irrelevant. However, many programs today are *event driven* (e.g., Visual Basic, C++), and event-driven programs include procedures that are executed in response to an event initiated by the user, system, or program code. After initialization, the program waits for some kind of event to happen, and when it does, the program carries out the appropriate task, then waits once again.

We have found that many programmers still use program specifications when programming in event-driven languages, and they include the event section on the

Program Specification 1.1 for ABC System

Module _____
Name:
Purpose:
Progammer:
Date due:

C   PowerScript   COBOL   Visual Basic

Events _____
_____
_____
_____
_____
_____

Input Name	Type	Used By	Notes

Output Name	Type	Used By	Notes

Pseudocode _____
_____
_____
_____
_____
_____

Other _____
_____
_____
_____
_____
_____

FIGURE 12-15
Program Specification Form

form to capture when the program will be invoked. Other programmers have switched to other design tools that capture event-driven programming instructions. One such tool, the state-transition diagram, is described in detail in Chapter 15.

Inputs and Outputs The next parts of the program specification describe the inputs and outputs to the program, which are identified by the data couples and control couples found on the structure chart. Programmers must understand what information is being passed and why because these ultimately will translate into variables and data structures within the actual program.

Pseudocode *Pseudocode* is a detailed outline of the lines of code that need to be written, and it is presented in the next section of the form. If you remember, when we had to describe the processes on the DFDs, we used a technique called structured English, a language with syntax based on English and structured programming. These DFD descriptions in structured English are now used as the primary input to produce pseudocode.

Pseudocode is a language that contains logical structures, including sequential statements, conditional statements, and iteration. It differs from structured English in that pseudocode contains details that are programming specific, such as initialization instructions or linking, and it also is more extensive so that a programmer can write the module by mirroring the pseudocode instructions. In general, pseudocode is much more like real code, and its audience is the programmer as opposed to the analyst. Its format is not as important as the information it conveys. Figure 12-16 shows a short example of pseudocode for a module that is responsible for finding CD information.

Writing good pseudocode can be difficult—imagine creating instructions that someone else can follow without having to ask for clarification or making a wrong assumption. For example, have you ever given a friend directions to your house, but your friend ended up getting lost? To you, the directions might have been very clear, but that is because of your personal assumptions. To you, the instruction "take the first left turn" may really mean "take a left turn at the first stoplight." Someone else's interpretation might be "take a left turn at the first road, with or without a light." (Barbara has a very bad sense of direction and has been known to make a first left turn into a driveway!)

Therefore, when writing pseudocode, pay special attention to detail and readability.

The last section of the program specification provides space for other information that must be communicated to the programmer, such as calculations, special business rules, calls to subroutines or libraries, and other relevant issues. This also

```
(Get_CD_Info module)
   Accept (CD.Title) {Required}
   Accept (CD.Artist) {Required}
   Accept (CD.Category) {Required}
   Accept (CD.Length)
Return
```

FIGURE 12-16
Pseudocode

can point out changes or improvements that will be made to the structure chart on the basis of problems that the analyst detected during the specification process.

Some project teams do not create program specifications using forms but instead input the specification information directly into a CASE tool. In these cases, the information is added to the description for the appropriate module on the structure chart (or to its corresponding process on the DFD). Figure 12-17 illustrates how the CASE repository can be used to capture program design information.

Applying the Concepts at CD Selections

Refer to the structure chart in Figure 12.9 for the following example. Each module on the diagram should have an associated program specification, but for now let's create one for module 1.1.1.1, *find CDs by title*.

The first part of the form (Figure 12-18) contains basic information about the specification, such as its name and purpose. Because an event-driven programming language will be used, we list the events that will trigger the program to run (i.e., a mouse click, a menu selection).

The inputs and outputs for the program correspond to the two couples on the structure chart: title that is sent to module 1.1.1.1 and CD that is passed by the module. We added these to the input and output sections of the form, respectively.

Next, we used the structured English description for the module that is found in the process description to develop the pseudocode that will communicate the code that should be written for the program. However, as we wrote the pseudocode and examined the process description, we discovered a problem—the structure chart does not appear to handle the situation in which a user's search request cannot be located by the system. Although the pseudocode includes a "not found" condition, there is no mechanism for the program to pass this result back to its calling program. At this point, we made a note in the last section of the program specification to add a control couple to the structure chart that passes a "not found" flag, and a second output was added to the current specification form.

Finally, we added a business rule to the specification to explain to the programmer what will happen when a CD is not found; however, the functionality of this rule will be handled elsewhere in the program. The program specification is now ready for the implementation phase, when the form will be passed off to a programmer who will develop the code that meets its requirements.

YOUR **12-2 PROGRAM SPECIFICATION**

TURN

Create a program specification for module 1.2.5.3, accept customer information, on the structure chart shown in Figure 12.9.

QUESTION:
On the basis of your specification, are there any changes to the structure chart that you would recommend?

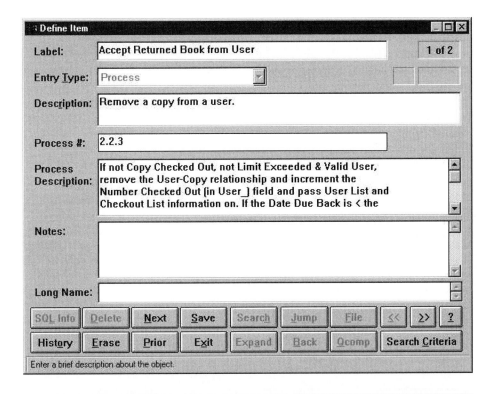

FIGURE 12-17
Process Description—Analysis and Design

Program Specification 1.3.1.1 for Internet Sales System

Module _____

Name: Display_CD_by_Title

Purpose: Display basic CD information, using a title input by the user

Progammer: John Smith

Date due: April 26, 2002

☐C   ■PowerScript   ☐COBOL   ☐FORTRAN

Events _____

search by title pushbutton is clicked

search by title menu choice is selected

_____

_____

_____

Input Name:	Type:	Provided by:	Notes:
CD title	String (50)	Program 1.3.1	

Output Name:	Type:	Used by	Notes:
CD id	String (10)	Program 1.3.1	
Not_found	Logical	Program 1.3.1	Used to communicate when CD is not found

Pseudocode _____

(Find_CD module)

  Find CD id via the Title

  If no CD is found

    Set not_found True

  Else

    Set not_found False

  End_If

  Return

Other _____

Business rule: if a CD id is not found, the "CD of the week" will appear to the user

Note: a control couple containing a not _found flag should be added from 1.3.1.1 to 1.3.1 to instruct 1.3.1 to display a not found message to the user and the CD of the week

FIGURE 12-18

Program Specification for Accept Customer Information

SUMMARY

Structure Chart

The structure chart shows all the functional components that must be included in the program at a high level, arranged in a hierarchical format that implies order and control. Lines that connect modules can contain a loop, which signifies that the subordinate module is repeated before other modules to its right are invoked. A diamond is placed over subordinate modules that are invoked conditionally. An arrow with a filled circle represents a control couple or flag, which passes system messages from one module to another. The data couple, an arrow with an empty circle, denotes the passing of records or fields.

Modules can be organized into one of two types of structures. The transaction structure contains a control module that calls subordinates that perform independent tasks. By contrast, transform structures convert some input into an output through a series of subordinate modules, and the control module describes the transformation that takes place.

Building Structure Charts

Creating a structure chart is usually a four-step process. First, the analyst identifies the top-level modules and then decomposes them into lower levels. Second, the analyst adds the control connections among modules, such as loops and conditional lines that show when modules call subordinates. Third, the analyst adds couples, the information that modules pass among themselves. Finally, the analyst reviews the structure chart and revises it again and again until it is complete.

Structure Chart Design Guidelines

There are several design guidelines that you should follow when designing structure charts. First, build modules with high cohesion so that each module performs only one function. There are seven kinds of cohesion, ranging from good to bad, and the instances of bad cohesion should be removed from the structure chart. Second, modules should not be interdependent but rather should be loosely coupled, using good types of coupling. There are five kinds of coupling, ranging from good to bad; as with cohesion, bad instances should be avoided. Finally, structure charts should display high fan-in and low fan-out, which means that modules should have many control modules but limited subordinates.

Program Specification

Program specifications provide more detailed instructions to the programmers about how to code the modules. The program specification contains several components that communicate basic module information (e.g., a name, calculations that must be performed, and the target programming language), inputs and outputs, special instructions for the programmer, and pseudocode.

Pseudocode

Pseudocode is a technique similar to structured English that communicates the code that must be written using programming structures and a generic language that is not program language specific. Pseudocode is much more like real code than is structured English, and its audience is the programmer as opposed to the analyst. Many programs today are event driven, meaning that their instructions are not necessarily invoked in a predefined order, as determined by the computer system. Instead, users control the order of modules through the way that they interact with the system. When event-driven programming is used, program specifications should include a section that describes the types of events that invoke the code.

KEY TERMS

Afferent process
Central process
Cohesion
Coincidental cohesion
Common coupling
Communicational cohesion
Conditional line
Connector
Content coupling
Control couple
Control coupling
Control module
Couple
Coupling
Data couple

Data coupling
Efferent processes
Event
Event driven
Factoring
Fan-in
Fan-out
Functional cohesion
Iteration
Library module
Logical cohesion
Loop
Module
Off-page connector
On-page connector

Procedural cohesion
Program design
Program specification
Pseudocode
Selection
Sequence
Sequential cohesion
Stamp coupling
Structure chart
Subordinate module
Temporal cohesion
Top-down modular approach
Transaction structure
Transform structure

QUESTIONS

1. What does it mean to use a top-down modular approach for program design? Describe the benefits of such an approach.
2. What is the deliverable called from the program design portion of the design phase? What does it include, and how is it used?
3. How is a structure chart used in program design?
4. What are the main components on the structure chart, and how are they depicted?
5. What analysis tools and techniques are used by analysts to create a structure chart?
6. What are the differences among control, subordinate, and library modules? Can a module be all three? Why or why not?
7. What is the difference between data couples and control couples?
8. In which direction should a control couple be passed? Why?
9. What is the difference between a transaction structure and a transform structure? Can a module be a part of both types of structures? Why or why not?
10. What are the five types of coupling? Give one example of good coupling and one example of bad coupling.

11. What are the seven types of cohesion? Give one example of good cohesion and one example of bad cohesion.
12. Why is it desirable for a structure chart to be highly cohesive and loosely coupled?
13. Describe three design guidelines for a structure chart and describe why they are important for a high-quality diagram.
14. Name two ways in which program specifications are useful during program design.
15. What are the main sections that are included in a program specification?
16. What is the difference between structured programming and event-driven programming? What section of the program specification is used for event-driven programming?
17. How can an analyst identify the inputs and outputs that belong on a program specification?
18. How can project teams capture information from the program specification without using a paper form?
19. Is program design more or less important when using event-driven languages such as Visual Basic?

EXERCISES

A. What symbols would you use to depict the following situations on a structure chart?

- A function occurs multiple times before the next module is invoked.

- A function is continued on the bottom of the page of the structure chart.
- A customer record is passed from one part of the program to another.
- The program will print a record either on screen or on a printer, depending on the user's preference.

- A customer's ID is passed from one part of the program to another.
- A function cannot fit on the current page of the structure chart.

B. Describe the differences in the meanings between the following two structure charts below. How have the symbols changed the meanings?

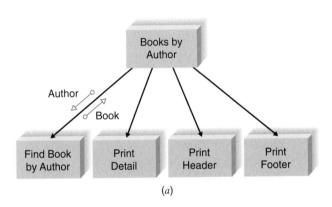

(a)

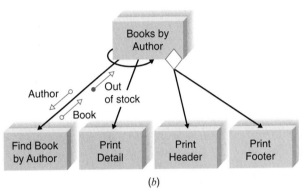

(b)

C. Create a structure chart based on the data flow diagrams (DFDs) that you created for the following exercises in Chapter 6:
- Question D
- Question E
- Question F
- Question G
- Question H

D. Critique the following structure chart that depicts a guest making a hotel reservation. Describe the chart in terms of fan-in, fan-out, coupling, and cohesion. Redraw the chart to improve the design.

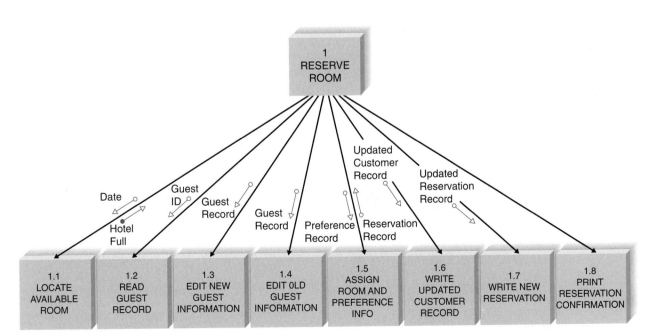

E. Identify the kinds of coupling that are represented in the following situations:

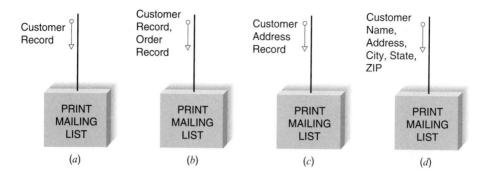

(a)   (b)   (c)   (d)

F. Identify the kinds of cohesion that are represented in the following situations:
   - accept customer address
   - print mailing label record
   - print customer address listing
   - print marketing address report

   - accept customer address
   - validate zip code and state
   - format customer address
   - print customer address
   - accept customer address
   - print mailing label record

   - accept customer address
   - print mailing label record or
   - print customer address listing or
   - print marketing address report or
   - validate customer address
   - print mailing label record
   - check customer balance
   - print marketing address report
   - record customer preference information

G. Identify whether the following structures are transaction or transform and explain the reasoning behind your answers.[2]

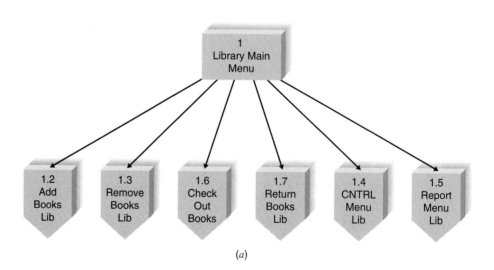

(a)

---

[2] The structure charts based on a library system are adapted from an example provided with the Visible Analyst Workbench software.

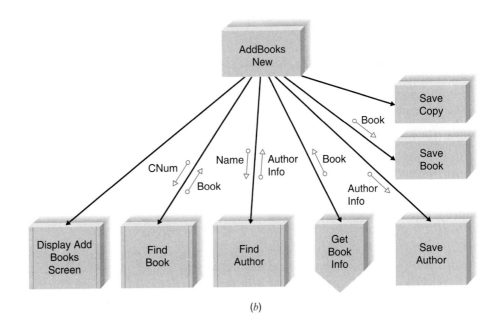

(b)

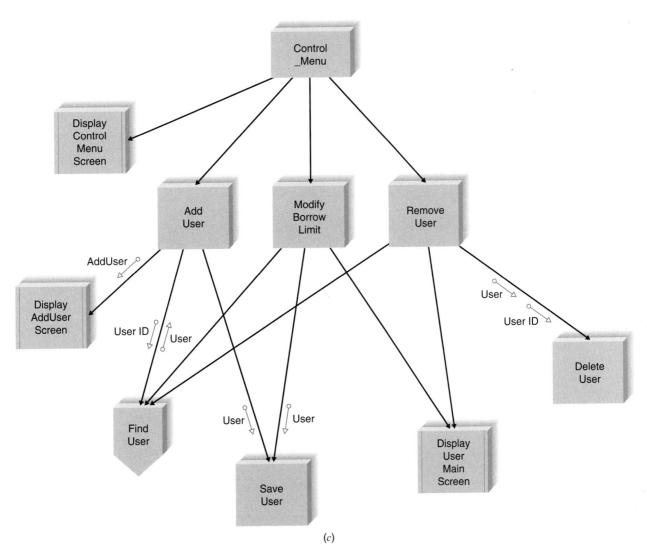

(c)

H. Create a program specification for module 1.1.1.2 on the structure chart in Figure 12-9.

I. Create a program specification for module 1.2.5.2 on the structure chart in Figure 12-9.

J. Create pseudocode for the program specification that you wrote in Exercise H.

K. Create pseudocode for the program specification that you wrote in Exercise I.

L. Create pseudocode that explains how to start the computer at your computer lab and open a file in the word processor. Exchange your pseudocode with a classmate and follow the instructions exactly as they appear. Discuss the results with your classmate. At what points were each set of instructions vague or unclear? How would you improve the pseudocode that you created originally?

MINICASES

1. In the new system for Holiday Travel Vehicles, the system users follow a two-stage process to record complete information on all of the vehicles sold. When an RV or trailer first arrives at the company from the manufacturer, a clerk from the Inventory Department creates a new vehicle record for it in the computer system. The data entered at this time include basic descriptive information on the vehicle such as manufacturer, name, model, year, base cost, and freight charges. When the vehicle is sold, the new vehicle record is updated to reflect the final sales terms and the dealer-installed options added to the vehicle. This information is entered into the system at the time of sale when the salesperson completes the sales invoice.

When it is time for the clerk to finalize the new vehicle record, the clerk will select a menu option from the system, which is called "Finalize New Vehicle Record." The tasks involved in this process are described next.

When the user selects the "Finalize New Vehicle Record" from the system menu, the user is immediately prompted for the serial number of the new vehicle. This serial number is used to retrieve the new vehicle record for the vehicle from system storage. If a record cannot be found, the serial number is probably invalid. The vehicle serial number is then used to retrieve the option records that describe the dealer-installed options that were added to the vehicle at the customer's request. There may be zero or more options. The cost of the option specified on the option record(s) is totaled. Then, the dealer cost is calculated using the vehicle's base cost, freight charge, and total option cost. The completed new vehicle record is passed back to the calling module.

   a. Develop a structure chart for this segment of the Holiday Travel Vehicles system.

   b. What type of structure chart have you drawn, a transaction structure or a transform structure? Why?

2. Develop a program specification for Module 4.2.5 *(Calculate Dealer Cost)* in minicase 1.

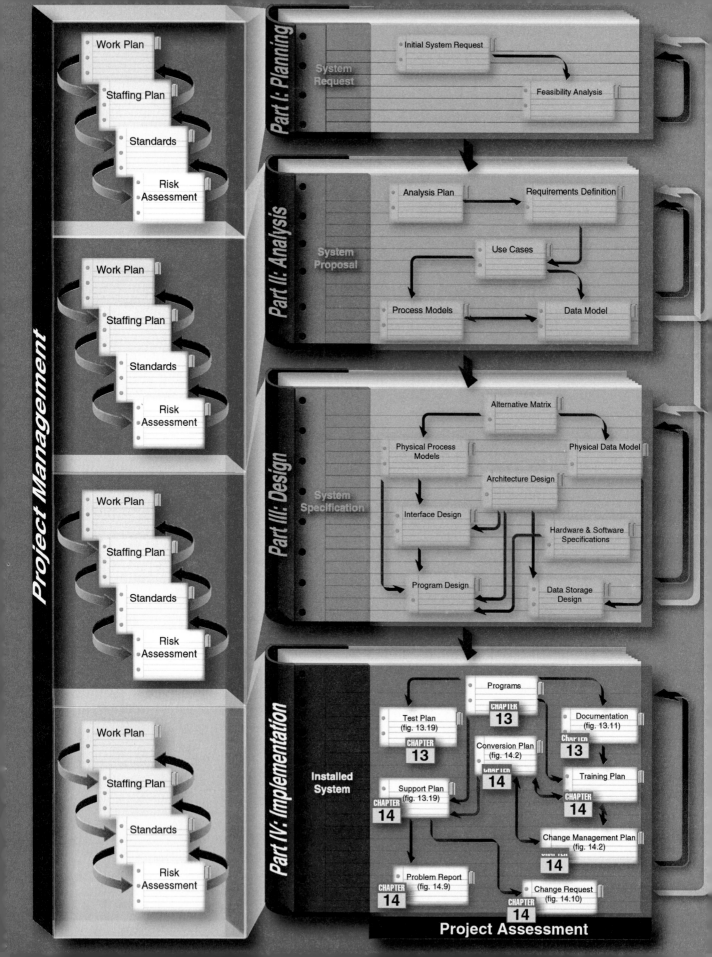

Programs

Test Plan

Documentation

Conversion Plan

Change Management Plan

Training Plan

Support Plan

Problem Report

Change Request

Project Assessment

PART FOUR

IMPLEMENTATION PHASE

PROJECT BINDER

Construction **CHAPTER 13**

Deployment **CHAPTER 14**

The final phase in the SDLC is the Implementation Phase, during which the system is actually built (or purchased, in the case of a packaged software design).

At the end of Implementation the final system is delivered to the project sponsor and approval committee.

PLANNING

ANALYSIS

DESIGN

IMPLEMENTATION

☐ **Program System**
☐ **Test Software**
☐ **Test Performance**
☐ Select System Conversion Style
☐ Train Users
☐ Select Support
☐ Maintain System
☐ Assess Project
☐ Conduct Post-Implementation Audit

T A S K C H E C K L I S T

PLANNING   ANALYSIS   DESIGN

CHAPTER 13

CONSTRUCTION

This chapter discusses the activities needed to successfuly build the information system: programming, testing, and documenting the system. Programming is time consuming and costly, but except in unusual circumstances, it is the simplest for the systems analyst because it is well understood. For this reason, the system analyst focuses on testing (proving that the system works as designed) and developing documentation during this part of the systems development life cycle.

OBJECTIVES

- Be familiar with the system construction process.
- Understand different types of tests and when to use them.
- Understand how to develop documentation.

CHAPTER OUTLINE

IMPLEMENTATION

INTRODUCTION

When people first learn about developing information systems, they usually immediately think about writing programs. Programming can be the largest single component of any systems development project in terms of both time and cost. However, it also can be the best understood component and therefore—except in rare circumstances—offers the least problems of all aspects of the SDLC. When projects fail, it is usually not because the programmers were unable to write the programs but because the analysis, design, installation, and/or project management were done poorly. In this chapter, we focus on the construction and testing of the software and the documentation.

Construction is the development of all parts of the system, including the software itself, documentation, and new operating procedures. Programming is often seen as the focal point of system development. After all, system development *is* writing programs. It is the reason why we do all the analysis and design. And it's fun. Many beginning programmers see testing and documentation as bothersome afterthoughts. Testing and documentation aren't fun, so they often receive less attention than does the creative activity of writing programs.

However, programming and testing are very similar to writing and editing. No professional writer (or student writing an important term paper) would stop after writing the first draft. Rereading, editing, and revising the initial draft into a good paper is the hallmark of good writing. Likewise, thorough testing is the hallmark of professional software developers. Most professional organizations devote more time and money to testing (and the subsequent revision and retesting) than to writing the programs in the first place.

The reasons are simple economics: downtime and failures caused by software bugs[1] are extremely expensive. Many large organizations estimate the costs of downtime of critical applications at $50,000 to $200,000 *per hour.*[2] One serious bug that causes an hour of downtime can cost more than one year's salary of a programmer—and how often are bugs found and fixed in one hour? Testing is therefore a form of insurance. Organizations are willing to spend a lot of time and money to prevent the possibility of major failures after the system is installed.

Therefore, a program is usually not considered finished until the test for that program has been passed. For this reason, programming and testing are tightly coupled, and since programming is the primary job of the programmer (not the analyst), testing (not programming) often becomes the focus of the construction stage for the systems analysis team.

In this chapter, we discuss three parts of the construction step: programming, testing, and writing the documentation. Because programming is primarily the job of programmers, not systems analysts, and because this is not a programming book, we devote less space here to programming than to testing and documentation.

---

[1] When I was an undergraduate, I had the opportunity to hear Admiral Grace Hopper tell how the term *bug* was introduced. She was working on one of the early U.S. Navy computers when suddenly it failed. The computer would not restart properly, so she began to search for failed vacuum tubes. She found a moth inside one tube and recorded in the log book that a bug had caused the computer to crash. From then on, every computer crash was jokingly blamed on a bug (as opposed to programmer error), and eventually the term *bug* entered the general language of computing.

[2] See Billie Shea, "Quality Patrol: Eye on the Enterprise," *Application Development Trends,* November 5, 1998, pp. 31–38.

13-A THE COST OF A BUG

My first programming job in 1977 was to convert a set of application systems from one version of COBOL to another version of COBOL for the government of Prince Edward Island. The testing approach was to first run a set of test data through the old system and then run it through the new system to ensure that the results from the two matched. If they matched, then the last three months of production data were run through both to ensure they, too, matched.

Things went well until I began to convert the gas tax system that kept records on everyone authorized to purchase gasoline without paying tax. The test data ran fine, but the results using the production data were peculiar. The old and new systems matched, but rather than listing several thousand records, the report listed only 50. I checked the production data file and found it listed only 50 records, not the thousands that were supposed to be there.

The system worked by copying the existing gas tax records file into a new file and making changes in the new file. The old file was then copied to tape backup. There was a bug in the program such that if there were no changes to the file, a new file was created, but no records were copied into it.

I checked the tape backups and found one with the full set of data that was scheduled to be overwritten three days after I discovered the problem. The government was only three days away from losing all gas tax records.

Alan Dennis

QUESTION:

What might have happened if this bug hadn't been caught and all gas tax records were lost?

MANAGING PROGRAMMING

In general, systems analysts do not write programs; programmers write programs. Therefore, the primary task of the systems analysts during programming is ... waiting. However, the project manager is usually very busy *managing* the programming effort by assigning the programmers, coordinating the activities, and managing the programming schedule.[3]

Assigning Programmers

The first step in programming is assigning modules to the programmers. As discussed in Chapter 12, each programming module should be as separate and distinct as possible from the other modules. The project manager first groups together modules that are related so that each programmer is working on related program modules. These groups of modules are then assigned to programmers.

One of the rules of system development is that the more programmers who are involved in a project, the longer the system will take to build. This is because as the size of the programming team increases, the need for coordination increases exponentially, and the more coordination that is required, the less time programmers can spend actually writing programs. The best size is the smallest possible programming team. When projects are so complex that they require a large team,

---

[3] One of the best books on managing programming (even though it was first written in 1975) is that by Frederick P. Brooks Jr, *The Mythical Man-Month,* 20th anniversary ed., Reading, MA: Addison-Wesley, 1995.

the best strategy is to try to break the project into a series of smaller parts that can function as independently as possible.

Coordinating Activities

Coordination can be done through both high-tech and low-tech means. The simplest approach is to have a weekly project meeting to discuss any changes to the system that have arisen during the past week—or just any issues that have come up. Regular meetings, even if they are brief, encourage the widespread communication and discussion of issues before they become problems.

Another important way to improve coordination is to create and follow standards that can range from formal rules for naming files to forms that must be completed when goals are reached to programming guidelines (see Chapter 3). When a team forms standards and then follows them, the project can be completed faster because task coordination is less complex.

The analysts also must put mechanisms in place to keep the programming effort well organized. Many project teams set up three "areas" in which programmers can work: a development area, a testing area, and a production area. These areas can be different directories on a server hard disk, different servers, or different physical locations, but the point is that files, data, and programs are separated on the basis of their status of completion. At first, programmers access and build files within the development area and then copy them to the testing area when the programmers are "finished." If a program does not pass a test, it is sent back to development. Once all programs and so forth are tested and ready to support the new system, they are copied into the production area—the location where the final system will reside.

Keeping files and programs in different places according to completion status helps manage *change control,* the action of coordinating a program as it changes through construction. Another change-control technique is keeping track of what programs are being changed by whom using a *program log.* The log is merely a form on which programmers sign out programs to write, and sign in when completed. Both the programming areas and program log help the analysts understand exactly who has worked on what and the program's status. Without these techniques, files can be put into production without the proper testing, two programmers can start working on the same program at the same time, files can be overlooked, and so on.

If a CASE tool is used during a construction step, it can be very helpful for change control because many CASE tools are set up to track the status of programs and help manage programmers as they work. In most cases, maintaining coordination is not conceptually complex. It just requires a lot of attention and discipline to track small details.

Managing the Schedule

The time estimates that were produced during the initial planning phase and refined during the analysis and design phases must almost always be refined as the project progresses during construction because it is virtually impossible to develop an exact assessment of the project's schedule. As we discussed in Chapter 3, a well-done set of time estimates will usually have a 10% margin of error by the time you reach the construction step. It is critical that the time estimates be revised as the construction step proceeds. If a program module takes longer to develop than expected, then the prudent response is to move the expected completion date later by the same amount.

One of the most common causes for schedule problems is *scope creep.* Scope creep occurs when new requirements are added to the project after the system design was finalized. Scope creep can be very expensive because changes made late in the SDLC can require much of the completed system design (and even programs already written) to be redone. Any proposed change during the construction phase must require the approval of the project manager and should be done only after a quick cost–benefit analysis has been done.

Another common cause is the unnoticed day-by-day slippages in the schedule. One module is a day late here; another one a day late there. Pretty soon these minor delays add up and the project is noticeably behind schedule. Once again, the key to managing the programming effort is to watch these minor slippages carefully and update the schedule accordingly.

Typically, a project manager will create a risk assessment that tracks potential risks along with an evaluation of their likelihood and potential impact. As the construction step moves to a close, the list of risks will change as some items are

PRACTICAL 13-1 AVOIDING CLASSIC IMPLEMENTATION MISTAKES

TIP

In previous chapters, we discussed classic mistakes and how to avoid them. Here, we summarize four classic mistakes in the implementation phase:

1. **Research-oriented development:** Using state-of-the-art technology requires research-oriented development that explores the new technology because "bleeding edge" tools and techniques are not well understood, are not well documented, and do not function exactly as promised.
   Solution: If you use state-of-the-art technology, you should significantly increase the project's time and cost estimates even if (some experts would say *especially if*) such technologies claim to reduce time and effort.

2. **Using low-cost personnel:** You get what you pay for. The lowest-cost consultant or staff member is significantly less productive than the best staff. Several studies have shown that the best programmers produce software six to eight times faster than the least productive (yet cost only 50% to 100% more).
   Solution: If cost is a critical issue, assign the best, most expensive personnel; never assign entry-level personnel in an attempt to save costs.

3. **Lack of code control:** On large projects, programmers must coordinate changes to the program source code (so that two programmers don't try to change the same program at the same time and one doesn't overwrite the other's changes). Although manual procedures appear to work (e.g., sending e-mail notes to others when you work on a program to tell them not to work on that program), mistakes are inevitable.
   Solution: Use a source code library that requires programmers to check out programs and prohibits others from working on them at the same time.

4. **Inadequate testing:** The number-one reason for project failure during implementation is ad hoc testing—in which programmers and analysts test the system without formal test plans.
   Solution: Always allocate sufficient time in the project plan for formal testing.

Source: Adapted from *Rapid Development*, Redmond, WA: Microsoft Press, 1996, pp. 29–50, by Steve McConnell.

removed and others surface. The best project managers, however, work hard to keep risks from having an impact on the schedule and costs associated with the project.

DESIGNING TESTS

There is often a temptation to rush into testing as soon as the very first program modules are complete and to spontaneously test different events and possibilities without spending time to develop a comprehensive test plan. This is dangerous because important tests may be overlooked, and if an error does occur, it may be difficult to reproduce the exact sequence of events that caused it. Instead, testing must be done systematically and the results must be documented so that the project team knows what has and has not been tested.

Test Planning

Testing starts with the tester's developing a *test plan* that defines a series of tests that will be conducted.[4] Figure 13-1 shows a typical test plan form. A test plan often has 20 to 30 pages, with a separate page for each individual test in the plan. Each individual test has a specific objective, describes a set of very specific *test cases* to examine, and defines the expected results and the actual results observed. The test objective is taken directly from the program specification or from the program source code. For example, suppose the program specification stated that the order quantity must be between 10 and 100 cases. The tester would develop a series of test cases to ensure that the quantity is validated before the system accepts it.

It is impossible to test every possible combination of input and situation; there are simply too many possible combinations. In this example of an order quantity that must be between 10 and 100 cases, the test requires a minimum of three test cases: one with a valid value (e.g., 15), one with an invalid value too low (e.g., 7), and one with invalid value too high (e.g., 110). Most tests would also include a test case with a nonnumeric value to ensure the data types were checked (e.g., ABCD). A really good test would include a test case with nonsensical but potentially valid data (e.g. 21.4).

In some cases, test cases cannot be conducted by entering data values but must instead be handled by selecting certain combinations of commands or menu choices. The script area on the test plan is used to describe the sequence of keystrokes or mouse clicks and movements for this type of test.

Not all program modules are likely to be finished at the same time, so the programmer usually writes *stubs* for the unfinished modules to enable the modules around them to be tested. A stub is placeholder for a module that usually displays a simple test message on the screen or returns some *hardcoded* value[5] when it is selected. For example, consider an application system that provides the five standard functions discussed in Chapter 6 for some data objects such as CDs, patients, or employees: creating, changing, deleting, finding, and printing (whether on the

---

[4] For more information on testing, see G. J. Myers, *The Art of Software Testing,* New York: Wiley-Interscience, 1979, and W. Hetzel, *The Complete Guide to Software Testing,* New York: John Wiley & Sons, 1993.

[5] The word *hardcoded* means "written into the program." For example, suppose you were writing a unit to calculate the net present value of a loan. The stub might be written to always display (or return to the calling module) a value of 100 regardless of the input values. In this case, we would say that the 100 was hardcoded.

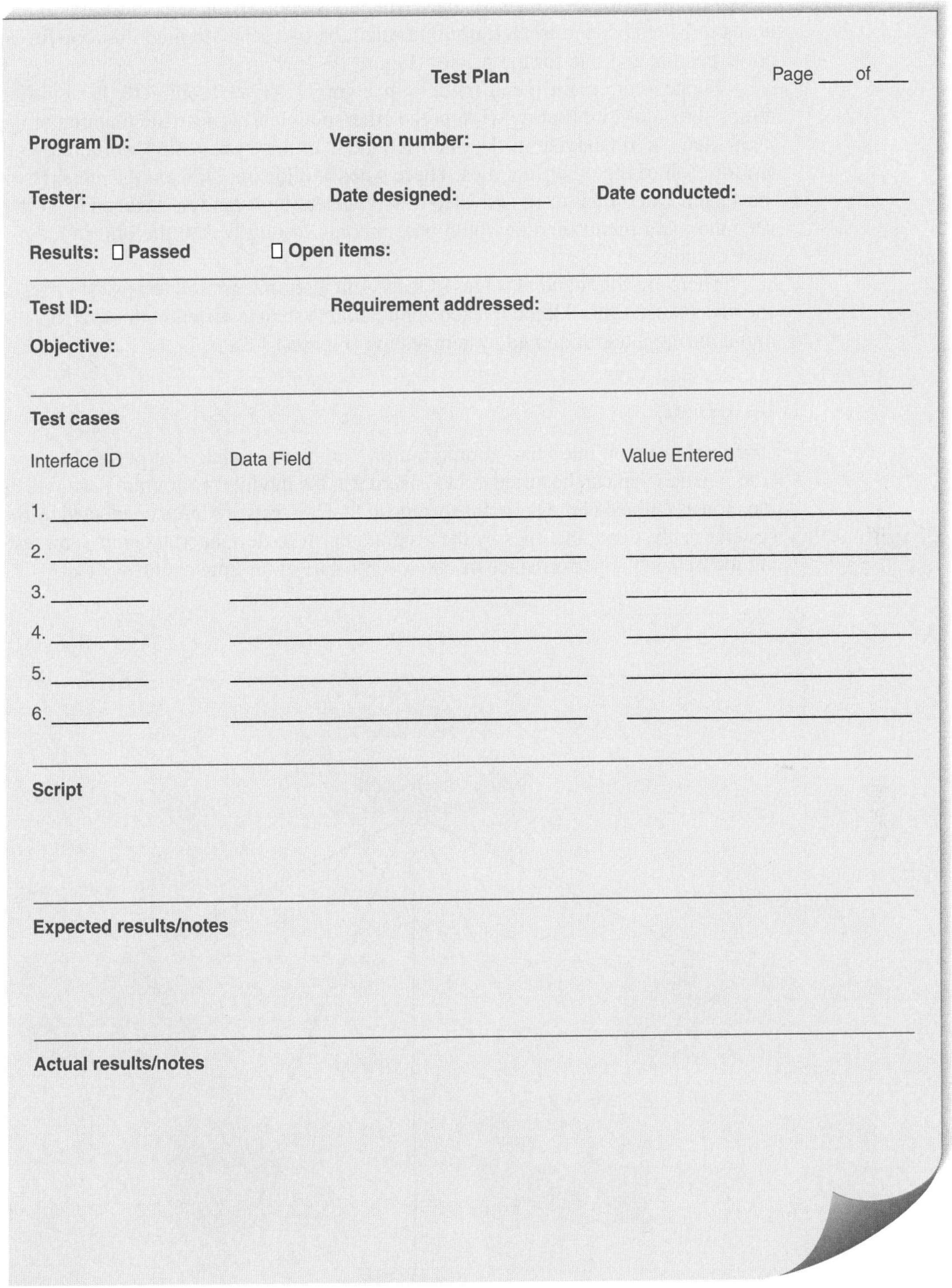

Test Plan   Page ____ of ____

Program ID: _____   Version number: _____

Tester: _____   Date designed: _____   Date conducted: _____

Results: ☐ Passed   ☐ Open items: _____

Test ID: _____   Requirement addressed: _____

Objective: _____

Test cases

Interface ID	Data Field	Value Entered
1. _____	_____	_____
2. _____	_____	_____
3. _____	_____	_____
4. _____	_____	_____
5. _____	_____	_____
6. _____	_____	_____

Script

Expected results/notes

Actual results/notes

FIGURE 13-1
Test Plan

screen or on a printer). Each of these functions could be a separate module that needs to be tested, and in fact, printing might be two separate modules, one for an on-screen list and one for the printer (Figure 13-2).

Suppose the main menu module in Figure 13-2 was complete. It would be impossible to test it properly without the other modules, because the function of the main menu is to navigate to the other modules. In this case, a stub would be written for each of the other modules. These stubs would simply display a message on the screen when they were activated (e.g., "Delete item module reached"). In this way, the main menu module could pass module testing before the other modules were completed.

There are four general stages of tests: unit tests, integration tests, system tests, and acceptance tests. Although each application system is different, most errors are found during integration and system testing (Figure 13-3).

Unit Tests

Unit tests focus on one unit—a program or a program module that performs a specific function that can be tested—and ensure that the module or program performs its function as defined in the program specification. Unit tests are often conducted by the systems analyst or sometimes by the programmer who developed the unit. Unit testing focuses on the performance of one specific part of the application system.

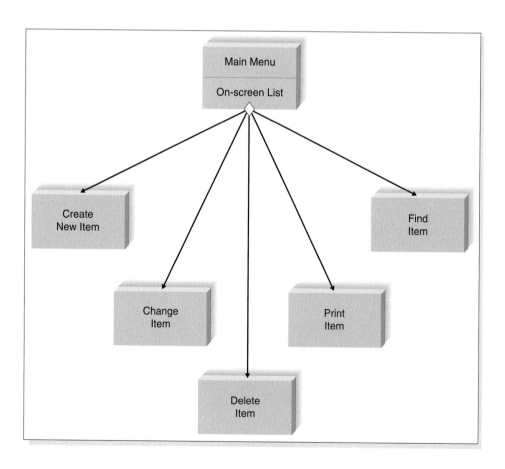

FIGURE 13-2
Testing Separate Modules

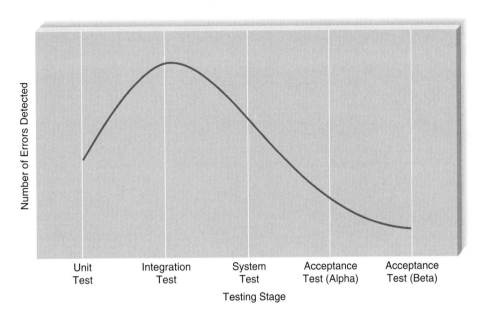

FIGURE 13-3
Error Discovery Rates for Different Stages of Tests

Programmers always test their code during development, so unit testing is performed only after the programmer believes the unit to be error free. There are two approaches to unit testing: *black-box* and *white-box* (Figure 13-4). Black-box testing is the most commonly used because it is driven by the program specification, not by the programmers' interpretation. In this case, the test plan is developed directly from the program specification: each item in the program specification becomes a test, and several test cases are developed for it.

Integration Tests

Integration tests assess whether a set of modules or programs that must work together do so without error. They ensure that the interfaces and linkages between different parts of the system work properly. At this point, the modules have passed their individual unit tests, so the focus now is on the flow of control among modules, and on the data exchanged among them. Integration testing follows the same general procedures as unit testing: the tester develops a test plan that has a series of tests that in turn have tests. Integration testing is often done by a set of programmers and/or systems analysts.

There are four approaches to integration testing: user interface testing, use scenario testing, data flow testing, and system interface testing (see Figure 13-4). Most projects use all four approaches.

System Tests

System tests are usually conducted by the systems analysts to ensure that all modules and programs work together without error. System testing is similar to integration testing is similar to integration testing but is much broader in scope. Whereas integration testing focuses on whether the modules work together without error, system tests examine how well the system meets business requirements and its usability, security, and performance under heavy load (see Figure 13-4). It also tests the system's documentation.

Stage	Types of Tests	Test Plan Source	When to Use	Notes
Unit Testing	Black-box testing: treats program as black box	Program specifications	For normal unit testing	The tester focuses on whether the unit meets the requirements stated in the program specifications.
	White-box testing: looks inside the program to test its major elements	Program source code	When complexity is high	By looking inside the unit to review the code itself, the tester may discover errors or assumptions not immediately obvious to someone treating the unit as a black box.
Integration Testing	User interface testing: the tester tests each interface function	Interface design	For normal integration testing	Testing is done by moving through each and every menu item in the interface either in a top-down or bottom-up manner.
	Use scenario testing: the tester tests each use scenario	Use scenario	When the user interface is important	Testing is done by moving through each use scenario to ensure it works correctly. Use scenario testing is usually combined with user interface testing because it does not test all interfaces.
	Data flow testing: tests each process in a step-by-step fashion	Physical DFDs	When the system performs data processing	The entire system begins as a set of stubs. Each unit is added in turn and the results of the unit are compared to the correct result from the test data; when a unit passes, the next unit is added and the test is rerun.
	System interface testing: tests the exchange of data with other systems	Physical DFDs	When the system exchanges data	Because data transfers between systems are often automated and not monitored directly by the users, it is critical to design tests to ensure they are being done correctly.
System Testing	Requirements testing: tests whether original business requirements are met	System design, unit tests, and integration tests	For normal system testing	This test ensures that changes made as a result of integration testing did not create new errors. Testers often pretend to be uninformed users and perform improper actions to ensure the system is immune to invalid actions (e.g., adding blank records).
	Usability testing: tests how convenient the system is to use	Interface design and use scenarios	When user interface is important	This test is often done by analysts with experience in how users think and in good interface design. This test sometimes uses the formal usability testing procedures discussed in Chapter 10.
	Security testing: tests disaster recovery and unauthorized access	Infrastructure design	When the system is important	Security testing is a complex task, usually done by an infrastructure analyst assigned to the project. In extreme cases, a professional firm may be hired.
	Performance testing: examines the ability to perform under high loads	System proposal and infrastructure design	When the system is important	High volumes of transactions are generated and given to the system. This test is often done by using special-purpose testing software.
	Documentation testing: tests the accuracy of the documentation	Help system, procedures, tutorials	For normal system testing	Analysts spot-check or check every item on every page in all documentation to ensure the documentation items and examples work properly.
Acceptance Testing	Alpha testing: conducted by users to ensure they accept the system	System tests	For normal acceptance testing	Alpha tests often repeat previous tests but are conducted by users themselves to ensure they accept the system.
	Beta testing: uses real data, not test data	No plan	When the system is important	Users closely monitor the system for errors or useful improvements.

DFD = data flow diagram.

FIGURE 13-4
Types of Tests

Acceptance Tests

Acceptance tests are done primarily by the users with support from the project team. The goal is to confirm that system is complete, meets the business needs that prompted the system to be developed, and is acceptable to the users. Acceptance testing is done in two stages: *alpha testing,* in which users test the system using made-up data, and *beta testing,* in which users begin to use the system with real data but are carefully monitored for errors (see Figure 13-4).

DEVELOPING DOCUMENTATION

There are two fundamentally different types of documentation. *System documentation* is intended to help programmers and systems analysts understand the application software and enable them to build it or maintain it after the system is installed. System documentation is a by-product of the systems analysis and design process and is created as the project unfolds. Each step and phase produces documents that are essential in understanding how the system is built or is to be built, and these documents are stored in the project binder(s).

User documentation (such as user's manuals, training manuals, and online help systems) is designed to help the user operate the system. Although most project teams expect users to have received training and to have read the user's manuals before operating the system, unfortunately this is not always the case. It is more common today—especially in the case of commercial software packages for microcomputers—for users to begin using the software without training or reading the user's manuals. In this section, we focus on user documentation.[6]

User documentation is often left until the end of the project, which is a dangerous strategy. Developing good documentation takes longer than many people expect because it requires much more than simply writing a few pages. Producing documentation requires designing the documents (whether paper or online), writing the text, editing them, and testing them. For good-quality documentation, this process usually takes about 3 hours per page (single-spaced) for paper-based documentation or 2 hours per screen for online documentation. Thus, a "simple" set of documentation such as a 10-page user's manual and a set of 20 help screens takes 70 hours. Of course, lower-quality documentation can be produced faster.

The time required to develop and test user documentation should be built into the project plan. Most organizations plan for documentation development to start once the interface design and program specifications are complete. The initial draft of documentation is usually scheduled for completion immediately after the unit tests are complete. This reduces—but doesn't eliminate—the chance that the documentation will need to be changed because of software changes, and it still leaves enough time for the documentation to be tested and revised before the acceptance tests are started.

Although paper-based manuals are still important, online documentation is becoming more important. Paper-based documentation is simpler to use because it is more familiar to users, especially novices who have less computer experience; online documentation requires the users to learn one more set of commands. Paper-based documentation also is easier to flip through to gain a general understanding of its organization and topics and can be used far away from the computer itself.

There are four key strengths of online documentation that all but guarantee it will be the dominant form for the next century. First, searching for information is often simpler (provided the help search index is well designed) because the user can type in a variety of keywords to view information almost instantaneously, rather than having to search through the index or table of contents in a paper document. Second, the same information can be presented several times in many different formats, so that the user can find and read the information in the most informative way. (Such redundancy is possible in paper documentation, but the cost and intimidating size of the resulting manual make it impractical.) Third, online documentation enables the user to interact with the documentation in many new ways that are not possible with static paper documentation. For example, it is possible to use links or "tool tips" (i.e., pop-up text; see Chapter 10) to explain unfamiliar terms, and programmers can write "show me" routines that demonstrate on the screen exactly what buttons to click and text to type. Finally, online documentation is significantly less expensive to distribute than paper documentation.

---

[6] For more information on developing documentation, see Thomas T. Barker, *Writing Software Documentation,* Boston: Allyn & Bacon, 1998.

Types of Documentation

There are three fundamentally different types of user documentation: reference documents, procedures manuals, and tutorials. *Reference documents* (also called the help system) are designed to be used when the user needs to learn how to perform a specific function (e.g., updating a field, adding a new record). Often, people read reference information when they have tried and failed to perform the function; writing reference documents requires special care because often users are impatient or frustrated when they begin to read them.

Procedures manuals describe how to perform business tasks (e.g., printing a monthly report, taking a customer order). Each item in the procedures manual typically guides the user through a task that requires several functions or steps in the system. Therefore, each entry is typically much longer than an entry in a reference document.

Tutorials teach people how to use major components of the system (e.g., an introduction to the basic operations of the system). Each entry in the tutorial is typically longer still than the entries in procedures manuals, and the entries are usually designed to be read in sequence, whereas enteries in reference documents and procedures manuals are designed to be read individually.

Regardless of the type of user documentation, the overall process for developing it is similar to the process of developing interfaces (see Chapter 10). The developer first designs the general structure for the documentation and then develops the individual components within it.

Designing Documentation Structure

In this section, we focus on the development of online documentation, because we believe it will become the most common form of user documentation. The general structure used in most online documentation, whether reference documents, procedures manuals, or tutorials, is to develop a set of *documentation navigation controls* that lead the user to *documentation topics*. The documentation topics are the material that users want to read, whereas the navigation controls are the way in which users locate and access a specific topic.

Designing the structure of the documentation begins by identifying the different types of topics and navigation controls that must be included. Figure 13-5 shows a commonly used structure for online reference documents (i.e., the help system). The documentation topics generally come from three sources. The first and most obvious source of topics is the set of commands and menus in the user interface. This set of topics is very useful if the user wants to understand how a particular command or menu is used.

However, users often don't know what commands to look for or where they are in the system's menu structure. Instead, users have tasks they want to perform and rather than thinking in terms of commands, they think in terms of their business tasks. Therefore, the second and often more useful set of topics focuses on how to perform certain tasks, usually those in the use scenarios from the user interface design (see Chapter 10). These topics walk the user through the set of steps (often involving several keystrokes or mouse clicks) needed to perform some task.

The third set of topics are definitions of important terms. These terms are usually the entities and data elements in the system, but sometimes they also include commands.

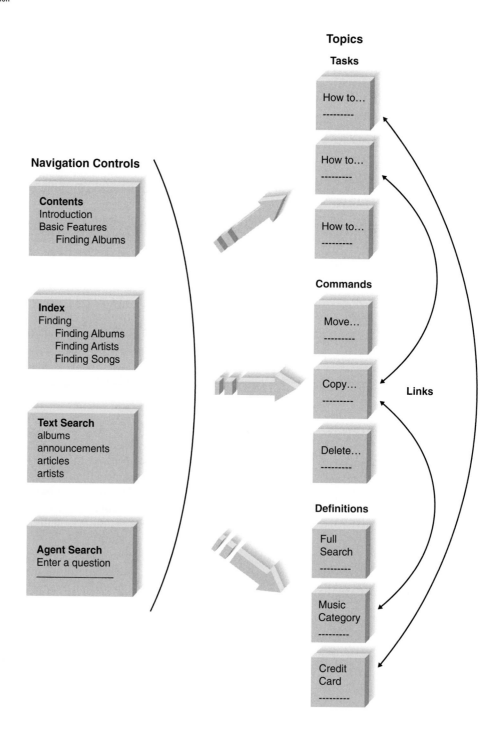

FIGURE 13-5
Organizing Online Reference
Documents

There are five general types of navigation controls for topics, but not all systems use all five types (see Figure 13-5). The first is the table of contents that organizes the information in a logical form, as though the users were to read the reference documentation from start to finish. The second, the index, provides access into the topics using important keywords, in the same way that the index at the back of

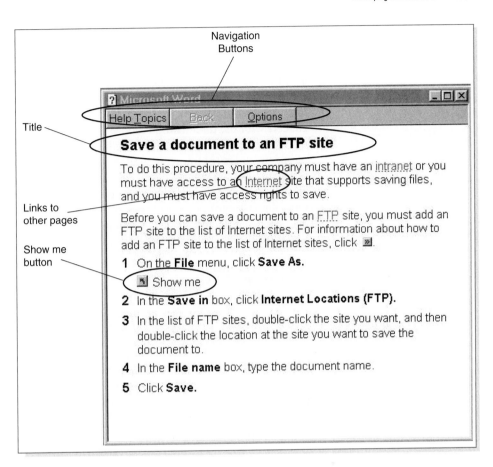

Navigation
Buttons

Title

Links to
other pages

Show me
button

FIGURE 13-6
A Help Topic in Microsoft Word

a book helps you find topics. Third, text search provides the ability to search through the topics either for any text the user types or for words that match a developer-specified set of words that is much larger than the words in the index. Unlike the index, text search typically provides no organization to the words (other than alphabetic). Fourth, some systems provide the ability to use an intelligent agent to help in the search (e.g., the Microsoft Office Assistant, also known as the paper clip guy). The fifth and final navigation control to topics are the Web-like links between topics that enable the user to click and move among topics.

Procedures manuals and tutorials are similar but often simpler in structure. Topics for procedures manuals usually come from the use scenarios developed during interface design and from other basic tasks the users must perform. Topics for tutorials are usually organized around major sections of the system and the level of experience of the user. Most tutorials start with basic, most commonly used commands and then move into more complex and less frequently used commands.

Writing Documentation Topics

The general format for topics is fairly similar across application systems and operating systems (Figure 13-6). Topics typically start with very clear titles, followed by some introductory text that defines the topic, and then provide detailed, step-by-step instructions on how to perform what is being described (where appropriate). Many topics include screen images to help the user find items on the screen; some also have "show me" examples in which the series of keystrokes and/or mouse movements and clicks needed to perform the function are demonstrated to the user. Most also include navigation controls to enable movement among topics, usually at the top of the window, plus links to other topics. Some also have links to related topics that include options or other commands and tasks the user may want to perform in concert with the topic being read.

Writing the topic content can be challenging. It requires a good understanding of the users (or, more accurately, the range of users) and a knowledge of what skills the users currently have and can be expected to import from other systems and tools they are using or have used (including the system the new system is replacing). Topics should always be written from the viewpoint of the user and describe what the user wants to accomplish, not what the system can do. Figure 13-7 provides some general guidelines to improve the quality of documentation text.[7]

Identifying Navigation Terms

As you write the documentation topics, you also begin to identify the terms that will be used to help users find topics. The table of contents is usually the most straightforward, because it is developed from the logical structure of the documentation topics, whether reference topics, procedure topics, or tutorial topics. The items for the index and search engine require more care because they are developed from the major parts of the system and users' business functions. Every time you write a topic, you must also list the terms that will be used to find the topic. Terms for the index and search engine can come from four distinct sources.

The first source for index terms is the set of the commands in the user interface, such as *open file, modify customer,* and *print open orders.* All commands contain two parts (action and object). It is important to develop the index for both parts because users could search for information using either part. A user looking for more information about saving files, for example, might search by using the term *save* or the term *files.*

The second source is the set of major concepts in the system, which are often the entities, data stores, and data elements in the data flow diagrams. In the case of CD Selections, for example, this might include *music category, album,* and *shipping costs.*

A third source is the set of business tasks the user performs, such as ordering replacement units or making an appointment. Often these will be contained in the command set, but sometimes these require several commands and use terms that do not always appear in the system. A good source for these terms is the use scenarios developed using interface design (see Chapter 10).

A fourth, often controversial, source is the set of synonyms for the three sets of items above. Users sometimes don't think in terms of the nicely defined terms used by the system. They may try to find information on how to *stop* or *quit* rather than *exit,* or on how to *erase* rather than *delete.* Including synonyms in the index increases the complexity and size of the documentation system but can greatly improve the value of the system to the users.

---

[7] One of the best books to explain the art of writing is that by William Strunk Jr. and E. B. White, *Elements of Style,* 3rd ed., Needham Heights, MA: Allyn & Bacon, 1995.

Guideline	Before the Guideline	After the Guideline
Use the active voice: The active voice creates more active and readable text by putting the subject at the start of the sentence, the verb in the middle, and the object at the end.	Finding albums is done using the album title, the artist's name, or a song title.	You can find an album by using the album title, the artist's name, or a song title.
Use e-prime style: E-prime style creates more active writing by omitting all forms of the verb *to be*.	The text you want to copy must be selected before you click on the copy button.	Select the text you want to copy before you click on the copy button.
Use consistent terms: Always use the same term to refer to the same items, rather than switching among synonyms (e.g., change, modify, update).	Select the text you want to copy. Pressing the *copy* button will copy the marked text to the new location.	Select the text you want to copy. Pressing the *copy* button will copy the selected text to the new location.
Use simple language: Always use the simplest language possible to accurately convey the meaning. This does not mean you should "dumb down" the text but that you should avoid artificially inflating its complexity. Avoid separating subjects and verbs and try to use the fewest words possible. (When you encounter a complex piece of text, try eliminating words; you may be surprised at how few words are really needed to convey meaning.)	The Georgia Statewide Academic and Medical System (GSAMS) is a cooperative and collaborative distance learning network in the state of Georgia. The organization in Atlanta that administers and manages the technical and overall operations of the currently more than 300 interactive audio and video teleconferencing classrooms throughout Georgia system is the Department of Administrative Service (DOAS). (56 words)	The Department of Administrative Service (DOAS) in Atlanta manages the Georgia Statewide Academic and Medical System (GSAMS), a distance learning network with more than 300 teleconferencing classrooms throughout Georgia. (29 words)
Use friendly language: Too often, documentation is cold and sterile because it is written in a very formal manner. Remember, you are writing for a person, not a computer.	Blank disks have been provided to you by Operations. It is suggested that you ensure your data is not lost by making backup copies of all essential data.	You should make a backup copy of all data that is important to you. If you need more diskettes, contact Operations.
Use parallel grammatical structures: Parallel grammatical structures indicate the similarity among items in list and help the reader understand content.	Opening files Saving a document How to delete files	Opening a file Saving a file Deleting a file
Use steps correctly: Novices often intersperse action and the results of action when describing a step-by-step process. Steps are always actions.	1. Press the *customer* button. 2. The customer dialogue box will appear. 3. Type the customer ID and press the *submit* button and the customer record will appear.	1. Press the *customer* button. 2. Type the customer ID in the customer dialogue box when it appears. 3. Press the *submit* button to view the customer record for this customer.
Use short paragraphs: Readers of documentation usually quickly scan text to find the information they need, so the text in the middle of long paragraphs is often overlooked. Use separate paragraphs to help readers find information more quickly.		

Source: Adapted from *Writing Software Documentation*, Boston: Allyn & Bacon, 1998, by T. T. Barker.

FIGURE 13-7
Guidelines for Crafting Documentation Topics

YOUR TURN

13-2 Documentation for an Automated Teller

Pretend you are a project manager for a bank developing software for automated teller machines. Develop an online help system.

APPLYING THE CONCEPTS AT CD SELECTIONS

Managing Programming

Three programmers were assigned by CD Selections to develop the three major parts of the Internet system. The first was the Web interface, both the client side (browser) and the server side. The second, the management system (managing the marketing materials databases), was client–server based. The third was the interfaces between the Internet system and CD Selections' existing special order system. Programming went smoothly and, despite a few minor problems, according to plan.

Testing

While the programmers were working, Alec—Senior Systems Analyst and Project Manager for CD Selections' Internet Order System—began developing the test plans and user documentation. The test plans for the three components were similar but slightly more intensive for the Web interface component (Figure 13-8). Unit testing, using black-box testing from program specifications, was planned for all components. Figure 13-9 shows part of one unit test for the Web interface component.

Integration testing for the Web interface and system management component would be subjected to all user interface and use scenario tests to ensure the interface worked properly. The system interface component would undergo system interface tests to ensure that the system performed calculations properly and was capable of exchanging data with CD Selections' other systems.

Systems tests are by definition tests of the entire system—all components together. However, not all parts of the system would receive the same level of testing. Requirements tests would be conducted on all parts of the system to ensure that all requirements were met. Security was a critical issue, so the security of all

Test Stage	Web Interface	System Management	System Interfaces
Unit tests	Black-box tests	Black-box tests	Black-box tests
Integration tests	User interface tests; use scenario tests	User interface tests; use scenario tests	System interface tests
System tests	Requirements tests; security tests; performance tests; usability tests	Requirements tests; security tests	Requirements tests; security tests; performance tests
Acceptance tests	Alpha test; beta test	Alpha test; beta test	Alpha test; beta test

FIGURE 13-8
CD Selections' Test Plan

Test Plan

Page: 12 of 32

Program ID: ORD56   Version Number: 3

Tester: Smith   Date Designed: 9/9   Date Conducted: 9/9

Results: ☑ Passed   ☐ Open Items

Test ID: 12   Required Addressed: Verify ordering information

Objective:
Ensure that the information entered by customer on the place request form
is valid

Test Cases

	Interface ID	Data Field	Value Entered
1)	REQ56-3.5	Zip code	Blank
2)	REQ56-3.5	Zip code	9021
3)	REQ56-3.5	Zip code	90210
4)	REQ56-3.5	Zip code	C1A 2C6
5)			
6)			

Script

Expected Results/Notes

Test 3 is a valid zip code. All others should be rejected.

Actual Results/Notes

Test 3 accepted. Tests 1, 2, and 4 were rejected with correct error message.

FIGURE 13-9
CD Selections Unit Test Plan Example

aspects of the system would be tested. Security tests would be developed by CD Selections' infrastructure team, and once the system passed those tests, an external security consulting firm would be hired to attempt to break in to the system.

Performance was an important issue for the parts of the system used by the customer (the Web interface and the system interfaces to the inventory system) but not as important for the management component that would be used by staff, not customers. The customer-facing components would undergo rigorous performance testing to see how many transactions (whether searching or placing a request) they could handle before they were unable to provide a response time of 2 seconds or less. Alec also developed an upgrade plan so that as demand on the system increased, there was a clear plan for when and how to increase the processing capability of the system.

Finally, formal usability tests would be conducted on the Web interface portion of the system with six potential users (both novice and expert Internet users).

Acceptance tests would be conducted in two stages, alpha and beta. Alpha tests would be done during the training of CD Selections' staff. The Marketing Manager would work together with Alec to develop a series of tests and training exercises to train staff on how to use the system. They would then load the real CD data into the system and begin adding marketing materials. These same staff and other CD Selections staff members would also pretend to be customers and test the Web interface.

Beta testing would be done by going live with the Web site but announcing its existence only to CD Selections employees. As an incentive to try the Web site employees would be offered triple their normal employee discount for the first three products requested from the Web site. The site would also have a prominent button on every screen that would enable employees to e-mail comments to the project team, and the announcement would encourage employees to report problems, suggestions, and compliments to the project team. After 1 month, assuming all went well, the beta test would be completed, and the site would be linked to the main Web site and advertised to the general public.

Developing User Documentation

There were three types of documentation (reference documents, procedures manuals, and tutorials) that could be produced for the Web interface and the management component. Since the number of CD Selections staff using the system management component would be small, Alec decided to produce only the reference documetation (an online help system). He believed that an intensive training program and a 1 month beta test period would be sufficient without tutorials and formal procedures manuals. Likewise, he felt that the process of requesting CDs and the V interface itself was simple enough to not require a tutorial on the Web—a help system would be sufficient, and a procedures manual didn't make sense.

Alec decided that the reference documents for both the Web interface and system management components would contain help topics for user tasks, commands, and definitions. He also decided that the documentation component would contain four types of navigation controls: a table of contents, an index, a finder, and links to definitions. He did not think that the system was complex enough to benefit from a search agent.

After these decisions were made, Alec assigned the development of the reference documents to a technical writer assigned to the project team. Figure 13-10

Tasks	Commands	Terms
Find an album	Find	Album
Add an album to my shopping cart	Browse	Artist
Placing a request	Quick search	Music type
Checkout	Full search	Special deals
What's in my shopping cart?		Cart
		Shopping cart

FIGURE 13-10
Sample Help Topics for CD Selections

shows examples of the topics the writer developed. The tasks and commands were taken directly from the interface design. The list of definitions was put together, once the tasks and commands were developed, on the basis of the writer's experience in understanding what terms might be confusing to the user.

Once the topic list was developed, the technical writer then began writing the topics themselves and the navigation controls to access. Figure 13-11 shows an example of one topic taken from the task list: how to place a request. This topic presents a brief description of what it is and then leads the user through the step-by-step process needed to complete the task. The topic also lists the navigation controls that will be used to find the topic, in terms of the table of contents entries, the index entries, and search entries. It also lists what words in the topic itself will have links to other topics (e.g., shopping cart).

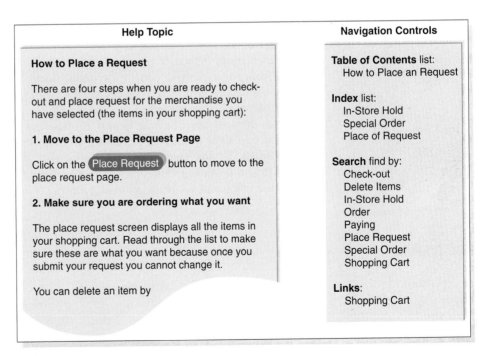

Help Topic

How to Place a Request

There are four steps when you are ready to check-out and place request for the merchandise you have selected (the items in your shopping cart):

1. Move to the Place Request Page

Click on the `Place Request` button to move to the place request page.

2. Make sure you are ordering what you want

The place request screen displays all the items in your shopping cart. Read through the list to make sure these are what you want because once you submit your request you cannot change it.

You can delete an item by

Navigation Controls

Table of Contents list:
　How to Place an Request

Index list:
　In-Store Hold
　Special Order
　Place of Request

Search find by:
　Check-out
　Delete Items
　In-Store Hold
　Order
　Paying
　Place Request
　Special Order
　Shopping Cart

Links:
　Shopping Cart

FIGURE 13-11
Example Documentation Topic for CD Selections

SUMMARY

Managing Programming

Programming is done by programmers, so systems analysts have few responsibilities during this stage. The project manager, however, is usually very busy. The first task is to assign the programmers—ideally the fewest possible to complete the project, because coordination problems increase as the size of the programming team increases. Coordination can be improved by having regular meetings, ensuring that standards are followed, implementing change control, and using computer-aided software engineering (CASE) tools effectively. One of the key functions of the project manager is to manage the schedule and adjust it for delays. Two common causes of delays are scope creep and minor slippages that go unnoticed.

Test Planning

Tests must be carefully planned because the cost of fixing one major bug after the system is installed can easily exceed the annual salary of a programmer. A test plan contains several tests that examine different aspects of the system. A test, in turn, specifies several test cases that will be examined by the testers. A unit test examines a module or program within the system; test cases come from the program specifications or the program code itself. An integration test examines how well several modules work together; test cases come from the interface design, use scenarios, and the physical data flow diagrams (DFDs). A system test examines the system as a whole and is broader than the unit and integration tests; test cases come from the system design, the infrastructure design, the unit tests, and the integration. Acceptance testing is done by the users to determine whether the system is acceptable to them; it draws on the system test plans (alpha testing) and the real work the users perform (beta testing).

Documentation

Documentation, both user documentation and system documentation, is moving away from paper-based documents to online documentation. There are three types of user documentation: reference documents are designed to be used when the user needs to learn how to perform a specific function (e.g., an online help system); procedures manuals describe how to perform business tasks; and tutorials teach people how to use the system. Documentation navigation controls (e.g., a table of contents, index, a "find" function, intelligent agents, or links between pages) enable users to find documentation topics (e.g., how to perform a function, how to use an interface command, an explanation of a term).

KEY TERMS

Acceptance test	Documentation navigation control	Requirements testing
Alpha test	Documentation topic	Scope creep
Beta test	Hardcoded	Security testing
Black-box testing	Integration test	Stub
Change control	Procedures manual	System documentation
Construction	Program log	System interface testing
Data flow testing	Reference document	System test

Test case	Unit test	User documentation
Test plan	Usability testing	User interface testing
Tutorial	Use scenario testing	White-box testing

QUESTIONS

1. Why is testing as important as programming?
2. What is the primary role of systems analysts during the programming stage?
3. In *The Mythical Man-Month*. Frederick Brooks argued that adding more programmers to a late project makes it later. Why?
4. Compare and contrast the terms *test, test plan,* and *test case.*
5. What is a stub, and why is it used in testing?
6. What is the primary goal of unit testing?
7. How are the test cases developed for unit tests?
8. What is the primary goal of integration testing?
9. How are the test cases developed for integration tests?
10. Compare and contrast black-box testing and white-box testing.
11. What is the primary goal of system testing?
12. How are the test cases developed for system tests?
13. What is the primary goal of acceptance testing?
14. How are the test cases developed for acceptance tests?
15. Compare and contrast alpha testing and beta testing.
16. Compare and contrast user documentation and system documentation.
17. Why is online documentation becoming more important?
18. What are the primary disadvantages of online documentation?
19. Compare and contrast reference documents, procedures manuals, and tutorials.
20. What are five types of documentation navigation controls?
21. What are the commonly used sources of documentation topics? Which is the most important? Why?
22. What are the commonly used sources of documentation navigation controls? Which is the most important? Why?
23. What do you think are three common mistakes that novice analysts make in programming and testing?
24. What do you think are three common mistakes that novice analysts make in developing documentation?
25. In our experience, documentation is often left to the very end of projects. Why do you think this happens, and how might you avoid this?
26. In our experience, few organizations conduct as thorough testing as they should (are the tests on your project as thorough as they should be?). Why do you think this happens, and how might you avoid this?
27. How can you develop good documentation? It might help if you think about developing bad documentation.

EXERCISES

A. Develop a unit test plan for the calculator program in Windows (or a similar program for the Mac or UNIX).
B. Develop a unit test plan for a Web site that enables you to perform some function (e.g., make travel reservations, order books).
C. If the registration system at your university does not have a good online help system, develop one for one screen of the user interface.
D. Examine and prepare a report on the online help system for the calculator program in Windows (or a similar program for the Mac or Unix). (You may be surprised at the amount of help for such a simple program).
E. Compare and contrast the online help at two different Web sites that enables you to perform the same function (e.g., make travel reservations, order books).

MINICASES

1. Pete is a project manager on a new systems development project. This project is Pete's first experience as a project manager, and he has led his team successfully to the programming phase of the project. The project has not always gone smoothly, and Pete has made a few mistakes, but he is generally pleased with the progress of his team and the quality of the system being developed. Now that programming has begun, Pete has been hoping for a little break in the hectic pace of his workday.

   Prior to beginning programming, Pete recognized that the time estimates made earlier in the project were too optimistic. However, he was firmly committed to meeting the project deadline because of his desire for his first project as project manager to be a success. In anticipation of this time pressure problem, Pete arranged with the Human Resources Department to bring in two new college graduates and two college interns to beef up the programming staff. Pete would have liked to find some staff with more experience, but the budget was too tight, and he was committed to keeping the project budget under control.

   Pete made his programming assignments, and work on the programs began about two weeks ago. Now, Pete has started to hear some rumbles from the programming team leaders that may signal trouble. It seems that the programmers have reported several instances where they wrote programs, only to be unable to find them when they went to test them. Also, several programmers

   have opened programs that they had written, only to find that someone had changed portions of their programs without their knowledge.

   a. Is the programming phase of a project a time for the project manager to relax? Why or why not?

   b. What problems can you identify in this situation? What advice do you have for the project manager? How likely does it seem that Pete will achieve his desired goals of being on time and within budget if nothing is done?

2. The systems analysts are developing the test plan for the user interface for the Holiday Travel Vehicles system. As the salespeople are entering a sales invoice into the system, they will be able to either enter an option code into a text box, or select an option code from a drop-down list. A combo box was used to implement this, since it was felt that the salespeople would quickly become familiar with the most common option codes and would prefer entering them directly to speed up the entry process.

   It is now time to develop the test for validating the option code field during data entry. If the customer did not request any dealer-installed options for the vehicle, the salesperson should enter "none"; the field should not be blank. The valid option codes are four-character alphabetic codes and should be matched against a list of valid codes.

   Prepare a test plan for the test of the option code field during data entry.

PLANNING

ANALYSIS

DESIGN

IMPLEMENTATION

- ☑ **Program System**
- ☑ **Test Software**
- ☑ **Test Performance**
- ☐ Select System Conversion Style
- ☐ Train Users
- ☐ Select Support
- ☐ Maintain System
- ☐ Assess Project
- ☐ Conduct Post-Implementation Audit

TASK CHECKLIST

PLANNING → ANALYSIS → DESIGN →

CHAPTER 14

INSTALLATION

This chapter examines the activities needed to install the information system and successfully convert the organization to using it. It also discusses postimplementation activities, such as system support, system maintenance, and project assessment. Installing the system and making it available for use from a technical perspective is relatively straightforward. However, the training and organizational issues surrounding the installation are more complex and challenging because they focus on people, not computers.

OBJECTIVES

- Be familiar with the system installation process.
- Understand different types of conversion strategies and when to use them.
- Understand several techniques for managing change.
- Be familiar with postinstallation processes.

CHAPTER OUTLINE

IMPLEMENTATION

INTRODUCTION

It must be remembered that there is nothing more difficult to plan, more doubtful of success, nor more dangerous to manage than the creation of a new system. For the initiator has the animosity of all who would profit by the preservation of the old institution and merely lukewarm defenders in those who would gain by the new.

—Machiavelli, *The Prince,* 1513

Although written almost 500 years ago, Machiavelli's comments are still true today. Managing the change to a new system—whether it is computerized or not—is one of the most difficult tasks in any organization. Because of the challenges involved, most organizations begin developing their conversion and change management plans while the programmers are still developing the software. Leaving conversion and change management planning to the last minute is a recipe for failure.

In many ways, using a computer system or set of work processes is much like driving on a dirt road. Over time with repeated use, the road begins to develop ruts in the most commonly used parts of the road. Although these ruts show where to drive, they make change difficult. As people use a computer system or set of work processes, those system/work processes begin to become habits or norms; people learn them and become comfortable with them. These systems or work processes then begin to limit people's activities and make it difficult for them to change because they begin to see their jobs in terms of these processes rather than of the final business goal of serving customers.

One of the earliest models for managing organizational change was developed by Kurt Lewin.[1] Lewin argued that change is a three-step process: unfreeze, move, refreeze (Figure 14-1). First, the project team must *unfreeze* the existing habits and norms (the as-is system) so that change is possible. Most of the SDLC to this point has laid the groundwork for unfreezing. Users are aware of the new system being developed, some have participated in an analysis of the current system (and so are aware of its problems), and some have helped design the new system (and so have some sense of the potential benefits of the new system). These activities have helped to unfreeze the current habits and norms.

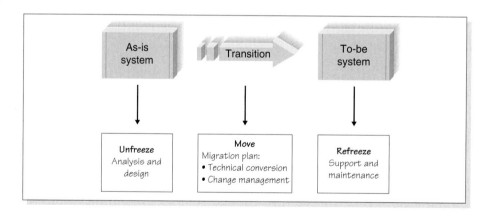

FIGURE 14-1

Implementing Change

---

[1] Kurt Lewin, "Fontiers in Group Dynamics," *Human Relations,* 1947, 1:5–41, and Kurt Lewin, "Group Decision and Social Change" in E.E. Maccoby, T.M. Newcomb, and E.L. Hartley (eds.), *Readings in Social Psychology,* New York: Holt, Rinehart & Winston, 1958, pp. 197–211.

The second step is to help the organization *transition* to the new system via a *migration plan.* The migration plan has two major elements. One is technical, which includes how the new system will be installed and how data in the as-is system will be moved into the to-be system; this is discussed in the "Conversion" section of this chapter. The second component is organizational, which includes helping users understand the change and motivating them to adopt it; this is discussed in the "Change Management" section of this chapter.

The third step is to *refreeze* the new system as the habitual way of performing the work processes—ensuring that the new system successfully becomes the standard way of performing the business function it supports. This refreezing process is a key goal of the "Postimplementation Activities" discussed in the final section of this chapter. By providing ongoing support for the new system and immediately beginning to identify improvements for the next version of the system, the organization helps solidify the new system as the new habitual way of doing business. Postimplementation activities include *system support,* which means providing help desk and telephone support for users with problems; *system maintenance,* which means fixing bugs and improving the system after it has been installed; and *project assessment,* which means the process of evaluating the project to identify what went well and what could be improved for the next system development project.

Change management is the most challenging of the three components, because it focuses on people, not technology, and because it is the one aspect of the project that is the least controllable by the project team. Change management means winning the hearts and minds of potential users and convincing them that the new system actually provides value.

Maintenance is the most costly aspect of installation process, because the cost of maintaining systems usually greatly exceeds the initial development costs. It is not unusual for organizations to spend 60% to 80% of their total IS development budget on maintenance. Although this may sound surprising initially, think about the software you use. How many software packages do you use that are the very first version? Most commercial software packages become truly useful and enter widespread use only in their second or third version. Maintenance and continual improvement of software is ongoing, whether it is a commercially available package or software developed in-house. Would you buy software if you knew that no new versions were going to be produced? Of course, commercial software is somewhat different from custom in-house software used by only one company, but the fundamental issues remain.

Project assessment is probably the least commonly performed part of the SDLC but is perhaps the one that has the most long-term value to the IS department. Project assessment enables project team members to step back and consider what they did right and what they could have done better. It is an important component in the individual growth and development of each member of the team, because it encourages team members to learn from their successes and failures. It also enables new ideas or new approaches to system development to be recognized, examined, and shared with other project teams to improve their performance.

CONVERSION

Conversion is the technical process by which the new system replaces the old system. Users are moved from using the as-is business processes and computer programs to the to-be business processes and programs. The migration plan specifies what activities will be performed when and by whom and includes both technical aspects (such as installing hardware and software and converting data from the as-is system to the to-be system) and organizational aspects (such as training and motivating the users to embrace the new system). Conversion refers to the technical aspects of the migration plan. Organizational aspects are discussed in the "Change Management" section.

There are three major steps to the conversion plan before commencement of operations: install hardware, install software, and convert data (Figure 14-2). Although it may be possible to do some of these steps in parallel, usually they must be done sequentially at any one location.

The first step in the conversion plan is to buy and install any needed hardware. In many cases, no new hardware is needed, but sometimes the project requires such new hardware as servers, client computers, printers, and networking equipment. It is critical to work closely with vendors who are supplying needed hardware and software to ensure that the deliveries are coordinated with the conversion schedule so that the equipment is available when it is needed. Nothing can stop a conversion plan in its tracks as easily as the failure of a vendor to deliver needed equipment.

Once the hardware is installed, tested, and certified as being operational, the second step is to install the software. This includes the to-be system under development, and sometimes additional software that must be installed to make the system operational. For example, the CD Selections Internet Order System needs Web server software. At this point, the system is usually tested again to ensure that it operates as planned.

The third step is to convert the data from the as-is system to the to-be system. Data conversion is usually the most technically complicated step in the migration plan. Often, separate programs must be written to convert the data from the as-is system to the new formats required in the to-be system and store it in the to-be system files and databases. This process is often complicated by the fact that the files and databases in the to-be system do not exactly match the files and databases in the as-is system (e.g., the to-be system may use several tables in a database to store customer data that was contained in one file in the as-is system). Formal test plans are always required for data conversion efforts (see Chapter 13).

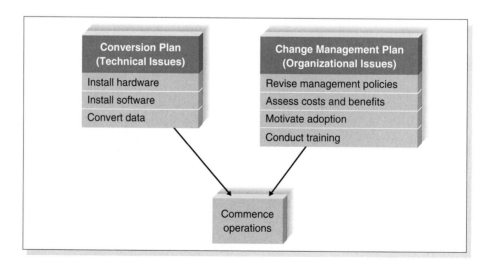

FIGURE 14-2
Elements of a Migration Plan

Conversion can be thought of along three dimensions: the style by which the conversion is done, what location or work groups are converted at what time, and what modules of the system are converted at what time (Figure 14-3).

Conversion Style

The *conversion style* is the way in which users are switched between the old and new systems. There are two fundamentally different approaches to the style of conversion: direct conversion and parallel conversion.

Direct Conversion   With *direct conversion* (sometimes called cold turkey, big bang, or abrupt cutover), the new system instantly replaces the old system. The new system is turned on, and the old system is immediately turned off. This is the approach that you probably use when you upgrade commercial software (e.g., Microsoft Word) from one version to another; you simply begin using the new version and stop using the old version.

Direct conversion is the simplest and most straightforward. However, it is the most risky, because any problems with the new system that have escaped detection during testing may seriously disrupt the organization.

Parallel Conversion   With *parallel conversion,* the new system is operated side by side with the old system; both systems are used simultaneously. For example, if a new accounting system is installed, the organization enters data into both the old system and the new system and then carefully compares the output from both systems to ensure that the new system is performing correctly. After some time period (often 1 to 2 months) of parallel operation and intense comparison between the two systems, the old system is turned off and the organization continues using the new system.

This approach is more likely to catch any major bugs in the new system and prevent the organization from suffering major problems. If problems are discovered in the new system, the system is simply turned off and fixed and then the conversion process starts again. The problem with this approach is the added expense of operating two systems that perform the same function.

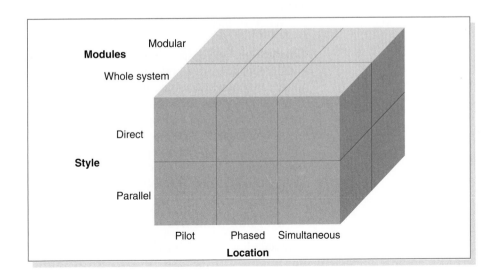

FIGURE 14-3
Conversion Strategies

Conversion Location

Conversion location refers to what parts of the organization are converted at what points in time. Often, parts of the organization are physically located in different offices (e.g., Toronto, Atlanta, Los Angeles). In other cases, location refers to different organizational units located in different parts of the same office complex (e.g., order entry, shipping, purchasing). There are at least three fundamentally different approaches to selecting the way in which different organizational locations are converted: pilot conversion, phased conversion, and simultaneous conversion.

Pilot Conversion   With a *pilot conversion,* one or more locations or units/work groups within a location are selected to be converted first as part of a pilot test. The locations participating in the pilot test are converted (using either direct or parallel conversion). If the system passes the pilot test, then the system is installed at the remaining locations (again using either direct or parallel conversion).

Pilot conversion has the advantage of providing an additional level of testing before the system is widely deployed throughout the organization, so that any problems with the system affect only the pilot locations. However, this type of conversion obviously requires more time before the system is installed at all organizational locations. Also, it means that different organizational units are using different versions of the system and business processes, which may make it difficult for them to exchange data.

Phased Conversion   With *phased conversion,* the system is installed sequentially at different locations. A first set of locations are converted, then a second set, then a third set, and so on, until all locations are converted. Sometimes there is a deliberate delay between the different sets (at least between the first and the second), so that any problems with the system are detected before too much of the organization is affected. In other cases, the sets are converted back-to-back so that as soon as those converting one location have finished, the project team moves to the next and continues the conversion.

Phased conversion has the same advantages and disadvantages of pilot conversion. Additionally, it means that a smaller set of people are required to perform the actual conversion (and any associated user training) than if all locations were converted at once.

Simultaneous Conversion   *Simultaneous conversion,* as the name suggests, means that all locations are converted at the same time. The new system is installed and made ready at all locations, and at a preset time, all users begin using the new system. Simultaneous conversion is often used with direct conversion, but it can also be used with parallel conversion.

Simultaneous conversion eliminates problems with having different organizational units using different systems and processes. However, it also means that the organization must have sufficient staff to perform the conversion and train the users at all locations simultaneously.

Conversion Modules

Although it is natural to assume that systems are usually installed in their entirety, this is not always the case.

CONCEPTS

14-A CONVERTING TO THE EURO (PART 1)

IN ACTION

When the European Union decided to introduce the euro, the European Central Bank had to develop a new computer system (called Target) to provide a currency settlement system for use by investment banks and brokerages. Prior to the introduction of the euro, settlement was performed between the central banks of the countries involved. After the introduction, the Target system, which consists of 15 national banking systems, would settle trades and perform currency conversions for cross-border payments for stocks and bonds.

Source: "Debut of Euro Nearly Flawless," *Computerworld*, 33(2) p. 16, January 11, 1999, by Thomas Hoffman.

QUESTION:
Implementing Target was a major undertaking for a number of reasons. If you were an analyst on the project, what kinds of issues would you have to address to make sure the conversion happened successfully?

Whole System Conversion   *Whole-system conversions,* in which the entire system is installed at one time, is the most common. It is simple and the easiest to understand. However, if the system is large and/or extremely complex (e.g., an enterprise resource planning system such as SAP or PeopleSoft) the whole system may prove too difficult for users to learn in one conversion step.

Modular Conversion   When the modules within a system are separate and distinct, organizations sometimes choose to convert to the new system one module at a time—that is, to use *modular conversion.* Modular conversion requires special care in developing the system (and usually adds extra cost) because each module must be written to work with both the old and new systems. When modules are tightly integrated, this is very challenging and therefore seldom done. However, when there is only a loose association between modules, this becomes easier. For example, consider a conversion from an old version of Microsoft Office to a new version. It is relatively simple to convert from the old version of Word to the new version without simultaneously having to change from the old to the new version of Microsoft Excel.

Modular conversion reduces the amount of training required to begin using the new system. Users need training in only the new module being implemented. However, modular conversion does take longer and has more steps than does the whole-system process.

Selecting the Appropriate Conversion Strategy

Each of the three dimensions in Figure 14-3 are independent, so that a conversion strategy can be developed to fit in any one of the boxes in this figure. Different boxes can also be mixed and matched into one conversion strategy. For example, one commonly used approach is to begin with a pilot conversion of the whole system using parallel conversion in a handful of test locations. Once the system has passed the pilot test at these locations, it is then installed in the remaining locations using phased conversion with direct cutover. There are three important factors to consider in selecting a conversion strategy: risk, cost, and the time required (Figure 14-4).

Characteristic	Conversion Style		Conversion Location			Conversion Modules	
	Direct Conversion	**Parallel Conversion**	**Pilot Conversion**	**Phased Conversion**	**Simultaneous Conversion**	**Whole-System Conversion**	**Modular Conversion**
Risk	High	Low	Low	Medium	High	High	Medium
Cost	Low	High	Medium	Medium	High	Medium	High
Time	Short	Long	Medium	Long	Short	Short	Long

FIGURE 14-4
Characteristics of Conversion Strategies

Risk   After the system has passed a rigorous battery of unit, system, integration, and acceptance testing, it should be bug free . . . maybe. Because humans make mistakes, nothing built by people is ever perfect. Even after all these tests, there may still be a few undiscovered bugs. The conversion process provides one last step in which to catch these bugs before the system goes live and the bugs have the chance to cause problems.

Parallel conversion is less risky than is direct conversion because it has a greater chance of detecting bugs that have gone undiscovered in testing. Likewise, pilot conversion is less risky than is phased conversion or simultaneous conversion because if bugs do occur, they occur in pilot test locations whose staff are aware that they may encounter bugs. Since potential bugs affect fewer users, there is less risk. Likewise, converting a few modules at a time lowers the probability of a bug because there is more likely to be a bug in the whole system than in any given module.

How important the risk is depends on the system being implemented—the combination of the probability that bugs remain undetected in the system and the potential cost of those undetected bugs. If the system has indeed been subjected to extensive methodical testing, including alpha and beta testing, then the probability of undetected bugs is lower than if the testing was less rigorous. However, there still might have been mistakes made in the analysis process, so that although there might be no software bugs, the software might fail to properly address the business needs.

Assessing the cost of a bug is challenging, but most analysts and senior managers can make a reasonable guess at the relative cost of a bug. For example, the cost of a bug in an automated stock market trading program or a heart–lung machine keeping someone alive is likely to be much greater than a bug in a computer game or word processing program. Therefore, risk is likely to be a very important factor in the conversion process if the system has not been as thoroughly tested as it might have been and/or if the cost of bugs is high. If the system has been thoroughly tested and/or the cost of bugs is not that high, then risk becomes less important to the conversion decision.

Cost   As might be expected, different conversion strategies have different costs. These costs can include things such as salaries for people who work with the system (e.g., users, trainers, system administrators, external consultants), travel expenses, operation expenses, communication costs, and hardware leases. Parallel conversion is more expensive than direct cutover because it requires that two systems (the old and the new) be operated at the same time. Employees must now perform twice the usual work because they have to enter the same data into both the old and the new systems. Parallel conversion also requires the results of the two systems to be completely cross-checked to make sure there are no differences between the two, which entails additional time and cost.

Pilot conversion and phased conversion have somewhat similar costs. Simultaneous conversion has higher costs because more staff are required to support all the locations as they simultaneously switch from the old to the new system. Modular conversion is more expensive than whole system conversion because it requires more programming. The old system must be updated to work with selected modules in the new system, and modules in the new system must be programmed to work with selected modules in both the old and new systems.

Time   The final factor is the amount of time required to convert between the old and the new system. Direct conversion is the fastest because it is immediate. Parallel conversion takes longer because the full advantages of the new system do not become available until the old system is turned off. Simultaneous conversion is fastest because all locations are converted at the same time. Phased conversion usually takes longer than pilot conversion because usually (but not always) once the pilot test is complete all remaining locations are simultaneously converted. Phased conversion proceeds in waves, often requiring several months before all locations are converted. Likewise, modular conversion takes longer than whole-system conversion because the models are introduced one after another.

CHANGE MANAGEMENT

In the context of a systems development project, change management is the process of helping people to adopt and adapt to the to-be system and its accompanying work processes without undue stress.[2] There are three key roles in any major organizational change. The first is the *sponsor* of the change—the person who wants the change. This person is the business sponsor who first initiated the request for the new system (see Chapter 2). Usually the sponsor is a senior manager of the part of the organization that must adopt and use the new system. It is critical that the sponsor be active in the change management process, because a change that is clearly being driven by the sponsor, not by the project team or the IS organization, has greater legitimacy. The sponsor has direct management authority over those who adopt the system.

The second role is that of the *change agent*—the person(s) leading the change effort. The change agent, charged with actually planning and implementing the change, is usually someone outside of the business unit adopting the system and therefore has no direct management authority over the potential adopters. Because the change agent is an outsider from a different organizational culture, he or she has less credibility than do the sponsor and other members of the business unit. After all, once the system has been installed, the change agent usually leaves and thus has no ongoing impact.

The third role is that of *potential adopter* or target of the change—the people who actually must change. These are the people for whom the new system is designed and who will ultimately choose to use or not use the system.

In the early days of computing, many project teams simply assumed that their job ended when the old system was converted to the new system at a technical level. The philosophy was "build it and they will come." Unfortunately, that happens only in the movies. Resistance to change is common in most organizations. Therefore, the change management plan is an important part of the overall installation plan that glues together the key steps in the change management process. Successful change requires that people want to adopt the change and are able to adopt the change. The change management plan has four basic steps: revising management policies, assessing the cost and benefit models of potential adopters, motivating adoption, and enabling people to adopt through training (see Figure 14-2). How-

---

[2] Many books have been written on change management. Some of our favorites are the following: Patrick Connor and Linda Lake, *Managing Organizational Change,* 2nd ed., Westport, CT: Praeger, 1994; Douglas Smith, *Taking Charge of Change,* Reading, MA: Addison-Wesley, 1996; and Daryl Conner, *Managing at the Speed of Change,* New York: Villard Books, 1992.

ever, before we can discuss the change management plan, we must first understand why people resist change.

Understanding Resistance to Change

People resist change—even change for the better—for very rational reasons.[3] What is good for the organization is not necessarily good for the people who work there. For example, consider an order-processing clerk who used to receive orders to be shipped on paper shipping documents but now uses a computer to receive the same information. Rather than typing shipping labels with a typewriter, the clerk now clicks on the print button on the computer and the label is produced automatically. The clerk can now ship many more orders each day, which is a clear benefit to the organization. The clerk, however, probably doesn't really care how many packages are shipped. His or her pay doesn't change; it's just a question of which the clerk prefers to use, a computer or typewriter. Learning to use the new system and work processes—even if the change is minor—requires more effort than continuing to use the existing well-understood, system and work processes.

So why do people accept change? Simply put, every change has a set of costs and benefits associated with it. If the benefits of accepting the change outweigh the costs of the change, then people change. And sometimes the benefit of change is avoidance of the pain that you would experience if you did not adopt the change (e.g., if you don't change, you are fired, so one of the benefits of adopting the change is that you still have a job).

In general, when people are presented with an opportunity for change, they perform a cost–benefit analysis (sometime consciously, sometimes subconsciously) and decide the extent to which they will embrace and adopt the change. They identify the costs of and benefits from the system and decide whether the change is worthwhile. However, it is not that simple, because most costs and benefits are not certain. There is some uncertainty as to whether a certain benefit or cost will actually occur; so both the costs of and benefits from the new system will need to be weighted by the degree of certainty associated with them (Figure 14-5). Unfortu-

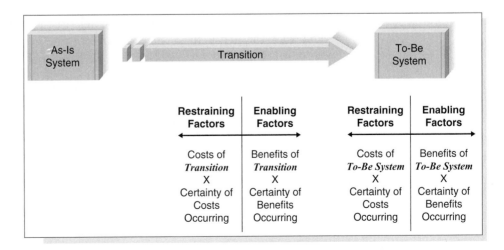

FIGURE 14-5
The Costs and Benefits of Change

---

[3] This section benefited from conversations with Dr. Robert Briggs, research scientist at the Center for the Management of Information at the University of Arizona.

nately, most humans tend to overestimate the probability of costs and underestimate the probability of benefits.

There are also costs and sometimes benefits associated with the actual transition process itself. For example, suppose you found a nicer house or apartment than your current one. Even if you liked it better, you might decide not to move simply because the cost of moving outweighed the benefits from the new house or apartment itself. Likewise, adopting a new computer system might require you to learn new skills, which could be seen as a cost to some people, or as a benefit to others, if they perceived that those skills would somehow provide other benefits beyond the use of the system itself. Once again, any costs and benefits from the transition process must be weighted by the certainty with which they will occur (see Figure 14-5).

Taken together, these two sets of costs and benefits (and their relative certainties) affect the acceptance of change or resistance to change that project teams encounter when installing new systems in organizations. The first step in change management is to understand the factors that inhibit change—the factors that affect the *perception* of costs and benefits and certainty that they will be generated by the new system. It is critical to understand that the "real" costs and benefits are far less important than the perceived costs and benefits. People act on what they *believe* to be true, not on what *is* true. Thus, any understanding of how to motivate change must be developed from the viewpoint of the people expected to change, not from the viewpoint of those leading the change.

Revising Management Policies

The first major step in the change management plan is to change the management policies that were designed for the as-is system to new management policies designed to support the to-be system. *Management policies* provide goals, define how work processes should be performed, and determine how organizational members are rewarded. No computer system will be successfully adopted unless management policies support its adoption. Many new computer systems bring changes to business processes; they enable new ways of working. Unless the policies that provide the rules and rewards for those processes are revised to reflect the new opportunities that the system permits, potential adopters cannot easily use it.

Management has three basic tools for structuring work processes in organizations.[4] The first is the *standard operating procedures (SOPs)* that become the habitual routines for how work is performed. The SOPs are both formal and informal. Formal SOPs define proper behavior. Informal SOPs are the norms that have developed over time for how processes are actually performed. Management must ensure that the formal SOPs are revised to match the to-be system. The informal SOPs will then evolve to refine and fill in details absent in the formal SOPs.

The second aspect of management policy is defining how people assign meaning to events. What does it mean to "be successful" or "do good work"? Policies help people understand meaning by defining *measurements* and *rewards*. Measurements explicitly define meaning because they provide clear and concrete evi-

---

[4] This section builds on the work of Anthony Giddons, *The Constitution of Society: Outline of the Theory of Structure,* Berkeley: University of California Press, 1984. A good summary of Giddon's theory that has been revised and adapted for use in understanding information systems is an article by Wanda Orlikowski and Dan Robey, "Information Technology and the Structuring of Organizations," *Information Systems Research,* 1991, 2(2):143–169.

Y O U R	**14-2 STANDARD OPERATING PROCEDURES**
T U R N	

Identify and explain three standard operating procedures for the course in which you are using this book. Discuss whether they are formal or informal.

dence about what is important to the organization. Rewards reinforce measurements because "what gets measured gets done" (an overused but accurate saying). Measurements must be carefully designed to motivate desired behavior. The IBM credit example ("Your Turn 4-2" in Chapter 4) illustrates the problem with flawed measurements' driving improper behavior (when the credit analysts became too busy to handle credit requests, they would find nonexistent errors so they could return them unprocessed).

A third aspect of management policy is *resource allocation*. Managers can have clear and immediate impacts on behavior by allocating resources. They can redirect funds and staff from one project to another, create an infrastructure that supports the new system, and invest in training programs. Each of these activities has both a direct and symbolic effect. The direct effect comes from the actual reallocation of resources. The symbolic effect shows that management is serious about its intentions. There is less uncertainty about management's long-term commitment to a new system when potential adopters see resources being committed to support it.

Assessing Costs and Benefits

The next step in developing a change management plan is to develop two clear and concise lists of costs and benefits provided by the new system (and the transition to it) compared with the as-is system. The first list is developed from the perspective of the organization, which should flow easily from the business case developed during the feasibility study and refined over the life of the project (see Chapter 2). This set of organizational costs and benefits should be distributed widely so that everyone expected to adopt the new system should clearly understand why the new system is valuable to the organization.

The second list of costs and benefits is developed from the viewpoints of the different potential adopters expected to change, or stakeholders in the change. For example, one set of potential adopters may be the frontline employees, another may be the first-line supervisors, and yet another might be middle management. Each of these potential adopters/stakeholders may have a different set of costs and benefits associated with the change—costs and benefits that can differ widely from those of the organization. In some situations, unions may be key stakeholders that can make or break successful change.

Many systems analysts naturally assume that front-line employees are the ones whose set of costs and benefits are the most likely to diverge from those of the organization and thus are the ones who most resist change. However, they usually bear the brunt of problems with the current system. When problems occur, they often experience them firsthand. Middle managers and first-line supervisors are the most likely to have a divergent set of costs and benefits and therefore resist change because new computer systems often change how much power they have. For example, a new computer system may improve the organization's control over a work process (a benefit to the organization) but reduce the decision-making power of middle management (a clear cost to middle managers).

An analysis of the costs and benefits for each set of potential adopter/stakeholders will help pinpoint those who will likely support the change and those who may resist the change. The challenge at this point is to try to change the balance of the costs and benefits for those expected to resist the change so that they support it (or at least do not actively resist it).

This analysis may uncover some serious problems that have the potential to block the successful adoption of the system. It may be necessary to reexamine the management policies and make significant changes to ensure that the balance of costs and benefits is such that important potential adopters are motivated to adopt the system.

Figure 14-6 summarizes some of the factors that are important to successful change. The first and most important reason is a compelling personal reason to change. All change is made by individuals, not organizations. If there are compelling reasons for the key groups of individual stakeholders to want the change, then the change is more likely to be successful. Factors such as increased salary, reduced unpleasantness, and—depending on the individuals—opportunities for promotion and personal development can be important motivators. However, if the change makes current skills less valuable, individuals may resist the change because they have invested a lot of time and energy in acquiring those skills and anything that diminishes those skills may be perceived as diminishing the individual (because important skills bring respect and power).

There must also be a compelling reason for the organization to need the change; otherwise, individuals become skeptical that the change is important and are less certain it will in fact occur. Probably the hardest organization to change is an organization that has been successful, because individuals come to believe that what worked in the past will continue to work. By contrast, in an organization that is on the brink of bankruptcy, it is easier to convince individuals that change is

---

CONCEPTS

14-C Understanding Resistance to a Decision Support System

IN ACTION

One of the first commercial software packages I developed was a decision support system to help schedule orders in a paper mill (see "Concepts in Action 4-B" in Chapter 4). The system was designed to help the person who scheduled orders decide when to schedule particular orders to reduce waste in the mill. This was a very challenging problem—so challenging, in fact, that it usually took the scheduler a year or two to really learn how to do the job well.

The software was tested by a variety of paper mills over the years and always reduced the amount of waste, usually by about 25% but sometimes by 75% when a scheduler new to the job was doing the scheduling. Although we ended up selling the package to most paper mills that tested it, we usually encountered significant resistance from the person doing the scheduling (except when the scheduler was new to the job and the package clearly saved a significant amount). At the time, I assumed that the resistance to the system was related to the amount of waste reduced: the less waste reduced, the more resistance because the payback analysis showed it took longer to pay for the software. *Alan Dennis*

Question:
1. What is another possible explanation for the different levels of resistance encountered at different mills?
2. How might this be addressed?

	Factor	Examples	Effects	Actions to Take
Benefits of to-be system	*Compelling* personal reason(s) for change	Increased pay, fewer unpleasant aspects, opportunity for promotion, most existing skills remain valuable	If the new system provides clear personal benefits to those who must adopt it, they are more likely to embrace the change.	Perform a cost–benefit analysis from the viewpoint of the stakeholders, make changes where needed, and actively promote the benefits.
Certainty of benefits	*Compelling* organizational reason(s) for change	Risk of bankruptcy, acquisition, government regulation	If adopters do not understand why the organization is implementing the change, they are less certain that the change will occur.	Perform a cost–benefit analysis from the viewpoint of the organization and launch a vigorous information campaign to explain the results to everyone.
	Demonstrated top management support	Active involvement, frequent mentions in speeches	If top management is not seen to actively support the change, there is less certainly that the change will occur.	Encourage top management to participate in the information campaign.
	Committed and involved business sponsor	Active involvement, frequent visits to users and project team, championing	If the business sponsor (the functional manager who initiated the project) is not seen to actively support the change, there is less certainty that the change will occur.	Encourage the business sponsor to participate in the information campaign and play an active role in the change management plan.
	Credible top management and business sponsor	Management and sponsor who do what they say instead of being members of the "management fad of the month" club	If the business sponsor and top management have credibility in the eyes of the adopters, the certainty of the claimed benefits is higher.	Ensure that the business sponsor and/or top management has credibility so that such involvement will help; if there is no credibility involvement will have little effect.
Costs of transition	Low personal costs of change	Few new skills needed	The cost of the change is not borne equally by all stakeholders; the costs are likely to be higher for some.	Perform a cost–benefit analysis from the viewpoint of the stakeholders, make changes where needed, and actively promote the low costs.
Certainty of costs	Clear plan for change	Clear dates and instructions for change, clear expectations	If there is a clear migration plan, it will likely lower the perceived costs of transition.	Publicize the migration plan.
	Credible change agent	Previous experience with change, does what he or she promises to do	If the change agent has credibility in the eyes of the adopters, the certainty of the claimed costs is higher.	If the change agent is not credible, then change will be difficult.
	Clear mandate for change agent from sponsor	Open support for change agent when disagreements occur	If the change agent has a clear mandate from the business sponsor, the certainty of the claimed costs is higher.	The business sponsor must actively demonstrate support for the change agent.

FIGURE 14-6

Major Factors in Successful Change

needed. Commitment and support from credible business sponsors and top management are also important in increasing the certainty that the change will occur.

The likelihood of successful change is increased when the cost of the transition to individuals who must change is low. The need for significantly different new skills or disruptions in operations and work habits may create resistance. A clear migration plan developed by a credible change agent who has support from the business sponsor is an important factor in increasing the certainty about the costs of the transition process.

Motivating Adoption

The single most important factor in motivating a change is providing clear and convincing evidence of the need for change. Simply put, everyone who is expected to adopt the change must be convinced that the benefits from the to-be system outweigh the costs of changing.

There are two basic strategies to motivating adoption: informational and political. Both strategies are often used simultaneously. With an *informational strategy,* the goal is to convince potential adopters that the change is for the better. This strategy works when the cost–benefit set of the target adopters has more benefits than costs. In other words, there really are clear reasons for the potential adopters to welcome the change.

Using this approach, the project team provides clear and convincing evidence of the costs and benefits of moving to the to-be system. The project team writes memos and develops presentations that outline the costs and benefits of adopting the system from the perspective of the organization and from the perspective of the target group of potential adopters. This information is disseminated widely throughout the target group, much like an advertising or public relations campaign. It must emphasize the benefits as well as increase the certainty in the minds of potential adopters that these benefits will actually be achieved. In our experience, it is always easier to sell painkillers than vitamins; that is, it is easier to convince potential adopters that a new system will remove a major problem (or other source of pain) than that it will provide new benefits (e.g., increase sales). Therefore, informational campaigns are more likely to be successful if they stress the reduction or elimination of problems, rather than focusing on the provision of new opportunities.

The other strategy to motivate change is a *political strategy.* With a political strategy, organizational power, not information, is used to motivate change. This approach is often used when the cost–benefit set of the target adopters has more costs than benefits. In other words, although the change may benefit the organization, there are no reasons for the potential adopters to welcome the change.

The political strategy is usually beyond the control of the project team. It requires someone in the organization who holds legitimate power over the target group to influence the group to adopt the change. This may be done in a coercive manner (e.g., "adopt the system or you're fired") or in a negotiated manner, in which the target group gains benefits in other ways that are linked to the adoption of the system (e.g., linking system adoption to increased training opportunities). Management policies can play a key role in a political strategy by linking salary to certain behaviors desired with the new system.

In general, for any change that has true organizational benefits, about 20% to 30% of potential adopters will be *ready adopters.* They recognize the benefits, quickly adopt the system, and become proponents of the system. Another 20% to 30% are *resistant adopters.* They simply refuse to accept the change and they fight against it, either because the new system has more costs than benefits for them personally or because they place such a high cost on the transition process itself that no amount of benefits from the new system can outweigh the change costs. The remaining 40% to 60% are *reluctant adopters.* They tend to be apathetic and will

14-D OVERCOMING RESISTANCE TO A DECISION SUPPORT SYSTEM

How would you motivate adoption if you were the developer of the decision support system described in "Concepts in Action 14-C" earlier in this chapter?

go with the flow to either support or resist the system, depending on how the project evolves and how their coworkers react to the system. Figure 14-7 illustrates the actors who are involved in the change management process.

The goal of change management is to actively support and encourage the ready adopters and help them win over the reluctant adopters. There is usually little that can be done about the resistant adopters because their set of costs and benefits may be divergent from those of the organization. Unless there are simple steps that can be taken to rebalance their costs and benefits or the organization chooses to adopt a strongly political strategy, it is often best to ignore this small minority of resistant adopters and focus on the larger majority of ready and reluctant adopters.

Enabling Adoption: Training

Potential adopters may want to adopt the change, but unless they are capable of adopting it, they won't. Adoption is enabled by providing the skills needed to adopt the change through careful *training*. Training is probably the most self-evident part of any change management initiative. How can an organization expect its staff members to adopt a new system if they are not trained? However, we have found that training is one of the most commonly overlooked part of the process. Many organizations and project managers simply expect potential adopters to find the system easy to learn. Since the system is presumed to be so simple, it is taken for granted that potential adopters should be able to learn with little effort. Unfortunately, this is usually an overly optimistic assumption.

Every new system requires new skills either because the basic work processes have changed (sometimes radically in the case of business process reengineering [BPR]; see Chapter 4) or because the computer system used to support the processes is different. The more radical the changes to the business processes, the more important it is to ensure the organization has the new skills required to operate the new business processes and supporting information sys-

Sponsor	Change Agent	Potential Adopters
The sponsor wants the change to occur.	The change agent leads the change effort.	Potential adopters are the people who must change.
		20–30% are ready adopters.
		20–30% are resistant adopters.
		40–60% are reluctant adopters.

FIGURE 14-7

Actors in the Change Management Process

tems. In general, there are three ways to get these new skills. One is to hire new employees who have the needed skills that the existing staff does not. Another is to outsource the processes to an organization that has the skills that the existing staff does not. Both of these approaches are controversial and are usually considered only in the case of BPR when the new skills needed are likely to be the most different from the set of skills of the current staff. In most cases, organizations choose the third alternative: training existing staff in the new business processes and the to-be system. Every training plan must consider what to train and how to deliver the training.

What to Train   What training should you provide to the system users? It's obvious: how to use the system. The training should cover all the capabilities of the new system so users understand what each module does, right?

Wrong. Training for business systems should focus on helping the users to accomplish their jobs, not on how to use the system. The system is simply a means to an end, not the end in itself. This focus on the performing the job (i.e., the business processes), not using the system, has two important implications. First, the training must focus on those activities around the system, as well as on the system itself. The training must help the users understand how the computer fits into the bigger picture of their jobs. The use of the system must be put in context of the manual business processes as well as of those that are computerized, and it must also cover the new management policies that were implemented along with the new computer system.

Second, the training should focus on what the user needs to do, not what the system can do. This is a subtle—but very important—distinction. Most systems will provide far more capabilities than the users will need to use (e.g., when was the last time you wrote a macro in Microsoft Word?). Rather than attempting to teach the users all the features of the system, training should instead focus on the much smaller set of activities that users perform on a regular basis and ensure that users are truly expert in those. When the focus is on the 20% of functions that the users will use 80% of the time (instead of attempting to cover all functions), users become confident about their ability to use the system. Training should mention the other little-used functions, but only so that users are aware of their existence and know how to learn about them when their use becomes necessary.

One source of guidance for designing training materials is the use cases. The use cases outline the common activities that users perform and thus can be helpful in understanding the business processes and system functions that are likely to be most important to the users.

How to Train   There are many ways to deliver training. The most commonly used approach is *classroom training* in which many users are trained at the same time by the same instructor. This has the advantage of training many users at one time with only one instructor and creates a shared experience among the users.

It is also possible to provide on *one-on-one training* in which one trainer works closely with one user at a time. This is obviously more expensive, but the trainer can design the training program to meet the needs of individual users and can better ensure that the users really do understand the material. This approach is typically used only when the users are very important or when there are very few users.

YOUR TURN

14-3 Developing a Training Plan

Suppose you are leading the conversion from one word processor to another in your organization. Develop an outline of topics that would be included in the training. Develop a plan for training delivery.

Another approach that is becoming more common is to use some form of *computer-based training (CBT),* in which the training program is delivered via computer, either on CD or over the Web. CBT programs can included text slides, audio, and even video and animation. CBT is typically more costly to develop but is cheaper to deliver because no instructor is needed to actually provide the training.

Figure 14-8 summarizes four important factors to consider in selecting a training method. CBT is typically more expensive to develop than one-on-one or classroom training, but it is less expensive to deliver. One-on-one training has the most impact on the user because it can be customized to the user's precise needs, knowledge, and abilities, whereas CBT has the least impact. However, CBT has the greatest reach—the ability to train the most users over the widest distance in the shortest time—because it is so much more simple to distribute, compared with classroom and one-on-one training, because no instructors are needed.

Figure 14-8 suggests a clear pattern for most organizations. If there are only a few users to train, one-on-one training is the most effective. If there are many users to train, many organizations turn to CBT. We believe that the use of CBT will increase in the future. Quite often, large organizations use a combination of all three methods. Regardless of which approach is used, it is important to leave the users with a set of easily accessible materials that can be referred to long after the training has ended (usually a quick reference guide and a set of manuals, whether on paper or in electronic form).

POSTIMPLEMENTATION ACTIVITIES

The goal of postimplementation activities is to *institutionalize* the use of the new system—that is to make it the normal, accepted, routine way of performing the

	One-on-One Training	Classroom Training	Computer-Based Training
Cost to develop	Low–medium	Medium	High
Cost to deliver	High	Medium	Low
Impact	High	Medium–high	Low–medium
Reach	Low	Medium	High

FIGURE 14-8
Selecting a Training Method

business processes. The postimplementation activities attempt to refreeze the organization after the successful transition to the new system. Although the work of the project team naturally winds down after implementation, the business sponsor and sometimes the project manager are actively involved in refreezing. These two—and ideally many other stakeholders—actively promote the new system and monitor its adoption and usage. They usually provide a steady flow of information about the system and encourage users to contact them to discuss issues.

In this section, we examine three key postimplementation activities; support (providing assistance in the use of the system), maintenance (continuing to refine and improve the system), and project assessment (analyzing the project to understand what activities were done well—and should be repeated—and what activities need improvement in future projects).

System Support

Once the project team has installed the system and performed the change management activities, the system is officially turned over to the *operations group*. This group is responsible for the operation of the system, whereas the project team is responsible for the development of the system. Members of the operations group usually are closely involved in the installation activities because they are the ones who must ensure that the system actually works. After the system is installed, the project team leaves but the operations group remains.

Providing system support means helping the users to use the system. Usually, this means providing answers to questions and helping users understand how to perform a certain function; this type of support can be thought of as *on-demand training*.

Online support is the most common form of on-demand training. This includes the documentation and help screens built into the system, as well as separate Web sites that provide answers to *frequently asked questions (FAQs)* that enable users to find answers without contacting a person. Obviously, the goal of most systems is to provide sufficiently good online support so that the user doesn't need to contact a person, because providing online support is much less expensive than is providing a person to answer questions.

Most organizations provide a *help desk* that provides a place for a user to talk with a person who can answer questions (usually over the phone, but sometimes in person). The help desk supports all systems, not just one specific system, so it receives calls about a wide variety of software and hardware. The help desk is operated by *level 1 support* staff who have very broad computer skills and are able to respond to a wide range of requests, from network problems and hardware problems to problems with commerical software and problems with the business application software developed in-house.

The goal of most help desks is to have the level 1 support staff resolve 80% of the help requests they receive on the first call. If the issue cannot be resolved by level 1 support staff, a *problem report* (Figure 14-9) is completed (often using a special computer system designed to track problem reports) and passed to a *level 2 support* staff member.

The level 2 support staff members are people who know the application system well and can provide expert advice. For a new system, they are usually selected during the implementation phase and become familiar with the system as it is being tested. Sometimes, the level 2 support staff members participate in training during

FIGURE 14-9
Elements of a Problem Report

- Time and date of the report
- Name, e-mail address, and telephone number of the support person taking the report
- Name, e-mail address, and telephone number of the person who reported the problem
- Software and/or hardware causing problem
- Location of the problem
- Description of the problem
- Action taken
- Disposition (problem fixed or forwarded to system maintenance)

the change management process to become more knowledgeable with the system, the new business processes, and the users themselves.

The level 2 support staff works with users to resolve problems. Most problems are successfully resolved by the level 2 staff. However, sometimes, particularly in the first few months after the system is installed, the problem turns out to be a bug in the software that must be fixed. In this case, the problem report becomes a *change request* that is passed to the system maintenance group (see the next section).

System Maintenance

System maintenance is the process of refining the system to make sure it continues to meet business needs. Substantially more money and effort is devoted to system maintenance than to the initial development of the system, simply because a system continues to change and evolve as it is used. Most beginning systems analysts and programmers work first on maintenance projects; usually only after they have gained some experience are they assigned to new development projects.

CONCEPTS
IN ACTION

14-E CONVERTING TO THE EURO (PART 2)

When the European Union decided to introduce the euro, the European Central Bank had to develop a new computer system (called Target) to provide a currency settlement system for use by investment banks and brokerages. The euro opened at an exchange rate of U.S. $1.167. However, a rumor that the Target system malfunctioned sent the value of the euro plunging 2 days later.

That evening, it was determined that the malfunction was not due to system problems. Instead, operators at some German banks had misunderstood how to use the system and had entered incorrect data. Once the problems were identified and the operators quickly retrained, the Target system continued to operate and the euro quickly regained its lost value.

Source: "Debut of Euro Nearly Flawless," *Computerworld,* January 11, 1999, 33(2), p. 16, by Thomas Hoffman.

QUESTION:
Target could be considered a high-risk system because of its effects on the European economy. What kinds of system support activities could be put in place to mitigate problems with Target?

Every system is "owned" by a project manager in the IS group (Figure 14-10). This individual is responsible for coordinating the systems maintenance effort for that system. Whenever a potential change to the system is identified, a change request is prepared and forwarded to the project manager. The change request is a "smaller" version of the *system request* discussed in Chapters 1 and 2. It describes the change requested and explains why the change is important.

Changes can be small or large. Change requests that are likely to require a significant effort are typically handled in the same manner as system requests: they follow the same process as the project described in this book, starting with project initiation in Chapter 2 and following through installation in this chapter. Minor changes typically follow a "smaller" version of this same process. There is an ini-

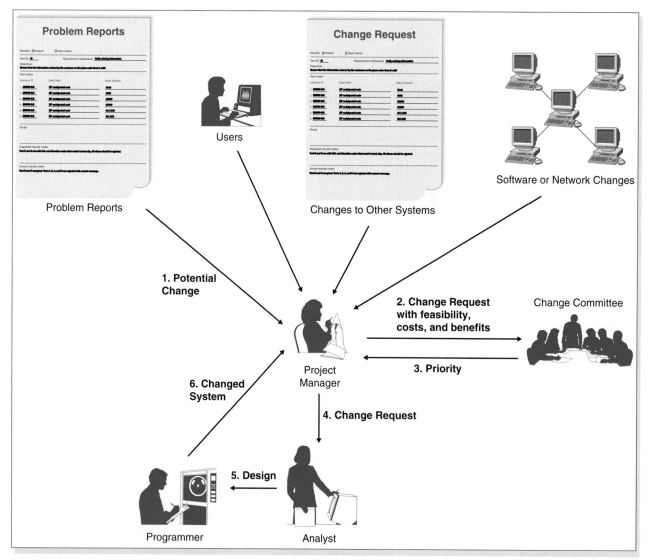

FIGURE 14-10
Processing a Change Request

CONCEPTS   **14-F SOFTWARE BUGS**

IN ACTION

The awful truth is that every operating system and application system is defective. System complexity, the competitive pressure to hurry applications to market, and simple incompetence contribute to the problem.

Will software ever be bug free? Not likely. Microsoft Windows Group General Manager Chris Jones believes that bigger programs breed more bugs. Each revision is usually bigger and more complex than its predecessor, which means new places for bugs to hide. Former Microsoft product manager Richard Freedman agrees that the potential for defects increases as software becomes more complex, but he believes users ultimately win more than they lose. "I'd say the features have gotten exponentially better, and the product quality has degraded a fractional amount."

Still, the majority of users who responded to our survey said they'd buy a software program with fewer features if it were bug free. This sentiment runs counter to what most software developers believe. "People buy features, plain and simple," explains Freedman. "There have been attempts to release stripped-down word processors and spreadsheets, and they don't sell." Freedman says a trend toward smaller, less-bug-prone software with fewer features will "never happen."

Eventually, the software ships and the bug reports start rolling in. What happens next is what separates the companies you want to patronize from the slackers. While almost every vendor provides bug fixes eventually, some companies do a better job of it than others. Some observers view Microsoft's market dominance as a roadblock to bug-free software. Todd Paglia, an attorney with the Washington, D.C.–based Consumer Project on Technology, says, "If actual competition for operating systems existed and we had greater competition for some of the software that runs on the Microsoft operating system, we would have higher quality than we have now."

Source: "Software Bugs Run Rampant," *PC World*, January 1999 17(1): p. 46, by Scott Spanbauer.

QUESTION:

If commercial systems contain the amount of bugs that this article suggests, what are the implications for systems developed in-house? Would in-house systems be more likely to have a lower or higher quality than commercial systems? Explain.

tial assessment of feasibility and of costs and benefits, and the changes request is prioritized. Then a systems analyst (or a programmer/analyst) performs the analysis, which may include interviewing users, and prepares an initial design before programming begins. The new (or revised) program is then extensively tested before the system is converted from the old system to the revised one.

Change requests typically come from five sources. The most common source is problem reports from the operations group that identify bugs in the system that must be fixed. These are usually given immediate priority because a bug can cause significant problems. Even a minor bug can cause major problems by upsetting users and reducing their acceptance of and confidence in the system.

The second most common source of change requests is enhancements to the system from users. As users work with the system, they often identify minor changes in the design that can make the system easier to use or identify additional functions that are needed. These enhancements are important in satisfying the users and are often key in ensuring that the system changes as the business requirements change. Enhancements are often given second priority after bug fixes.

A third source of change requests is other system development projects. For example, as part of CD Selections' Internet Order System project, CD Selections likely had to make some minor changes to the distribution system to ensure that the two systems would work together. These changes required by the need to integrate two systems are generally rare but are becoming more common as system integration efforts become more common.

A fourth source of change requests are those that occur when underlying software or networks change. For example, new versions of Windows often require an application to change the way they interact with Windows, or enable application systems to take advantage of new features that improve efficiency. While users may never see these changes (because most changes are inside the system and do not affect its user interface or functionality), these changes can be among the most challenging to implement because analysts and programmers must learn about the new system characteristics, understand how application systems use (or can use) those characteristics, and then make the needed programming changes.

The fifth source of change requests is senior management. These change requests are often driven by major changes in the organization's strategy (e.g., the CD Selections Internet Order project) or operations. These significant change requests are typically treated as separate projects, but the project manager responsible for the initial system is often placed in charge of the new project.

Project Assessment

The goal of *project assessment* is to understand what was successful about the system and the project activities (and therefore should be continued in the next system or project) and what needs to be improved. Project assessment is not routine in most organizations, except for military organizations, which are accustomed to preparing after-action reports. Nonetheless, assessment can be an important component in organizational learning because it helps organizations and people understand how to improve their work. It is particularly important for junior staff members because it helps promote faster learning. There are two primary parts to project assessment—project team review and system review.

Project Team Review *Project team review* focuses on the way in which the project team carried out its activities. Each project member prepares a short two- to three-page document that reports and analyzes his or her performance. The focus is on performance improvement, not penalties for mistakes made. By explicitly identifying mistakes and understanding their causes, project team members will, it is hoped, be better prepared for the next time they encounter a similar situation—and less likely to repeat the same mistakes. Likewise, by identifying excellent performance, team members will be able to understand why their actions worked well and how to repeat them in future projects.

The documents prepared by each team member are assessed by the project manager, who meets with the team members to help them understand how to improve their performance. The project manager then prepares a summary document that outlines the key learnings from the project. This summary identifies what actions should be taken in future projects to improve performance but is careful not to identify team members who made mistakes. The summary is widely circulated among all project managers to help them understand how to manage their projects better. Often, it is also circulated among regular staff members who did not work on the project so that they, too, can learn from other projects.

System Review   The focus of the *system review* is understand the extent to which the proposed costs and benefits from the new system that were identified during project initiation were actually recognized from the implemented system. Project team review is usually conducted immediately after the system is installed, while key events are still fresh in team members' minds, but system review is often undertaken several months after the system is installed because it often takes a while before the system can be properly assessed.

System review starts with the system request and feasibility analysis prepared at the start of the project. The detailed analyses prepared for the expected business value (both intangible and intangible) as well as the economic feasibility analysis are reexamined and a new analysis is prepared after the system has been installed. The objective is to compare the anticipated business value against the actual realized business value from the system. This helps the organization assess whether the system actually provided the value it was planned to provide. Whether or not the system provides the expected value, future projects can benefit from an improved understanding of the true costs and benefits.

A formal system review also has important behavior implications for project initiation. Since everyone involved with the project knows that all statements about business value and the financial estimates prepared during project initiation will be evaluated at the end of the project, they have an incentive to be conservative in their assessments. No one wants to be the project sponsor or project manager for a project that goes radically overbudget or fails to deliver promised benefits.

PRACTICAL TIP   **14-1 BEATING BUGGY SOFTWARE**

How do you avoid bugs in the commercial software you buy? Here are six tips:

1. **Know your software:** Find out if the few programs you use day in and day out have known bugs and patches, and track the Web sites that offer the latest information on them.
2. **Back up your data:** This dictum should be tattooed on every monitor. Stop reading right now and copy the data you can't afford to lose onto a floppy disk, second hard disk or Web server. We'll wait.
3. **Don't upgrade—yet:** It's tempting to upgrade to the latest and greatest version of your favorite software, but why chance it? Wait a few months, check out other users' experiences with the upgrade on Usenet newsgroups or the vendor's own discussion forum, and then go for it. But only if you must.

4. **Upgrade slowly:** If you decide to upgrade, allow yourself at least a month to test the upgrade on a separate system before you install it on all the computers in your home or office.
5. **Forget the betas:** Installing beta software on your primary computer is a game of Russian roulette. If you really have to play with beta software, get a second computer.
6. **Complain:** The more you complain about bugs and demand remedies, the more costly it is for vendors to ship buggy products. It's like voting—the more people participate, the better the results.

Source: "Software Bugs Run Rampant," *PC World*, January, 1999, 17(1): p. 46, Scott Spanbauer.

APPLYING THE CONCEPTS AT CD SELECTIONS

Installation of the Internet system at CD Selections was somewhat simpler than the installation of most systems because the system was new; there were only minor changes to the as-is system. Most changes would be felt by the staff in the stores who would need to respond to the in-store holds. Since this functioned in a very similar manner to the existing special order system, Alec anticipated few major problems.

Conversion

The team was faced with the problem of installing the in-store system and training the staff in all 50 of CD Selections stores. Because the new in-store functions were new and not changing any old functions, it was less important to make sure the conversion was completely synchronized across all stores, although the in-store system needed to be updated before the store could be listed in the Web search. Because the risk associated with the system was low, and because both cost and time were somewhat important, Alec chose to use direct conversion with a whole-system conversion, but with a pilot phase first. You will recall from last chapter, that the system was beta tested with employees. Alec chose to use the beta test as the pilot conversion.

The in-store system was installed in the 18 stores in a greater Los Angeles area for the beta test; system was modified so it would only locate inventory in these 18 stores. The manager and a small set of employees in each of these 18 stores were trained in the use of the new in-store system; it was expected that the manager of these employees would then train the remaining employees.

Conversion during the pilot phase went smoothly. First, the new hardware was purchased and installed. Then the software was installed on the Web server and on the client computers to be used by the staff of the Internet group. There was no data conversion per se, although the system started receiving data from the inventory system and sending data to the special order system.

At the end of the beta test and pilot conversion phase assistant training procedures were considered appropriate. The system was then installed in the remaining 32 stores in the CD Selections chain and their staff trained. The system was technically ready for operation.

Change Management

There were few change management issues, because there were few existing staff members who had to change. New staff were hired, most by internal transfer from other groups within CD Selections. The most likely stakeholders to be concerned by the change would be managers and employees in the traditional retail stores who might see the Internet system as a threat to their stores. Alec therefore developed an information campaign (distributed through the employee newsletter and internal Web site) that discussed the reasons for the change and explained that the Internet system was seen as a complement to the existing stores, not as a competitor. The system was instead targeted at Web-based competitors, such as Amazon.com as well as their traditional store-based competitors such as Tower Records.

The new management policies were developed, along with a training plan that encompassed both the manual work procedures and computerized procedures. Alec decided to use classroom training for the Internet system personnel because there was a small number of them and it was simpler and more cost effective to train them all together in one classroom session.

Postimplementation Activities

Support of the system was turned over to the CD Selections operations group, who had hired four additional support staff members with expertise in networking and Web-based systems. System maintenance began almost immediately, with Alec designated as the project manager responsible for maintenance of this version of the system plus the development of the next version. Alec began the planning to develop the next version of the system.

Project team review uncovered several key learnings, mostly involving Web-based programming and the difficulties in linking to existing Structured Query Language (SQL) database. The project was delivered on budget, with the exception that more was spent on programming than was anticipated.

A preliminary system review was conducted after the 2 month of operations. Thanks to an advertising campaign, sales via the Internet system were $120,000 for the first month and $162,000 for the second, showing a gradual increase (remember that the goal for the first year of operations was $2.6 million). Operating expenses averaged $47,000 per month, a bit higher than the projected average, owing to startup costs. Nonetheless, Margaret Mooney, Vice President of Marketing and the project sponsor, was quite pleased. She approved the feasibility study for the follow-on project to develop the second version of the Internet system, and Alec began the SDLC all over again.

SUMMARY

Conversion

Conversion, the technical process by which the new system replaces the old system, has three major steps: install hardware, install software, and convert data. Conversion style, the way in which users are switched between the old and new systems, can be via either direct conversion (in which users stop using the old system and immediately begin using the new system) or parallel conversion (in which both systems are operated simultaneously to ensure the new system is operating correctly). Conversion location, what parts of the organization are converted when, can be via a pilot conversion in one location; via a phased conversion, in which locations are converted in stages over time; or via simultaneous conversion, in which all locations are converted at the same time. The system can be converted module by module or as a whole at one time. Parallel and pilot conversion are less risky because they have a greater chance of detecting bugs before the bugs have widespread effect, but parallel conversion can be expensive.

Change Management

Change management is the process of helping people to adopt and adapt to the to-be system and its accompanying work processes. People resist change for very rational reasons, usually because they perceive the costs to themselves of the new system (and the transition to it) to outweigh the benefits. The first step in the change management plan is to change the management policies (such as standard operating procedures), devise measurements and rewards that support the new system, and allocate resources to support it. The second step is to develop a concise list of costs and benefits to the organization and all relevant stakeholders in the change, which will point out who is likely to support and who is likely to resist the change. The third step is to motivate adoption both by providing information and by using political strategies—using power to induce potential adopters to adopt the new system. Finally, training, whether classroom, one-on-one, or computer based, is essential to enable successful adoption. Training should focus on the primary functions the users will perform and look beyond the system itself to help users integrate the system into their routine work processes.

Postimplementation Activities

System support is performed by the operations group, which provides online and help desk support to the users. System support has both a level 1 support staff, which answers the phone and handles most of the questions, and level 2 support staff, which follows up on challenging problems and sometimes generates change requests for bug fixes. System maintenance responds to change requests (from the system support staff, users, other development project teams, and senior management) to fix bugs and improve the business value of the system. The goal of project assessment is to understand what was successful about the system and the project activities (and therefore should be continued in the next system or project) and what needs to be improved. Project team review focuses on the way in which the project team carried out its activities and usually results in documentation of key lessons learned. System review focuses on understanding the extent to which the proposed costs and benefits from the new system that were identified during project initiation were actually recognized from the implemented system.

KEY TERMS

Change agent	Management policies	Refreeze
Change management	Measurements	Reluctant adopters
Change request	Migration plan	Resistant adopters
Classroom training	Modular conversion	Resource allocation
Computer-based training (CBT)	On-demand training	Rewards
Conversion	One-on-one training	Simultaneous conversion
Conversion location	Online support	Sponsor
Conversion modules	Operations group	Standard operating procedures
Conversion strategy	Parallel conversion	(SOPs)
Conversion style	Phased conversion	System maintenance
Direct conversion	Pilot conversion	System request
Frequently asked questions (FAQs)	Political strategy	System review
Help desk	Postimplementation	System support
Informational strategy	Potential adopter	Training
Installation	Problem report	Transition
Institutionalization	Project assessment	Unfreeze
Level 1 support	Project team review	Whole-system conversions
Level 2 support	Ready adopters	

QUESTIONS

1. What are the three basic steps in managing organizational change?
2. What are the major components of a migration plan?
3. Compare and contrast direct conversion and parallel conversion.
4. Compare and contrast pilot conversion, phased conversion, and simultaneous conversion.
5. Compare and contrast modular conversion and whole-system conversion.
6. Explain the trade-offs among selecting between the types of conversion in questions 3, 4, and 5.
7. What are the three key roles in any change management initiative?
8. Why do people resist change? Explain the basic model for understanding why people accept or resist change.
9. What are the three major elements of management policies that must be considered when implementing a new system?
10. Compare and contrast an information change management strategy with a political change management strategy. Is one better than the other?
11. Explain the three categories of adopters you are likely to encounter in any change management initiative.
12. How should you decide what items to include in your training plan?
13. Compare and contrast three basic approaches to training.
14. What is the role of the operations group in the systems development life cycle (SDLC)?
15. Compare and contrast two major ways to provide system support.
16. How is a problem report different from a change request?
17. What are the major sources of change requests?
18. Why is project assessment important?
19. How is project team review different from system review?
20. What do you think are three common mistakes that novice analysts make in migrating from the as-is to the to-be system?
21. Some experts argue that change management is more important than any other part of the SDLC. Do you agree or not? Explain.
22. In our experience, change management planning often receives less attention than conversion planning. Why do you think this happens?

EXERCISES

A. Suppose you are installing a new accounting package in your small business. What conversion strategy would you use? Develop a conversion plan (i.e., technical aspects only).

B. Suppose you are installing a new room reservation system for your university that tracks which courses are assigned to which rooms. Assume that all the rooms in each building are "owned" by one college or department and only one person in that college or department has permission to assign them. What conversion strategy would you use? Develop a conversion plan (i.e., technical aspects only).

C. Suppose you are installing a new payroll system in a very large multinational corporation. What conversion strategy would you use? Develop a conversion plan (i.e., technical aspects only).

D. Consider a major change you have experienced in your life (e.g., taking a new job, starting a new school). Prepare a cost–benefit analysis of the change in terms of both the change and the transition to the change.

E. Suppose you are the project manager for a new library system for your university. The system will improve the way in which students, faculty, and staff can search for books by enabling them to search over the Web, rather than using only the current text-based system available on the computer terminals in the library. Prepare a cost–benefit analysis of the change in terms of both the change and the transition to the change for the major stakeholders.

F. Suppose you are the project manager for a new library system for your university. The system will improve the way in which students, faculty, and staff can search for books by enabling them to search over the Web, rather than using only the current text-based system available on the computer terminals in the library. Prepare a plan to motivate the adoption of the system.

G. Suppose you are the project manager for a new library system for your university. The system will improve the way in which students, faculty, and staff can search for books by enabling them to search over the Web, rather than using only the current text-based system available on the computer terminals in the library. Prepare a training plan that includes both what you would train and how the training would be delivered.

H. Suppose you are leading the installation of a new decision support system to help admissions officers manage the admissions process at your university. Develop a change management plan (i.e., organizational aspects only).

I. Suppose you are the project leader for the development of a new Web-based course registration system for your university that replaces an old system in which students had to go to the coliseum at certain times and stand in line to get permission slips for each course they wanted to take. Develop a migration plan (including both technical conversion and change management).

J. Suppose you are the project leader for the development of a new airline reservation system that will be used by the airline's in-house reservation agents. The system will replace the current command-driven system designed in the 1970s that uses terminals. The new system uses PCs with a Web-based interface. Develop a migration plan (including both conversion and change management) for your telephone operators.

K. Suppose you are the project leader for the development of a new airline reservation system that will be used by the airline's in-house reservation agents. The system will replace the current command-driven system designed in the 1970s that uses terminals. The new system uses PCs with a Web-based interface. Develop a migration plan (including both conversion and change management) for the independent travel agencies who use your system.

MINICASES

1. Nancy is the IS department head at MOTO Inc., a human resources management firm. The IS staff at MOTO Inc. completed work on a new client management software system about a month ago. Nancy was impressed with the performance of her staff on this project because the firm had not previously undertaken a project of this scale in-house. One of Nancy's weekly tasks is to evaluate and prioritize the change requests that have come in for the various applications used by the firm.

   Right now, Nancy has five change requests for the client system on her desk. One request is from a system user who would like some formatting changes made to a daily report produced by the system. Another request is from a user who would like the sequence of menu options changed on one of the system menus to more closely reflect the frequency of use for those options. A third request came in from the Billing Department. This department performs billing through the use of a billing software package. A major upgrade of this software is being planned, and the interface between the client system and the bill system will need to be changed to accommodate the new software's data structures. The fourth request seems to be a system bug that occurs whenever a client cancels a contract (a rare occurrence, fortunately). The last request came from Susan, the company president. This request confirms the rumor that MOTO Inc. is about to acquire another new business. The new business specializes in the temporary placement of skilled professional and scientific employees, and represents a new business area for MOTO Inc. The client management software system will need to be modified to incorporate the special client arrangements that are associated with the acquired firm.

   How do you recommend that Nancy prioritize these change requests for the client/management system?

2. Sky View Aerial Photography offers a wide range of aerial photographic, video, and infrared imaging services. The company has grown from its early days of snapping pictures of client houses to its current status as a full-service aerial image specialist. Sky View now maintains numerous contracts with various governmental agencies for aerial mapping and surveying work. Sky View has its offices at the airport where its fleet of specially equipped aircraft are hangared. Sky View contracts with several freelance pilots and photographers for some of its aerial work and also employs several full-time pilots and photographers.

   The owners of Sky View Aerial Photography recently contracted with a systems development consulting firm to develop a new information system for the business.

As the number of contracts, aircraft flights, pilots, and photographers increased, the company experienced difficulty keeping accurate records of its business activity and the utilization of its fleet of aircraft. The new system will require all pilots and photographers to swipe and ID badge through a reader at the beginning and conclusion of each photo flight, along with recording information about the aircraft used and the client served on that flight. These records would be reconciled against the actual aircraft utilization logs maintained and recorded by the hangar personnel.

The office staff was eagerly awaiting the installation of the new system. Their general attitude was that the system would reduce the number of problems and errors that they encountered and would make their work easier. The pilots, photographers, and hangar staff were less enthusiastic, being unaccustomed to having their activities monitored in this way.

a. Discuss the factors that may inhibit the acceptance of this new system by the pilots, photographers, and hangar staff.
b. Discuss how an informational strategy could be used to motivate adoption of the new system at Sky View Aerial Photography.
c. Discuss how a political strategy could be used to motivate adoption of the new system at Sky View Aerial Photography.

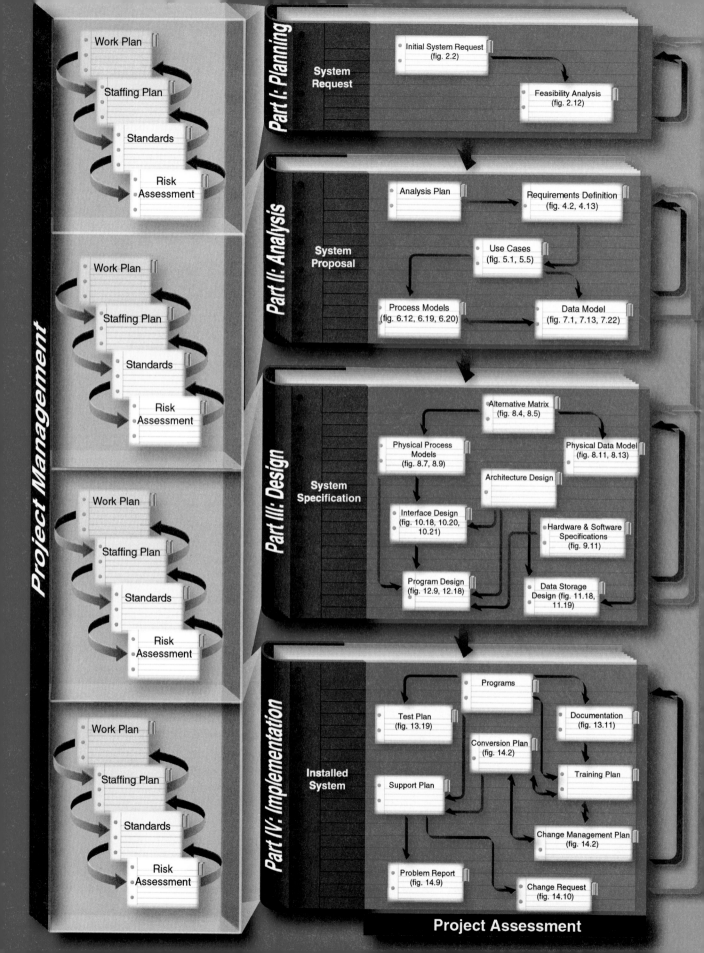

CHAPTER 15

THE MOVEMENT
TO OBJECTS

T he field of Systems Analysis and Design now incorporates object-oriented concepts and techniques, which view a system as a collection of self-contained objects that include both data and processes. Objects can be built as individual pieces and then put together to form a system, leading to modular, reusable project efforts. In 1997, the Unified Modeling Language (UML) was accepted as the standard language for object development. This chapter describes the four most effective UML models: the use-case diagram, class diagram, sequence diagram, and statechart diagram.

OBJECTIVES

- Understand basic concepts of the object approach and UML.
- Be able to create a use-case diagram.
- Be able to create a class diagram.
- Be able to create a sequence diagram.
- Be able to create a statechart diagram.

CHAPTER OUTLINE

INTRODUCTION

By this point we have presented the important skills that you will need for a real-world systems development project. You can be certain that all projects move through the four phases of planning, analysis, design, and implementation; all projects require you to gather requirements, model the business needs, and create blueprints for how the system should be built; and all projects require an understanding of organizational behavior concepts like change management and team building. This is true for large and small projects; custom built and packaged; local and international.

These underlying skills remain largely unchanged over time, but the actual techniques and approaches that analysts and developers use do change—often dramatically—over time. As we implied in Chapter 1, the field of systems analysis and design still has a lot of room for improvement: projects still run overbudget, users often cannot get applications when they need them, and some systems still fail to meet important user needs. Thus, the state of the systems analysis and design field is one of constant transition and continuous improvement. Analysts and developers learn from past mistakes and successes and evolve their practices to incorporate new techniques and new approaches that work better.

Today, the most exciting change to systems analysis and design is the move to object-oriented techniques. Object-oriented techniques view a system as a collection of self-contained objects, which include both data and processes. These objects can be built as individual pieces and then put together to form a system. The beauty of objects is that they can be reused over and over again in many different systems and changed without affecting other components.

Although some people feel that the move to objects will radically change the field of systems analysis and design, and the SDLC, we see the incorporation of objects as an evolving process in which object-oriented techniques are gradually integrated into the mainstream SDLC. Therefore, it is important for you as an analyst to understand what object orientation is and why it is causing such a stir in industry, as well as some popular object-oriented techniques that you may need to use on projects.

This chapter will first provide a background of object orientation and explain several key object concepts supported by the *Unified Modeling Language (UML)*, which has become the standard set of object-modeling techniques used by systems analysts and developers. Then, we will explain how to draw four of the most effective models in UML: the use-case diagram, the class diagram, the sequence diagram, and the statechart diagram.

THE OBJECT APPROACH AND UML

Until recent years, analysts focused on either data or business processes when developing systems. As they moved through the SDLC performing the techniques presented in Chapters 1–14, they emphasized either the data for the system (data approach) or the processes that it would support (process approach). In the mid-1980s, developers realized that building systems could be more efficient if the analyst worked with a system's data and processes simultaneously and focused on integrating the two. Project teams began using an *object approach,* whereby self-contained modules called objects (containing both data and processes) were used as the building blocks for systems. The following sections first describe the key characteristics of the object approach, object-oriented analysis and design, the benefits of the object approach, and then object-oriented analysis and design using UML.

Object Concepts

Objects and Classes   An *object* is a person, place, event, or thing about which we want to capture data and define processes. If we were building an appointment system for a doctor's office, objects might include doctors, patients, and appointments.

A *class* is the template we use to define and create objects. Every object is an *instance* of a class—that is, one specific member of the class. For example, we could create a class called "patient" with certain data attributes (e.g., names, addresses, and birthdates) and processes (e.g., insert new patients, maintain information, and delete entries). The specific patients like Jim Maloney, Mary Wilson, and Theresa Marks are considered *instances,* or objects, of the patient class (see Figure 15-1).

Each object has *attributes* that describe information about the object, such as a patient's name, birthdate, address, and phone number. The *state* of an object is defined by the value of its attributes and its relationships with other objects at a particular point in time. For example, a patient might have a state of "new" or "current" or "former."

Each object also has *behaviors*. The behaviors, implemented by *methods,* specify what the object can do. A method is nothing more than an action or process that an object can perform. For example, an appointment object likely can schedule a new appointment, delete an appointment, and locate the next available appointment. See Figure 15-2 for an illustration of a class and its objects.

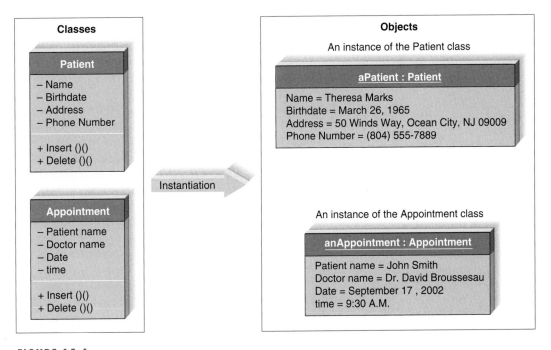

FIGURE 15-1
Classes and Objects

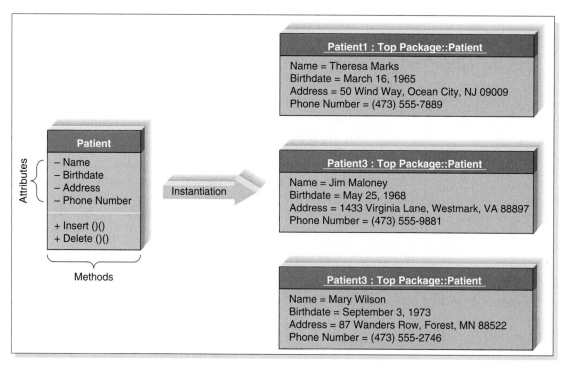

FIGURE 15-2
Classes and Their Objects

Objects do not have to include *primary keys* or *foreign keys* as are required when building relational data models.[1] Instead, each instance of an object is assigned a *unique identifier (UID)* by the system when it is created. This UID is often hidden from the user and is only used behind the scenes by the system to differentiate one object instance from another, even if they were to have the same values and methods. Therefore, if the system created two patients, both with the name John Smith and with the same exact address (twins?), it would still be able to distinguish one instance from another because the UIDs would be different.

One of the more confusing aspects of object-oriented systems development is the fact that in most object-oriented programming languages, both classes and instances of classes can have attributes and methods. Class attributes and methods tend to be used to model attributes (or methods) that deal with issues related to all instances of the class. For example, to create a new patient object, a message is sent to the patient class to create a new instance of itself. However, from a systems analysis and design point of view, we will focus primarily on attributes and methods of objects and not of classes.

Methods and Messages   As stated earlier, methods implement the object's behavior. Methods are very much like a function or procedure in a traditional programming language such as C, COBOL, or Pascal. *Messages* are the information

---

[1] Nonetheless, based on our experiences, we recommend that you include them.

sent to objects to trigger methods. A message is essentially a function or procedure call from one object to another object. For example, if a patient is new to the doctor's office, the system will send an insert message to the patient object. The patient object will receive a message and do what it needs to do to insert the new patient into the system (see Figure 15-3).

Encapsulation and Information Hiding   *Encapsulation* means to combine process and data into a single object. *Information hiding* is a key principle of object-oriented systems that means that only the information required to use an object is visible to the user of the object. Exactly how the object stores data or how it performs a method is not important, as long as the object behaves as expected. Typically, this means that only the information needed for the message that triggers a method and the information returned from the object is known. As such, objects are treated like black boxes.

The fact that we can use an object by calling methods is the key to reusability because it shields the internal workings of the object from changes in the outside system, and it keeps the system from being affected when changes are made to an object. In Figure 15-3, notice how a message (insert new patient) is sent to an object yet the internal methods needed to respond to the message are hidden from other parts of the system. All objects need to know is the set of operations, or methods, that other objects can perform and the format of messages that trigger them.

Inheritance   *Inheritance* allows developers to define new classes by reusing existing classes and adding new data and/or methods to them. Classes are arranged in a hierarchy so the *superclasses* (the most general classes) are at the top, and the *subclasses* (the more detailed classes that reuse them) are at the bottom. In Figure 15-4, person is a superclass to the classes doctor and patient. Doctor, in turn, is a superclass to general practitioner and specialist. Notice how a class (e.g., doctor) can serve as a superclass and subclass concurrently. The relationship between the class and its superclass is known as the *A-Kind-Of(AKO)* relationship. For example, in Figure 15-4, a general practitioner is A-Kind-Of doctor which is A-Kind-Of person.

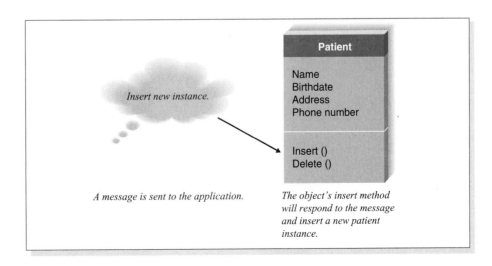

FIGURE 15-3
Messages and Methods

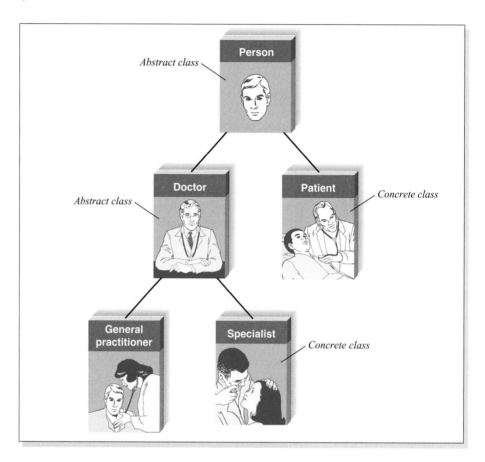

FIGURE 15-4
Class Hierarchy

Subclasses *inherit* attributes and methods from the superclasses above them. That is, each subclass contains attributes and methods from its parent superclass. For example, Figure 15-4 shows that both doctor and patient are subclasses of person and therefore will inherit the attributes and methods of the person class. Inheritance makes it simpler to define classes. Instead of repeating the attributes and methods in the doctor and patient classes separately, the attributes and methods that are common to both are placed in the person class and inherited by those classes below it. Notice how much more efficient hierarchies of object classes are than the same objects without a hierarchy in Figure 15-5.

Most classes throughout a hierarchy will lead to instances, and any class that has instances is called a *concrete class.* For example, if Mary Wilson and Jim Maloney were instances of the patient class, patient would be considered a concrete class. Some classes do not produce instances because they are used merely as templates for other more specific classes (especially those classes located high up in a hierarchy). They are *abstract classes.* Person would be an example of an abstract class. Instead of creating objects from person, we would create instances representing the more specific classes of doctor and patient, both types of "person" (see Figure 15-4). What kind of class is the general practitioner class? Why?

Polymorphism   *Polymorphism* means that the same message can be interpreted differently by different classes of objects. For example, inserting a patient means

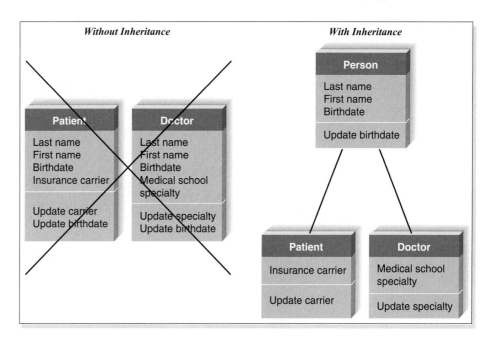

FIGURE 15-5
Inheritance

something different than inserting an appointment. As such, different pieces of information need to be collected and stored. Luckily, we do not have to be concerned with *how* something is done when using objects. We can simply send a message to an object, and that object will be responsible for interpreting the message appropriately. For example, if we sent the message "draw yourself" to a square object, a circle object, and a triangle object, the results would be very different even though the message is the same. Notice in Figure 15-6 how each object responds appropriately (and differently) even though the message is identical.

An Object-Oriented Approach to Analysis and Design

Object-oriented approaches to developing information systems can use any of the previous methodologies presented in Chapter 1: waterfall development, parallel development, phased development, prototyping, and so on. However, the object-oriented approaches are most often associated with a phased development RAD methodology or an extreme programming agile methodology. Typically, in object-oriented development, each version of the system goes through the entire SDLC in a time period of 1 to 2 weeks or 1 to 2 months depending on the size of the problem. To be capable of delivering systems in this very short amount of time, any object-oriented approach to developing information systems must be (1) use-case driven, (2) architecture centric, and (3) iterative and incremental.

Use-case driven means that *use cases* are the primary modeling tool to define the behavior of the system. They are also used to develop the criteria to verify and validate the system throughout the system's development. Use cases are transformed into *use-case diagrams,* that are graphical specification of the system's behavior from the perspective of the user(s). They are used to identify and communicate the high-level business requirements for the system in much the same way as DFDs are used to graphically illustrate the high-level business requirements in a non-objected-oriented world. All other details of the system are derived from the system's use cases.

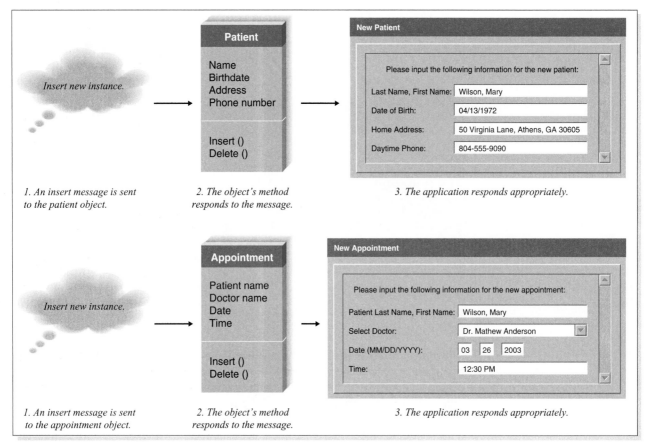

FIGURE 15-6
Polymorphism and Encapsulation

Architecture centric means that the underlying architecture of the evolving system drives the specification, construction, and documentation of the system. There are three separate but interrelated architectural views of a system: functional, static, and dynamic. The *functional view* describes the external behavior of the system from the perspective of the user. On the surface, this view is closely related to process-modeling approaches in structured analysis and design. However, as we will see, there are important differences between them. The *static view* describes the structure of the systems in terms of attributes, methods, classes, relationships, and messages. This view is very similar to data-modeling approaches in structured analysis and design. The *dynamic view* describes the internal behavior of the system in terms of messages passed between objects and state changes within an object. This view, in many ways combines the process and data-modeling approaches, in that, the execution of a method can cause state changes in the receiving object.

Object-oriented approaches emphasize *iterative* and *incremental* development that undergoes continuous testing throughout the life of the project. Each iteration of the system brings the system closer and closer to the final needs of the users.

The primary difference between structured analysis and design methodologies and object-oriented approaches is where emphasis is placed in decomposing a problem. In structured analysis and design approaches, the emphasis is on decomposing the problem along process or functional lines. That is, problems are decomposed into a set of subprocesses, which are in turn decomposed into further subprocesses. This goes on until no further process decomposition makes sense. In object-oriented approaches, the emphasis is on object or class decomposition. In structured techniques such as DFDs, the information for the entire system is tightly coupled on one—sometimes very complex—diagram, while with object-oriented approaches, each object can be viewed separately thereby reducing some complexity.

The Benefits

So far we have described several major concepts that permeate any object-oriented approach to systems development, but you may be wondering how these concepts affect the performance of a project team. The answer is simple. Concepts like polymorphism, encapsulation, and inheritance taken together allow analysts to break a complex system into smaller, more manageable components, to work on the components individually, and to more easily piece the components back together to form a system. This modularity makes system development easier to grasp, easier to share among members of a project team, and easier to communicate to users who are needed to provide requirements and confirm how well the system meets the requirements throughout the SDLC.

By modularizing system development, the project team is actually creating reusable pieces that can be plugged into other systems efforts, or used as starting points for other projects. Ultimately this can save time because new projects do not have to start from scratch, and learning curves are not as steep.

Finally, many people argue that "object think" is a much more realistic way to think about the real world. Users typically do not think in terms of data or process; instead they see their business as a collection of logical units that contain both—so communicating in terms of objects improves the interaction between the user and the analyst or developer. Figure 15-7 summarizes the major concepts of the object approach and how each contributes to the benefits that have just been described.

Unified Modeling Language

Until 1990, object concepts were implemented in different of ways by different companies and tool vendors, so object-oriented methods were slow to become widely used. Then in 1990 Rational Software brought three industry leaders together to create a single approach to object development. Grady Booch, Ivar Jacobson, and James Rumbaugh worked with others to create a standard set of diagramming techniques known as the *Unified Modeling Language, or UML.*[2] The objective of UML is to provide a common vocabulary of object-based terms and diagramming techniques that is rich enough to model any systems development project from analysis through implementation.

In November 1997, the Object Management Group (OMG) formally accepted UML as the standard for all object developers. A variety of CASE soft-

---

[2] For a complete description of all UML diagramming techniques see http://www.rational.com/uml.

Concept	Supports...	Leads to...
Classes, objects, methods, and messages	• A more realistic way for people think about their business • Highly cohesive units that contain both data and processes	• Better communication between user and analyst or developer • Reusable objects • Benefits from having a highly cohesive system (see cohesion in Chapter 12
Encapsulation and information hiding	• Loosely coupled units	• Reusable objects • Fewer ripple effects from changes within an object or in the system itself • Benefits from having a loosely coupled system design (see coupling in Chapter 12)
Inheritance	• Allows us to use classes as standard templates from which other classes can be built	• Less redundancy • Faster creation of new classes • Standards and consistency within and across development efforts • Ease in supporting exceptions
Polymorphism	• Minimal messaging that is interpreted by objects themselves	• Simpler programming of events • Ease in replacing or changing objects in a system • Fewer ripple effects from changes within an object or in the system itself
Use-case driven	• Allows users and analysts to focus on how a user will interact with the system to perform a single activity	• Better understanding and gathering of user needs • Better communication between user and analyst
Architecture centric and functional, static, and dynamic views	• Viewing the evolving system from multiple points of view	• Better understanding and modeling of user needs • More complete depiction of information system
Iterative and incremental development	• Continuous testing and refinement of the evolving system	• Meeting real needs of users • Higher-quality systems

FIGURE 15-7
Benefits of the Object Approach

ware packages, such as Rational Software Corporation's Rational Rose, Platinum Technology's Paradigm Plus, Visible System's Visible Analyst, Microsoft Corporation's VISIO, and Togethersoft's Together Control Center, support all or some of the UML components.

UML Diagramming Techniques UML defines a set of nine object diagramming techniques used to model a system (see Figure 15-8). All nine diagramming techniques use the same syntax and notation, making it easier for analysts and developers to learn the language. The same diagramming techniques are used throughout all phases of the SDLC (although some diagrams become more important in some phases), with the diagrams becoming more detailed as they move from the logical design to the physical design and include implementation details of the system. Overall, the consistent notation, integration among diagramming techniques, and the application of the diagrams across all phases of the SDLC makes UML a powerful language for analysts and developers.

The key building block of UML is the use case. UML requires analysts and developers to break the system into use cases, small logical pieces of the system,

Diagram Name	What Diagram Shows	What Diagram Is Used To Do	What Diagram Is Similar To	Systems Development Life Cycle Phases
Use-case diagram	The interaction between external users and the system	Capture business requirements for the system	Context diagram	Use cases drive the entire development process
Class diagram	The static nature of a system at the class level	Illustrate the relationships between classes modeled in the system	Data model	Analysis, design
Object diagram	The static nature of a system at the object level	Illustrate the relationships between objects modeled in the system; used when actual instances of the classes will better communicate the model	Data model	Analysis, design
Sequence diagram	The interaction between classes for a given use case, arranged by time sequence	Model the behavior of classes within a use case	Process model	Analysis, design
Collaboration diagram	The interaction between classes for a given use case, *not* arranged by time sequence	Model the behavior of classes within a use case	Process model	Analysis, design
Statechart diagram	Sequence of states that an object can assume, the events that cause an object to transition from state to state, and significant activities and actions that occur as a result	Examine the behavior of one class within a use case		Analysis, design
Activity diagram	A specific business process, or the dynamics of a group of objects; provides a view of flows and what is going on inside a use case or among several classes	Illustrate the flow of activities in a use case		Analysis, design
Component diagram	The physical components (i.e., exe files, dll files) in a design and where they are located	Illustrate the physical structure of the software		Architectural analysis, design, implementation
Deployment diagram	The structure of the run-time system; for example, it can show how physical modules of code are distributed across various hardware platforms	Show the mapping of software to hardware components		Architectural analysis, design, implementation

FIGURE 15-8
Unified Modeling Language Diagrams

and deal with each separately (in contrast, the traditional SDLC approach requires the analysts to develop DFDs and ERDs that attempt to encompass the entire system in one diagram). This use of many small use cases makes UML ideal for representing complex systems because the analysts and designers can focus on small views of the system, without getting overwhelmed by the details of the big picture. However, because the diagrams are so tightly integrated syntactically and conceptually, the underlying UML model represented by all of the diagrams depicts the system as an integrated whole.

UML is not a methodology in that it does not formally mandate *how* to apply the diagramming techniques. Many organizations are experimenting with UML and trying to understand how to incorporate its diagramming techniques into their system analysis and design methodologies. In many cases, the UML diagrams simply replace the older structured techniques (e.g., class diagrams replace ERDs). That is, the basic SDLC stays the same, but one step is simply performed using a different diagramming technique. However, we recommend that if you are going to use UML, you should follow a methodology suited to object-oriented techniques, such as the *rational unified process (RUP)*.

The Rational Unified Process   Other organizations are adopting new methodologies created especially for UML. Rational Software Corporation, for example, has created a methodology called the *rational unified process (RUP)* that defines *how* to apply UML. RUP is a rapid application development approach to building systems that is similar to the phased development approach or extreme programming described in Chapter 1 (see Figure 15-9 and Figure 1-5). The first step of the methodology is building use cases for the system, which identify and communicate the high-level business requirements for the system. This step drives the rest of the SDLC. Next analysts draw analysis diagrams, later building on the analysis efforts through design and development. The UML diagrams start off conceptual and abstract, and then include details that ultimately will lead to code generation and development. The diagrams move from showing the *what* to showing the *how*.

RUP emphasizes iterative, incremental development and prototyping that undergoes continuous testing throughout the life of the project. In Figure 15-9, each iteration of the system brings the system closer and closer to the real needs of the users. The nine UML diagrams are drawn and changed throughout the process.

Key Aspects   One could spend an entire book explaining how to use UML to develop systems,[3] but we don't have that much space here. Fortunately, four UML diagramming techniques have come to dominate object-oriented projects: use-case diagrams, class diagrams, sequence diagrams, and statechart diagrams. The other diagramming techniques are useful for their particular purposes, but these four techniques form the core of UML as used in practice today and will be the focus of this chapter.

The four diagramming techniques are integrated and used together to replace DFDs and ERDs in the traditional SDLC (see Figure 15-10). The use-case diagram is typically used to summarize the set of use cases for a logical part of the system (or the whole system). Then class diagrams, sequence diagrams, and statechart diagrams are used to further define the evolving system from various perspectives. The use-case diagram is always created first, but the order in which the other diagrams are created depends upon the project and the personal preferences of the analysts. Most analysts start either with the class diagrams (which show what objects contain and how they are related, much like ERDs) or the sequence diagrams (which show

---

[3] In fact, we have: see Alan Dennis, Barbara Haley Wixom, David Tegarden, *Systems Analysis & Design: An Object-Oriented Approach with UML,* New York: John Wiley & Sons, 2002.

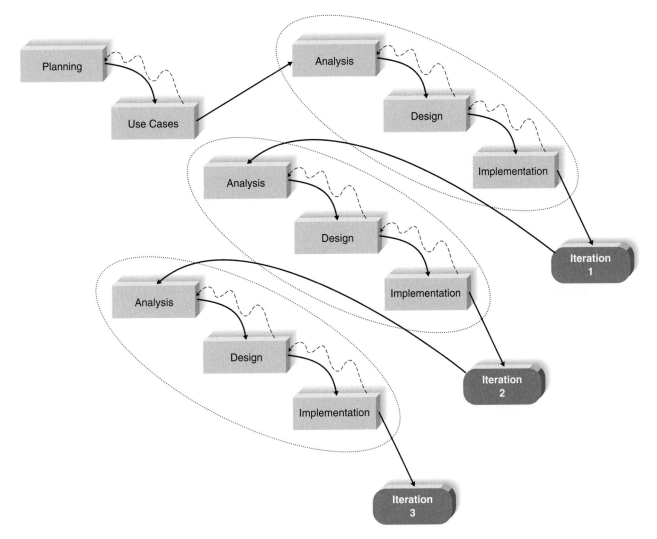

FIGURE 15-9
An Adaptation of the Unified Process Phased Development Methodology

how objects dynamically interact, much like DFDs), but in practice, the process is iterative. Developing sequence diagrams often leads to changes in the class diagrams and vice versa, so analysts often move back and forth between the two, refining each in turn as they define the system. Generally speaking, statecharts are developed later, after the class diagrams have been refined. In this chapter, we will start with the use-case diagram, move to the class diagrams, and finish up with the sequence diagrams and statechart diagrams.

USE-CASE DIAGRAM

Use cases are the primary drivers for all of the UML diagramming techniques. The use case communicates at a high level what the system needs to do, and each of the

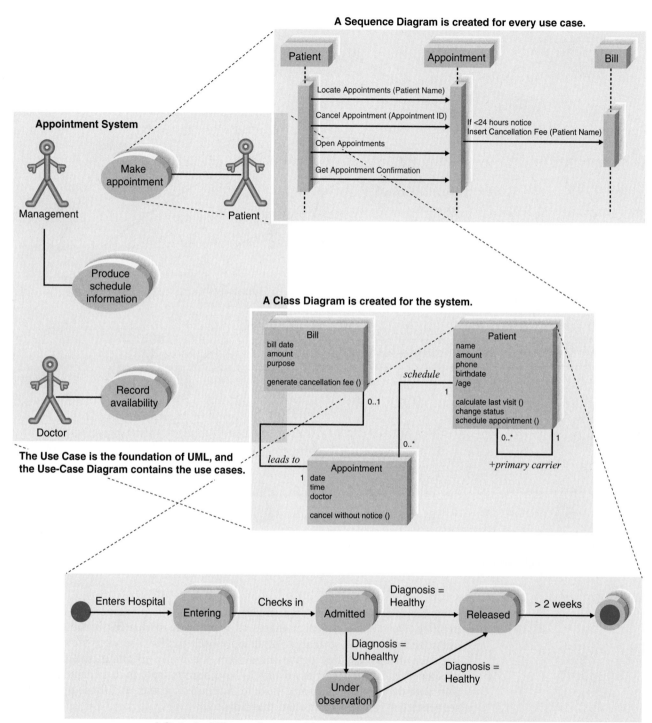

A Sequence Diagram is created for every use case.

A Class Diagram is created for the system.

The Use Case is the foundation of UML, and the Use-Case Diagram contains the use cases.

A Statechart Diagram is created for every complex class on the Class Diagram.

FIGURE 15-10
The Integration of Four Unified Modeling Language (UML) Diagrams

UML diagramming techniques build upon this by presenting the functionality in different ways, each view having a different purpose (as described in Figure 15-8). In the early stages of analysis, the analyst first identifies one use case for each major part of the system and creates accompanying documentation, the use-case report, to describe each function in detail. A use case may represent several "paths" that a user can take while interacting with the system; each path is referred to as a *scenario.* Use cases and use-case diagrams support the functional view just described. You may want to take a moment and review Chapter 5 on use cases before continuing with the rest of the chapter.

For now, we will learn how the use case is the building block for the use-case diagram, which summarizes all of the use cases (for the part of the system being modeled) together in one picture. An analyst can use the use-case diagram to better understand the functionality of the system at a very high level. Typically, the use-case diagram is drawn early on in the SDLC, when the analyst is gathering and defining requirements for the system, because it provides a simple, straightforward way of communicating to the users exactly what the system will do (i.e., at the same point of the SDLC as we would create a DFD).

This section will first describe the syntax for the use-case diagram and then demonstrate how to build one using an example from CD Selections.

Elements of a Use-Case Diagram

A use-case diagram illustrates in a very simple way the main functions of the system and the different kinds of users that will interact with it. For example, Figure 15-11 presents a use-case diagram for a doctor's office appointment system. We can see from the diagram that patients, doctors, and management personnel will use the Appointment System to make appointments, record availability, and produce schedule information, respectively.

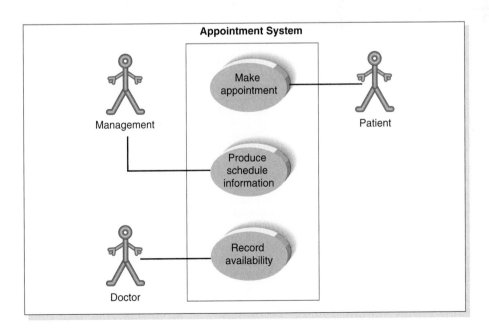

FIGURE 15-11

Use-Case Diagram for Appointment System

Actor   The labeled stick figures on the diagram represent actors (see Figure 15-12). An *actor* is similar to an external entity found in DFDs—it is a person or another system that interacts with and derives value from the system. An actor is not a specific user, but a *role* that a user can play while interacting with the system. Actors are external to the system, so if we were modeling a doctor's office appointment system, a data-entry clerk (or a nurse entering patient information) would not be considered an actor because he or she would fall within the scope of the system itself (this is the same rule for DFD external entities). The diagram in Figure 15-11 shows that three actors will interact with the appointment system (a patient, a doctor, and management).

Sometimes an actor plays a specialized role of a more general type of actor. For example, there may be times when a new patient interacts with the system in a way that is somewhat different than a general patient. In this case, a *specialized actor* (i.e., *new patient*) can be placed on the model, shown using a line with a hollow triangle at the end of the more general superclass of actor (i.e., *patient*). The specialized actor will inherit the behavior of the superclass and extend it in some way (see Figure 15-13). Can you think of some ways in which a new patient may behave differently than an existing patient?

Use Case   A use case, depicted by an oval, is a major process that the system will perform that benefits an actor(s) in some way (see Figure 15-12), and it is labeled using a descriptive verb phrase (much like a DFD process). We can tell from Figure 15-13 that the system has three primary use cases: make appointment, produce schedule information, and record availability.

Term and Definition	Symbol
An actor 　Is a person or system that derives benefit 　from and is external to the system 　Is labeled with its role 　Can be associated with other actors using a 　specialization/superclass association, 　denoted by an arrow with a hollow arrowhead 　Are placed outside the system boundary	Actor role name
A use case 　Represents a major piece of system 　functionality 　Can extend another use case 　Can use another use case 　Is placed inside the system boundary 　Is labeled with a descriptive verb–noun 　phrase	Use case name
A system boundary 　Includes the name of the system inside or on top 　Represents the scope of the system	System name
An association 　Links an actor with the use case(s) with 　which it interacts	

FIGURE 15-12
Syntax for Use-Case Diagram

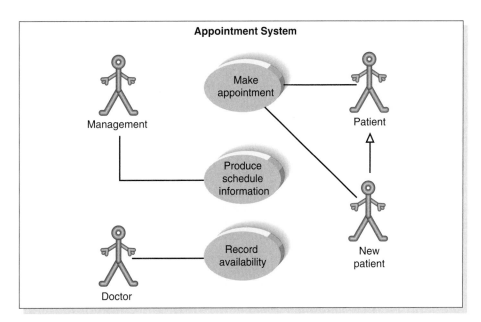

FIGURE 15-13
Use-Case Diagram with Specialized Actor

There are times when one use case will either use the functionality or extend the functionality of another use case on the diagram, and these are shown using *includes* or *extends associations*. It may be easier to understand these associations with the help of examples. Let's assume that every time patients make an appointment, they are asked to confirm their contact and basic patient information to ensure that the system always contains the most up-to-date information on its patients. Therefore, we may want to include a use case called *update patient information* that extends the *make appointment* use case to include the functionality just described. Notice how a hollow arrow was drawn in Figure 15-14 between "update patient information" and "make appointment" to denote the *extends* relationship.

Similarly, there are times when a single use case contains common functions that are used by other use cases. For example, suppose there is a use case called *manage schedule* that performs some routine tasks needed to maintain the doctor's office appointment schedule, and the two use cases *record availability* and *produce schedule information* both perform the routine tasks. Figure 15-14 shows how we can design the system so that *manage schedule* is a shared use case that is used by others. A hollow arrow again is used to denote the *includes* association.

System Boundary   The use cases are enclosed within a *system boundary,* which is a box that represents the system and clearly delineates what parts of the diagram are external or internal to it (see Figure 15-12). The name of the system can appear either inside or on top of the box.

Association   Finally, use cases are connected to actors through *associations,* which show with which use cases the actors interact (see Figure 15-12). An association is depicted by a line drawn from an actor to a use case, and it is normally shown as a one-to-one relation with no direction.

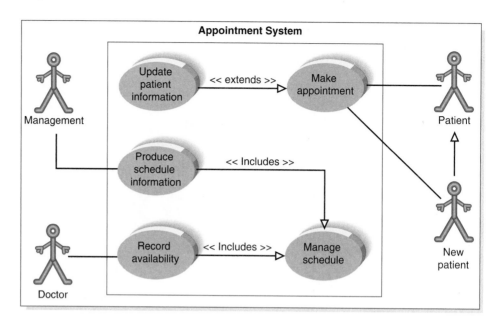

FIGURE 15-14
Extends and Includes Associations

Creating a Use-Case Diagram

Let's demonstrate how to draw a use-case diagram by using the CD Selections Internet system. You should note that the use-case diagram communicates information that is similar to information found in DFD context and level 0 diagrams. In fact, you may want to compare the use-case diagram that we are about to draw with the diagrams that were created in Chapter 6.

Identify Use Cases Before a use-case diagram can be created, it is helpful to go through the process of identifying the use cases that correspond to the system's major functionality and putting together the use-case documentation for each of them. The way to create use cases was explained in Chapter 5, and we refer you to that chapter now if you would like to refresh your memory. If you recall, we found that the CD Selections Internet system needs to support the three use cases that are presented in Figure 15-15.

Draw the System Boundary First, place a box on the use-case diagram to represent the system, and place the system's name either inside or on top of the box. This will form the border of the system, separating use cases (i.e., the system's functionality) from actors (i.e., the roles of the external users).

Place the Use Cases on the Diagram The next step is to add the use cases inside the system boundary. There should be no more than six to eight use cases on the model, so if you identify more than eight, you should group together the use cases into *packages* (i.e., logical groups of use cases) to make the diagrams easier to read and keep the models at a reasonable level of complexity.

At this point special use-case associations (includes or extends) should be added to the model. These are identified by looking for use cases that may include common functionality that other use cases require (i.e., includes association) or use cases that add additional functionality to others (i.e., extends association). The current model does not include examples of these associations.

Use Case 1

Use case name: Take requests for CDs **ID number:** 1

Short description: This describes how customers can search the Web site and place requests to hold CDs in stock or place special orders

Trigger: Customer searches Web and places request to hold a CD or to special order it

Type: (External) Temporal

Major Inputs: Description	Source	Major Outputs: Description	Destination
Search request	Customer	Special Order	Special order DBs
CDs selected for request	Customer	Hold for in-stock CD	In-store hold DB
Customer information	Customer	CDs matching search request	Customer
Marketing materials	Marketing DB	CDs requested	Customer
CD information request	Customer	CD information	Customer
CD inventory	Inventory DB	Marketing materials	Customer

Major Steps

Use Case 2

Use case name: Maintain marketing materials **ID number:** 2

Short description: This adds, deletes, and modifies the additional marketing material from vendors (e.g, reviews, musics clips)

Trigger: Materials from vendors, distributors, wholesalers, record companies, and articles in trade magazines

Type: (External) Temporal

Major Inputs: Description	Source	Major Outputs: Description	Destination
Marketing materials	Vendor	Marketing materials	Marketing DB
Marketing materials	Marketing manager	Marketing material report	Marketing manager
CD information	CD DB		
Vendor information	Vendor		

Major Steps

Use Case 3

use case name: Process in-store holds **ID number:** 3

Short description: This alerts the store staff to pull a request CD from the shelves and place it in the special order section

Trigger: Hold request from take request use case

Type: (External) Temporal

Major Inputs: Description	Source	Major Outputs: Description	Destination
Hold request	In-store hold DB	Hold Label	In-store staff
Hold confirmation	In-store staff	Hold request alert	In-store staff
		Hold confirmation	In-store hold DB
		Inventory adjustment	Inventory DB

Major Steps Performed	Information for Steps

FIGURE 15-15
Final Use Cases for CD Selections

15-1 Identifying Use-Case Associations and Specialized Actors

The use-case diagram for the Internet system does not include special use-case associations (e.g., extends or includes) or specialized actors. See if you can come up with one example for each of these spe- cial cases that may be helpful for CD Selections to add to the use-case diagram. Describe how the development effort may benefit from including your examples.

Identify the Actors   Once the use cases are placed on the diagram, you will need to identify the actors. We recommend that you look at the sources and destinations to major inputs and outputs that you identified in the use cases. Although some sources and destinations refer to internal system components (e.g., *marketing materials database, CD information database*), many others refer to actors (e.g., *customer, vendor*). Look at the use-case reports in Figure 15-15 and see if you can identify the actors that belong on the use-case diagram.

You likely have listed *special order system, marketing manager, in-store staff customer,* and *vendor.* At this point, there are no specialized actors that need to be included.

Add Associations   The last step is to draw lines connecting the actors with the use cases with which they interact. No order is implied by the diagram, and the items

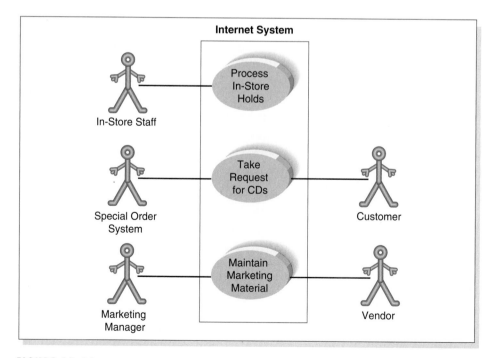

FIGURE 15-16
CD Selections Internet System

15-2 DRAWING A USE-CASE DIAGRAM

Create a use-case diagram for the system described next:

Owners of apartments fill in information forms about the rental units they have available (e.g., location, number of bedrooms, monthly rent) which are entered into a database. Students can search through this database via the Web to find apartments that meet their needs (e.g., a two-bedroom apartment for $400 or less per month within $1/2$ mile of campus). They then contact the apartment owners directly to see the apartment and possibly rent it. Apartment owners call the service to delete their listing when they have rented their apartment(s).

you have added along the way do not have to be placed in a particular order; therefore, you may want to rearrange the symbols a bit to minimize the number of lines that cross so the diagram is less confusing. Figure 15-16 is the use-case diagram that we have created.

CLASS DIAGRAM

The next major diagramming technique is the *class diagram.* The class diagram is a *static model* that supports the static view of the evolving system. It shows the classes and the relationships among classes that remain constant in the system over time. The class diagram is very similar to the entity relationship diagram (ERD) in Chapter 7; however, the class diagram depicts classes, which include attributes, behaviors and states, while entities in the ERD only include attributes. The scope of a class diagram, like the ERD, is systemwide. The following sections will first present the syntax of the class diagram, followed by the way in which a class diagram is drawn.

Elements of a Class Diagram

Figure 15-17 shows a class diagram that was created to reflect the classes and relationships that are needed for the set of use cases which describe the appointment system in Chapters 5 and 6.

Class   The main building block of a class diagram is the class, which stores and manages information in the system (see Figure 15-18). During analysis, classes refer to the people, places, events, and things about which the system will capture information. Later, during design and implementation, classes can refer to implementation-specific artifacts like windows, forms, and other objects used to build the system. Each class is drawn using three part-rectangles with the class's name at the top, attributes in the middle, and methods (also called operations) at the bottom. You should be able to identify that *Person, Employee, Doctor, Nurse, Administrative Staff, Health Team, Patient, Medical History, Appointment, Bill, Treatment, Illness,* and *Symptom* are classes in Figure 15-17. The attributes of a class and their values define the state of each object that is created from the class, and the behavior is represented by the methods.

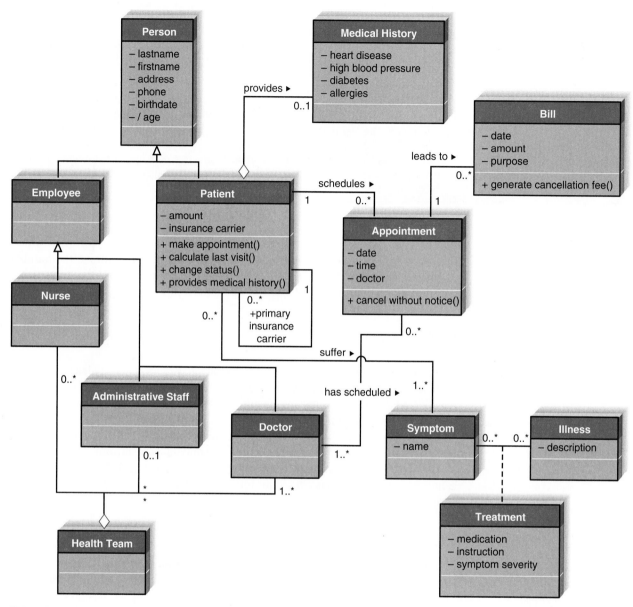

FIGURE 15-17
Class Diagram for Manage Appointment

Attributes are properties of the class about which we want to capture information (see Figure 15-18). Notice that the *Person* class in Figure 15-17 contains the attributes lastname, firstname, address, phone, birthdate, and age. At times, you may want to store *derived attributes,* which are attributes that can be calculated or derived from other attributes and therefore are not stored. Derived attributed are indicated by placing a slash (/) in front of the attribute's name. Notice how the person class contains a derived attribute called "/age," which can be derived by subtracting the person's birthdate from the current date. It is also possible to show the

Term and Definition	Symbol
A class Represents a kind of person, place, or thing about which the system must capture and store information Has a name typed in bold and centered in its top compartment Has a list of attributes in its middle compartment Has a list of operations in its bottom compartment Does not explicitly show operations that are available to all classes	**Class name** Attribute name /derived attribute name Operation name ()
An attribute Represents properties that describe the state of an object Can be derived from other attributes, shown by placing a slash before the attribute's name	Attribute name /derived attribute name
A method Represents the actions or functions that a class can perform Can be classified as a constructor, query, or update operation Includes parentheses that may contain special parameters or information needed to perform the method	Method name ()
An association Represents a relationship between multiple classes, or a class and itself Is labeled using a verb phrase or a role name, whichever better represents the relationship Can exist between one or more classes Contains multiplicity symbols, which represent the minimum and maximum times a class instance can be associated with the related class instance	1..* verb phrase 0..1

FIGURE 15-18
Class Diagram Syntax

visibility of the attribute on the diagram. Visibility relates to the level of information hiding to be enforced for the attribute. The visibility of an attribute can either be public (+), protected (#), or private (−). A *public* attribute is one that is not hidden from any other object. As such, other objects can modify its value. A *protected* attribute is one that is hidden from all other classes except its immediate subclasses. A *private* attribute is one that his hidden from all other classes. The default visibility for an attribute is normally private.

Methods are actions or functions that a class can perform (see Figure 15-18). The functions that are available to all classes (e.g., *create a new instance, return a value for a particular attribute, set a value for a particular attribute,* or *delete an instance*) are not explicitly shown within the class rectangle. Instead, only those methods that are unique to the class are included, such as the "cancel without notice" and "generate cancellation fee" methods in Figure 15-17. Notice that both of the operations are followed by parentheses. Methods should be shown with parentheses that are either empty or filled with some value, which represents a parameter that the method needs for it to act. As with attributes, the visibility of an operation can be designated as public, protected, or private. The default visibility for an operation is normally public.

There are three kinds of methods that a class can contain: constructor, query, and update. A *constructor method* creates a new instance of a class. For example, the patient class may have a method called *insert ()* that creates a new patient instance. Because a method available to all classes (e.g., create a new instance) is not shown on class diagrams, you will not see constructor methods in Figure 15-17.

A *query method* makes information about the state of an object available to other objects, but it will not change the object in any way. For instance, the "calculate last visit ()" method that determines when a patient last visited the doctor's office will result in the object being accessed by the system, but it will not make any change to its information. If a query method merely asks for information from attributes in the class (e.g., a patient's name, address, or phone), then it is not shown on the diagram because we assume that all objects have operations that produce the values of their attributes.

An *update method* will change the value of some or all of the object's attributes, which may result in a change in the object's state. Consider changing the status of a patient from new to current with a method called "change status ()," or associating a patient with a particular appointment with *schedule appointment (appointment)*.

Relationships   A primary purpose of the class diagram is to show the relationships, or associations, that classes have with one another. These are depicted on the diagram by drawing lines between classes (see Figure 15-18). These associations are very similar to the relationships that are found on the ERD. Relationships are maintained by *references,* which are similar to pointers and maintained internally by the system (unlike in the relational models where relationships are maintained using foreign and primary keys).

When multiple classes share a relationship (or a class shares a relationship with itself), a line is drawn and labeled with either the name of the relationship or the roles that the classes play in the relationship. For example, in Figure 15-17 the two classes *patient* and *appointment* are associated with one another whenever a patient schedules an appointment. Thus, a line labeled *schedules* connects *patient* and *appointment,* representing exactly how the two classes are related to each other. Also, notice that there is a small solid triangle beside of the name of the relationship. The triangle allows a direction to be associated with the name of the relationship. In Figure 15-17, the schedules relationship includes a triangle, indicating that the relationship is to be read as *patient schedules appointment.* Inclusion of the triangle simply increases the readability of the diagram.

Sometimes a class is related to itself, as in the case of a patient being the primary insurance carrier for other patients (e.g., their spouse, children). In Figure 15-17, notice that a line was drawn between the patient class and itself and called *primary insurance carrier* to depict the role that the class plays in the relationship. Notice that a plus "+" sign is placed before the label to communicate that it is a role as opposed to the name of the relationship. When labeling an association, use either a relationship name or a role name (not both), whichever communicates a more thorough understanding of the model.

Relationships also have *multiplicity,* which shows how an instance of an object can be associated with other instances. Numbers are placed on the association path to denote the minimum and maximum instances that can be related through the association in the format *minimum number maximum number* (see Figure 15-19). This is identical to the modality and cardinality of a relationship on an ERD. The numbers specify the relationship from the class at the far end of the relationship line to the end with the number. For example, in Figure 15-17, there is a "0.." on the appointment end of the *patient schedules appointment* relationship. This means that a patient can be associated with zero through many different appointments. At the patient end of this same relationship there is a "1," meaning that an appointment must be associated with one and only one (1) patient.

There are times when a relationship itself has associated properties, especially when its classes share a many-to-many relationship. In these cases, a class is formed, called an *association class* that has its own attributes and methods. This is very similar to the intersection entity that is placed on an ERD. It is shown as a rectangle attached by a dashed line to the association path, and the rectangle's name matches the label of the association e.g., see the *treatment* association in Figure 15-17. Think about the case of capturing information about illnesses and symptoms. An illness (e.g., the flu) can be associated with many symptoms (e.g., sore throat, fever), and a symptom (e.g., sore throat) can be associated with many illnesses (e.g., the flu, strep throat, the common cold). Figure 15-17 shows how an associa-

Instance(s)	Representation of Instance(s)	Diagram Involving Instance(s)	Explaination of Diagram
Exactly one	1	Department — 1 — Boss	A department has one and only one boss.
Zero or more	0..*	Employee — 0..* — Child	An employee has zero to many children.
One or more	1..*	Boss — 1..* — Employee	A boss is responsible for one or more employees.
Zero or one	0..1	Employee — 0..1 — Spouse	An employee can be married to zero or no spouse.
Specified range	2..4	Employee — 2..4 — Vacation	An employee can take between two to four vacations each year.
Multiple, disjoint ranges	1..3, 5	Employee — 1..3, 5 — Committee	An employee is a member of one to three or five committees.

FIGURE 15-19
Multiplicity

tion class can capture information about treatments that change depending on the various combinations. For example, a sore throat caused by strep throat will require antibiotics, while treatment for a sore throat from the flu or a cold could be throat lozenges or hot tea. Most times, classes are related through a "normal" association, but there are two special cases of an association that you will see appear quite often: generalization and aggregation.

Generalization and Aggregation   *Generalization* shows that one class (subclass) inherits from another class (superclass), meaning that the properties and operations of the superclass are also valid for objects of the subclass. The generalization path is shown with a solid line from the subclass to the superclass and a hollow arrow pointing at the superclass. For example, Figure 15-17 communicates that *doctors, nurses,* and *administrative staff* are all kinds of *employees* and those *employees* and *patients* are kinds of *persons*. The generalization relationship occurs when you need to use words like "is a kind of" to describe the relationship.

Aggregation is used when classes actually comprise other classes. For example, think about a doctor's office that has decided to create health care teams that include doctors, nurses, and administrative personnel. As patients enter the office, they are assigned to a health care team that cares for their needs during their visits. Figure 15-17 shows how this relationship is denoted on the class diagram. A diamond is placed nearest the class representing the aggregation *(health team),* and lines are drawn from the arrow to connect the classes that serve as its parts *(doctors, nurses,* and *administrative staff)*. Aggregation relationships are typically identified when you need to use words like "is a part of" or "is made up of" to describe the relationship.

Simplifying Class Diagrams

When a class diagram is drawn with all of the classes and relationships for a real-world system, the class diagram can become very complex. When this occurs, it is sometimes necessary to simplify the diagram by using a *view* to limit the amount of information displayed. Views are simply subsets of information contained in the entire model. For example, a use-case view shows only the classes and relationships that are needed for a particular use case. Another view could show only a particular type of relationship, such as aggregation, association, or generalization. A third type of view could restrict the information shown to just that associated with a specific class, such as its name, attributes, and/or methods.

A second approach to simplifying class diagrams is through the use of packages (i.e., logical groups of classes). To make the diagrams easier to read and keep the models at a reasonable level of complexity, you can group related classes together into packages. Packages are general constructs that can be applied to any of the elements in UML models. We discussed them previously as a way to simplify use-case diagrams, and they can also be used to simplify class diagrams.

Creating a Class Diagram

Creating a class diagram (like any UML diagram) is an iterative process whereby the analyst starts by drawing a rough version of the diagram and then refines it over time. We will take you through one iteration of class diagram creation, but we would expect that the diagram would change dramatically as you communicate with users and fine-tune your understanding of the system.

Identify Classes   The steps when creating a class diagram are quite similar to the steps that we learned to create an ERD. First, you will need to identify what classes should be placed on the diagram. Remember, like ERDs, class diagrams show the classes that are needed for the system as a whole. However, for demonstration purposes, we only investigate a single use case: customer places order.

Many different approaches have been suggested to aid the analyst in identifying a set of candidate classes for the class diagram. The most common approach is *textual analysis,* the analysis of the text in the use cases. The analyst starts by reviewing the use cases and the use-case diagrams. The text descriptions in the use cases are examined to identify potential objects, attributes, methods, and relationships. The nouns in the use case suggest possible classes, while the verbs suggest possible methods or relationships. Figure 15-20 presents a summary of guidelines we have found useful.

Identify Attributes and Methods   Hopefully you determined that the class diagram will need to include classes that represent *CD, Inventory, Marketing Material, In-Store Hold, Vendor,* and *Customer.* The next step is to define the kinds of information that we want to capture about each class. The use case also provides insight into the kind of information that needs to be captured, under the section

---

Nouns → imply *objects* or *classes*

A common or improper noun implies a class of objects

For example, "an employee serves the customer" implies two class of objects, employee and customer

A proper noun or direct reference implies an instance of a class

For example, "John addressed the issues raised by Susan" implies two instances of a object, John and Susan

A collective noun implies a class of objects made up of groups of instances of another class

For example, "The list of students was not verified" implies that a list of students is an object that has its own attributes and methods

Verbs → imply *relationships* or *operations*

A doing verb implies an operation

For example, "Don files purchase orders" implies a "file" operation

A being verb implies a classification relationship between an object and its class

For example, "Joe is a dog" implies that Joe is an instance of the dog class

A having verb implies an aggregation or association relationship

For example, "the car has an engine" implies an aggregation relationship between car and engine

A transitive verb implies an operation

For example, "Frank sent Donna a order" implies that Frank and Donna are instances of some class that has an operation related to sending an order

A predicate or descriptive verb phrase implies an operation

For example, "if the two employees are in different departments, then…" implies an an operation to test whether employees are or are not in different departments

Adjectives → imply *attributes* of a class

An adjective implies an attribute of an object

For example, "All 55-year-old customers are now eligible for the senior discount" implies age is an attribute

Adverbs → imply an attribute of a *relationship* or *operation*

An adverb implies an attribute of a relationship or an attribute of an operation

For example "John drives very fast" implies a speed attribute associated with the driving operation

---

These guidelines are based on Russell J. Abbott, "Program Design by Informal English Descriptions," *Communications of the ACM,* 26(11), 1983, pp. 882–94; Peter P-S Chen, "English Sentence Structure and Entity-Relationship Diagrams" *Information Sciences: An International Journal,* 29(2–3), 1983, pp. 127–149; and Ian Graham, *Migrating to Object Technology,* Reading, MA: Addison-Wesley, 1995.

FIGURE 15-20
Textual Analysis Guidelines

labeled "Information for Steps." Sometimes additional requirements-gathering techniques in Chapter 4 are also needed. At this point, try to list some pieces of information that we likely will want to capture for each class—it may help to reread the description of the Take Request use case in Chapter 5 (Figure 5-5) and the overviews of the Maintain Marketing Material and Process In-Store Holds use cases in Chapter 5 (Figure 5-6)[4] One issue, of course, is that the use-case report in Figure 5-5 was written to be used in a structured environment, not an object-oriented environment, but you should be able to make the translation. Take a moment and add attributes to the classes.

At this point, we also want to consider what special methods each class will need to contain (remember that all classes can peform basic methods like inserting a new instance, so these are *not* placed on the diagram). Figure 15-21 shows a "first cut" at the attributes for the class diagram. Can you think of other attributes that we could have included in any of these classes?

Draw Relationships Between the Classes   Relationships are added to the class diagram by drawing association lines. Go through each class and determine other classes to which it is related, the name of the relationship (or the role it plays), and the number of instances that can participate in the association. For example, the *customer* class is related to the *in-store hold* class because a customer *places* an in-store hold. A customer can place zero or many in-store holds (multiplicity

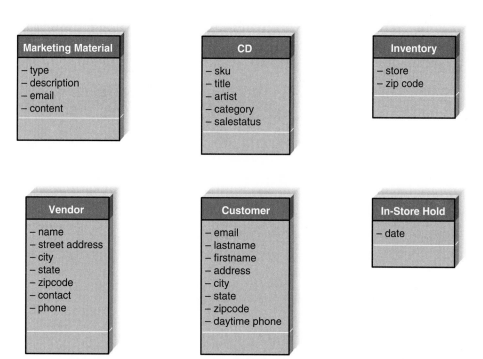

FIGURE 15-21
Initial Attributes for Class Diagram

---

[4] Since we did not complete the use case descriptions for these two use cases, you should also review the Revised Functional Requirements (Figure 5-8).

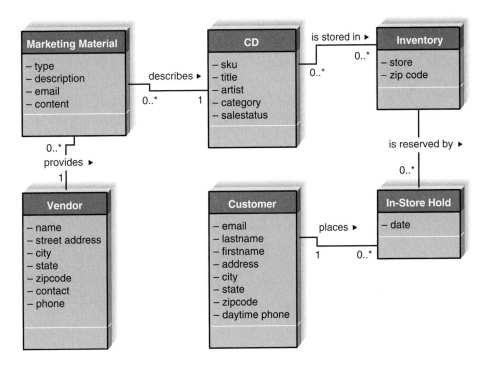

FIGURE 15-22
Revised Attributes and Methods

= 0..*), and an in-store hold can be associated with one and only one customer (multiplicity = 1). Take time now to formulate the associations for *In-Store Hold, Inventory, CD, Marketing Material,* and *Vendor.* Figure 15-22 shows the diagram that we have created so far. Notice that we included relationships between *CD* and *Marketing Material,* and *CD* and *Inventory, Inventory* and *In-Store Hold, In-Store Hold* and *Customer,* and *Vendor* and *Marketing Material.*

Finally, you should examine the model for opportunities to use aggregation or generalization associations. For example, if CD Selections decided that marketing material is best viewed as a collections of different kinds of promotional materials (e.g., artist information, sample clips, and reviews), it may be best to model *marketing material* as an aggregation that includes other classes. Also, CD Selections may want to leave room for the opportunity to sell items other then CDs. The class diagram could include a class called *product* that is a generalization of CD. In this way, other kinds of products (e.g., books, video, clothing) could easily be incorporated into the system by adding additional subclasses to the *product* superclass. Figure 15-23 shows the aggregation and generalization associations that we have included.

YOUR TURN

15-3 DRAWING A CLASS DIAGRAM

In "Your Turn 15-2" you created a use-case diagram for the campus housing service that helps students find apartments. Based on the use cases and the use-case diagram, create a class diagram for the campus housing service. See if you can identify at least one potential derived attribute, aggregation association, and generalization association for the diagram.

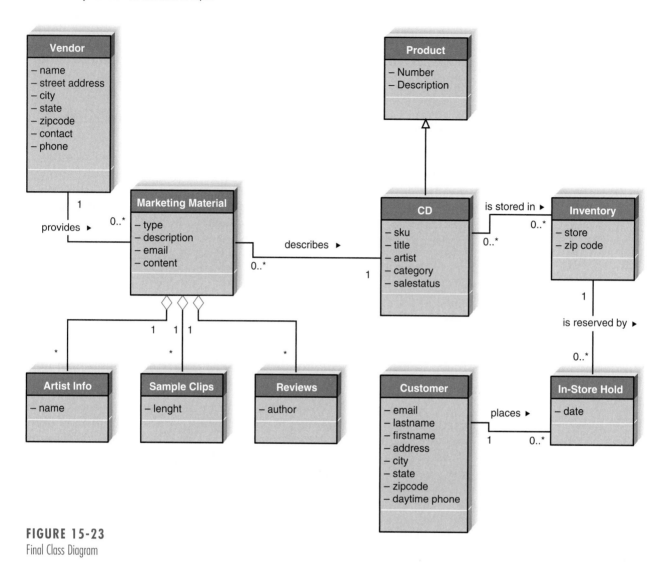

FIGURE 15-23
Final Class Diagram

SEQUENCE DIAGRAM

The next major UML diagramming technique is the *sequence diagram.* A sequence diagram illustrates the objects that participate in a use case and the messages that pass between them over time for *one* use case. A sequence diagram is a *dynamic model* that supports a dynamic view of the evolving systems. It shows the explicit sequence of messages that are passed between objects in a defined interaction. Since sequence diagrams emphasize the time-based ordering of the activity that takes place among a set of objects, they are very helpful for understanding real-time specifications and complex use cases.

The sequence diagram can be a *generic sequence diagram* that shows all possible scenarios[5] for a use case, but usually each analyst develops a set of *instance*

---

[5] Remember that a scenario is a single executable path through a use case.

sequence diagrams each of which depicts a single scenario within the use case. If you are interested in understanding the flow of control of a scenario by time, you should use a sequence diagram to depict this information. The diagrams are used throughout both the analysis and design phases; however, the design diagrams are very implementation specific, often including database objects or specific GUI components as the classes. The following sections first present the syntax of a sequence diagram and then demonstrate how one should be drawn.

Elements of a Sequence Diagram   Figure 15-24 presents an instance sequence diagram that depicts the objects and messages for the "Make Appointment" use case that describes the process by which a patient creates a new appointment, cancels, or reschedules an appointment for the doctor's office appointment system. In this specific instance, the create appointment process is portrayed.

Objects that participate in the sequence are placed across the top of the diagram using unlabeled rectangles (see Figure 15-25). Notice that the objects in Figure 15-24 are *aPatient, aReceptionist, Patients, UnpaidBills, Appointments,* and *anAppt.* They are not placed in any particular order, although it is nice to organize them in some logical way, such as the order in which they participate in the sequence. For each of the objects, the name of the class that they are an instance of is given after the object's name (e.g., *Patients:List* means that *Patients* is an instance of the *List* class that contains individual patient objects).

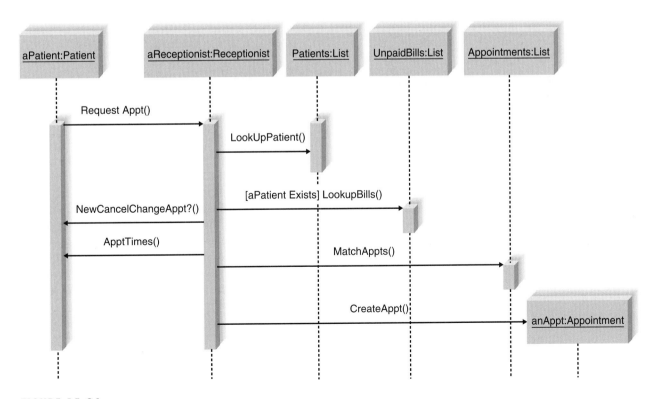

FIGURE 15-24
Sequence Diagram

Term and Definition	Symbol

An actor
    Is a person or system that derives benefit
    from and is external to the system
    Participates in a sequence by sending
    and/or receiving messages
    Is placed across the top of the diagram
	 anActor

An object:
    Participates in a sequence by sending
    and/or receiving messages
    Is placed across the top of the diagram
	anObject:aClass

A lifeline:
    Denotes the life of an object during a
    sequence
    Contains an X at the point at which the
    class no longer interacts	

A focus of control:
    Is a long narrow rectangle placed atop a
    lifeline
    Denotes when an object is sending or
    receiving messages	

A message:
    Conveys information from one object to
    another one
	aMessage()

Object destruction:
    An X is placed at the end of an object's
    lifeline to show that it is going out of
    existence
	X

FIGURE 15-25
Sequence Diagram Syntax

A dotted line runs vertically below each actor and object to denote the *lifeline* of the actors/objects over time (see Figure 15-24). Sometimes an object creates a *temporary object,* and in this case an X is placed at the end of the lifeline at the point the object is destroyed (not shown). For example, think about a shopping cart object for a Web commerce application. The shopping cart is used for temporarily capturing line items for an order, but once the order is confirmed, the shopping cart is no longer needed. In this case, an X would be located at the point at which the shopping cart object is destroyed. When objects continue to exist in the system after they are used in the sequence diagram then the lifeline continues to the bottom of the diagram (this is the case with all of the objects in Figure 15-24).

A thin rectangular box, called the *focus of control,* is overlaid onto the lifeline to show when the classes are sending and receiving messages (see Figure 15-25). A *message* is a communication between objects that conveys information with the expectation that activity will ensue, and messages passed between objects are shown using solid lines connecting two objects, called *links* (see Figure 15-25). The arrow on the link shows which way the message is being passed, and any argument values for the message are placed in parentheses next to the message's name. The order of messages goes from the top to the bottom of the page, so messages located higher on the diagram represent messages that occur earlier on in the sequence, versus the lower messages that occur later. In Figure 15-24, *LookUpPatient* is a message sent from the object *aReceptionist* to the object *Patients,* which is a container for the current patients to determine whether the *aPatient* object is a current patient.

There are times that a message is only sent if a *condition* is met. In those cases, the condition is placed between a set of [ ], e.g., *[aPatient Exists] Lookup-Bills()*. The condition is placed in front of the message name. However, when using a sequence diagram to model a specific scenario, conditions typically are not shown on any single sequence diagram. Instead, conditions are implied only through the existence of different sequence diagrams. Finally, it is possible to explicitly show the return from a message with a return link, a dashed message. However, adding return links tends to clutter the diagram. As such, unless the return links add a lot of information to the diagram, they should be omitted.

Sometimes, an object will create another object. This is shown by the message being sent directly to an object instead of its lifeline. In Figure 15-24, the object *aReceptionist* creates an object *anAppt.*

Creating a Sequence Diagram

The best way to learn how to create a sequence diagram is to draw one. We will use the scenario that results in the placing of a special order from the use case "Take Requests for CD" that was created in Chapter 5 and illustrated in Figure 5-5. Figure 15-26 lists the main steps that this "special order" sequence diagram will need to communicate. The steps when creating a sequence diagram are somewhat similar to the steps that we learned to create DFD.

Identify Objects   The first step is to identify instances of the classes that participate in the sequence being modeled; that is, the objects that interact with each other during the use-case sequence. Think of the major kinds of information that needs to be captured by the system. Objects typically can be taken from the use-case report created during the development of the use-case diagram (see Chapter 5). The sources and destinations on the use case (i.e., the external entities or data stores) are usually a good starting point for identifying the classes. Also, classes can be external actors that are represented on the use-case diagram.

For example, the instances of classes used for the Customer Places Order scenario include *Customer, CD, Marketing Material, Inventory,* and *Special Order System.* All of these should be placed in boxes and listed across the top of the drawing (see Figure 15-27). The instances of *Customer, CD, Inventory,* and *Marketing Material* correspond to data stores that we ultimately will need in the system, whereas the instances of *Special Order System* represent external actors that appeared on the use-case diagram. Because the latter instances interact in the sequence, we want to include them in the diagram.

1. User requests CD information
2. User requests information about what store(s) have the CD
3. User inserts CD into shopping cart
4. User checks out and provides customer information
5. A Special Order is placed

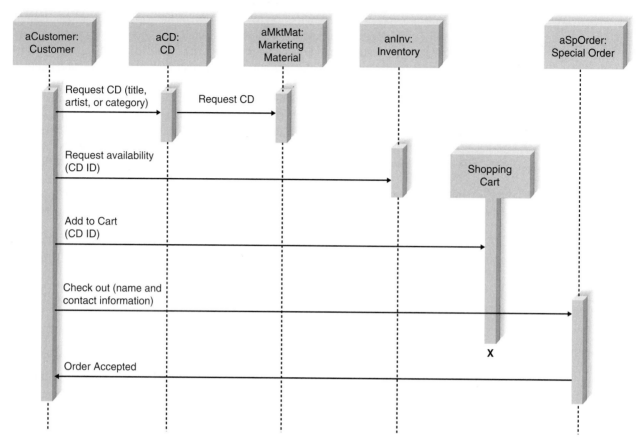

FIGURE 15-27
Sequence Diagram for *Customer Places Order* Scenario

Remember, at this point in time, you are only trying to identify the objects that take part in a specific scenario of a use case. As such, don't worry too much about perfectly identifying all the objects of the use case. Other scenario-based sequence diagrams may uncover additional objects. Also, the class diagram created in the previous section of this chapter described how the classes are defined and refined. However, based on the sequence diagrams created, the class diagram usually is revised since analysts have a better understanding of the classes after they develop them.

Add Messages   Next, draw arrows to represent the messages being passed from object to object, with the arrow pointing in the message's transmission direction. The arrows should be placed in order from the first message (at the top) to the last (at the bottom) to show time sequence. Any parameters passed along with the messages should be placed in parentheses next to the message's name. If a message is expected to be returned as a response to a message, then the return message is not explicitly shown on the diagram. Examine the steps in Figure 15-26 and see if you can determine the way in which the messages should be added to the sequence diagram. Figure 15-27 shows our results. Notice how we did not include messages back to Customer in response to *Request CD* and *Request Avail-*

YOUR

TURN

15-4 DRAWING A SEQUENCE DIAGRAM

In "Your Turn 15-3" you were asked to draw a use-case diagram for the campus housing system. Select one of the use cases from the diagram and create a sequence diagram that represents the interaction among objects in the use case.

ability. In these cases, it is assumed that the Customer will receive response messages about the requested CD and its availability, respectively.

Place Lifeline and Focus of Control   Last, you will need to show when objects are participating in the sequence. A vertical dotted line is added below each object to represent the object's existence during the sequence, and an X should be placed below objects at the point on the lifeline where they are no longer interacting with other objects. You should draw a narrow rectangle box over top of the lifelines to represent when the objects are sending and receiving messages. See Figure 15-27 for the completed sequence diagram.

STATECHART DIAGRAM

Some of the classes in the class diagrams are quite dynamic in that they pass through a variety of states over the course of their existence. For example, a patient can change over time from being "new" to "current," "former," and so forth based on his or her status with the doctor's office. A *statechart diagram* is a dynamic model that shows the different states that a single class passes through during its life in response to events, along with its responses and actions. Typically, statechart diagrams are not used for all classes, but just to further define complex classes to help simplify the design of algorithms for their methods. The statechart diagram shows the different states of the class and what events cause the class to change from one state to another. In comparison to the sequence diagrams, statechart diagrams should be used if you are interested in understanding the dynamic aspects of a single class and how its instances evolve over time, and not with how a particular use-case scenario is executed over a set of classes.

Elements of a Statechart Diagram

Figure 15-28 presents an example of a statechart diagram representing the patient class in the context of a hospital environment. From this diagram we can tell that a patient enters a hospital and is admitted after checking in. If a doctor finds the patient to be healthy, he or she is released and is no longer considered a patient after 2 weeks elapses. If a patient is found to be unhealthy, he or she remains under observation until the diagnosis changes.

State   A state is a set of values that describe an object at a specific point in time, and it represents a point in an object's life in which it satisfies some condition,

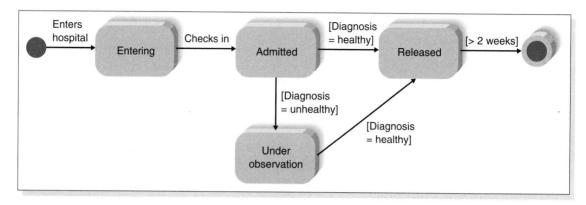

FIGURE 15-28
Statechart Diagram for a Hospital Patient

Term and Definition	Symbol

A state
    Is shown as a rectangle with rounded corners
    Has a name that represents the state of an object	

An initial state
    Is shown as a small filled-in circle
    Represents the point at which an object begins to exist	

A final state
    Is shown as a circle surrounding a small solid filled-in circle (bull's-eye)
    Represents the completion of activity	

An event
    Is a noteworthy occurrence that triggers a change in state
    Can be a designated condition becoming true, the receipt of an explicit signal from one object to another, or the passage of a designated period of time
    Is used to label a transition
	Event name

A transition
    Indicates that an object in the first state will enter the second state
    Is triggered by the occurrence of the event labeling the transition
    Is shown as a solid arrow from one state to another, labeled by the event name	

FIGURE 15-29
Statechart Syntax

performs some action, or waits for something to happen (see Figure 15-29). In Figure 15-28, states include *entering, admitted, released,* and *under observation.* A state is depicted by a *state symbol,* which is a rectangle with rounded corners with a descriptive label that communicates a particular state. There are two exceptions. An *initial state* is shown using a small solid filled circle, and an object's *final state* is shown as a circle surrounding a small solid filled-in circle. These exceptions depict when an object begins and ceases to exist, respectively.

The attributes or properties of an object affect the state that it is in; however, not all attributes or attribute changes will make a difference. For example, think about a patient's address. In Figure 15-28, those attributes make very little difference as to changes in a patient's state. However, if states were based upon a patient's geographic location (e.g., in-town patients were treated differently than out-of-town patients), changes to the patient's address may influence state changes. In the current diagram, the attributes that influence state transitions are the patient's hospital status and diagnoses.

Event An *event* is something that takes place at a certain point in time and changes a value(s) that describes an object, which in turn changes the object's state. It can be a designated condition becoming true, the receipt of the call for a method by an object, and the passage of a designated period of time. The state of the object determines exactly what the response will be. In Figure 15-28, checking in to the hospital, a healthy diagnosis, and a 2-week time lapse are events that cause changes to the patient's state.

Arrows are used to connect the state symbols, representing the transitions between states. A *transition* is a relationship that represents the movement of an object from one state to another state. Some transitions will have a guard condition. A *guard condition* is a Boolean expression that includes attribute values, which allows a transition only if the condition is true. Each arrow is labeled with the appropriate event name, and any parameters or conditions that may apply. For example, the two transitions from *admitted* to *released* and *under observation* contain guard conditions.

Creating a Statechart Diagram

Statechart diagrams are drawn to depict a single class from a class diagram. Typically the classes are very dynamic and complex, requiring a good understanding of their states over time and events triggering changes. You should examine your class diagram to identify which classes will need to undergo a complex series of state changes and draw a diagram for each of them. Let us investigate the state of the *special order* class from the CD Selections Internet system.

Identify the States The first step is to identify the various states that an order will have over its lifetime. This information is gleaned from reading the use-case reports, talking with users, and relying on the requirements-gathering techniques that you learned about in Chapter 4. You should begin by writing the steps of what happens to an order over time, from start to finish, similar to how you would create the "major steps performed" section of a use-case report. Figure 15-30 shows an order from start to finish, from an order's perspective.

Identify the Transitions The next step is to identify the sequence of states that an order object will pass through during its lifetime and then determine exactly what

1. The customer adds items into the shopping cart, some of which may end up being special orders, some of which may be in-store holds
2. The customer checks out and submits the special order once he or she is finished
3. The special order is either transferred from another store (if another store has it in stock) or a special order is placed with the CD vendor
4. The special order is shipped to the store from the owning store or from the vendor
5. The store receives the special order and holds it for the customer
6. The customer picks up the special order and pays for it
7. The special order is closed

FIGURE 15-30
The Life of a Special Order

causes each state to occur. Place state figures on the diagram to represent the states and label the transitions to describe the events that are taking place to cause state changes. For example, the event *Customer checks out, creating special order* moves the order from the *initial* state to the *submitted* state (see Figure 15-31). During the *submitted* state, the inventory database is checked to see if any CD Selections store has the special order items. The guard condition *In inventory= yes* prevents the order from moving from the *submitted* state to the *transfer requested* state unless the CD is in CD Selections' inventory at some store. Also, the guard condition *In inventory= no* prevents the order from moving from the *submitted* state to the *on-order* state unless the CD is not in inventory at any store. As such, between the two guard conditions, the order is stuck in the *submitted* state until the inventory check has been completed.

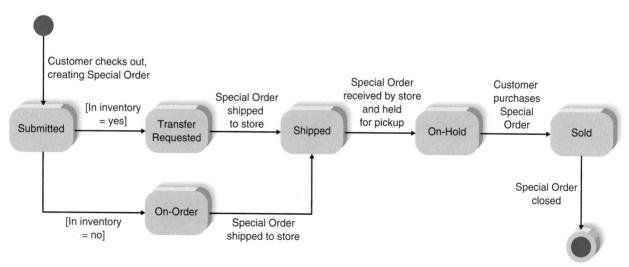

FIGURE 15-31
Statechart Diagram for a Special Order

15-5 DRAWING A STATECHART DIAGRAM

You have been working with the system for the campus housing service that helps students find apartments. One of the dynamic classes in this system likely is the apartment class. Draw a statechart diagram to show the various states that an apartment class transitions to throughout its lifetime. Can you think of other classes that would make good candidates for a statechart diagram?

SUMMARY

Today, the most exciting change to systems analysis and design is the move to object-oriented techniques, which view a system as a collection of self-contained objects that have both data and processes. However, the ideas underlying object-oriented techniques are simply ideas that have either evolved from traditional approaches to systems development or they are old ideas that have now become practical due to the cost-performance ratios of modern computer hardware in comparison to the ratios of the past. Today, the cost of developing modern software is composed primarily of the cost associated with the developers themselves and not the computers. As such, object-oriented approaches to developing information systems hold out much promise in controlling these costs.

An object is a person, place, or thing about which we want to capture information. Each object has attributes that describe information about it and behaviors, which are described by methods that specify what an object can do. Objects are grouped into classes, which are collections of objects that share common attributes and methods. The classes can be arranged in a hierarchical fashion in which low-level classes, or subclasses, inherit attributes and methods from superclasses above them to reduce the redundancy in development. Objects communicate with each other by sending messages, which trigger methods. Polymorphism allows a message to be interpreted differently by different kinds of objects. This form of communication allows us to treat objects like black boxes and ignore the inner workings of the objects. Encapsulation and information hiding allow an object to conceal its inner processes and data from the other objects.

Object-oriented analysis and design using UML allows the analyst to decompose complex problems into smaller, more manageable components using a commonly accepted set of notation. UML is a standard set of diagramming techniques that provide a graphical representation that is rich enough to model any systems development project from analysis through implementation. Typically, today most object-oriented analysis and design approaches use the UML to depict an evolving system. Finally, many people believe that users do not think in terms of data or processes but instead in terms of a collection of collaborating objects. As such, object-oriented analysis and design using UML allows the analyst to interact with the user employing objects from the user's environment instead of a set of separate processes and data.

Use-Case Diagram

A use-case diagram illustrates the main functions of a system and the different kinds of users that interact with it. The diagram includes actors, which are people or things that derive value from the system, and use cases that represent the functionality of the system. The actors and use cases are separated by a system boundary and connected by lines representing associations. At times, actors are specialized versions of more general actors. Similarly, use cases can extend or include other use cases. Building use-case diagrams is a five-step process whereby the analyst identifies the use cases, draws the system boundary, adds the use cases to the diagram, identifies the actors, and finally adds appropriate associations to connect use cases and actors together.

Class Diagram

The class diagram shows the classes and relationships among classes that remain constant in the system over time. The main building block of the class diagram is a class, which stores and manages information in the system. Classes have attributes that capture information about the class and methods, which are actions that a class can perform. There are three kinds of methods: constructor, query, and update. The classes are related to each other through an association, which has a name and a multiplicity that denotes the minimum and maximum instances that participate in the relationship. Two special associations, aggregation and generalization, are used when classes comprise other classes or when one subclass inherits properties and behaviors from a superclass, respectively. Class diagrams are created by first identifying classes, along with their attributes and operations. Then relationships are drawn among classes to show associations. Special notations are used to depict the aggregation and generalization associations.

Sequence Diagram

The sequence diagram is a dynamic model that illustrates instances of classes that participate in a use case and messages that pass between them over time. Objects are placed horizontally across the top of a sequence diagram, each having a dotted line called a lifeline vertically below it. The focus of control, represented by a thin rectangle, is placed over the lifeline to show when the objects are sending or receiving messages. A message is a communication between objects that conveys information with the expectation that activity will ensue, and the messages are shown using an arrow connecting two objects that points in the direction that the message is being passed. To create a sequence diagram, first identify the classes that participate in the use case and then add the messages that pass among them. Finally, you will need to add the lifeline and focus of control. Sequence diagrams are helpful for understanding real-time specifications and for complex scenarios of a use case.

Statechart Diagram

The statechart diagram shows the different states that a single instance of a class passes through during its life in response to events, along with responses and actions. A state is a set of values that describe an object at a specific point in time, and it represents a point in an object's life in which it satisfies some condition, performs some action, or waits for something to happen. An event is something that takes place at a certain point in time and changes a value(s) that describes an object, which in turn changes the object's state. As objects move from state to state, they undergo transitions. When drawing a statechart diagram, rectangles with rounded corners are first placed on the model to represent the various states that the class will take on. Next, arrows are drawn between the rectangles to denote the transitions, and event labels are written above the arrows to describe the event that causes the transition to occur.

KEY TERMS

A-Kind-Of (AKO)
Abstract class
Actor
Aggregation association
Architecture centric
Association
Association class
Attribute
Behavior
Class
Class diagram
Condition
Concrete class
Constructor method
Derived attribute
Dynamic model
Dynamic view
Encapsulation
Event
Extends association
Final state
Focus of control
Foreign key
Functional view
Generalization association
Generic sequence diagram

Guard condition
Includes association
Incremental
Information hiding
Inherit
Inheritance
Initial state
Instance
Instance sequence diagram
Iterative
Lifeline
Links
Message
Method
Multiplicity
Object
Object approach
Package
Polymorphism
Primary key
Private
Protected
Public
Query method
Rational Unified Process (RUP)
References

Role
Scenario
Sequence diagram
Specialized actor
State
State symbol
Statechart diagram
Static Model
Static view
Subclass
Superclass
System boundary
Temporary object
Textual analysis
Transitions
Unified Modeling Language
   (UML)
Unique identifier (UID)
Update method
Use case
Use-case diagram
Use-case driven
View
Visibility

QUESTIONS

1. Contrast the items in the following sets of terms:
   - Object; class; instance; entity relationship diagram (ERD) entity
   - Property; method; attribute
   - State: behavior
   - Unique identifier; primary key; foreign key
   - Superclass; subclass
   - Concrete class; abstract class
   - Method; message
   - Encapsulation; inheritance; polymorphism

2. How is the object approach different from the data and process approaches to systems development?
3. How can the object approach improve the systems development process?
4. Describe how the object approach supports the program design concepts of cohesion and coupling that were presented in Chapter 12.
5. What is Unified Modeling Language (UML)? How does it support the object approach to systems development?

6. Describe the steps in creating a use-case diagram.

7. How can you employ the use-case report to develop a use-case diagram?

8. How is the use-case diagram similar to the context and level 0 data flow diagrams (DFDs)? How is it different?

9. Give two examples of the extends associations on a use-case diagram. Give two examples for the includes association.

10. Which of the following could be an actor found on a use-case diagram? Why?
    • Ms. Mary Smith
    • Supplier
    • Customer
    • Internet customer
    • Mr. John Seals
    • Data-entry clerk
    • Database administrator

11. Consider a process called *validate credit history,* which is used to validate the credit history for customers who want to take out a loan. Explain how it can be an example of an includes association on a use-case diagram. Describe how it is an example of an extends association. As an analyst, how would you know which interpretation is correct?

12. Give examples of a static model and a dynamic model in UML. How are the two kinds of models different?

13. Describe the main building blocks for the sequence diagram and how they are represented on the model.

14. Do lifelines always continue down the entire page of a sequence diagram? Explain.

15. Describe the steps used to create a sequence diagram.

16. How is a class diagram different from an ERD?

17. Give three examples of derived attributes that may exist on a class diagram. How would they be denoted on the model?

18. Identify the following methods as constructor, query, or update. Which operations would not need to be shown in the class rectangle?
    • Calculate employee raise (raise percent)
    • Insert employee ()
    • Insert spouse ()
    • Calculate sick days ()
    • Locate employee name ()
    • Find employee address ()
    • Increment number of employee vacation days ()
    • Change employee address ()

• Place request for vacation (vacation day)

19. Draw the associations that are described by the following business rules. Include the multiplicities for each relationship.
    • A patient must be assigned to only one doctor and a doctor can have one or many patients.
    • An employee has one phone extension, and a unique phone extension is assigned to an employee.
    • A movie theater shows at least one movie, and a movie can be shown at up to four other movie theaters around town.
    • A movie either has one star, two costars, or more than 10 people starring together. A star must be in at least one movie.

20. What are the two kinds of labels that a class diagram can have for each association? When is each kind of label used?

21. Why is an association class used for a class diagram? Give an example of an association class that may be found in a class diagram that captures students and the courses that they have taken.

22. Give two examples of aggregation associations and generalization associations. How is each type of association depicted on a class diagram?

23. Are states always depicted using rounded rectangles on a statechart diagram? Explain.

24. What three kinds of events can lead to state transitions on a statechart diagram?

25. Describe the type of class that is best represented by a statechart diagram. Give two examples of classes that would be good candidates for statechart diagrams.

26. Compare and contrast the rational unified process (RUP) with UML.

27. Describe the way in which the RUP is implemented on a systems project.

28. Identify the model(s) that contains each of the following components:
    • Aggregation association
    • Class
    • Derived attributes
    • Extends association
    • Focus of control
    • Guard condition
    • Initial state
    • Links
    • Message
    • Multiplicity
    • Specialized actor

- System boundary
- Update method

29. What do you think are three common mistakes that novice analysts make in using UML techniques?

30. Do you think that UML will become more popular than the traditional structured techniques discussed previously? Why or why not?

31. Some experts argue that object-oriented techniques are simpler for novices to understand and use than are DFDs and ERDs. Do you agree? Why or why not?

EXERCISES

A. Investigate the Web site for Rational Software (www.rational.com) and its repository of information about Unified Modeling Language (UML). Write a paragraph news brief on the current state of UML (the current version and when it will be released, future improvements, etc.).

B. Investigate the Object Management Group (OMG). Write a brief memo describing what it is, its purpose, and its influence on UML and the object approach to systems development. (Hint: a good resource is: www.omg.org.)

C. Investigate rational unified process (RUP). Describe the major benefits of RUP and the steps that it contains. Compare the methodology to one of the other methodologies described in Chapter 1.

D. Investigate computer-aided software engineering (CASE) tools that support UML (e.g., Rational Software's Rational Rose, Microsoft's VISIO) and describe how well they support the language. What CASE tool would you recommend for a project team about to embark on a project using the object approach? Why?

E. Consider a system that is used to run a small clothing store. Its main functionality is maintaining inventory of stock, selling items to customers, and producing sales reports for management. List examples for each of the following items that may be found on a use-case diagram that models such a system: use case; extends use case: includes use case; actor; specialized actor.

F. Create a use-case diagram that would illustrate the use cases for the following dentist office system. Whenever new patients are seen for the first time, they complete a patient information form that asks their name, address, phone number, and brief medical history, which is stored in the patient information file. When a patient calls to schedule a new appointment or change an existing appointment, the receptionist checks the appointment file for an available time. Once a good time is found for the patient, the appointment is scheduled. If the patient is a new patient, an incomplete entry is made in the patient file: the full information will be collected when the patient arrives for the appointment. Because appointments are often made far in advance, the receptionist usually mails a reminder postcard to each patient 2 weeks before his or her appointment.

G. Create a use-case diagram that would illustrate the use cases for the following online university registration system. The system should enable the staff members of each academic department to examine the courses offered by their department, add and remove courses, and change the information about courses (e.g., the maximum number of students permitted). It should permit students to examine currently available courses, add and drop courses to and from their schedules, and examine the courses for which they are enrolled. Department staff should be able to print a variety of reports about the courses and the students enrolled in them. The system should ensure that no student takes too many courses and that students who have any unpaid fees are not permitted to register (assume that a fees data store is maintained by the university's financial office, which the registration system accesses but does not change).

H. Create a use-case diagram that would illustrate the use cases for the following system. A Real Estate Inc. (AREI) sells houses. People who want to sell their houses sign a contract with AREI and provide information on their house. This information is kept in a database by AREI and a subset of this information is sent to the citywide multiple listing service used by all real estate agents. AREI works with two types of potential buyers. Some buyers have an interest in one specific house. In this case, AREI prints information from its database, which the real estate agent uses to help show the house to the buyer (a process beyond the scope of the system to be modeled). Other buyers seek AREI's advice in finding a house that meets their needs. In this case, the buyer completes a buyer information form that is entered into a buyer database, and AREI real estate agents use its information to search AREI's database and the multiple listing service for houses that meet their needs. The results of these searches are printed and used to help the real estate agent show houses to the buyer.

I. Create a sequence diagram for each of the following scenario descriptions for a video store system. A Video Store (AVS) runs a series of fairly standard video stores:

- Every customer must have a valid AVS customer card to rent a video. Customers rent videos for 3 days at a time. Every time a customer rents a video, the system must ensure that he or she does not have any overdue videos. If there are overdue videos, they must be returned and an overdue fee must be paid before the customer can rent more videos.

- If the customer has returned overdue videos but has not paid the fee for overdue videos, the fee must be paid before new videos can be rented. If the customer is a premier customer, the first two overdue fees can be waived, and the customer can rent the video.

- Every morning, the store manager prints a report that lists overdue videos; if a video is two or more days overdue, the manager calls the customer to remind him or her to return the video.

J. Create a sequence diagram for each of the following scenario descriptions for a health club membership system:

- When members join the health club, they pay a fee for a certain length of time. The club wants to mail out reminder letters to members asking them to renew their memberships 1 month before their memberships expire. About half of the members do not renew their memberships. These members are sent follow-up surveys to complete about why they decided not to renew so that the club can learn how to increase retention. If the member did not renew because of cost, a special discount is offered to that customer. Typically, 25% of accounts are reactivated because of this offer.

- Every time a member enters the club, an attendant takes his or her card and scans it to make sure the

person is an active member. If the member is not active, the system presents the amount of money it costs to renew the membership. The customer is given the chance to pay the fee and use the club, and the system makes note of the reactivation of the account so that special attention can be given to this customer when the next round of renewal notices is dispensed.

K. Create class diagrams that describe the classes and relationships depicted in the following scenarios:

- Researchers are placed into a database that is maintained by the state of Georgia. Information of interest includes researcher name, title, position, date began current position, number of years at current position; university name, location, enrollment; and research interests. Researchers are associated with one institution, and each researcher can have up to five research interests. More than one researcher can have the same interest, and the system tracks the ranking of the best researchers for each interest. The system should be able to insert new researchers, universities, and research interests; produce information, such as the number of researchers at each university, contact information for the researchers, and research interests that do not have associate researchers; and change researcher rankings for the various research interests.

- A department store has a bridal registry. This registry keeps information about the bride, the products that the store carries, and the products for which each customer registers. Some products include several related items; for example, dish sets, include plates, specialty dishes, and serving bowls. Customers typically register for a large number of products, and many customers register for the same products. Draw the class diagram and give at least two examples of query and update operations that could be placed somewhere on the model.

- Jim Smith's dealership sells Fords, Hondas, and Toyotas. The dealership keeps information about each car manufacturer with whom employees deal so that they can get in touch with manufacturers easily. The dealership also keeps basic information about the models of cars that it carries from each manufacturer. The dealership keeps such information as list price, the price the dealership paid to obtain the model, and the model name and series (e.g., Honda Civic LX). It also keeps information about all sales that employees have made (for instance, the dealership will record the buyer's name, the car he or she bought, and the amount the buyer paid for the car). To contact the buyers in the future, the dealership also keeps contact information (e.g., address, phone number).

L. Think about sending a first-class letter to an international pen pal. Describe the process that the letter goes through to get from your initial creation of the letter to being read by your friend, from the letter's perspective. Draw a statechart diagram that depicts the states that the letter moves through.

M. Consider the video store that is described in question I. Draw a statechart diagram that describes the various states that a video goes through from the time it is placed on the shelf through the rental and return process.

N. Draw a statechart diagram that describes the various states that a travel authorization can have through its approval process. A travel authorization form is used in most companies to approve travel expenses for employees. Typically, an employee fills out a blank form and sends it to his or her boss for a signature. If the amount is fairly small (under $300), then the boss signs the form and routes it to Accounts Payable to be input into the accounting system. The system cuts a check that is sent to the employee for the right amount, and after the check is cashed, the form is filed away with the canceled check. If the check is not cashed within 90 days, the travel form expires. When the amount of the travel voucher is a large amount (over $300), then the boss signs the form and sends it to the Chief Financial Officer (CFO) along with a paragraph explaining the purpose of the travel, and the CFO will sign the form and pass it along to Accounts Payable. Of course, both the boss and the CFO can reject the travel authorization form if they do not feel that the expenses are reasonable. In this case, the employee can change the form to include more explanation, or decide to pay the expenses.

O. Identify the use cases for the following system. Picnics R Us (PRU) is a small catering firm with five employees. During a typical summer weekend, PRU caters 15 picnics for 20 to 50 people each. The business has grown rapidly over the past year and the owner wants to install a new computer system for managing the ordering and buying process. PRU has a set of 10 standard menus. When potential customers call, the receptionist describes the menus to them. If the customer decides to book a picnic, the receptionist records the customer information (name, address, phone number, etc.) and the information about the picnic (e.g., place, date, time, which one of the standard menus, total price) on a contract. The customer is then faxed a copy of the contract and must sign and return it along with a deposit (often by credit card or check) before the picnic is officially booked. The remaining money is collected when the picnic is delivered. Sometimes, the customer wants something special (e.g., birthday cake). In this case, the receptionist takes the information and gives it to the owner, who determines the cost; the receptionist then calls the customer back with the price information. Sometimes the customer accepts the price; other times, the customer requests some changes, which have to go back to the owner for a new cost estimate. Each week, the owner looks through the picnics scheduled for that weekend and orders the supplies (e.g., plates) and food (e.g., bread chicken) needed to make them. The owner would like to use the system for marketing as well. It should be able to track how customers learned about PRU and to identify repeat customers, so that PRU can mail special offers to them. The owner also wants to track the picnics on which PRU sent a contract but the customer neither signed the contract nor actually booked a picnic.

- Create the use-case diagram for the PRU system.
- Choose one use-case diagram and create a sequence diagram.
- Create a class diagram for the PRU system.
- Create a statechart diagram to depict one of the classes on the class diagram above.

P. Identify the use cases for the following system. Of-the-Month Club (OTMC) is an innovative young firm that sells memberships to people who have an interest in certain products. People pay membership fees for one year and each month receive a product by mail. For example, OTMC has a coffee-of-the-month club that sends members one pound of special coffee each month. OTMC currently offers six memberships (coffee, wine, beer, cigars, flowers, and computer games), each of which costs a different amount. Customers usually belong to just one, but some belong to two or more. When people join OTMC, the telephone operator records the name, mailing address, phone number, e-mail address, credit card information, start date, and membership service(s) (e.g., coffee). Some customers request a double or triple membership (e.g., two pounds of coffee, three cases of beer). The computer game membership operates a bit differently from the others. In this case, the member must also select the type of game (action, arcade, fantasy/science-fiction, educational, etc.) and age level. OTMC is planning to greatly expand the types of memberships it offers (e.g., video games, movies, toys, cheese, fruit, vegetables), so the system must accommodate this future expansion. OTMC is also planning to offer 3-month and 6-month memberships.

- Create the use-case diagram for the OTMC system.
- Choose one use-case diagram and create a sequence diagram.
- Create a class diagram for the OTMC system.
- Create a statechart diagram to depict one of the classes on the class diagram above.

Q. Think about your school or local library and the processes involved in checking out books, signing up new borrowers, and sending out overdue notices from the library's perspective. Describe three use cases that represent these three functions.

- Create the use-case diagram for the library system.
- Choose one use-case diagram and create a sequence diagram.
- Create a class diagram for the library system
- Create a statechart diagram to depict one of the classes on the class diagram above.

R. Think about the system that handles student admissions at your university. The primary function of the system should be to track a student from the request for information through the admission process until the student is either admitted or rejected for attendance at the school. Write the use-case report that can describe an "admit student" use case.

- Create the use-case diagram for the one use case. Pretend that students of alumni are handled differently from other students. Also, a generic *update student information* use case is available for your system to use.
- Choose one use-case diagram and create a sequence diagram. Pretend that a temporary student object is used by the system to hold information about people before they send in an admission form. After the form is sent in, these people are considered students.
- Create a class diagram for the student admission system. An admissions form includes the contents of the form, SAT information, and references. Additional information is captured about students of alumni, such as their parents' graduation year(s), contact information, and college major(s).
- Create a statechart diagram to depict a person as he or she moves through the admissions process.

MINICASES

1. The new information system at Jones Legal Investigation Services will be developed using objects. The data to be managed by this system will be complex, consisting of large amounts of text, dates, numbers, graphical images, video clips, and audio clips. The primary functions of the system will be to establish an investigation when requests come in from client-attorneys, record the investigative procedures that are conducted and the information that is gathered during an investigation, and produce bills for investigative services. Develop a use-case diagram for this new system.

2. The case investigation undergoes several states in the Jones Legal Investigation Services system. The case

investigation is first established when the attorney requests an investigation be conducted. When the investigator begins to perform the various investigative techniques, the case investigation becomes active. The client-attorney can begin settlement negotiations, or the case can go to trial. Settlement negotiations may result in a settlement, or the case may have to go to trial if settlement negotiations fail. Ultimately, the case investigation is closed when the case is closed by settlement agreement or judicial verdict. Develop a statechart diagram for this situation.

INDEX